New Developments in Singularity Theory

T0137731

NATO Science Series

A Series presenting the results of scientific meetings supported under the NATO Science Programme.

The Series is published by IOS Press, Amsterdam, and Kluwer Academic Publishers in conjunction with the NATO Scientific Affairs Division

Sub-Series

I. Life and Behavioural Sciences	IOS Press
II. Mathematics, Physics and Chemistry	Kluwer Academic Publishers
III. Computer and Systems Science	IOS Press
IV. Earth and Environmental Sciences	Kluwer Academic Publishers

The NATO Science Series continues the series of books published formerly as the NATO ASI Series.

The NATO Science Programme offers support for collaboration in civil science between scientists of countries of the Euro-Atlantic Partnership Council. The types of scientific meeting generally supported are "Advanced Study Institutes" and "Advanced Research Workshops", and the NATO Science Series collects together the results of these meetings. The meetings are co-organized bij scientists from NATO countries and scientists from NATO's Partner countries – countries of the CIS and Central and Eastern Europe.

Advanced Study Institutes are high-level tutorial courses offering in-depth study of latest advances in a field.
Advanced Research Workshops are expert meetings aimed at critical assessment of a field, and identification of directions for future action.

As a consequence of the restructuring of the NATO Science Programme in 1999, the NATO Science Series was re-organized to the four sub-series noted above. Please consult the following web sites for information on previous volumes published in the Series.

http://www.nato.int/science
http://www.wkap.nl
http://www.iospress.nl
http://www.wtv-books.de/nato-pco.htm

Series II: Mathematics, Physics and Chemistry – Vol. 21

New Developments in Singularity Theory

edited by

D. Siersma

Department of Mathematics,
University Utrecht,
Utrecht, The Netherlands

C.T.C. Wall

Mathematical Sciences,
Liverpool, United Kingdom

and

V. Zakalyukin

Moscow Aviation Institute,
Moscow, Russian Republic

Kluwer Academic Publishers

Dordrecht / Boston / London

Published in cooperation with NATO Scientific Affairs Division

Proceedings of the NATO Advanced Study Institute on
New Developments in Singularity Theory
Cambridge, United Kingdom
31 July–11 August 2000

A C.I.P. Catalogue record for this book is available from the Library of Congress.

ISBN 0-7923-6996-3 (HB)
ISBN 0-7923-6997-1 (PB)

Published by Kluwer Academic Publishers,
P.O. Box 17, 3300 AA Dordrecht, The Netherlands.

Sold and distributed in North, Central and South America
by Kluwer Academic Publishers,
101 Philip Drive, Norwell, MA 02061, U.S.A.

In all other countries, sold and distributed
by Kluwer Academic Publishers,
P.O. Box 322, 3300 AH Dordrecht, The Netherlands.

Printed on acid-free paper

Table of Contents

Preface

The primary purpose of the Advanced Study Institute was the introduction of new developments in singularity theory to a wide audience, including many young mathematicians from all over the world. The event was held at the Isaac Newton Institute in Cambridge, which provided an excellent venue and a very pleasant atmosphere and surroundings; it attracted over a hundred participants.

Singularities arise naturally in a huge number of different areas of mathematics and science. As a consequence, singularity theory lies at the crossroads of paths connecting the most important areas of applications of mathematics with some of its most abstract parts. For example, it connects the investigation of optical caustics with simple Lie algebras and the theory of regular polyhedra, while also relating the wavefronts of hyperbolic PDE to knot theory, and the theory of solid shape to commutative algebra.

The main goal in most problems of singularity theory is to understand the dependence of some objects of analysis, geometry, physics, or some other science (functions, varieties, mappings, vector or tensor fields, differential equations, models and so on) on parameters. For generic points in the parameter space, their exact values influence only the quantitative aspects of the phenomena, their qualitative, topological features remaining stable under small changes of parameter values.

However for certain exceptional values of the parameters, the qualitative, topological features of the phenomena may suddenly change under a small variation of the parameter. This change is called a perestroika, bifurcation, metamorphosis, catastrophe etc in different branches of science.

The meeting consisted of several courses consisting of a couple of survey lectures, a large number of research lectures, and a problem session. This volume consists of articles amplifying the survey lectures given at the meeting. We can group them under three headings.

A: Singularities of real maps

The classification of singularities and applications is discussed by Bruce, and plays a central rôle in the contribution of Zakalyukin on flag contact singularities. The relation between combinatorics, Stokes sets and bifurcation sets is studied in the paper of Baryshnikov. Interesting developments on topological invariants dual to singular sets form the theme of Vassiliev's work. He gives an outline of his programme for studying discriminant spaces and their complements, with numerous examples. The work of Kazarjan (reported by Vassiliev at the meeting) uses a direct construction of a 'classifying space' for singularities to develop a new method for calculating Thom polynomials. Brasselet writes about generalized index theorems on singular varieties, relating them to non-commutative geometry.

B: Singular complex varieties

The geometry of families of plane curves, a classical subject with many new results, is the topic of the article of Greuel, Lossen and Shustin. Parusinski proves in his article a preparation theorem for subanalytic functions. Interesting developments on the geometry of arc spaces were presented by Denef[1]. This subject is related to questions in Hodge theory and mirror symmetry. The computation of Hodge theoretic invariants is the subject of the paper by Steenbrink and Schultze. The relation between Frobenius manifolds and singularities was explained by Hertling. In his article Dimca combines the theory of Hodge structures with the study of polynomials at infinity. These can be regarded as giving examples of meromorphic functions; an account of these, focussing on the zeta function, is given by Luengo, Melle and Gusein-Zade. The relation between simple singularities and complex reflection groups is developed in the articles of Goryunov and Slodowy.

C: Singularities of holomorphic maps

Discriminants and vector fields on singular hypersurfaces form the topic of the article by du Plessis and Wall. The transition from geometrical questions to algebraic ones relates this subject to the contribution of Gaffney. He presents the theory of integral closure of ideals and modules. Algebra also plays a role in Damon's work on nonlinear sections of non-isolated complete intersections. Moreover, these results are related to Sierma's article about the vanishing topology of non-isolated singularities.

Dirk Siersma
Terry Wall
Volodya Zakalyukin

February 2001

---

[1] this talk is not represented here; the speaker's transparencies are posted on http://www.wis.kuleuven.ac.be/wis/algebra/NotesCambridge/default.htm

Part A: Singularities of real maps

Classifications in Singularity Theory and Their Applications

James William Bruce
Mathematical Sciences, University of Liverpool
(jwbruce@liv.ac.uk)

1. Introduction

Classifying objects in mathematics is a fundamental activity. Each branch has its own natural notion of equivalence and it is equally natural to list the objects in question up to that equivalence. So, for example, we have the notions of isomorphism for vector spaces or for groups, diffeomorphism for manifolds, conjugacy for dynamical systems.

In this article we consider classifications which arise in singularity theory. We start with some straightforward classifications in linear algebra, and elementary algebraic geometry. We then move to the classification of jets under the natural equivalence relations in singularity theory, using the idea of a complete transversal. We also establish some results on classification in infinite dimensional spaces, that is discuss (formal) determinacy theorems.

Prerequisites We assume that the reader is familiar with the notion of smooth manifold and mapping, the inverse function theorem, the idea of a germ, some elementary facts concerning projective spaces, projective curves and Lie groups.

There is a brief discussion of the history of the subject in the final section. I make no great claim to originality, but much of this material emerged from joint work with Andrew du Plessis, Terry Wall and Neil Kirk to whom I am grateful.

EXAMPLE 1.1.   We start by considering some familiar objects in linear algebra and their associated notions of equivalence.
(1) Two linear maps $f, g : V \to W$ are equivalent if there are isomorphisms $\phi : V \to V$ and $\psi : W \to W$ with $\psi \circ f \circ \phi^{-1} = g$.
(2) Two symmetric bilinear forms α, β on a vector space V are isomorphic if there is an isomorphism $\phi : V \to V$ such that $\alpha(\phi-, \phi-) = \beta(-, -)$.
(3) Two automorphisms $f, g : V \to V$ are are equivalent if there is an isomorphism $\phi : V \to V$ with $\phi \circ f \circ \phi^{-1} = g$.

3

D. Siersma et al. (eds.), New Developments in Singularity Theory, 3–33.
© 2001 Kluwer Academic Publishers. Printed in the Netherlands.

Usually to make classifications one has to work with the objects in fairly explicit/concrete form. Assuming that the vector spaces above are finite-dimensional (as we shall) each of the above problems have familiar matrix formulations. In what follows $M(p, n, \mathbb{K})$ denotes the set of $p \times n$ matrices, $SM(n, \mathbb{K})$ the set of $n \times n$ real symmetric matrices, and $Gl(n, \mathbb{K})$ denotes the set of real invertible $n \times n$ matrices, all with entries lying in the field $\mathbb{K} = \mathbb{R}$ or $\mathbb{C}$. We sometimes omit $\mathbb{K}$. The above notions of equivalence then reduce to the following.

(1) Two matrices X, $Y \in M(p, n)$ are equivalent if there are matrices $A \in Gl(n)$ and $B \in Gl(p)$ with $X = BYA^{-1}$.
(2) Two $n \times n$ symmetric matrices X, Y are equivalent if there is a matrix $A \in Gl(n)$ with $X = AYA^{T}$.
(3) Two matrices X, $Y \in M(n, n)$ are equivalent if there is a matrix $A \in Gl(n)$ with $X = AYA^{-1}$. Since this equivalence, and that in (1), both apply to square matrices we distinguish them by calling this notion *conjugacy*.

The above is familiar, as are the following results of the classification.

(1) Every $p \times n$ matrix is equivalent to one of the form

$$\begin{pmatrix} I_k & 0 \\ 0 & 0 \end{pmatrix}$$

where I_k is the identity $k \times k$ matrix, and the zeros are zero matrices. This is the so-called Smith normal form.
(2) Over the complex numbers every symmetric matrix can be reduced to the same (square) form. Over the real numbers the identity block is replaced by a diagonal matrix whose entries are each ± 1.
(3) For conjugacy the normal forms (over the complex numbers) are direct sums of square matrices of the form

$$\begin{pmatrix} \lambda & 1 & 0 & 0 & \dots & 0 \\ 0 & \lambda & 1 & 0 & \dots & 0 \\ 0 & 0 & \lambda & 1 & \dots & 0 \\ . & . & . & . & . & . \\ 0 & 0 & 0 & 0 & \dots & \lambda \end{pmatrix}$$

with varying λ's and size of block.

REMARK 1.2. The first question to ask is why is it useful to classify? Some answers are given in (1)-(5) below.

(1) Firstly classifications yield examples to study; mathematics is an experimental science!

(2) Classifications lead to an understanding of the structure of the space of objects from simple to complex; they also expose common themes or patterns.

(3) Natural equivalences arise for objects involving more abstract structures. So, for example, the notion of right-equivalence we shall meet below is absolutely necessary if we are to discuss singularities of functions on manifolds.

(4) Good classifications should yield normal forms which are (in some sense) simple; generally this means they contain few non-zero terms. This facilitates their subsequent study. In each case above the classifications arose for quite explicit reasons. The first is essential in the study of linear equations. The second tells us something about the number and form of quadrics in real or complex projective space, and also determines the nature of generic critical points of several variables. The third has applications to the study of singular points of ordinary differential equations. Here one can often reduce the study of ODE's to the special case of a linear system. These simple normal forms then aid our understanding of such systems.

(5) Sometimes reduction to normal form is hard or impossible. However partial reduction to a 'pre-normal form' may have many of the advantages of a full classification.

(6) A key question is whether a given reduction can be done constructively or not: if two objects are equivalent can we find the change of bases/invertible matrices which yield the equivalence? Sometimes this is vital, often the fact that one can identify the normal form is enough. The first two examples from linear algebra have associated algorithms which allow reduction to normal form. The third depends on the solution of polynomial equations to determine the eigenvalues. Classically an alternative 'rational normal form' circumvents this problem.

(7) A related question is that of deciding which normal form a given object reduces to without carrying out the reduction. The ideal here is a list of invariants, that is properties unchanged under the equivalence, which are computable and determine the normal form. So, for example, rank is invariant under all three notions of equivalence above, and determines the equivalence class in the first case.

(8) One immediate difference between classifications (1), (2) and that in (3) is that in the first two cases there are only finitely many normal forms to consider. In the third the eigenvalues are 'continuous' invariants. Classifications without continuous invariants (or *moduli* as they are called) are particularly satisfying, and often important.

We now formalise some of the common features of the examples studied above.

DEFINITION 1.3. *Let G be a group and X a set. We say that G acts on X if there is a map $\Phi : G \times X \to X$ for which the following is true:*
  (i) $\Phi(e, x) = x$ (where e is the identity element of G);
  (ii) $\Phi(g, \Phi(h, x)) = \Phi(gh, x)$.
In our examples usually G is a Lie group, that is, G is a manifold and the group operations (product and inverse) are smooth, X is a smooth manifold, and the map Φ is smooth.
We usually write $g.x$ in place of $\Phi(g, x)$ so the above become $e.x = x$, $g.(h.x) = (gh).x$. The set $G.x = \{g.x : g \in G\}$ is the orbit *of x.*

REMARK 1.4. (1) In the three cases above we have a group (respectively $Gl(n) \times Gl(p), Gl(n), Gl(n)$) acting on the sets $M(p, n)$, $SM(n)$, $M(n, n)$. Note that we needed the inverses in the formulae above to ensure that we had a group action in the above sense.
(2) So in our examples 'equivalence' is synonymous with 'lies in the same orbit'; classification reduces to listing orbits; a normal form is a representative of an orbit; an invariant is some property or function constant on orbits.

2. Groups and Singularity Theory

The equivalences we shall be interested in concern some infinite dimensional spaces and groups. Throughout we can work over the reals or complexes; generally we consider the former, but one of the surprising features about classifications in singularity theory is the extent to which the processes in the two cases coincide. We start with some notation.

Let $\mathcal{E}_n$ denote the space of germs of functions $f : (\mathbb{R}^n, 0) \to \mathbb{R}$, and $\mathcal{E}(n, p)$ the space of germs of mappings $f : (\mathbb{R}^n, 0) \to \mathbb{R}^p$. We shall also need to consider finite-dimensional analogues of these spaces provided by the notion of a k-jet.

DEFINITION 2.1. *Given a smooth germ $f : (\mathbb{R}^n, 0) \to (\mathbb{R}^p, 0)$ its k-jet is the Taylor series of f at 0 truncated to degree k, and is denoted by $j^k f$. The set of such jets (polynomial mappings of degree $\leq k$ with zero constant term) is denoted $J^k(n, p)$.*

This has an algebraic formulation. Clearly $\mathcal{E}_n$ is a ring under the obvious operations of addition and multiplication (inherited from those of $\mathbb{R}$). This ring contains a unique maximal ideal, consisting of functions vanishing at the origin, denoted by $\mathcal{M}_n$. It is not hard to prove that its k-th power $\mathcal{M}_n^k$ is the set of germs whose derivatives of order $\leq k - 1$ vanish at the origin. Now $\mathcal{E}(n, p)$ is an $\mathcal{E}_n$-module, so we can consider the product $\mathcal{M}_n^{k+1}.\mathcal{E}(n,p)$, and the quotient $\mathcal{M}_n.\mathcal{E}(n,p)/\mathcal{M}_n^{k+1}.\mathcal{E}(n,p)$, and this quotient can be identified with $J^k(n,p)$.

The group of diffeomorphisms $(\mathbb{R}^n,0) \to (\mathbb{R}^n,0)$ is denoted by $\mathcal{D}(n)$, and consists of those germs in $\mathcal{M}_n.\mathcal{E}(n,n)$ whose linear part at 0 is invertible. It is clearly a group under map-composition. The group $\mathcal{D}(n)$ acts on $\mathcal{M}_n.\mathcal{E}(n,p)$: given $\phi \in \mathcal{D}(n)$ and $f \in \mathcal{M}_n.\mathcal{E}(n,p)$ define $\phi.f = f \circ \phi^{-1}$. The corresponding equivalence is called *right equivalence*, denoted by $\mathcal{R}(n)$-equivalence, since ϕ appears to the right of f.

Similarly the group $\mathcal{D}(p)$ acts on $\mathcal{M}_n.\mathcal{E}(n,p)$: given $\psi \in \mathcal{D}(p)$ and $f \in \mathcal{M}_n.\mathcal{E}(n,p)$ define $\psi.f = \psi \circ f$; this is also a group action. Since ψ appears to the left of f the corresponding equivalence is called *left equivalence*, or $\mathcal{L}(p)$-equivalence.

The product group $\mathcal{D}(n) \times \mathcal{D}(p)$ acts on $\mathcal{M}_n.\mathcal{E}(n,p)$: given $(\phi, \psi) \in \mathcal{D}(n) \times \mathcal{D}(p)$ and $f \in \mathcal{M}_n.\mathcal{E}(n,p)$ define $(\phi, \psi).f = \psi \circ f \circ \phi^{-1}$. The corresponding equivalence is called (for by now obvious reasons) *right-left equivalence*, or $\mathcal{A}(n,p)$-equivalence.

Finally let $\mathcal{C}(n,p)$ denote the group of germs $\Psi : (\mathbb{R}^n,0) \to \mathcal{D}(p)$, where the map $(\mathbb{R}^n \times \mathbb{R}^p, (0,0)) \to (\mathbb{R}^p,0)$, $(x,y) \mapsto \Psi(x)(y)$ is smooth and the group structure is inherited from the target. Since Ψ is a map into the group $\Psi(x,0) = 0$ for all x. Now $\mathcal{C}$ acts on $\mathcal{M}_n.\mathcal{E}(n,p)$: given $\Psi \in \mathcal{C}(n,p)$ and $f \in \mathcal{M}_n.\mathcal{E}(n,p)$ define $\Psi.f(x) = \Psi(x, f(x))$. This is so-called $\mathcal{C}(n,p)$-equivalence. If we use the product of this group with the $\mathcal{R}(n)$-action of $\mathcal{D}(n)$, then the corresponding equivalence is called *contact equivalence*, denoted $\mathcal{K}(n,p)$-equivalence. (Note that we can replace $\mathcal{D}(p)$ by $Gl(p, \mathbb{R})$ and obtained the same equivalence; exercise!)

REMARK 2.2.   (1) Strictly speaking the above symbols are labelling the groups involved and the action. So, for example, we use $\mathcal{R}(n)$ or simply $\mathcal{R}$ to denote the group of germs of diffeomorphisms in the target acting in the above way.

(2) The group $\mathcal{R}$ is the easiest to deal with, and contains enough of the salient features to provide a good introduction.

(3) Classifications up to $\mathcal{A}$-equivalence are generally rather hard. If we expand the group to $\mathcal{K}$ the classifications become very much easier, although we lose a good deal of information along the way. When the dimension of the source exceeds the target $\mathcal{K}$-equivalence is the natural notion preserving the zero fibre of the mapping $(\mathbb{R}^n,0) \to (\mathbb{R}^p,0)$, while $\mathcal{A}$-equivalence preserves all of the fibres of the mapping.

(4) Clearly all of the above makes perfect sense in the real analytic and complex analytic cases.

(5) There are many important subgroups of these groups which also arise e.g. groups of diffeomorphisms $\phi : (\mathbb{K}^n,0) \to (\mathbb{K}^n,0)$ preserv-

ing a variety $(V, 0) \subset (\mathbb{K}^n, 0)$. There is a beautiful general theory due to Damon [16] of 'geometric subgroups' which shows that most of the subgroups which arise naturally behave as well as $\mathcal{R}$, $\mathcal{A}$, $\mathcal{K}$. For example consider diffeomorphisms $\phi : (\mathbb{K}^2, 0) \to (\mathbb{K}^2, 0)$ preserving the x-axis. Such diffeomorphisms clearly take the form $(x, y) \mapsto (\alpha(x, y), y\beta(x, y))$. Note however that it is generally impossible to write down the diffeomorphisms preserving most varieties explicitly.

(6) Clearly the group $\mathcal{D}(n)$ acts on the space of germs $(\mathbb{R}^n, 0) \to (\mathbb{R}^n, 0)$ by $\phi.f = \phi \circ f \circ \phi^{-1}$. This is a useful action in the theory of dynamical systems. The linearised case (which we considered above) already provided a relatively complicated example, and this is one situation we shall not pursue. Our techniques give useful results when constructing normal forms, but there are no straightforward 'determinacy' results.

This article explains various techniques used to classify map-germs under the above equivalence relations. One such (rather unusual) classification which we shall use below is the following which is a consequence of the Inverse Function Theorem. (See [8] for a proof.) This provides the technical tool required to prove our key results on group actions.

LEMMA 2.3. *Let $f : (\mathbb{R}^n, 0) \to (\mathbb{R}^p, 0)$ be a smooth map germ with the property that its derivative df has rank k at every point near $0 \in \mathbb{R}^n$. Then f is $\mathcal{A}$-equivalent to the germ $x \mapsto (x_1, \ldots, x_k, 0, \ldots, 0)$.*

3.   Some Elementary Algebraic Geometry

Linear mathematics is the key tool in geometry and much of differential analysis. Where possible we linearise problems; indeed what lies at the heart of this article is an attempt to classify nonlinear objects using linear techniques. Sometimes however the problems one faces are fundamentally non-linear. Our first examples of this type are some basic classification problems in algebraic geometry. They also provide the starting point for some classifications we shall make later on.

In what follows we denote the space of homogeneous polynomials (or forms) of degree k in n variables, over the field $\mathbb{K}$, by $H^k(n)$. Each such (non-zero) homogeneous polynomial determines a hypersurface in projective space $\mathbb{K}P^{n-1}$. Any element of $Gl(n, \mathbb{K})$ induces an automorphism of this projective space, and we regard two hypersurfaces as equivalent if one is taken to the other by such an automorphism. This corresponds to the action on $H^k(n)$ given by $(f, \phi) \mapsto f \circ \phi^{-1}$. Our wish is to classify the orbits of this action. Generally this is impossible, and a discussion of invariants very complicated. We will, however, carry out a few elementary classifications which will illustrate the problems and provide results of some use later. One general remark: one (characteristic) property of a form of degree k is that if we replace each x_i by tx_i the form is multiplied by t^k. So over C non-zero multiples of the same form are equivalent. (What happens over R?) We shall generally work over the complexes, although real classifications are possible (but more difficult). Note that $H^k(n)$ is a vector space of dimension $N = (n + k - 1)!/(n - 1)!k!$.

EXAMPLE 3.1.   (1) The case $k = 2$. Here we are dealing with quadrics (when $n = 3$ the classical case of conics). There is a one-to-one correspondence between $n \times n$ symmetric matrices and homogeneous polynomials of degree 2 in n variables, and it is not difficult to check that two such polynomials are equivalent if and only if the corresponding matrices X and Y are related as above via $X = A^T Y A$ for some $A \in Gl(n)$. So we have a list of normal forms $\sum_{i=1}^r x_i^2$ in the complex case and $\sum_{i=1}^r x_i^2 - \sum_{i=1}^s x_i^2$ in the real case.
(2) When $n = 2$ we have a so-called binary form. Over the complexes any such form will factor, most of them as a product $\Pi(x - \alpha_i y)$. The α_i are referred to as roots; the only other type of factor which can occur is y, corresponding to the root ∞. It is not difficult to produce normal forms when k is small. So for $k = 3$ there are three types $xy(x - y)$, $x^2 y$, x^3 according as the form has 3, 2, 1 distinct roots. The problems start with the case $k = 4$. If the quartic form has three distinct roots or less we are reduced to the previous case. If it has four distinct roots factor as a linear times a cubic; the cubic we can fix as $xy(x - y)$, so the quartic can be written $xy(x - y)(x - \alpha y)$, $\alpha \neq 0, 1$. Suppose given another such quartic $xy(x - y)(x - \beta y)$. If we can change co-ordinates to take one to the other then, for example, the change might preserve $x = 0$, $y = 0$, $x = y$. You can check that any such linear change is of the form $(x, y) \mapsto (ax, ay)$ for some $a \neq 0$, and this will only take the one form to the other if $\alpha = \beta$. There are other possible changes of co-ordinates preserving the union of the lines $x = 0$, $y = 0$, $x = y$ and it is not hard to show that a linear map exists interchanging the above two forms if and only if $\beta = \alpha$, $1/\alpha$, $1 - \alpha$, $1/(1 - \alpha)$, $1 - (1/\alpha)$ or $1 - 1/(1 - \alpha)$. Clearly we have moduli. (If we look at real forms with 4 real roots, so take α real, then for $0 < \alpha < 1/2$, all of the quartics $xy(x - y)(x - \alpha y)$ are inequivalent. It is not hard to believe that there will be 2 moduli for quintics.
(3) The next simplest case to look at is $n = 3$, $k = 3$, the case of cubic curves in the projective plane; here $\dim H^3(3) = 10$. Consider first the case of a singular curve; by a change of co-ordinates we may

suppose that one of the singular points is at $(0:0:1)$ in $\mathbb{C}P^2$, so we can write our cubic in the form $zQ(x,y) + C(x,y)$ where Q and C are binary forms of degree 2 and 3 respectively. Straightforward changes of co-ordinates produces the normal forms $zxy + x^3 + y^3$, $zxy + x^3$, zxy, (respectively a nodal cubic, a line meeting the residual conic in two points and three distinct non-concurrent lines) $zx^2 + y^3$, $zx^2 + xy^2, zx^2$, (respectively a cuspidal cubic, conic plus tangent, line plus repeated line). The final (non-zero) forms are $xy(x-y)$, x^2y and x^3, respectively three concurrent lines, a line plus repeated line (again), and a triple line.

If our cubic is non-singular then it can be shown that one can reduce to the form $x^3 + y^3 + z^3 + 3axyz$. Neither reaching this normal form nor deciding which of these are equivalent is an entirely trivial task.

Although the classification is elementary it is, even in this simple case, a little messy. The reduction in the singular case relied on finding a singular point. Given a general singular cubic this involves solving 3 non-linear equations. In other words it is not easy to reduce even a singular cubic to normal form.

We finish by widening our study to the product $H^k(n)^p$ of p copies of $H^k(n)$. Given an element of this space we can consider the subspace of $H^k(n)$ spanned by the corresponding p forms. There is a natural action of the linear group $Gl(p)$ on $H^k(n)^p$ obtained by applying a matrix A to a vector consisting of p forms. Since we are only interested in the space spanned by the components of this vector any pair related in this way will be deemed equivalent. We also continue to allow the action of $Gl(n)$ on the forms themselves and hence on the product $H^k(n)^p$.

DEFINITION 3.2. *An element $f = (f_1, \ldots, f_p)$ is referred to as a* linear system *of dimension p. Any two linear systems related by the actions of the group $Gl(n) \times Gl(p)$ are said to be* equivalent.

Our definition of a linear system of dimension p is slightly anomolous, since the space spanned by the components of f may be of dimension $< p$. Naturally linear systems are even harder to classify than forms. In the case when $p = 2$ a linear system is referred to as a pencil. We shall consider just one case, in the form of an extended exercise.

EXERCISE 3.3.   Consider the case of pencils of real binary forms, so $n = 2$, $k = 2$, $p = 2$, and $\mathbb{K} = \mathbb{R}$. By a change of x, y variables reduce the first form of the pair to $x^2 \pm y^2$, x^2, or 0. Now the other change of co-ordinates allows us to subtract multiples of the first entry from the second. Using this reduce any pencil to one of the forms

$$(x^2, y^2),\ (x^2 - y^2, xy),\ (x^2, xy)\ (0, x^2 \pm y^2),\ (0, x^2),\ (0, 0).$$

The first 3 cases are clearly different from the last 3 (where the dimension of the space spanned by the components has dropped). Distinguish the initial cases as follows. The space of binary quadratics is of dimension 3, the corresponding projective space is a plane. If $(a : b : c)$ corresponds to some (non-zero) multiple of $ax^2 + 2bxy + cy^2$ then the forms with a repeated root are those for which $b^2 = ac$, a conic in the $(a : b : c)$ plane. Now these three genuine pencils each yield a line in this plane. Show that the line determined by the above three types respectively meets the conic in 2, 0, 1 points (in the last case the line is tangent to the conic). Hence show that the pencils are distinct.

4. Lie Groups and Actions

Most of the groups so far are Lie, the exceptions are those arising in singularity theory. Even here to carry out efficient classifications of smooth map-germs it will be necessary to work with their finite dimensional analogues, the jet-spaces. So we will need to prove some results on Lie groups and their actions. Our approach (and that generally taken within the subject) is inductive. So we start with 1-jets and classify them. For the inductive step suppose given a list of k-jets. We seek to classify all $(k+1)$-jets with a given k-jet. In the case of functions if the k-jet happens to be zero then the classification corresponding to the group $\mathcal{R}$ reduces to the $Gl(n)$-classification of forms of degree $k + 1$. (See the remark below.)

We shall illustrate our need to establish the results we do with the following extended example.

EXAMPLE 4.1.   Consider 3-jets of functions $f : (\mathbb{C}^3, 0) \to (\mathbb{C}, 0)$ with 2-jet 0. The classification reduces to that of cubic curves under projective equivalence. One possibility is the triangle xyz. Now we consider all 4-jets with 3-jet xyz. There are, of course, many terms of degree 4 to add (a total of 15). However when considering

$$xyz + ax^4 + bx^3y + cx^3z + \ldots$$

one notes that one can kill off any terms of degree 4 involving at least two variables. For example to kill off the x^3y term replace z by $z - ax^2$. What this establishes is that every 4-jet with 3-jet xyz is equivalent to a jet of the form $xyz + ax^4 + by^4 + cz^4$. Such an affine subspace of $J^4(3, 1)$ is called a *complete transversal*. The orbit of every 4-jet with 3-jet xyz meets this space. So instead of having to look at a 15-dimensional space we are reduced to working in a space of dimension 3. Suppose now that we extend the group of equivalences and allow multiplication by a non zero constant, i.e. consider the action of the group $\mathcal{R} \times \mathbb{C}^*$. If any of a, b, c are non-zero we can scale to ensure that they are 1 (± 1 over $\mathbb{R}$). Suppose for example that $ab \neq 0$ but $c = 0$. Then we can reduce to $xyz + x^4 + y^4$. Then consider all 5-jets with this 4-jet. The same argument shows that we need only consider the additional terms x^5, y^5, z^5. However some ad-hoc changes of co-ordinates suggest that the x^5 and y^5 terms are now redundant, so a complete transversal is given by $xyz + x^4 + y^4 + cz^5$. If $c \neq 0$ we can scale to make $c = 1$. This analysis suggests that (in a very strong sense) most germs with 3-jet xyz are equivalent to $xyz + x^p + y^q + z^r$ for some $p, q, r \geq 4$. This is indeed the case, but we seek tools which leads us to this conclusion in a systematic way. The final step in this inductive approach is to decide when the given k-jet is *determined*, that is when is *any* germ with this k-jet equivalent to it? We shall also discuss this issue. For dynamical systems determinacy is generally very delicate. For most of the natural equivalences in singularity theory it can be shown to reduce to a certain number of 'consecutive transversals' being empty.

REMARK 4.2.   (1) The set of germs of diffeomorphisms $(\mathbb{R}^n, 0) \rightarrow (\mathbb{R}^n, 0)$ which we have denoted by $\mathcal{D}(n)$ is not a Lie group; neither is the space of all germs $(\mathbb{R}^n, 0) \rightarrow (\mathbb{R}^p, 0)$ a manifold. However the jet-spaces, the finite-dimensional approximations alluded to above and denoted $J^k(n, p)$, are clearly finite-dimensional vector spaces, and hence smooth manifolds. The set $\mathcal{D}^k(n)$ of k-jets of invertible mappings forms an open subset of $J^k(n, n)$. Such mappings do not form a group under map composition, but we can define a group structure using $j^k \phi_1 * j^k \phi_2 = j^k(\phi_1 \circ \phi_2)$. With this group operation of truncated composition $\mathcal{D}^k(n)$ *is* a Lie group; when $k = 1$ this is just the general linear group again.
(2) The same construction works for the group $\mathcal{C}(n, p)$.
(3) One can see that for each $1 \leq r \leq k$ the subset of $\mathcal{D}^k(n)$ consisting of polynomial mappings with r-jet the identity also forms a Lie group, denoted $\mathcal{D}^k_r(n)$. The case $r = 1$ is particularly important for us.
(4) There is an action of the Lie group $\mathcal{D}^k(n)$ on the manifold $J^k(n, p)$ via $(\phi, f) \mapsto j^k(f \circ \phi^{-1})$. This is one of the key examples. Of course there is also an action of the group $\mathcal{D}^k(p)$ on this jet-space via $(\psi, f) \mapsto j^k(\psi \circ f)$, that is by changes of co-ordinates on the target. In keeping with the notation used above these groups will be denoted by $\mathcal{R}^k(n)$, $\mathcal{L}^k(p)$ respectively, though we often omit the n and p. There is an action of the product group by $(\phi, \psi, f) \mapsto j^k(\psi \circ f \circ \phi^{-1})$, where we are changing co-ordinates in the source and target; this group is denoted by $\mathcal{A}^k(n, p)$. Finally there are corresponding actions for the groups $\mathcal{C}$ and $\mathcal{K}$, which we will leave the reader to define.

(5) Note that $H^k(n)^p$ can be thought of as a subset (indeed vector subspace) of $J^k(n,p)$, and then the action of $\mathcal{A}^k$ or $\mathcal{K}^k$ coincides with that of $Gl(n, \mathbb{R}) \times Gl(p, \mathbb{R})$ discussed above.

If G acts smoothly on X, and $x \in X$, then the map $\phi^x : G \to X$ defined by $\phi^x(g) = g.x$ has constant rank. So it plausible that orbits of smooth actions should be smooth manifolds, and they *are* locally. The proof essentially uses the rank theorem.

PROPOSITION 4.3. *Let G act smoothly on X with $x \in X$, and $y \in G.x$. There is some neighbourhood W of $e \in G$ such that $W.y = \{g.y : g \in G\}$ is a smooth manifold. (Of course $W.y \subset G.x$.)*

REMARK 4.4. Clearly the dimension of the orbit is at most the dimension of the group. It follows that there must be moduli in any classification for which dim $G <$ dim X. So for example the space of cubic curves is 10-dimensional and the group $Gl(3, \mathbb{R})$ of dimension 9. Similarly the space of quartic binary forms is of dimension 5, while $Gl(2, \mathbb{R})$ is of dimension 4. This explains why both have moduli. The converse is, of course, not true. For example consider the action of $Gl(n, \mathbb{C})$ on $M(n, n, \mathbb{C})$ by conjugation.

The tangent space to the orbit $G.x$ at the point y is the image of the differential $d\phi^y_e : T_eG \to T_yY$. We need to compute these tangent spaces, and give some sample calculations using the following trivial result.

PROPOSITION 4.5. *Let $X \subset \mathbb{R}^n$ and $Y \subset \mathbb{R}^m$ be smooth manifolds with $f : X \to Y$ a smooth map. Given any vector $v \in T_xX$ there is a smooth curve $\alpha : (-\epsilon, \epsilon) \to \mathbb{R}^n$ with $\alpha(0) = x$, $\alpha'(0) = d\alpha_0(1) = v$. Moreover $df_x(v) = \lim_{t \to 0} \frac{f(\alpha(t)) - f(\alpha(0))}{t}$.*

EXAMPLE 4.6. (1) Consider the action of $Gl(n, \mathbb{R})$ on $H^k(n)$, the space of homogeneous polynomials of degree k in n variables, given by $(A, h(x)) \mapsto h(A^{-1}(x))$. The orbit of an element h is the set $\{h(A^{-1}(x)) : A \in Gl(n, \mathbb{R})\}$. The inverse, though needed for this example to satisfy the definition of a group action, is an embarrassment. In fact clearly the tangent space to the orbit is given by the image of the derivative of the map $Gl(n, \mathbb{R}) \to H^k(n)$, $A \mapsto h(A(x))$ at I.

We now use some obvious curves in $Gl(n, \mathbb{R})$ and the previous Lemma to compute the tangent space to the orbit at a point h. Let E_{ij} denote the matrix with a 1 in the $(i, j)^{th}$ place and zeros elsewhere. So we obtain

$$\lim_{t \to 0} \frac{h(I + tE_{ij}) - h(I)}{t}$$

which is $x_j \partial h / \partial x_i$. For example in $H^2(3) = Sp\{x_1^2, x_2^2, x_3^2, x_2 x_3, x_1 x_3,$ $x_1 x_2\}$ the tangent space to the orbit $Gl(3, \mathbb{R}).h$ at $h = x_1 x_2$ is spanned by the vectors $x_1 x_2, x_2^2, x_2 x_3, x_1^2, x_1 x_3$.

(2) Let G be the group $\mathcal{R}^k(n)$ (or $\mathcal{R}^k(n)_1$) acting on $X = J^k(n, 1)$, the action being given by $(\phi, f) \mapsto j^k(f \circ \phi^{-1})$. As before we want to compute the tangent space to the orbit of a k-jet $f \in J^k(n, 1)$. As in the previous example we can ignore the inverse and consider paths γ in the groups $\mathcal{R}^k(n)$ or $\mathcal{R}^k(n)_1$ through the identity and then compute the relevant limits. If h is a polynomial of degree $\leq k$ with $h(0) = 0$, i.e. a k-jet, and i is an integer, with $1 \leq i \leq n$ consider the path $\gamma_{(h,i)} : (-\epsilon, \epsilon) \to \mathcal{R}^k(n)$ defined by

$$\gamma_{(h,i)}(t) = ((x_1, \ldots, x_n) \mapsto (x_1, \ldots, x_{i-1}, x_i + th(x), x_{i+1}, \ldots, x_n)).$$

This corresponds to the tangent vector $he_i \in T_e \mathcal{R}^k(n) \subset J^k(n, n)$. Note that this is a path in $\mathcal{R}^k(n)_1$ if and only if $h \in \mathcal{M}_n^2$. A short calculation shows that the corresponding tangent vector in $J^k(n, 1)$ at f is $j^k(h \partial f / \partial x_i)$, so the tangent space to the $\mathcal{R}^k$-orbit of f is spanned by k-jets of elements of $\mathcal{M}_n \langle \partial f / \partial x_1, \ldots, \partial f / \partial x_n \rangle$. The ideal generated by the partial derivatives of f, the Jacobian ideal, is denoted by J_f and this product $\mathcal{M}_n.J_f$ is referred to as *the $\mathcal{R}$-tangent space of f*. Similarly the tangent space to the $\mathcal{R}_1^k$-orbit is spanned by k-jets of elements of $\mathcal{M}_n^2.J_f$, this latter being the $\mathcal{R}_1$-*tangent space*.

EXAMPLE 4.7. Consider the k-jet $f = xyz \in J^k(3, 1)$. The tangent space to the $\mathcal{R}^k$-orbit of f is spanned by the vectors $j^k(h \partial f / \partial x) = j^k(hyz)$ and $j^k(hxz)$, $j^k(hxy)$ for $h \in J^k(n, 1)$. In other words we have $j^k\{\mathcal{M}_n \langle yz, xz, xy \rangle\}$.

We often need to decide when a subset of a space is contained in a single orbit, i.e. to answer the question 'when are a set of elements all equivalent?' So we needed to know above if $xyz + x^4 + y^4 + ax^5 + by^5 + z^5$ lay in a single orbit of the action of $\mathcal{R}^5 \times \mathbb{C}^*$ in $J^5(3, 1)$. Let G be a Lie group acting on a manifold X with Y a submanifold. When is Y contained in a single orbit of G? We seek infinitesimal criteria for such an inclusion. Clearly if Y is contained in a single orbit then $T_y G.y \supset T_y Y$ and $\dim T_y G.y$ is constant for all $y \in Y$. So the hypotheses of the following important result of Mather are clearly natural.

LEMMA 4.8. *Let G be a Lie group acting on a manifold X. Let $Y \subset X$ be a connected manifold with*
   (1) $\dim T_y G.y$, $y \in Y$ is independent of y,
   (2) $T_y G.y \supset T_y Y$ for all $y \in Y$.
Then Y is contained in a single orbit; the converse is obviously true.

Proof. Using the rank theorem one shows that for each $y \in Y$ the set $G.y \cap Y$ is open in Y. Since distinct orbits are disjoint, and Y is connected, the result follows. Let $k = \dim T_y G.y$; we claim that the action map $\psi : G \times Y \to X$, $\psi(g, y) = g.y$ has constant rank k. This is clearly the case at the points (e, y); for $d\psi_{(e,y)}(T_{(e,y)}G \times Y) = T_y G.y + T_y Y = T_y G.y$. Now given $g_0 \in G$ consider the diagram

$$
\begin{array}{ccc}
G \times Y & \xrightarrow{\gamma} & G \times Y \\
\psi \downarrow & & \downarrow \psi \\
X & \xrightarrow{\quad\delta\quad} & X
\end{array}
$$

where $\gamma(g, y) = (g_0 g, y)$ and $\delta(x) = g_0^{-1}.x$. It clearly commutes, the maps γ and δ are diffeomorphisms, so the rank of $d\psi$ at (g_0, y) is its rank at (e, y), as required.

Let U be an open neighbourhood of $0 \in \mathbb{R}^m$ with $\phi : U \to G \times Y$ a parametrisation of a neighbourhood of (e, y). The map $\psi \circ \phi : U \to X$ has constant rank k, so taking a chart at $y \in Y$ for X, that is an open neighbourhood V of y in X say and a diffeomorphism $\theta : V \to V_1$ where V_1 is an open subset of $\mathbb{R}^n$, the composite $H = \theta \circ \psi \circ \phi : U \to \mathbb{R}^n$ has constant rank k. Shrinking U if necessary we can choose co-ordinates on $\mathbb{R}^m$ and $\mathbb{R}^n$ so that $H(x_1, \ldots, x_m) = (x_1, \ldots, x_k, 0, \ldots, 0)$, so we have $\psi(\phi(U))$ a manifold of dimension k, which coincides with $G.y$ locally. However $\psi(\phi(U))$ contains a neighbourhood of $y \in Y$, and the result follows.

REMARK 4.9. This result provides a crucial tool in the classifications of germs. In practice, the second condition reduces to checking that one set of vectors are in the span of another set (which depend on certain parameters). Establishing the first condition involves checking that the dimension of the space spanned by this set is independent of the parameters. While the second hypothesis is reasonable to check, and in some sense the key issue, the first is generally rather difficult in practice, and our use of the lemma below avoids it.

We now establish our main technical tool (taken from [12]).

THEOREM 4.10. *Let G be a Lie group acting smoothly on a vector space V, with H a subspace of V satisfying*

$$
g.(\alpha + \beta) = g.\alpha + \beta
$$

for all $\alpha \in V$ and $\beta \in H$.

Given any $\alpha \in V$ we have

$$G.\alpha \cap (\{\alpha\} + H) \supseteq \{\alpha\} + (T_\alpha G.\alpha \cap H).$$

Moreover if W is a vector subspace of H with $H \subseteq W + T_\alpha G.\alpha$ then for any $\beta \in H$ we have $\alpha + \beta$ G-equivalent to (that is, in the same orbit as) some $\alpha + \beta'$ for some $\beta' \in W$.

Proof. We clearly need to show that $\{\alpha\} + (T_\alpha G.\alpha \cap H)$ is contained in a single orbit (that of α), so we need to check the hypotheses of Mather's Lemma. We shall establish them both simultaneously by showing that $T_{\alpha+\beta} G.(\alpha+\beta) = T_\alpha G.\alpha$ for all $\beta \in H$. Let $g : (-\epsilon, \epsilon) \to G$ be a smooth path through e. Then

$$\lim_{t \to 0} \frac{g.(\alpha + \beta) - (\alpha + \beta)}{t} = \lim_{t \to 0} \frac{g.(\alpha) - (\alpha)}{t}$$

so that $T_{\alpha+\beta} G.(\alpha + \beta) = T_\alpha G.\alpha$ and the result is proved.

To prove the second assertion note that since, for any $\beta \in H$,

$$G.(\alpha + \beta) \cap (\{\alpha + \beta\} + H) \supseteq \{\alpha + \beta\} + (T_{\alpha+\beta} G.(\alpha + \beta) \cap H),$$

which is equal to $\{\alpha + \beta\} + (T_\alpha G.\alpha \cap H)$, we have

$$\cup_{\beta \in W} G.(\alpha + \beta) \supseteq \cup_{\beta \in W} (\alpha + \beta + T_\alpha G.\alpha \cap H) = \{\alpha\} + W + T_\alpha G.\alpha \cap H$$

which contains $\{\alpha\} + H$ as required.

We shall, for example, apply this result when $V = J^k(n,1)$, $G = \mathcal{R}_1^k$ and $H = H^k(n)$ is the subspace of homogeneous polynomials of degree k. One easily checks that for any $f \in J^k(n,1)$, $F \in H^k(n)$, $\phi \in \mathcal{R}_1^k$ we have $\phi.(f + F) = f \circ \phi + F \circ \phi = f \circ \phi + F = \phi.f + F$. Actually it is easier to check this at an infinitesimal level. For this we need a little new notation.

DEFINITION 4.11. *Let G be a Lie group acting on a vector space V. Given $\alpha \in V$ and $l \in T_e G$ we define $l.\alpha$ to be the vector $d\phi_e^\alpha(l) \in V$. If we write $T_e G.\alpha$ for $\{l.\alpha : l \in T_e G\}$ then $T_e G.\alpha = T_\alpha(G.\alpha)$.*

We now have the following

THEOREM 4.12. *(Addendum to Main Theorem) If we replace the condition $g.(\alpha + \beta) = g.\alpha + \beta$ for all $\alpha \in V$, $\beta \in H$ and $g \in G$ by $l.(\alpha + \beta) = l.\alpha$ for all $\alpha \in V$, $\beta \in H$ and $l \in T_e G$ then the theorem continues to hold.*

Proof. The point is that

$$T_{\alpha+\beta}G.(\alpha+\beta) = T_eG.(\alpha+\beta) = T_eG.\alpha = T_\alpha G.\alpha$$

for all $\alpha \in V$, $\beta \in H$, and this was all that was used in the proof above.

EXAMPLE 4.13. If f is a k-jet, and $F \in H^k(n)$ the space of homogeneous polynomials of degree k then given a vector of the form $h(x)e_i$ in $T_e\mathcal{R}_1^k$, so $h \in \mathcal{M}_n^2$, we have $he_i.(f+F) = j^k(h.\partial(f+F)/\partial x_i) = j^k(h.\partial f/\partial x_i) = he_i.f$ as required.

DEFINITION 4.14. *A space of the form W in Theorem 4.9 is called a complete transversal.*

The key challenge is to try to make the space W as small as possible.

5. Classification of Functions

The results from the previous section yield the following tool for the classification of functions.

THEOREM 5.1. *Let $f : (\mathbb{R}^n, 0) \to (\mathbb{R}, 0)$ be a polynomial germ of degree k, and let $\{G_1, \ldots, G_r\}$ be a collection of homogeneous polynomials of degree $k+1$ with the property that*

$$\mathcal{M}_n^2.J_f + sp\{G_1, \ldots, G_r\} + \mathcal{M}_n^{k+2} \supseteq \mathcal{M}_n^{k+1}.$$

Then any polynomial germ $g \in J^{k+1}(n, 1)$ with $j^k g = j^k f$ is $\mathcal{R}^{k+1}$-equivalent to a jet of the form

$$f(x) + \sum_{i=1}^{r} u_i G_i(x).$$

Moreover this $\mathcal{R}$-equivalence has 1-jet the identity.
If $r = 0$, and $N \geq k$ then any N-jet g with $j^k(g) = f$ is $\mathcal{R}^N$-equivalent to f.

   Proof. The first assertion is the content of the transversal result. For the second we note that if $r = 0$ then $\mathcal{M}_n^2.J_f + \mathcal{M}_n^{k+2} \supseteq \mathcal{M}_n^{k+1}$. Multiplying through by powers of the maximal ideal it follows that if we take f as our N-jet, then the complete transversal at the $(N+1)$-jet level is also empty.

DEFINITION 5.2. *If we have a k-jet $f \in J^k(n,p)$ with the property that for any $N \geq k$ and N-jet g, with $j^k g = f$, g is $\mathcal{G}^N$-equivalent to f we say that f is k-$\mathcal{G}$-formally-determined. (Here $\mathcal{G}$ is one of $\mathcal{R}, \mathcal{L}, \mathcal{A}, \mathcal{C}, \mathcal{K}$.)*

With this definition we see we have proved:

COROLLARY 5.3. *Let $f : (\mathbb{R}^n, 0) \to (\mathbb{R}, 0)$ be a smooth function-germ with $\mathcal{M}_n^2.J_f + \mathcal{M}_n^{k+2} \supseteq \mathcal{M}_n^{k+1}$. Then f is k-$\mathcal{R}$-formally-determined.*

REMARK 5.4. (1) A preferable property of f would be that any germ g with $j^k g = f$ is actually $\mathcal{R}$-equivalent to f. We then say that f is k-$\mathcal{R}$-determined. It turns out that if f is k-$\mathcal{R}$-formally-determined it is actually k-$\mathcal{R}$-determined. (See [8].) This means that in $\mathcal{R}$-classifications as soon as we reach a k-jet with empty transversal we can end the classification: all germs with this k-jet are $\mathcal{R}$-equivalent. We shall see that the same results hold true if we replace $\mathcal{R}$ by $\mathcal{K}$. We shall also see that the inclusion of Corollary 5.3 implies that $\mathcal{M}_n^2.J_f \supset \mathcal{M}_n^{k+1}$.
(2) Note that if the above inclusion holds then $\mathcal{M}_n^{s+1}.J_f + \mathcal{M}_n^{k+s+1} \supseteq \mathcal{M}_n^{k+s}$ for each $s \geq 1$. It follows that the $(k+s)$-transversal of f is empty for the group of diffeomorphisms with s-jet the identity. Let g be a germ with the same k-jet as f. We deduce that the sequence of diffeomorphisms which one constructs at the jet-level to obtain an equivalence between f and the N-jets of g can be chosen to converge as $N \to \infty$ (though this does not prove that the limit is differentiable).

We now apply Theorem 5.1 inductively to yield a classification of an interesting collection of function germs. (What we will list, inefficiently, is a tiny subset of the collection of germs obtained by Arnold who developed very powerful techniques to cover this case. Our methods however also yield, for example, $\mathcal{A}$-classifications of mappings.)

EXAMPLE 5.5. (1) Note that by the rank theorem any germ $f : (\mathbb{R}^n, 0) \to (\mathbb{R}, 0)$ of rank 1 at 0 is $\mathcal{R}$-equivalent to $f(x) = x_1$. So we need only deal with germs whose 1-jets are 0, that is singular germs.
(2) Consider next homogeneous 2-jets. We know that the action of the group $\mathcal{R}^2$ is simply that of the general linear group. So we have normal forms at the 2-jet level $\sum_{i=1}^r \epsilon_i x_i^2$, with $\epsilon_i = \pm 1$. If $r = n$ then the 3-transversal is empty, and the germ is 2-$\mathcal{R}$-determined. (These yield the so-called Morse singularities, labelled by Arnold A_1.)
(3) In the case $n = 1$ the result shows that any germ $f : (\mathbb{R}, 0) \to (\mathbb{R}, 0)$ is either $\mathcal{R}$-equivalent to some power x^{k+1}, or for each N is $\mathcal{R}$-equivalent to a germ whose N-jet is 0. (This is easy to see directly.)

(4) Suppose now that $f : (\mathbb{R}^n, 0) \to (\mathbb{R}, 0)$ is singular and the Hessian (quadratic part) has rank r. Then by a linear change of co-ordinates we can reduce the 2-jet of f to $\sum_{i=1}^r \epsilon_i x_i^2$. Now suppose that f is a k-jet of the form $\sum_{i=1}^r \epsilon_i x_i^2 + g(x_{r+1}, \ldots, x_n)$; in other words a k-jet where the variables $x_1, \ldots, x_r$ only occur in the initial sum of squares. Then since $\mathcal{M}_n^2 \langle \partial f / \partial x_1, \ldots, \partial f / \partial x_n \rangle$ contains $\mathcal{M}_n^2 \langle x_1, \ldots, x_r \rangle$ we can always choose our G_i to be polynomials in the variables $x_{r+1}, \ldots, x_n$. A little thought shows that the classification procedure only produces $(k+1)$-jets of the form $f(x) = \sum_{i=1}^r \epsilon_i x_i^2 + g(x_{r+1}, \ldots, x_n)$. In other words the $x_1, \ldots, x_r$ variables have been 'split off' at the 2-jet level, and play no further part. This is a version of an important result called the *Splitting Lemma*. The upshot is that when carrying out our classifications we may always suppose that the 2-jet of our germ is identically zero.

(5) We deduce that any germ f whose Hessian has rank $(n-1)$ is $\mathcal{R}$-equivalent to a germ of the form $\sum_{i=1}^{n-1} \epsilon_i x_i^2 \pm x_n^{k+1}$, (an A_k-singularity), or for each N is $\mathcal{R}$-equivalent to a germ whose N-jet is $\sum_{i=1}^{n-1} \epsilon_i x_i^2$. (The second type being infinitely rare in a certain precise sense.)

(6) Consider next 3-jets of germs $(\mathbb{R}^2, 0) \to (\mathbb{R}, 0)$ with 2-jet 0. There are just 5 types: $x(x^2 \pm y^2)$, $x^2 y$, x^3 and 0. The first pair are 3-$\mathcal{R}$-determined. Regarding $x^2 y$ as a k-jet it is easy to see that a complete transversal is $x^2 y + a y^{k+1}$, and that if $a \neq 0$ this is equivalent to $x^2 y \pm y^{k+1}$ which is $(k+1)$-$\mathcal{R}$-determined, a so-called D_{k+2} singularity.

(7) We now illustrate one of the shortcomings of this method. Consider the 3-jet x^3. A 4-transversal is given by $x^3 + a x y^3 + b y^4$. If $b \neq 0$ we may suppose that $b = \pm 1$ and one can apply Mather's lemma to show that this family is contained in a single $\mathcal{R}^4$-orbit, and a representative $x^3 \pm y^4$ is 4-$\mathcal{R}$-determined. If $b = 0$, $a \neq 0$ we can reduce to the 4-jet $x^3 + x y^3$; a complete 5-transversal is now given by $x^3 + x y^3 + a y^5$; again one has to employ Mather's lemma to show that we may take $a = 0$. In other words the complete transversals produced are not always minimal.

EXAMPLE 5.6. (1) We revisit an earlier example. Consider functions of 3 variables. As we have seen we may suppose that the 2-jet is zero. Then we have to start by classifying homogeneous germs of degree 3, up to $Gl(3)$-equivalence. If, for example, we start with 3 distinct and non-concurrent lines, say xyz, then you can check that a complete 4-transversal is given by $xyz + a x^4 + b y^4 + c z^4$. Use the theorem to establish the result mentioned above, namely that this 3-jet yields a triply indexed series $\lambda xyz + x^p + y^q + z^r$ where $p, q, r \geq 4$, $\lambda \neq 0$. (Of course if we extend the group as before to $\mathcal{R} \times \mathbb{C}^*$ we may suppose $\lambda = 1$.)

REMARK 5.7. Note that we begin all classifications with a homogeneous jet. The complete transversal method requires some initial jet to 'get started' before it begins to give information. In many instances the main problem is in locating the initial jet or 'stem' to work with.

6. Classification of Mappings

In this section we illustrate how to apply these results to the case of mappings, that is germs with target dimension > 1. We shall start with a quick look at a (contact) $\mathcal{K}$-classification, and then move to the more difficult case of $\mathcal{A}$-classifications. The basic technique remains that of complete transversals, and the key is finding a group that satisfies the hypotheses of the complete transversal result. Determinacy results are considerably more difficult to prove in the $\mathcal{A}$-case, as we shall see.

For $\mathcal{A}$-equivalence the groups we are interested in are respectively $\mathcal{A}^k$ and $\mathcal{A}_1^k$. We need to compute the tangent space to the orbits of their actions. This will be made up of two parts, one emerging from the group of changes of co-ordinates in the source, and one from those in the target. We start with the former.

Recall that the tangent space to $\mathcal{R}^k$ at the identity is spanned by $h(x)e_i$ where $h(x)$ is polynomial of degree $\leq k$ and $h(0) = 0$. A path $\gamma(t)$ in $\mathcal{R}^k$ through the identity with this as its tangent vector is given by $\gamma(t) = ((x_1, \ldots, x_n) \mapsto (x_1, \ldots, x_{i-1}, x_i + th(x), x_{i+1}, \ldots, x_n))$. Given a jet $f \in J^k(n,p)$ we need to consider the limit $\lim_{t \to 0} \frac{\gamma(t).f(x) - f(x)}{t}$. This yields $j^k(h(x)\partial f(x)/\partial x_i)$. So this part of the tangent space to the orbit is spanned by k-jets of germs of the form $h(x)\partial f(x)/\partial x_i$ where $h(x) \in \mathcal{M}_n$ and $1 \leq i \leq n$. In the case of the group $\mathcal{A}_1$ we know that $h(x)$ must lie in $\mathcal{M}_n^2$. Now consider the changes of co-ordinates in the target. Again the tangent space to $\mathcal{L}^k$ at the identity is spanned by $g(y)e_i$ where $g(y)$ is polynomial of degree $\leq k$ and $g(0) = 0$. A path γ in $\mathcal{L}^k$ through the identity with this as its tangent vector is given by $\gamma(t) = ((y_1, \ldots, y_p) \mapsto (y_1, \ldots, y_{i-1}, y_i + tg(y), y_{i+1}, \ldots, y_p))$. Given a jet $f \in J^k(n,p)$ we need to consider the limit $\lim_{t \to 0} \frac{\gamma(t).f(x) - f(x)}{t}$. This time we get $j^k(g(f(x))e_i)$ where $1 \leq i \leq p$. Again in the case of the group $\mathcal{A}_1$ we must have $g(y)$ in $\mathcal{M}_p^2$. Summing up, we find:

PROPOSITION 6.1. *The tangent space to the $\mathcal{A}^k$-orbit of a k-jet $f \in J^k(n,p)$ is the sum of the set of k-jets of germs in $\mathcal{E}(n,p)$ of the form $h(x)\partial f(x)/\partial x_i$, with $h(x) \in \mathcal{M}_n$ and $1 \leq i \leq n$, and the set of k-jets of the form $g(f(x))e_i$, with $g(y) \in \mathcal{M}_p$ and $1 \leq i \leq p$. For the $\mathcal{A}_1^k$-orbit the tangent space is as above, but with $h(x) \in \mathcal{M}_n^2$ and $g(y) \in \mathcal{M}_p^2$. These can be expressed (respectively) as the set of k-jets of elements of*

$$\mathcal{M}_n.J_f + f^*\mathcal{M}_p\{e_1, \ldots, e_p\} \quad \text{and} \quad \mathcal{M}_n^2.J_f + (f^*\mathcal{M}_p)^2\{e_1, \ldots, e_p\},$$

where J_f is the Jacobian module generated by the partial derivatives of f, and $f^*(I)$ denotes, for some ideal $I \subset \mathcal{E}_p$, the set of elements of the form $g \circ f$ where $g \in I$. We refer to these as the $\mathcal{A}$- and $\mathcal{A}_1$-tangent spaces of f.

We can similarly prove the following.

PROPOSITION 6.2. *The tangent space to the $\mathcal{K}^k$-orbit of a k-jet $f \in J^k(n,p)$ is the sum of the set of k-jets of germs in $\mathcal{E}(n,p)$ of the form $h(x)\partial f(x)/\partial x_i$, with $h(x) \in \mathcal{M}_n$ and $1 \leq i \leq n$ together with set of k-jets of the form $g(x)f_j(x)e_i$, with $g(x) \in \mathcal{E}_n$ and $1 \leq i \leq p$. For the $\mathcal{K}_1^k$-orbit the tangent space is as above, but with $h(x) \in \mathcal{M}_n^2$ and $g(x) \in \mathcal{M}_n$.*
These are k-jets of elements of $\mathcal{M}_n.J_f + \mathcal{E}_n.f^*\mathcal{M}_p\{e_1,\ldots,e_p\}$ and $\mathcal{M}_n^2.J_f + \mathcal{M}_n.f^*\mathcal{M}_p\{e_1,\ldots,e_p\}$ respectively, which we refer to as the $\mathcal{K}$- and $\mathcal{K}_1$-tangent spaces of f.

REMARK 6.3. (1) We shall use the notation $T\mathcal{G}f$ for the $\mathcal{G}$ tangent space to f, where $\mathcal{G}$ is one of our usual groups $\mathcal{R}, \mathcal{L}, \mathcal{C}, \mathcal{A}, \mathcal{K}, \mathcal{R}_1$ etc. So $T\mathcal{R}f = \mathcal{M}_n.J_f, T\mathcal{R}_1f = \mathcal{M}_n^2.J_f$.
(2) Note an important difference here: the $\mathcal{K}$ and $\mathcal{K}_1$-tangent spaces are $\mathcal{E}_n$-modules, while the $\mathcal{A}$ and $\mathcal{A}_1$-tangent spaces generally are not.

We see how this all works out in some simple examples.

EXAMPLE 6.4. (1) Consider the k-jet $(x_1,0)$ in $J^k(2,2)$. Differentiating with respect to x_1 we obtain $(1,0)$ so the tangent space to the $\mathcal{A}^k$-orbit contains everything of the form $j^k(h(x_1,x_2),0)$ where $h(x_1,x_2) \in \mathcal{M}_2$. On the other hand from changes of co-ordinates in the target we obtain $j^kg(x_1,0)(1,0)$ and $j^kg(x_1,0)(0,1)$ for any $g \in \mathcal{M}_2$. So, for example, taking $k=2$ the tangent space is spanned by $(x_1,0), (x_2,0), (x_1^2,0), (x_1x_2,0), (x_2^2,0), (0,x_1), (0,x_1^2)$.
(2) In the case when we have a k-jet (x_1,x_2^2) in $J^k(2,2)$ we obtain everything of the form $j^k(h(x_1,x_2),0)$ and of the form $j^k(0,x_2h(x_1,x_2))$ where $h(x_1,x_2) \in \mathcal{M}_2$. On the other hand, from changes of co-ordinates in the target we obtain $j^kg(x_1,x_2^2)(1,0)$ and $j^kg(x_1,x_2^2)(0,1)$ for any $g \in \mathcal{M}_2$. A little thought shows that we obtain all monomials in both slots apart from $(0,x_2)$.

If we are to use the complete transversal method we need to check that the hypotheses of Theorem 4.20 hold.

LEMMA 6.5. *For the action of the Lie group $\mathcal{A}_1^k$ on the jet space $J^k(n,p)$ the following holds. Given $f \in J^k(n,p)$ and $F \in H^k(n,p)$, the subspace of homogeneous germs of degree k, we have $l.(f+F) = l.f$ for all $l \in T_e\mathcal{A}_1^k$. The same is true if we replace $\mathcal{A}_1^k$ by $\mathcal{K}_1^k$.*

Proof. Consider the tangent vectors from changes of co-ordinates in the source: given $l = h(x)e_i$, with $h \in \mathcal{M}_n^2$, $1 \leq i \leq n$ we have

$$l.(f+F) = j^k(h(x)(\partial(f+F)/\partial x_i)) = j^k(h(x)(\partial f(x)/\partial x_i + \partial F(x)/\partial x_i))$$

this is $j^k(h(x)\partial f(x)/\partial x_i) = l.f$. (Since F is homogeneous of degree k, its derivatives will be homogeneous of degree $k-1$, and multiplying by an element of $\mathcal{M}_n^2$, and then taking k-jets we lose this term.) For the target given $l = g(y)e_i$, with $g \in \mathcal{M}_p^2$, $1 \leq i \leq p$ we have

$$l.(f+F) = j^k(g(f+F)(x)e_i) = j^k(g(f(x))e_i) = l.f.$$

Again the key is that $g \in \mathcal{M}_p^2$. The $\mathcal{K}$ case is left as an exercise.

The fact that the $\mathcal{K}$-tangent space is an $\mathcal{E}_n$-module means that we have a formal determinacy result.

THEOREM 6.6. *Let $f : (\mathbb{R}^n, 0) \to (\mathbb{R}^p, 0)$ be a germ. Then if*

$$T\mathcal{K}_1.f + \mathcal{M}_n^{k+2}.\mathcal{E}(n,p) \supset \mathcal{M}_n^{k+1}.\mathcal{E}(n,p)$$

then f is k-$\mathcal{K}$-formally-determined.

Proof. The k-jet of f has an empty $(k+1)$-transversal. But it then follows that all subsequent transversals are also empty. (Why?)

EXAMPLE 6.7. Consider the $\mathcal{K}$-classification of map-germs $(\mathbb{R}^2, 0) \to (\mathbb{R}^2, 0)$ with zero 1-jet. The classification of homogeneous 2-jets reduces to that of pencils of binary quadratics which we completed earlier. So, for example, we have the types (x^2, y^2) and $(x^2 - y^2, xy)$ both of which are 2-$\mathcal{K}$-determined. Consider next the jet (x^2, xy); thinking of this as a k-jet we find that a complete transversal is given by $(x^2 + ay^{k+1}, xy)$, and if $a \neq 0$ we obtain $(x^2 \pm y^{k+1}, xy)$ which is $(k+1)$-$\mathcal{K}$-determined. Next consider the 2-jets $(x^2 \pm y^2, 0)$. In the minus case we can use the alternative normal form $(xy, 0)$. Thinking of this as a k-jet a complete transversal is given by $(xy, ax^{k+1} + by^{k+1})$. If $ab \neq 0$ we reduce to $(xy, \pm x^{k+1} \pm y^{k+1})$, which is $(k+1)$-$\mathcal{K}$-determined. If say $a \neq 0$, $b = 0$ we reduce to (xy, x^{k+1}), which regarded as an l-jet has complete transversal $(xy, x^{k+1} + by^{l+1})$. If $b \neq 0$ we reduce to the normal form $(xy, x^{k+1} \pm y^{l+1})$, which is $(l+1)$-determined. Investigate the germs emerging from $(x^2 + y^2, 0)$ and the other (more degenerate) pencils in Exercise 3.3. Note that over $\mathbb{C}$ we can eliminate the $\pm$ signs, over $\mathbb{R}$ we can sometimes do this depending on the parity of the powers.

We now consider a simple example of an $\mathcal{A}$-classification, namely that of germs $f : (\mathbb{R}, 0) \to (\mathbb{R}^2, 0)$.

EXAMPLE 6.8. (1) If the 1-jet is non-zero we can reduce to $(x,0)$ by the rank theorem.

(2) There is one class of non-zero 2-jets, namely $(x^2,0)$. A complete transversal for this as a k-jet in $J^{k+1}(1,2)$ is (x^2, ax^{k+1}) if k is even, and empty if k is odd. If $a \neq 0$ we scale to (x^2, x^{2m+1}), $k = 2m$, and all subsequent transversals are empty.

(3) Now consider homogeneous 3-jets. Again there is only one non-zero type, $(x^3,0)$. Thinking of this as a $3k$-jet (resp. $(3k+1)$-jet, $(3k+2)$-jet) a complete transversal is (x^3, ax^{3k+1}) (resp. (x^3, ax^{3k+2}), $(x^3,0)$), and this yields new jets (x^3, x^{3k+1}), (x^3, x^{3k+2}). Consider the first: if this is a $3l$ or $(3l+2)$-jet the corresponding transversals are empty. As a $(3l+1)$-jet we obtain $(x^3, x^{3k+1} + ax^{3l+2})$ as a transversal if $k \leq l \leq 2k - 1$ and an empty transversal otherwise. If $a \neq 0$ we reduce to $(x^3, x^{3k+1} \pm x^{3l+2})$, all subsequent transversals are empty, and we obtain this and (x^3, x^{3k+1}) as a list of normal forms. From the second case we obtain the family $(x^3, x^{3k+2} \pm x^{3l+1})$ where $k+1 \leq l \leq 2k$, and (x^3, x^{3k+2}).

(4) Finally consider the 4-jet $(x^4,0)$. At the 5-jet level we obtain either (x^4, x^5) or $(x^4, 0)$. In the former case the 6-transversal is empty, and the 7-transversal $(x^4, x^5 + ax^7)$. If $a \neq 0$ we reduce to $(x^4, x^5 \pm x^7)$ and one can check that all subsequent transversals are empty.

In the above cases we could complete a (formal) classification because it was easy to see when future transversals were empty, even though it was no longer true that one transversal empty implies the same is true for all subsequent ones. To carry out more complicated $\mathcal{A}$-classifications however we need to have a determinacy result to use. (In the curve case above we can in fact apply a rather easy determinacy result for the $\mathcal{L}$-group; see [14].)

DEFINITION 6.9. *A germ* $f : (\mathbb{R}^n, 0) \to (\mathbb{R}^p, 0)$ *is* k-$\mathcal{A}$-*determined if every germ* $g : (\mathbb{R}^n, 0) \to (\mathbb{R}^p, 0)$ *with the same* k-*jet as* f *is* $\mathcal{A}$-*equivalent to* f. *We say that* f *is* finitely-$\mathcal{A}$-*determined if it is* k-$\mathcal{A}$-*determined for some* k. *The same terminology is used for the group* $\mathcal{A}_1$, *and indeed the other groups* $\mathcal{R}$, $\mathcal{L}$, $\mathcal{C}$, $\mathcal{K}$, $\mathcal{R}_1$ *etc.*

THEOREM 6.10. *(1) A germ* $f : (\mathbb{R}^n, 0) \to (\mathbb{R}^p, 0)$ *is finitely* $\mathcal{A}$-*determined if and only if for some* N *we have* $\mathcal{M}_n^N.\mathcal{E}(n,p) \subset T\mathcal{A}f$.

(2) A germ $f : (\mathbb{R}^n, 0) \to (\mathbb{R}^p, 0)$ *is* $(2r+1)$-$\mathcal{A}$-*determined if we have* $\mathcal{M}_n^{r+1}.\mathcal{E}(n,p) \subset T\mathcal{A}f + \mathcal{M}_n^{2r+2}.\mathcal{E}(n,p)$.

(3) A germ $f : (\mathbb{R}^n, 0) \to (\mathbb{R}^p, 0)$ *is* r-$\mathcal{A}_1$-*determined if and only if we have* $\mathcal{M}_n^{r+1}.\mathcal{E}(n,p) \subset T\mathcal{A}_1f$.

(4) A germ $f : (\mathbb{R}^n, 0) \to (\mathbb{R}^p, 0)$ *is* r-$\mathcal{A}_1$-*determined if and only if we have* $\mathcal{M}_n^{r+1}.\mathcal{E}(n,p) \subset T\mathcal{A}_1f + \mathcal{M}_n^{r+1}.(f^*M_p.\mathcal{E}_n + \mathcal{M}_n^{r+1}).\mathcal{E}(n,p)$.

REMARK 6.11. Of these criteria for determinacy (4) is the most useful, but looks rather complicated. Statement (3) is the easiest to understand. This is the analogue of our determinacy result for the group $\mathcal{R}_1$. However to establish k-$\mathcal{R}$-determinacy we only had to check that the homogeneous terms of degree $(k+1)$ were in the ideal $\mathcal{M}_n^2.J(f)$. Result (4) provides similar help; the extra term on the right hand side makes checking the inclusion much easier. We illustrate with some examples before discussing the proofs of these results. Sadly the degrees of $\mathcal{A}_1$- and $\mathcal{A}$-determinacy do not coincide, as we shall see in our second example. (See [15], p 530 for this and 'best possible' determinacy results using the ideas discussed in Section 7.)

EXAMPLE 6.12.   (1) To prove, for example that $f = (x^3, x^{3k+1} + x^{3l+2})$ is $(3l+2)$-$\mathcal{A}$-determined, for $k \leq l \leq 2k - 1$ we need to check that $\mathcal{M}_1^{3l+3}.\mathcal{E}(1,2) \subset T\mathcal{A}_1 f + \mathcal{M}_1^{3l+6}.\mathcal{E}(1,2)$ which is straightforward. (2) One can easily check that the fold map (x, y^2) is 2-$\mathcal{A}$-determined using the above criteria. Consider the cusp singularity $(x, y^3 + xy)$; this is 4-$\mathcal{A}_1$-determined. Before computing the $\mathcal{A}_1$-tangent space consider the extra term referred to above. The first thing to note is that $f^*(\mathcal{M}_p).\mathcal{E}_n$ is the *ideal* generated by the components of f, in this case generated by x, y^3. So when checking the inclusion we can work modulo $\mathcal{M}_2^5(\mathcal{E}_2.\langle x, y^3 \rangle + \mathcal{M}_2^5).\mathcal{E}(2,2)$, and can ignore all terms of degree ≥ 8, all terms of degree ≥ 6 divisible by x. Establishing the inclusion is now straightforward but tedious. An application of the transversal method and Mather's Lemma shows that this germ is, in fact, 3-$\mathcal{A}$-determined.

Proof of Part of Theorem 6.11.

We shall prove part (3) assuming (1). So suppose that $\mathcal{M}_n^{r+1}.\mathcal{E}(n,p) \subset T\mathcal{A}_1 f$. Since $T\mathcal{A}_1 f \subset T\mathcal{A}f$ it follows from (1) that f is N-$\mathcal{A}$-determined for some N. Now let g be a germ with the same r-jet as f. We shall prove for each $k \geq 0$ that g is $\mathcal{A}$-equivalent to a germ g_k with the same $(r+k)$-jet as f. Since f is N-$\mathcal{A}$-determined (3) follows.

The result is established for $k = 0$. So suppose that g_{k+r} is a germ which is $\mathcal{A}$-equivalent to g and with the same $(k+r)$-jet as f. Then in the jet-space $J^{k+r+1}(n,p)$ the $(k+r)$-jet of f has an empty transversal, so $j^{k+r+1}g_k$ is $\mathcal{A}_1^{k+r+1}$-equivalent to $j^{k+r+1}f$. Applying this equivalence to g_k we obtain g_{k+1} with the required properties. Note that this proves that any germ f for which $\mathcal{M}_n^{r+1}.\mathcal{E}(n,p) \subset T\mathcal{A}_1 f$ holds is formally r-$\mathcal{A}_1$-determined (no need to assume result (1)).

Full references for proofs of the above can be found in [15]. However we can also establish (4) for formal determinacy. We first need a simple result in (linear) commutative algebra. This has direct applications to $\mathcal{R}$- and $\mathcal{K}$-determinacy estimates, as we see in the corollary which follows.

LEMMA 6.13. *(Nakayama's Lemma.) Suppose that R is a commutative ring with identity which has an ideal M for which $1+m$ is invertible for all $m \in M$. (The examples we have in mind are of course $R = \mathcal{E}_n$, $M = \mathcal{M}_n$.) Let A be a finitely-generated R-module, and B an R-module. Suppose that $A \subset B + M.A$. Then $A \subset B$.*

Proof. Let $e_1, \ldots, e_p$ be generators for A. By hypothesis each can be written $e_i = b_i + \sum m_{ij} e_j$, with the $b_i \in B$, and $m_{ij} \in M$. So writing e for the vector $(e_1, \ldots, e_p)^T$, $b = (b_1, \ldots, b_r)^T$ and m for the matrix (m_{ij}) we find $e = me + b$, so $(I - m)e = b$. Now $\det(I - m)$ is invertible, by hypothesis, so multiplying by the adjoint of this matrix we see that each e_i lies in B.

COROLLARY 6.14. *(i) If $\mathcal{M}_n^{r+1} \subset T\mathcal{R}_1 f + \mathcal{M}_n^{r+2}$ then $\mathcal{M}_n^{r+1} \subset T\mathcal{R}_1 f$ (and f is r-$\mathcal{R}$-determined).*

(ii) If $\mathcal{M}_n^{r+1}.\mathcal{E}(n,p) \subset T\mathcal{K}_1 f + \mathcal{M}_n^{r+2}.\mathcal{E}(n,p)$ then $\mathcal{M}_n^{r+1}.\mathcal{E}(n,p) \subset T\mathcal{K}_1 f$ (and f is r-$\mathcal{K}$-determined).

PROPOSITION 6.15. *Suppose that f is as above and $\mathcal{M}_n^{r+1}.\mathcal{E}(n,p) \subset TA_1 f + \mathcal{M}_n^{r+1}.(f^*\mathcal{M}_p.\mathcal{E}_n + \mathcal{M}_n^{r+1}).\mathcal{E}(n,p)$. Then f is formally r-$\mathcal{A}$-determined.*

Proof. We need to show that if the above inclusion holds then for any $N \geq 2r+2$ we have $\mathcal{M}_n^{r+1}.\mathcal{E}(n,p) \subset TA_1 f + \mathcal{M}_n^N.\mathcal{E}(n,p)$ (this was all that was used in the argument above). Note that the quotient $\mathcal{M}_n^{r+1}.\mathcal{E}(n,p)/\mathcal{M}_n^N.\mathcal{E}(n,p)$ is a finite dimensional $\mathbb{R}$-vector space, so clearly a finitely generated $\mathcal{E}_p$-module. If we take $N = 2r+2$ then by Nakayama's Lemma the given inclusion implies that $\mathcal{M}_n^{r+1}.\mathcal{E}(n,p) \subset TA_1 f + \mathcal{M}_n^{2r+2}.\mathcal{E}(n,p)$.

We shall prove by induction on k that $\mathcal{M}_n^{r+1}.\mathcal{E}(n,p) \subset TA_1 f + (f^*\mathcal{M}_p.\mathcal{M}_n^{r+1} + \mathcal{M}_n^{k(r+1)}).\mathcal{E}(n,p)$. The case $k = 2$ is our hypothesis, and suppose the result holds for k. Let $\alpha \in \mathcal{M}_n^{r+1}.\mathcal{E}(n,p)$; then $\alpha = a+b+c$ where $a \in TA_1 f$, $b \in f^*\mathcal{M}_p.\mathcal{M}_n^{r+1}.\mathcal{E}(n,p)$ and $c \in \mathcal{M}_n^{k(r+1)}.\mathcal{E}(n,p)$.

By induction we can write $c = \sum \gamma_i a_i$, with $\gamma_i \in \mathcal{M}_n^{(k-1)(r+1)}$, $a_i \in TA_1 f + \mathcal{M}_n^{2r+2}$. Using the fact that the $\mathcal{R}_1$-tangent space is an $\mathcal{E}_n$-module, and the $\mathcal{L}_1$-tangent space is a subset of $f^*\mathcal{M}_p.\mathcal{E}(n,p)$, we see that α lies in $TA_1 f + f^*\mathcal{M}_p.\mathcal{M}_n^{r+1}.\mathcal{E}(n,p) + \mathcal{M}_n^{(k-1)(r+1)}.(f^*\mathcal{M}_p.\mathcal{E}_n + \mathcal{M}_n^{2r+2}).\mathcal{E}(n,p)$ and the inclusion follows. Applying Nakayama's Lemma again we deduce that $\mathcal{M}_n^{r+1}.\mathcal{E}(n,p) \subset TA_1 f + \mathcal{M}_n^{k(r+1)}.\mathcal{E}(n,p)$ which establishes the result.

EXAMPLE 6.16.   Classifications up to $\mathcal{A}$-equivalence are quite involved and there is much to recommend an approach using these methods implemented on a computer! Here we briefly discuss aspects of a
classification of germs $(I\!\!R^2, 0) \to (I\!\!R^2, 0)$. We use (x, y) co-ordinates
in the source and (u, v) co-ordinates in the target. Any change of coordinates obtained by multiplying the x, y, u, v variables by non-zero
constants is called *scaling*. Consider the germs of rank 1 so that we
have a 1-jet $(x, 0)$. A complete 2-transversal is given by $(x, axy + by^2)$.
If $b \neq 0$ then we can, by scaling, suppose that $b = 1$. A change of
co-ordinates $(x, y) \mapsto (x, y - ax/2)$ kills the xy term, and introduces an
x^2 term which can be killed off by a change of co-ordinates of the form
$(u, v) \mapsto (u, v + \alpha u^2)$; we reduce to (x, y^2). Note that we must expect to
have to use scalings when simplifying the jets; the other ad-hoc changes
of co-ordinates are what we are trying to avoid! If $b = 0, a \neq 0$ reduce
to (x, xy); if $a = b = 0$ reduce to $(x, 0)$. So starting with one 1-jet we
now have three 2-jets to consider. The jet (x, y^2) is 2-$\mathcal{A}$-determined, so
we need only consider the other two.

A complete transversal for (x, xy) is $(x, xy + ay^3)$; if $a \neq 0$ this is
$\mathcal{A}$-equivalent to $(x, xy + y^3)$, otherwise (x, xy). Again the former is 3-$\mathcal{A}$-
determined. The complete transversal for $(x, 0)$ is $(x, ax^2y + bxy^2 + cy^3)$.
If $c \neq 0$ we may suppose it is 1 (by scaling again) and completing the
cube to kill the xy^2 term, killing off the resulting x^3 term, we obtain
three possibilities $(x, y^3 \pm x^2y)$ or (x, y^3). The former pair are 3-$\mathcal{A}$-
determined. If $c = 0$ similar changes of co-ordinates yield (x, xy^2) and
(x, x^2y). Of these we consider (x, y^3) as a k-jet; a complete transversal
in the $(k + 1)$-jet space is $(x, y^3 + ax^ky)$; if $a \neq 0$ we reduce to $(x, y^3 \pm
x^ky)$ which is k-$\mathcal{A}$-determined. We refer to (x, y^3) as the 'stem' for this
family. We can obviously continue with the other jets: the classification
has a clear tree-like structure.

The determinacy calculations are lengthy and we refer the reader to
[15] for samples. This brief discussion illustrates many of the features of
$\mathcal{A}$-classifications however. The aim of the complete transversal method
is to minimise the use of ad-hoc changes of co-ordinates. Scaling is
needed, and this is where one finds $\pm$ signs occuring in normal forms
over $I\!\!R$ and avoids them over $\mathbb{C}$. Determinacy results are obtained using
the same sort of calculations, by showing that a number of consecutive
complete transversals are empty. The aim throughout is to avoid a
direct use of Mather's Lemma. One can reduce the use of ad-hoc
changes of co-ordinates, Mather's Lemma and obtain more accurate
determinacy estimates using the finer filtrations discussed below.

7. Other Filtrations

The key to the results above was the rather simple theorem on Lie group actions which yields complete transversals. The crucial idea was that the tangent vectors to the Lie group pushed homogeneous terms of top degree out of the jet-space. Put another way, we have been using the filtration

$$\mathcal{M}_n.V \supset \mathcal{M}_n^2.V \supset \mathcal{M}_n^3.V \supset \mathcal{M}_n^4.V \supset \mathcal{M}_n^5.V \supset \ldots$$

where $V = \mathcal{E}(n,p)$, and our vectors in $T_e\mathcal{G}_1$ shift the filtration to the right, when $\mathcal{G} = \mathcal{R}, \mathcal{L}, \mathcal{A}, \mathcal{C}, \mathcal{K}$. This provided a powerful tool for classification, but one still often has to resort to Mather's lemma, where the calculations are complicated and tiresome. There are other filtrations however beyond the standard one by degree. For example if we assign integral weights α_j to the variables $x_j, 1 \leq j \leq n$ and β_j to the variables $y_j, 1 \leq j \leq p$ we can filter $\mathcal{M}_n.V$: set $wt\ x^a.e_j = \sum a_i\alpha_i - \beta_j$ and considering the resulting filtration

$$\ldots W_{-1} \supset W_0 \supset W_1 \supset W_2 \supset \ldots \supset W_k \supset W_{k+1} \supset \ldots$$

where W_k is spanned by those terms of $wt \geq k$. One can then seek Lie subgroups of the groups $\mathcal{G}$ whose Lie algebras shift this filtration to the right. We see that such filtrations emerge naturally. (See [12] for further details of the ideas discussed here.)

EXAMPLE 7.1.   Consider functions $(\mathbb{C}^2,0) \to (\mathbb{C},0)$ up to diffeomorphisms in the source preserving the cusp $u^2 + v^3 = 0$. As mentioned above it is difficult to explicitly display the diffeomorphisms preserving this set, but one can write down the corresponding elements of the Lie algebra. These are, roughly speaking, the vector fields tangent to the cusp (i.e. those tangent to the cusp at its smooth points), which form a module spanned by $\xi_1 = 3u\partial/\partial u + 2v\partial/\partial v$ and $\xi_2 = 3v^2\partial/\partial u - 2u\partial/\partial v$. If we assign weights by $wt\ u = 3$, $wt\ v = 2$ then we have a natural filtration of $\mathcal{E}_n$ and $u\xi_1, v\xi_1, \xi_2$ shift this filtration.
(2) Consider the group of diffeomorphisms $(\mathbb{R}^3,0) \to (\mathbb{R}^3,0)$ whose differential (at 0) is an upper unitriangular matrix. The corresponding Lie algebra $T_e\mathcal{R}_\Delta$ is $T_e\mathcal{R}_1$ augmented by the vectors $y\partial/\partial x, z\partial/\partial x, z\partial/\partial y$. Now assign weights: $wt\ x = 1$, $wt\ y = 2$, $wt\ z = 3$. We can define a filtration on $\mathcal{E}_3$ by setting $F_{r,s} = \mathcal{M}_3^{r+1} + \mathcal{M}_3^r \cap \{wt \geq s\}$. Clearly $F_{r,r-1} = \mathcal{M}_n^r$ and $F_{r,3r+1} = \mathcal{M}_n^{r+1}$ and we have a filtration

$$F_{1,1} \supset F_{1,2} \supset F_{1,3} \supset F_{2,1} \supset F_{2,2} \supset \ldots F_{r,3r} \supset F_{r+1,r} \supset F_{r+1,r+1} \supset \ldots.$$

Again the vector fields in $T_e\mathcal{R}_\Delta$ shift the filtration to the right: one simply checks that all of the operators increase weight. This clearly generalises to the group $\mathcal{R}_\Delta$ of diffeomorphisms $(\mathbb{R}^n, 0) \to (\mathbb{R}^n, 0)$ whose differential (at 0) is upper unitriangular; assign weights by $wt\, x_i = i$.

How can we exploit these filtrations? For functions we have to date been working in the jet-spaces $J^k(n, 1) = \mathcal{M}_n/\mathcal{M}_n^{k+1}$; instead we can work in the quotients $J_s^k(n, 1) = \mathcal{M}_n/F_{k,s}$. Given a germ g we define $j_s^k g$ to be the projection of g to the quotient $J_s^k(n, 1)$.

EXAMPLE 7.2.   Consider the case $n = 2$, where by increasing the subscript s by 1 the spaces J_s^k acquire one additional term. Our group $\mathcal{R}_\Delta$ also yields one additional tangent vector, namely $y\partial/\partial x$. We consider the classification of function germs $(\mathbb{R}^2, 0) \to (\mathbb{R}, 0)$ with 3-jet y^3. We work in $J_4^4(2, 1) = J_4^4$ the vector space spanned by monomials of degree ≤ 3 and x^4. For this space and group a complete transversal for y^3 is $y^3 + ux^4$ and if $u \neq 0$ this can be scaled to $y^3 \pm x^4$. We now consider those jets in J_5^4 whose j_4^4-jet is $y^3 \pm x^4$. The next complete transversal is empty, because we can get the $x^3 y$ term from $y\partial/\partial x$. It is easy to check that all subsequent transversals are empty and that these germs are 4-$\mathcal{R}$-determined. If $u = 0$ then we consider the complete transversal of y^3 in J_5^4, namely $y^3 + ux^3 y$. Again if $u \neq 0$ we can scale to get $y^3 + x^3 y$. One can now check that all subsequent transversals in the J_s^4 spaces are empty. Indeed we can show that the same is true for the J_s^5 spaces too, and deduce that this germ is also 4-$\mathcal{R}$-determined. (Our previous approach involved a use of Mather's lemma; the 5-transversal to $y^3 + x^3 y$ is $y^3 + x^3 y + ux^5$ and we needed to show that these jets were all $\mathcal{R}$-equivalent. But using $y\partial/\partial x$ we can obtain x^5, since $x^5 = (x^2\partial/\partial y - 3y\partial/\partial x)(y^3 + x^3 y)$; it was this that showed that the J_5^5 transversal was empty.)

This illustrates in a relatively trivial case the advantages of finer filtrations; the disadvantage is that there are far more steps in the process. This type of filtration does however give the following result.

THEOREM 7.3.   *Let $f : (\mathbb{R}^n, 0) \to (\mathbb{R}, 0)$ be a smooth function germ, and suppose that*

$$sp\{x_i \partial f/\partial x_j : 1 \leq i < j \leq n\} + \mathcal{M}_n^2.J_f + \mathcal{M}_n^{k+2} \supset \mathcal{M}_n^{k+1}$$

then f is k-$\mathcal{R}$-determined.

Similar results hold for the other groups $\mathcal{G}$ we have been considering. The space on the left hand side is not intrinsic, but a formulation of this result in terms of unipotent groups is contained in [15].

8. Final Remarks and References

Our treatment of classification problems has, of course, been rather elementary and somewhat idiosyncratic and has given no sense of the origin or history of the problems and methods involved. We attempt to give a broader picture in the following short overview.

The classification of singularities has always been very important within the subject. They started with Whitney in the early 1950's who listed the stable map germs (roughly those which remain unchanged after small deformation) $(\mathbb{R}^2, 0) \to (\mathbb{R}^2, 0)$ and $(\mathbb{R}^n, 0) \to (\mathbb{R}^{2n+1}, 0)$; see [41], [42]. The normal form for Morse (function) singularities was also known at that time. Towards the end of the 1960's Mather uncovered a massive (and astonishing) generalisation of Whitney's work and devised a method of producing normal forms for stable map-germs for any source and target dimension, making crucial use of the concept of $\mathcal{K}$-equivalence; it is this that makes the classification amenable. There were also lists of Morin, a little before this, of certain special but important types of stable germ. See [22] and [24] and the references therein, and the interesting article by Arnold, [1].

Around the same time Thom made a classification of singularities of function germs, required for his catastrophe theory. These were in turn superseded by the work of Arnold who produced his famous list of simple singularities (and, of course, related them to so many other phenomena). Siersma also codified and extended Thom's work around the same time in his thesis; see also [38]. Arnold hugely expanded the scope of the listing process for functions and introduced many crucial tools, which we have not discussed above, such as the ideas of weighted homogeneous functions and mappings, the use of spectral sequences, and of Newton polyhedra. See [2], [4], [7]. These provide much more powerful tools than those given above for $\mathcal{R}$-equivalence.

These classifications contrasted sharply with those available under $\mathcal{A}$-equivalence. Although in theory Mather provided the tools required for classification, in practice they proved inadequate. In our terms a key problem was the provision of a good determinacy result. The major breakthrough was made by Gaffney who produced criteria which allowed practical classifications; [19], [20], [21]. An early success here was the description of all germs of mappings which arise when a generic surface is orthogonally projected (from any direction), work completed by Gaffney and Ruas. Around the same time Arnold made the same classification [3], showing that his techniques could be applied to a wide range of problems. Martinet made some important contributions to $\mathcal{A}$-determinacy, and du Plessis sharpened up many of the ideas and techniques, producing several significant new determinacy estimates and classifications in his paper [33].

Using these and other techniques various lists up to $\mathcal{A}$-equivalence were produced, by amongst others Gaffney, Mond, du Plessis, Reiger and Bruce including the list of simple singularities for mappings $(\mathbb{C}, 0) \to (\mathbb{C}^2, 0)$, $(\mathbb{C}^2, 0) \to (\mathbb{C}^3, 0)$, $(\mathbb{C}^2, 0) \to (\mathbb{C}^2, 0)$. (See [9], [30], [33], [37].)

Other important list makers included Giusti (simple singularities of complete intersections), Goryunov (simple projections of complete intersections), Wall, who motivated by his long-term project with topological stability made extensive lists up to $\mathcal{K}$-equivalence, and more recently Zakalyukin who produced an important generalisation of contact equivalence. See the references [25], [40], [34], [43].

It was clear, from early on, that classifications of different classes of mappings were possible (e.g. equivariant mappings, mappings on manifolds with boundary etc.). Lists were also developed here, with a general framework provided by the Geometric Subgroups of Damon discussed in [16]. See for example [5], but there are many other papers here we do not cite.

In [15] an approach to determinacy was taken related to the ideas exposed above, which provided criteria which are very effective in practice (and indeed in some sense best possible). The tools in this paper have been used in a number of subsequent works to classify, for example, orthogonal projections of surfaces with boundary to planes [11]. There is an elegant formulation in [15] of determinacy criteria using the notion of a unipotent group. However this can also be understood in terms of the finer filtrations discussed in Section 7.

The notion of complete transversal first appeared in a paper of Dimca and Gibson where it was used in the classification of $\mathcal{K}$-simple singularities; see [17] and [18]. Its generalisation, to cover the other groups used here, is due to Bruce, du Plessis and Kirk and appeared in [12]. The methods employed can be automated, and this has been done by Kirk who has produced a Maple package which supplies all of the tools required to classify map-germs in a wide variety of situations. These techniques allowed the classification of simple germs $(\mathbb{C}, 0) \to (\mathbb{C}^3, 0)$, $(\mathbb{C}^3, 0) \to (\mathbb{C}^3, 0)$, $(\mathbb{C}^2, 0) \to (\mathbb{C}^4, 0)$, $(\mathbb{C}^3, 0) \to (\mathbb{C}^4, 0)$ and from a surface with boundary to 3-space. See [23], [27], [13], [26], [32]. A version of these results, and some extensive lists of mappings from $(\mathbb{C}^2, 0) \to (\mathbb{C}^3, 0)$ were also obtained by Ratcliffe, see [35] and [36].

Finally the determinacy results which underpin the classifications above all depend on a different set of techniques, namely the trivialisation of families using integration of vector fields. This in turn requires some mildly complicated algebra (hinging in the case of $\mathcal{A}$-equivalence on the Malgrange Preparation Theorem). For a good introduction see the first part of Wall's survey article [39], and [28]. It would appear however that the best approach is to use these results to ensure that the problem is finite-dimensional and then apply the sort of methods discussed above.

For further reading we recommend [7], [8] and [10] for accounts of the classification of function germs, and [6] for an explanation of why these classifications are of interest. For stable mappings and a fuller account of the contact group see [22] and [28]. For global properties of stable mappings see [24].

References

1. V. I. Arnold, Singularities of smooth mappings, Uspekhi Mat. Nauk. **23**:1 (1968) 3–44; Russian Math. Surveys **23**:1 (1968) 1–43.

2. V. I. Arnold, Normal forms for functions near a degenerate critical point, the Weyl groups A_n, D_n, E_n, and Legendrian singularities, Funct. Anal. Appl. **6** (1972) 254–272.

3. V. I. Arnold, Indices of 1-forms on a manifold, Uspekhi Mat. Nauk., **34**:1 (1979) 3–38; Russian Math. Surveys **34** (1979), 1–42.

4. V. I. Arnold, Normal forms for functions in neighbourhoods of degenerate critical points, Russ. Math. Surveys **29** (1974) 11–49.

5. V. I. Arnold, Wave front evolution and equivariant Morse lemma, Commun. Pure and Appl. Math. **29** (1976) 557–582.

6. V. I. Arnold, *Catastrophe Theory*, Springer-Verlag, New York-Heidelberg-Berlin, 1984, 1986; third edition 1992.

7. V. I. Arnold, S. Gusein-Zade and A. N. Varchenko, *Singularities of Differentiable Maps, Volume I*, Monographs in Math. **82**, Birkhäuser, Boston-Basel, 1985.

8. Th. Brocker and L. C. Lander, *Differentiable germs and catastrophes*, London Math. Soc. Lecture Notes **17**, Cambridge Univ. Press, 1975.

9. J. W. Bruce and T. J. Gaffney, Simple singularities of mappings $(C, 0) \rightarrow (C^2, 0)$, J. London Math. Soc.(2) **26** (1982) 465–474.

10. J. W. Bruce and P. J. Giblin, *Curves and singularities*, Cambridge Univ. Press, 1994.

11. J. W. Bruce and P. J. Giblin, Projections of surfaces with boundary, Proc. London. Math. Soc. (3) **60** (1990) 392–416.

12. J. W. Bruce, N. P. Kirk and A. A. du Plessis, Complete transversals and the classification of singularities, Nonlinearity, **10**:1 (1997) 253–275.

13. J. W. Bruce, N. P. Kirk and J. M. West, Classification of map-germs from surfaces to four-space, preprint, University of Liverpool, 1995.

14. J. W. Bruce, T. J. Gaffney and A. A. du Plessis, On left equivalence of map-germs, Bull. London Math. Soc. **16** (1984) 301–306.

15. J. W. Bruce, A. A. du Plessis and C. T. C. Wall, Determinacy and unipotency, Invent. Math. **88** (1987) 521–554.

16. J. N. Damon, *The unfolding and determinacy theorems for subgroups of A and K*, Memoirs Amer. Math. Soc. **50**, no. 306, 1984.

17. A. Dimca and C. G. Gibson, Contact unimodular germs from the plane to the plane, Quart. J. Math. Oxford (2) **34** (1983) 281–295.

18. A. Dimca and C. G. Gibson, Classification of equidimensional contact unimodular map-germs, Math. Scand. **56** (1985) 15–28.

19. T. J. Gaffney, On the order of determination of a finitely determined germ, Invent. Math. **37** (1976) 83–92.

20. T. J. Gaffney, A note on the order of determination of a finitely determined germ, Invent. Math. **52** (1979) 127–130.

21. T. J. Gaffney, The structure of $T\mathcal{A}.f$, classification, and an application to differential geometry, *Singularities* (ed Peter Orlik), Proc. Symp. pure math. **40**, part 1, (Amer. Math. Soc., 1983) pp 409–428.

22. C. G. Gibson, *Singular points of smooth mappings*, Research Notes in Maths. **25**, Pitman, London, 1973.

23. C. G. Gibson and C. A. Hobbs, Simple singularities of space curves, Math. Proc. Camb. Phil. Soc **113** (1993) 297–310.

24. M. Golubitsky and V. Guillemin, *Stable mappings and their singularities*, Graduate Texts in Maths. **14**, Springer Verlag, Berlin, 1973.

25. V. V. Goryunov, Geometry of bifurcation diagrams of simple projections onto the line, Funct. Anal. Appl. **15** (1981) 77–82.

26. K. Houston and N. P. Kirk, On the classification and geometry of map-germs from 2-space to 4-space, *Singularity Theory* (eds Bill Bruce and David Mond), London Math. Soc. Lecture Notes **263**, (Cambridge Univ. Press, 1999) pp 325–352.

27. W. L. Marar and F. Tari, On the geometry of simple germs of corank 1 maps from $\mathbb{R}^3$ to $\mathbb{R}^3$, Math. Proc. Camb. Phil. Soc. **119** (1996) 469–481.

28. J. Martinet, *Singularities of smooth functions and maps*, London Math. Soc. Lecture Notes **58**, Cambridge Univ. Press, 1982.

29. J. N. Mather, Stability of C^∞-mappings III: Finitely determined map-germs, Publ. Math. IHES **35** (1969) 127–156.

30. D. M. Q. Mond, On the classification of germs of maps from $\mathbb{R}^2$ to $\mathbb{R}^3$, Proc. London Math. Soc. **50** (1985) 333–369.

31. B. Morin, Formes canoniques des singularités d'une application différentiable, Comptes Rendus Acad. Sci. Paris **260** (1965) 5662–5665, 6503–6506.

32. A. C. Nogueira, thesis, Sao Carlos, 1998.

33. A. A. du Plessis, On the determinacy of smooth map germs, Invent. Math. **58** (1980) 107–160.

34. A. A. du Plessis and C. T. C. Wall, *The geometry of topological stability*, London Math. Soc. Monographs **9**, Oxford Univ. Press, 1995.

35. D. Ratcliffe, Stems and series in $\mathcal{A}$-classification, Proc. London Math. Soc. (1) **70** (1995) 181–213.

36. D. Ratcliffe, A classification of map-germs $(\mathbf{C}^2, 0) \rightarrow (\mathbf{C}^3, 0)$ up to $\mathcal{A}$-equivalence, preprint, Univ. of Warwick, 1994.

37. J. H. Rieger, Families of maps from the plane to the plane, J. London Math. Soc (2) **36** (1987) 351–369.

38. D. Siersma, The singularities of C^∞-functions of right codimension smaller than or equal to eight, Indag. Math. **35** (1973) 31–37.

39. C. T. C. Wall, Finite determinacy of smooth map-germs, Bull. London Math. Soc. **13** (1981) 481–539.

40. C. T. C. Wall, Notes on the classification of singularities, Proc. London Math. Soc. (3) **48** (1984) 461–513.

41. H. Whitney, The singularities of smooth n-manifolds in $(2n-1)$-space, Ann. of Math. **45** (1944) 247–293.

42. H. Whitney, On singularities of mappings of Euclidean spaces I, mappings of
     the plane to the plane, Ann. of Math. **62** (1955) 374–410.

43. V. M. Zakalyukin, Flag contact singularities, *Real and Complex Singularities*
     (eds J.W. Bruce and F. Tari), Research Notes in Math. **412** (Chapman and
     Hall/CRC 2000) pp 134–146. See also article in this volume.

Applications of Flag Contact Singularities

Vladimir Zakalyukin*

Moscow Aviation Institute
(vladimir@zakal.mccme.ru)

1. Introduction

Discriminants of isolated hypersurface singularities arise in various problems of geometry, physics and differential equations. Mostly they occur as the projections (wavefronts) of Legendre submanifolds of projectivized cotangent bundles. Arnold's list of simple classes (related to A,D,E Lie algebras) provides normal forms of generic wavefront singularities in small dimensions.

Later V. Arnold proved [6] that the simple singularity classes of functions with respect to the group of diffeomorphisms of the source manifold which preserve a regular hypersurface (boundary) provide the discriminants of the Weyl groups B, C, F_4.

According to F. Pham and N. H. Duc [16] the theory of these boundary singularities corresponds to the theory of Lagrangian (or Legendre) projections of a pair of Lagrangian (Legendre) submanifolds which intersect in a submanifold of codimension one in each of them and are transversal in the remaining directions.

Certain physical applications require the study of singularities on a manifold with corners [33] and in general on the spaces of orbits of certain group actions. The singularities of function-germs defined on arrangements of hyperplanes [37, 38, 39] represent a generalization of corner singularities. In other words all these are singularity classes of functions with respect to certain subgroups of the group of right equivalences. Another example considered by O. Lyashko [25] shows that the discriminant of the Coxeter group H_3 is isomorphic to the bifurcation diagram of a function on the space of invariants of the dihedral group I_5.

We show that these constructions have natural counterparts in terms of singularities of the contact between a submanifold germ and a germ of a flag (a sequence of nested submanifolds) embedded in an ambient space.

* Partially supported by the NWO-047008005,RBRF-99010147,UR-992365 and INTAS-1644 grants

These *flag contact singularities* (FCS) naturally occur in various other contexts of singularity theory [44].

They are also related to the projections of complete intersections onto a line studied by V. Goryunov [18], to the flattenings of curves [22], to the singularities of fractions [5], and to the singularities of diagrams of mappings [15].

Their applications to differential geometry give various classification results. For example, singularities of contact of generic surfaces with spheres and circles are related to the discriminants of Coxeter groups [40].

We describe the generic singularities of the relative minimum arising in problems of parametric optimization [13, 14], using flag contact classes. Among them we distinguish those which represent the generic singularities of the envelopes of moving wavefronts in control systems [42].

Finally, other possible applications are mentioned. In particular, the stability of the wave front of an exponential mapping in the simplest case in subriemannian geometry [1, 2] was proved in [3, 4] using flag contact group techniques.

2. Preliminaries and examples

2.1. BASIC DEFINITIONS

By a flag in an affine space $\mathbf{A}^n$ (here $\mathbf{A}$ is either $\mathbb{R}$ or $\mathbb{C}$) we mean a set F_s of nested affine subspaces (or germs of submanifolds)

$$\mathbf{L}_1^{k_1} \subset \mathbf{L}_2^{k_2} \subset \ldots \subset \mathbf{L}_s^{k_s} \subset \mathbf{A}^n$$

of dimensions $k_1 < k_2 < \ldots < k_s \leq n$. We study the local contact of a distinguished flag F_s with the image of a smooth mapping f of a smooth manifold M^m to $\mathbf{A}^n$, considering the appropriate spaces of germs.

The product of the group of diffeomorphism-germs of the source space and the group of diffeomorphisms $\theta : \mathbf{A}^n \to \mathbf{A}^n$ of the target which preserve the flag F_s (that is $\theta(\mathbf{A}^{k_i}) = \mathbf{A}^{k_i}$) is called the group of F_s-*equivalences*. It acts on the space of germs of mappings f.

Consider a pair consisting of a map-germ f with $f(0) = 0$ and a flag F_s. Define the r-*extension* $\tilde{f}, \tilde{F}$ of this pair as follows.
The mapping $\tilde{f} : M \times \mathbf{A}^r \to \mathbf{A}^n \times \mathbf{A}^r$ is defined by $\tilde{f}(x, z) = (f(x), z)$; and

$$\tilde{F} = \{\tilde{\mathbf{L}}_1^{k_1} \subset \ldots \subset \tilde{\mathbf{L}}_s^{k_s} \subset \mathbf{A}^n \times \mathbf{A}^r\},$$

where $\tilde{\mathbf{L}}_i^{k_i} = \mathbf{L}_i^{k_i} \times \{0\}, i = 1, \ldots, s$.

Denote by π_i a linear projection $\pi_i : \mathbf{A}^n \to \mathbf{A}^{n-k_i}$, with kernel subspace $\mathbf{L}_i$.

If the mapping germ $\pi_s \circ f$ has rank r, then evidently f is F-equivalent to the r-extension of the restriction $\hat{f} = f|_{M_0^{m-r}} : M_0^{m-r} \to \mathbf{A}^{n-r}$ of f to a submanifold germ M_0, whose tangent space is the kernel of the differential of $\pi_s \circ f$.

In particular, if f is transversal to the largest element $\mathbf{L}_s$ of the flag, than the largest element $\mathbf{L}_s = \mathbf{A}^{n-r}$ of the reduced pair $\hat{f}, \hat{F}$ coincides with the ambient target space of $\hat{f}$. Hence, the F_s-equivalence class of the initial mapping is determined by the F_{s-1}-class of the reduced mapping with respect to a reduced flag $\hat{\mathbf{L}}_1 \subset \ldots \subset \hat{\mathbf{L}}_{s-1} \subset \mathbf{A}^{n-r}$ with $s - 1$ elements.

Define also an r-*lifting* $(\bar{f}, \bar{F})$ of a pair (f, F) as follows:

$$\bar{f} : M^m \to \mathbf{A}^n \times \mathbf{A}^r, \quad \bar{f} : x \mapsto (f(x), 0),$$

$$\bar{F} = \{\mathbf{L}_1 \times \mathbf{A}^r \subset \ldots \subset \mathbf{L}_s \times \mathbf{A}^r \subset \mathbf{A}^n \times \mathbf{A}^r\}.$$

Obviously, the r-lifting of an embedding FCS is F-equivalent to the map $M^m \to \mathbf{A}^n \times \mathbf{A}^r$ given by $x \mapsto (f(x), h(x))$ for h an arbitrary map.

A germ of FCS is called *simple*, if the germ at any near-by point of any mapping close to a representative of the given map is equivalent to one of a finite set of germs.

Let $y_1, \ldots, y_n$ be coordinates on A^n and the elements of the flag F be the coordinate subspaces

$$\mathbf{L}_s = \{y_1 = \ldots y_{n-k_s} = 0\}, \ldots, \mathbf{L}_1 = \{y_1 = \ldots y_{n-k_1} = 0\}.$$

Denote by $g_i = y_i(f)$, $i = 1, \ldots, n - k_s$ the compositions of the map f with the coordinate functions on $\mathbf{A}^n$.

EXAMPLE 2.1.   If the flag consists of one element only ($s = 1$) then the FCS of mappings $f : M \to \mathbf{A}^n$, which are embeddings coincide with equivalence classes of complete intersection $X \subset M$ defined by equations $g_1(x) = \ldots = g(x)_{n-k_1} = 0$. In particular, when the codimension $n - k_1 = 1$, one gets a singularity class of hypersurfaces in M (or, equivalently, in $f(M)$). In particular, Giusti's list [20] of simple complete intersection germs provides the list of simple FCS of this type.

On the other hand, the problem of F-classification of embeddings of complete intersections M of dimension m_1 in $\mathbf{A}^{m_2}$ is equivalent to that of F_{s+1}- classification of regular embeddings of $\mathbf{A}^{m_2}$ into $\mathbf{A}^{2m_2 - m_1}$ with respect to the enlarged flag in this space with $\mathbf{A}^{m_2}$ as an additional largest element.

EXAMPLE 2.2. Suppose that $n = 3$, $s = 1$, $k_1 = 2$ and f is a generic mapping of a surface into three-space. The germs of f are either germs of embedding or cross-caps.

The FCS in this case are the singularities of hypersurfaces and the contact classes of functions on the cross-cap. The simple FCS were classified in [9], and the list consists of the following three series:

$$I : (x_1, x_2) \mapsto (x_1 \pm x_2^{2k+2}, x_1 x_2, x_2^2),$$

$$II : (x_1, x_2) \mapsto (x_2^2 \pm x_1^{k+1}, x_1 x_2, x_1),$$

$$III : (x_1, x_2) \mapsto (x_1 x_2 \pm x_2^{2k+2}, x_1, x_2^2),$$

where k is a non-negative integer, x_1, x_2 are coordinates in the sourse and the unique flag element is $y_1 = 0$.

Denote by V_F the Lie algebra of vector fields $v = \sum v_i \frac{\partial}{\partial y_i}$ on A^n tangent to the flag F. Their components have the following block lower-diagonal form. For $i \in \{1, \ldots, n - k_1\}$ denote by $J(i) \subset \{1, \ldots, n - k_1\}$ the set of subscripts j such that either $j \leq i$, or the vectors $\frac{\partial}{\partial y_i}$, $\frac{\partial}{\partial y_j}$ are tangent to the same flag element. The component v_i for $i \in \{1, \ldots, n - k_1\}$ belongs to the ideal generated by the coordinate functions y_j, $j \in J(i)$. The components v_i with $i > n - k_1$ are arbitrary.

The tangent space T_f to the F-equivalence class is the f^*-module of vertical vector fields on the graph of the mapping f :

$$T_f = df(\Gamma(TM^m)) + f^*(V_F).$$

If f is a germ of embedding, then this space is a module over the ring $\mathcal{O}_m$ of function-germs on M^m.

The right action of diffeomorphisms on M composed with multiplication of the column of coordinate functions g_i by non-singular block lower-triangular matrices

$$\begin{pmatrix} E_{11} & 0 & \cdots & 0 \\ E_{21} & E_{22} & \cdots & 0 \\ \cdots & & & \\ E_{s1} & E_{s2} & \cdots & E_{ss} \end{pmatrix}$$

whose blocks E_{ij} are $(k_{i-1} - k_i) \times (k_{j-1} - k_j)$ matrices depending on x form a subgroup T of the contact equivalences acting on the space of mappings $G : x \mapsto (g_1, \ldots, g_{n-k_s})$.

The coordinate function g_i corresponding to the ith element of the flag is determined by the flag contact singularity up to addition of a function in the ideal $\mathcal{O}_x\{g_m\}, 1 \leq m \leq n - k_{i-1}$ generated by the coordinate functions determining flag elements of higher dimensions.

PROPOSITION 2.3. *The FCS of embeddings are correspond to the orbits of the action of T on the space of mappings G.*

To any map-germ f one can associate the graph embedding $Gr(f)$: $M^m \to M^m \times \mathbf{A}^n$, $Gr(f) : x \mapsto (x, f(x))$, which is a certain desingularisation of the image of f. Then to the problem of classifying the F-orbits of f corresponds the problem of classifying the F_{gr}-orbits of the graph mappings with respect to the lifted flag whose elements are $M^m \times \mathbf{L}_i$. This relation sends F-orbits to F_{gr}-orbits, while different orbits may glue together. Thus the associated problem, which is much easier to treat (the tangent spaces are modules over the ring of function germs on the source) yields a preliminary classification of singularities.

For germs of embeddings these classes coincide.

2.2. BOUNDARY SINGULARITIES AND FLAGS WITH TWO ELEMENTS

Consider an embedding $f : M \to \mathbf{A}^n$ into a space equipped with a flag $F_2 :$ $\mathbf{A}^{n-2} \subset \mathbf{A}^{n-1} \subset \mathbf{A}^n$ of two elements of neighbouring dimensions and suppose that the second coordinate function g_2 is nonsingular.

PROPOSITION 2.4. *The FCS of such mappings coincide with the contact boundary singularities of the first component g_1 provided that the equation $g_2 = 0$ determines a boundary.*

For a T-orbit in this case is uniquely determined by the contact class of the zero level hypersurface of g_1 and its intersection with $g_2 = 0$.

There is an involution (studied by I. Scherback [11]) on the space of boundary singularities. To a pair consisting of a function-germ f and its restriction f_0 to the boundary corresponds the germ of the function f^* (in new variables) and its restriction f_0^* to another boundary, such that f^* is stably equivalent to f_0, and f_0^* is stably equivalent to f. It was proven in [11] that the intersection forms on the vanishing homology of dual boundary singularities are dual.

In terms of flags this involution is defined as follows.

Consider the pencil of hyperplanes $\mathbf{A}_2(a)$, $a \in \mathbf{A}$ containing the subspace $\mathbf{A}_1 : y_1 = y_2 = 0$ (that is, determined by the equation $y_1 - ay_2 = 0$).

To the contact of the mapping f with the family of flags $A_1^{n-2} \subset A_2^{n-1}(a) \subset A^n$ corresponds the contact of the suspended mapping $f^* :$ $M \times A \to A^n \times A$, $f^* : (x, a) \mapsto (f(x), a)$ with the *polar* flag $A_2 \times \{0\} \subset \bigcup A_2(a) \times \{a\} \subset A^n \times A$,

PROPOSITION 2.5. *The associated pair* g_1^*, $g_1^*\big|_{g_2^*=0}$ *is dual to the original pair* g_1; $g_1\big|_{g_2=0}$.

For $g^*(x,a) = g_1(x) - ag_2(x)$ and $g_2(x)$ is regular. Hence g_1^* is stably right equivalent ($\approx$) to the restriction $g_1(x)\big|_{g_2(x)=0}$. The restriction $g_1^*(x,a)\big|_{g_2^*(x,a)=0} = g_1^*(x,a)\big|_{a=0} = g_1(x)$.

2.3. CORNER SINGULARITIES

The same polar construction establishes an isomorphism between the singularity classes (up to stabilization) of a function on a manifold with a 2-corner (union of two transversal hypersurfaces) and singularities of contact of an embedding with a flag of three elements $A_1^{n-3} \subset A_2^{n-3} \subset A_3^{n-3} \subset A^n$, provided that the coordinate functions g_2 and g_3 are regular and independent.

Namely, to the triple of coordinate functions (g_1, g_2, g_3) there corresponds the zero level hypersurface of the function $h(x,a) = g_1 - ag_2$ $\approx g_1(x)\big|_{g_2(x)=0}$ and its restrictions $h\big|_{a=0} = g_1$, $h\big|_{g_3=0} \approx g_1\big|_{g_2=g_3=0}$.

An embedding $i : \Sigma \to A^n$ of a hypersurface germ into a space equipped with a flag $A_1^{k_1} \subset A_2^{k_2} \subset \ldots \subset A_s^{k_s} \subset A^n$ determines a sequence $h, h_1, \ldots, h_s$ of contact classes of function singularities, where h generates the ideal of functions vanishing on Σ and $h_j = h\big|_{A_{s-j}}$.

On the other hand the singularity class of i determines the contact of the embedded graph $f : A^n \to A^n \times A$, $f : x \mapsto (x, h(x))$ with the flag $A_1^{k_1} \times 0 \subset A_2^{k_2} \times 0 \subset \ldots \subset A_s^{k_s} \times 0 \subset A^n \times 0 \subset A^n \times A$.

These classes are also related to singularities of functions on a manifold with corners (transversal intersections of several smooth boundaries). The group of transformations preserving the corner preserves in particular a flag, which consists of a component of the boundary, its intersection with some other component, the intersection of three components, and so on. Since the flag contact group is larger that the corner contact group, certain families of corner orbits may glue into a single flag contact orbit. This actually happens for unimodal singularities on corners with two branches (compare the list of simple flag classes from Theorem 3.5 and classifications from [34]).

Denote by $A_j^* \subset B = (A^{n+1})^\wedge$ the set of hyperplanes in A^{n+1} which contain A_j. The set $\{A_j^*\}$ forms a flag dual to $\{A_j\}$. The suspension $\tilde{f} :$ $A^n \to A^{n+1} \times B$ of the embedding f determines a (dual) hypersurface embedding, whose sequence of functions $h^*, h_1^*, \ldots, h_s^*$ is dual to the initial one: $h_i^* \approx h_{s-i}$. This operation induces a duality involution on the sequence of relative vanishing homology of restrictions of g_1 to pairs of neighbouring flag elements [30, 31].

2.4. FUNCTIONS ON SINGULAR BOUNDARY

The group of flag contact equivalences is a geometrical subgroup (in the sense of J. Damon [12]) the group of right-left equivalences. So the notion and properties of versal deformations of FCS are standard.

For example, a mini-versal deformation of an F_2-flag singularity with coordinate functions g_1, g_2 can be given as a family $G = (\hat{g}_1, \hat{g}_2)$ of embeddings depending on parameters $\Lambda = (\lambda_1, ..., \lambda_{\tau_1}, \varepsilon_1, ..., \varepsilon_{\tau_2})$

$$\hat{g}_1(x, \lambda) = g_1(x) + \sum_{i=1}^{\tau_1} \lambda_i \varphi_i(x), \quad \hat{g}_2(x, \varepsilon) = g_2(x) + \sum_{j=1}^{\tau_2} \varepsilon_j \psi_j(x),$$

where the φ_i span the A-module $Q_1 = \mathcal{O}_x / \mathcal{O}_x \{g_1, \frac{\partial g_1}{\partial x}\}$, ψ_j span the A-module $Q_2 = \mathcal{O}_x / \mathcal{O}_x \{g_1, g_2, St_{g_1} \cdot g_2\}$ (provided that these modules have finite dimensions τ_1 and τ_2) and $St_{g_1} \cdot g_2$ is the ideal of derivatives of g_2 along vector fields v in the stabiliser $\{v \mid D_v g_1 \in \mathcal{O}_x \{g_1\}\}$ of the function g_1.

This deformation yields a contact versal deformation of the hypersurface germ $g_1 = 0$ (if one forgets the second component of the flag). Therefore the submersion $S : (\lambda, \varepsilon) \mapsto \lambda$ is well defined and the intrinsically determined subdeformation $\hat{g}_2(x, 0, \varepsilon)$ is the contact versal deformation of the restriction of g_2 to the zero level hypersurface X_1 of the function g_1.

The parameter values of the deformation G corresponding to non-transversal intersections of the embedding with the flag form the *discriminant* $D(G, F)$.

The intersection of $D(G, F_2)$ with the fiber of S over the origin is the discriminant $D^r(\hat{g}_2, g_1)$ of the restriction of the hypersurface $g_2 = 0$ to the (singular) boundary $g_1 = 0$.

In the case of a smooth boundary this construction shows that τ_2 is equal to the dimension of the contact mini-versal deformation of the restriction of g_1 to the boundary.

Our restricted discriminants D^r are a contact counterpart of the discriminants of functions on manifolds with singular boundaries studied by O. Lyashko [25, 6]. The main objective of those works was to relate non-crystallographic Coxeter groups to discriminants of singularities. Our contact classification is easier. In particular, the discriminant of the Coxeter group H_3 occurs in our list as the discriminant of a simple class, while in [25] it is a hyperplane section of a discriminant of a unimodal singularity. The latter is equisingular along a $\mu = $ constant stratum since it belongs to a single contact orbit.

DEFINITION 2.6. *The pair (g_1, g_2) is called pseudo-simple if $g_1 = 0$ is a germ of simple ICIS germ and the contact class of the restriction* *of g_2 to $g_1 = 0$ is simple within the space of all restrictions of functions to this variety.*

A pseudo-simple class can be non-simple as a flag contact class since it can be adjacent to a non-simple class of the restriction of g_2 for a perturbed germ g_1 which itself remains simple.

Pseudo-simple classes with singular germs g_2 can exist only if $M = \mathbf{A}^2$ and g_1 has an A_k singularity.

However, pseudo-simple pairs with g_2 regular (equivalent to special classes of boundary singularities) occur in higher dimensions as well as for some other simple plane curves.

THEOREM 2.7. *The pseudo-simple classes on the plane curves of type A_k are the following (here $\mathbf{A} = \mathbb{C}$ and (x, y) are coordinates on the plane):*

Class	Boundary	g_1	g_2	Conditions
A_{k-1}	A_0	x	y^k	$k \geq 1$
$B_{l,m}$	A_1	xy	$x^l + y^m$	$l \geq m \geq 0$
$\widetilde{F}_{2k-1}$	A_2	$x^2 + y^3$	y^k	$k \geq 1$
$\widetilde{F}_{2k}$	A_2	$x^2 + y^3$	xy^{k-1}	$k \geq 1$
$\widetilde{U}_{l,m}(k)$	A_{2k+1}	$x^2 + y^{2k+2}$	$xy^l + y^m$	$k \geq m > l+1, l \geq 0$
$U_m(k)$	A_{2k+1}	$x^2 + y^{2k+2}$	y^m	$k \geq m \geq 1$
$\widetilde{V}_q(k)$	A_{2k}	$x^2 + y^{2k+1}$	y^q	$q \geq 1, k \geq 2$
$V_{l,m}(k)$	A_{2k}	$x^2 + y^{2k+1}$	$xy^l + y^m$	$m \geq l+1 \geq 1, k \geq 2$
$W_{p+1}(k)$	A_{2k}	$x^2 + y^{2k+1}$	xy^p	$p \geq 0, k \geq 2$

The discriminants of these classes are of some interest (see also [19]). Of course, the discriminant of A_{k-1} is the usual one.

The miniversal deformation of a $B_{l,m}$ singularity is the sum of versal deformations $h_1(x, \varepsilon) = \varepsilon_0 + \sum_{i=1}^{l-1} \varepsilon_i x^i + x^l$ and $h_2(y, e) = \varepsilon_0 + \sum_{i=1}^{m-1} \varepsilon_{i+l-1} y^i + y^m$ of the monomials x^l and y^m in the space of functions in one variable with respect to the group of diffeomorphisms preserving the origin, where the parameter ε_0 is common.

The discriminant of a $B_{l,m}$ singularity is a collection of three hypersurfaces in $\mathbb{C} \times \mathbb{C}^{l-1} \times \mathbb{C}^{m-1}$ with coordinates $(\varepsilon_0, \ldots, \varepsilon_{l-1}, \varepsilon_l, \ldots, \varepsilon_{l+m-2})$, being the union of {discriminant of B_l} $\times \mathbb{C}^{m-1}$ and $\mathbb{C}^{l-1} \times$ {discriminant of B_m}. Each of these has two components, but they have one (the hyperplane $\varepsilon_0 = 0$) in common. When $m = 0$ one has the usual discriminant of the Weyl group B_l. The adjacency table of these classes is generated by $B_{l,m-1} \leftarrow B_{l,m} \rightarrow B_{l-1,m}$.

There exists another real normal form $g_1 = x^2 + y^2$ of the boundary, and corresponding normal forms of the restrictions.

The adjacency diagram of the classes $\tilde{F}_n$ is a string $\tilde{F}_1 \leftarrow \tilde{F}_2 \leftarrow \tilde{F}_3 \leftarrow \ldots$. Their miniversal deformations are equivalent to

$$\varepsilon_0 + \sum_1^{k-1}(\varepsilon_{2i-1} + \varepsilon_{2i}x)y^i + y^k \quad \text{for } n = 2k-1$$

and to

$$\varepsilon_0 + \sum_1^{k-1}(\varepsilon_{2i-1} + \varepsilon_{2i}x)y^i + \varepsilon_{2k-1}y^k + xy^{k-1} \quad \text{for } n = 2k.$$

The uniformization $u : \mathbb{C} \to \mathbb{C}^2 = \{(x,y)\}$, $u : t \mapsto (t^3, -t^2)$ of the cuspidal boundary provides the following description of these discriminants: $D^r(\tilde{F}_n) = \left\{\varepsilon \,\middle|\, P_{n+1}(t, \varepsilon) = 0, \frac{\partial P_{n+1}}{\partial t} = 0\right\}$, where $P_{n+1}(x, t) = t^{n+1} + \varepsilon_{n-1}t^n + \ldots + \varepsilon_1 t^2 + \varepsilon_0$ is a polynomial in t with the linear term missing.

The discriminant $D^r(F_2)$ is isomorphic to the discriminant of the Weyl group G_2 (two smooth curves with second order tangency at the origin).

The discriminant $D^r(F_3)$ is diffeomorphic to the folded Whitney umbrella with a tangent plane at the singular point [6]. This variety occurs in many contexts. In particular it is diffeomorphic to:

– the bifurcation set of the boundary function singularity C_4 (without the cut-locus);

– the envelope of moving wavefronts (or caustics) with semicubical edges [42];

– the bifurcation diagram of a C_4 singularity of projections of ICIS onto a line [18]

– its real part (without the tangent plane) is diffeomorphic to the union of straight lines in $\mathbb{R}^3$ tangent to a space curve near a simple flattening point on this curve [32].

All these settings essentially represent (as is shown below) various applications of flag contact singularities.

The discriminants $D^r(F_n)$ form a series of images of stable simple Legendre projections of singular Legendre varieties isomorphic to open Whitney umbrellas [21], which naturally arise as singularities of the solutions of the Cauchy problem for a Hamilton-Jacobi equation when the initial data fail to be transversal to the characteristic Hamilton vector field.

There is only one simple class $U_{1,0}(3)$ with boundary A_3. It has codimension one.

The adjacency diagram of the classes $\tilde{V}_*(2), V_{*,*}(2), W_*(2)$ on the A_4 boundary starts with the following table (where we omit (2) in the notations for these classes).

$$
\begin{array}{ccccccc}
\tilde{V}_1 & \leftarrow & V_{0,2} & \leftarrow & V_{0,3} & \leftarrow & W_1 \\
\uparrow & & \uparrow & & \uparrow & & \uparrow \\
& \tilde{V}_2 & \leftarrow & V_{1,3} & \leftarrow & V_{1,4} & \leftarrow & W_2 \\
& \uparrow & & \uparrow & & \uparrow & & \uparrow \\
& & \tilde{V}_3 & \leftarrow & V_{2,4} & \leftarrow & V_{2,5} & \leftarrow & W_3 \\
& & \uparrow & & \uparrow & & \uparrow & & \uparrow
\end{array}
$$

Here again a uniformization of the boundary curve simplifies the study of the discriminants. For example, the restricted discriminant $D^r(V_{0,2})$ is diffeomorphic to the *wave front* curve of the family of polynomials $t^5 + t^4 + \varepsilon_1 t^2 + \varepsilon_0$, that is, to the subset of $(\varepsilon_0, \varepsilon_1)$ -parameter space corresponding to polynomials with zero critical value. This curve is diffeomorphic to the union of a 2/5-cusp curve (the discriminant of the dihedral group $I_2(5)$) and its tangent line at the origin.

The discriminant of $V_{0,3}$ is isomorphic to the wave front surface of the 3-parameter family of polynomials

$$
t^6 + t^5 + \varepsilon_2 t^4 + \varepsilon_1 t^2 + \varepsilon_0.
$$

This surface is diffeomorphic to the union of the discriminant of the Coxeter group H_3 generated by reflections (symmetries of an icosahedron) and its tangent plane at the origin. To see this it is enough to compare this family of polynomials with the standard description of the discriminant of H_3 as the graph of a multivalued length function on the germ of the evolute at an inflection point of a smooth curve.

The discriminant of $\tilde{V}_2$ is isomorphic to the wave front surface H_3^* of the 3-parameter family of polynomials

$$
t^4 + \varepsilon_2 t^5 + \varepsilon_1 t^2 + \varepsilon_0.
$$

This surface is birationally equivalent to the preceding one.

Finally, the discriminant of W_1 is isomorphic to the wave front hypersurface $\tilde{H}_4$ of the 4-parameter family of polynomials

$$
t^5 + \varepsilon_3 t^6 + \varepsilon_2 t^4 + \varepsilon_1 t^2 + \varepsilon_0,
$$

and the discriminant $\tilde{H}_4^*$ of the class $V_{1,3}$ is the wave front of

$$
t^7 + t^6 + \varepsilon_3 t^5 + \varepsilon_2 t^4 + \varepsilon_1 t^2 + \varepsilon_0.
$$

These two varieties are related to the discriminant of the Coxeter group H_4.

The discriminant of H_4 is diffeomorphic [8] to the bifurcation diagram of zeros of the family H of functions of x, y with parameters a, b, c, d given by

$$H = x^3 + \int_0^y (u^2 + ax + b)^2 du + cx + d$$

which is a special non-versal deformation of the simple function singularity E_8. Let $p : \mathbf{C}^4 \to \mathbf{C}^4$ be the meromorphic mapping $p : (a, b, c, d) \mapsto (\varepsilon_3, \dots, \varepsilon_0)$ with components:

$$\varepsilon_3 = a^{-3}, \ \varepsilon_2 = ba^{-3}, \ \varepsilon_1 = ca^{-1} + 3b^2 a^{-3}, \varepsilon_0 = d - cba^{-1} - b^3 a^{-3}.$$

This has poles along $a = 0$, and is a 3-fold covering of the complement of the hyperplane $\varepsilon_3 = 0$ in the target space.

PROPOSITION 2.8. *The mapping p sends the intersection of the H_4 discriminant with the open set $a \neq 0$ onto the intersection of $\tilde{H}_4$ with the open set $\varepsilon_3 \neq 0$.*

The plane curve A_k is itself the discriminant of the dihedral group $I_2(k)$. It is diffeomorphic to the locus of critical values of the mapping $\Psi : \mathbf{C}^2 \to \mathbf{C}^2$ whose components are the basic invariants of the $I_2(k)$ action which is the complexification of the group of symmetries of a regular k-gon in $\mathbf{R}^2$.

The pullback under Ψ of a pseudo-simple germ $g_2 = 0$ is a germ of isolated $I_2(k)$-invariant singularity of a plane curve. Consider the one-dimensional invariant stabilization $Y \subset \mathbf{C}^3$ determined by the stabilized equation $Y = \{g_2 \cdot \Psi^{-1} + z^2 = 0\}$, $z \in \mathbf{C}$. Extend the $I_2(k)$ action to be trivial on the z axis. Then the $I_2(k)$ action on the vanishing homology of the Milnor fiber $\tilde{Y}$ of Y commutes with the monodromy action.

PROPOSITION 2.9. [28] *The isotypic decomposition (by $I_2(k)$-characters) of the monodromy action on the vanishing homology of the Milnor fiber of the stabilization contains:*
  – the standard representations of the Coxeter group H_3 in the $\tilde{V}_2$ and $V_{0,3}$ cases;
  – the standard representation of the Coxeter group H_4 in the W_1 and $V_{1,3}$ cases.

The classification of pseudo-simple singularities changes when a boundary of some A_k type is stabilized. For example to a single class $V_{0,2}$ on a curve A_4 corresponds an infinite series of pseudo-simple classes $x + z^k + y^2$, $k \geq 1$ on the boundary $xz + y^5 = 0$.

2.5. CONTACT OF SURFACES WITH SPHERES AND CIRCLES

Singularities of the contact of special families of circles with a given generic surface in Euclidean 3-space [40] (which are useful in computer vision and numerical geometry, see e.g. [23]) supply a geometrical application of Theorem 2.7.

Let S be a 2-sphere in Euclidean space $\mathbb{R}^3$, and let M_S^3 be the 3-manifold of all circles (including those of zero radius) contained in S.

Let Γ be a smooth surface in $\mathbb{R}^3$. The *bifurcation set* $D(\Gamma, S) \subset M_S$ is the set of circles which are not transversal to Γ.

PROPOSITION 2.10. *For a generic surface Γ and an arbitrary sphere S, the germ of $D(\Gamma, S)$ at any of its points is diffeomorphic either to the germ at the origin of one of the sets listed below, to the union of such germs intersecting transversely, or to the intersection of such germs with the half-space defined by the boundary of the manifold M_S. This list consists of :*

    – a smooth surface (A_1 case);

    – cylinders over the discriminants of the groups A_2, B_2, G_2, generated by reflections;

    – the union of the cylinder over the discriminant of the group H_2 (the curve with a 2/5-cusp point) and its tangent plane;

    – the discriminants of the groups A_3, B_3;

    – the union of the discriminant of the group H_3 of symmetries of the icosahedron and its tangent plane;

    – a surface birationally equivalent to the preceding one (case H_3^*) (geometrically the H_3, H_3^* cases arise when the sphere S passes through a point on Γ whose normal contains the vertex of a swallow tail of the focal set, and the center of S coincides with this vertex);

    – the union of the folded Whitney umbrella and its tangent plane;

    – the union of three smooth surfaces such that the first and second are each tangent to the third along a line, and these two lines are transversal;

    – the union of three smooth surfaces which are tangent to each other along one smooth curve (this set has moduli).

Proof. Stereographic projection reduces the problem to the study of the restricted bifurcation sets of contacts of a surface Γ' embedded in $\mathbb{R}^4$ with the flag formed by the affine 3-space corresponding to the sphere S, and its 2-subspaces corresponding to circles from M_S (that is, to the study of FSC).

A suitable application of the transversality theorem shows that only those singularities can arise which have codimension at most 3 on a singular boundary which is a plane curve germ of codimension at most

5. These singularities (besides the nonsimple A_2 and D_4 cases, which can be treated similarly) are classified in theorem 2.7.

2.6. LEGENDRE THEORY

A germ of a family of hypersurface singularities $h(x, q) = 0$ in $\mathbf{A}^m$ depending on parameters $q \in \mathbf{A}^s$ is a local generating family [43] of the Legendre variety $\mathcal{L}_h$ in $PT^*\mathbf{A}^s$ given by

$$\mathcal{L}_h = \left\{ (q, \bar{p}) \,\middle|\, \exists x : h(x, q) = 0, \ \frac{\partial h}{\partial x} = 0, p = \frac{\partial h}{\partial q} \right\}$$

Here p are homogeneous coordinates of the point $\bar{p}$ in the projectivized fiber.

If the rank of the mapping $J^1 : (x, q) \mapsto \left(h, \frac{\partial h}{\partial x}\right)$ is maximal (Morse condition), then $\mathcal{L}_h$ is non singular. If the regular points of the projection of $\mathcal{L}_h$ to the base $\mathbf{A}^s$ are dense in $\mathcal{L}_h$, its image (bifurcation diagram of zeros or wave front) determines the hypersurface $h = 0$ up to stabilization. This holds for versal deformations of isolated hypersurface singularities.

Let $F(x, q)$ be a contact versal deformation of a map-germ $F(\cdot, 0)$ defining a complete intersection of codimension k with isolated singularity. The set Δ_F of points q such that the variety $F(\cdot, q) = 0$ is singular is the *bifurcation diagram of zeros of F*.

Consider the germ $\mathcal{F}$ given by $\mathcal{F} = \sum_{i=1}^m \lambda_i F_i$, where the F_i are the components of F. This depends on the variables x, the parameters q, and the auxiliary variables (Lagrange multipliers) $\lambda \in \mathbf{A}^k$.

Note that the singularity theory of complete intersections coincides with the singularity theory of hypersurfaces (with nonisolated singularities) determined by equations of the form $\mathcal{F} = 0$ (linear in the variables λ) with respect to the right action of the diffeomorphisms preserving the vector bundle structure $(x, \lambda) \mapsto x$. This type of function often occurs as the Hamiltonian of a variational problem with constraints.

The family $\mathcal{F}$ is the generating family of the Legendre variety $\mathcal{L}_F \subset PT^*\mathbf{A}^s$, which is the Nash modification of Δ_F. In fact, $\mathcal{L}_F$ is a submanifold [24].

The projection of $\mathcal{L}_F$ to the base is a finite-to-one mapping only if the rank r of the mapping $F(\cdot, 0)$ equals $k - 1$ (hypersurface case). Otherwise the intersection of $\mathcal{L}_F$ with the fiber over the origin is a projective subspace $P\left(\mathbf{A}^{k-r}\right)$.

Let $g_1(x, q), \ldots, g_n(x, q)$ be a contact versal deformation of an embedding with respect to the flag $\mathbf{A}^n$ formed by the coordinate subspaces

L_j. For each element L_j^* of the dual flag in $(A^n)^*$, consider the Legendre submanifold $\mathcal{L}_j \subset PT^*A^s$ determined by the generating family $\sum_{i=1}^m \lambda_i g_i$, where the vector λ belongs to L_j. The collection of all these submanifolds forms a Legendre variety $\mathcal{L}_G$ associated to the flag contact singularity. Its projection is the set of parameters corresponding to non-transversal intersection of the embedding with the flag (the bifurcation diagram of the FCS). The variety $\mathcal{L}_G$ is the Nash modification of the bifurcation diagram.

The bifurcation diagram of zeros uniquely determines the complete intersection singularity [36] (up to stabilization in the hypersurface case). A similar result holds for flag contact bifurcation diagrams.

THEOREM 2.11. *Let U_1 and U_2 be germs of holomorphic (or real analytic) flag contact singularities with isomorphic bifurcation diagrams. Then either these germs are isomorphic, or they are generalized boundary singularities, determined by stably equivalent equations, up to trivial extensions.*

3. Simple flag singularities

3.1. CLASSIFICATION

The complete classification of simple flag contact classes is enormous. In the case of embeddings, if the generic intersection of the mapping with flag elements has negative dimension the simple classes correspond to nested systems of ideals in the local algebras of simple isolated singularities of complete intersections.

For singular mappings the classification is much more difficult [9, 29].

Below we concentrate on cases which have applications in geometry and analysis. Other cases which may arise in specific problems can be treated by similar methods [44].

Let F_2 be the flag in $\mathbb{C}^m$ formed by two affine subspaces $\mathbb{C}^k \hookleftarrow \mathbb{C}^{k-1}$ of neighbouring dimensions. Let $i : X \to \mathbb{C}^m$ be an embedding of a complete intersection germ. Suppose that $\dim(X) + k - 1 \geq m$, that is, the generic intersection of X with $\mathbb{C}^{k-1}$ is non-empty.

THEOREM 3.1. [40] *The simple F_2CS of i-germs are as follows (up to extensions and liftings):*
   i. *The set X is a smooth hypersurface; the intersection $X \cap \mathbb{C}^k$ and the boundary $\mathbb{C}^{k-1}$ define in $\mathbb{C}^k$ a simple boundary singularity (of type A, B, C, D, E, F_4).*

ii. *The set X is a smooth 2-surface embedded in $\mathbb{C}^4$, $k = 3$; the restrictions to X of the coordinate functions f and g, defining the equations of $\mathbb{C}^k$ in $\mathbb{C}^{k+1}$ and $\mathbb{C}^{k-1}$ in $\mathbb{C}^k$ respectively, have the form (in some local coordinates x, y on X):*

$$
\begin{aligned}
C_{n,l} &: f = xy, & g = x^n + y^l \quad 2 \le n \le l \\
F_{2n+1} &: f = x^2 + y^3, & g = y^n \quad\quad\; n \ge 1 \\
F_{2n+4} &: f = x^2 + y^3, & g = xy^n \quad\; n \ge 1.
\end{aligned}
$$

These normal forms appeared originally in Goryunov's list of simple singularities of projections of complete intersections onto a line [18]. However the group of equivalences for the projections is smaller than the group of flag-contact equivalences. The coincidence of the lists follows from the weighted homogeneity of Goryunov's simple classes and the list of bounding non-simple classes.

The corresponding adjacency diagrams for the classes $C_{*,*}, F_*$ are as follows.

$$
\begin{array}{ccccccccccc}
A_1 & \leftarrow & A_2 & \leftarrow & A_3 & \leftarrow & A_4 & \leftarrow & A_5 & \leftarrow & \cdots \\
\uparrow & & \uparrow & & \uparrow & & \uparrow & & \uparrow & & \\
B_2 & \leftarrow & C_3 & \leftarrow & C_4 & \leftarrow & C_5 & \leftarrow & C_6 & \leftarrow & \cdots \\
\uparrow & & \uparrow & & \uparrow & & \uparrow & & \uparrow & & \\
B_3 & & C_{2,2} & \leftarrow & C_{3,2} & \leftarrow & C_{4,2} & \leftarrow & C_{5,2} & \leftarrow & \cdots \\
\uparrow & & \uparrow & & \uparrow & & \uparrow & & \uparrow & & \\
F_4 & \leftarrow & F_5 & & C_{3,3} & \leftarrow & C_{4,3} & \leftarrow & C_{5,3} & \leftarrow & \cdots \\
& & \uparrow & & \uparrow & & \uparrow & & \uparrow & & \\
& & F_6 & \leftarrow & F_7 & & C_{4,4} & \leftarrow & C_{5,4} & \leftarrow & \cdots \\
& & & & \uparrow & & \uparrow & & \uparrow & & \\
& & & & F_8 & \leftarrow & F_9 & & \vdots & & \\
& & & & & & \uparrow & & & & \\
& & & & & & \vdots & & & &
\end{array}
$$

$$
F_{p+q+1} \to C_{p,q} \to A_{p+q-1}
$$

REMARK 3.2. The bifurcation sets of these singularities and the intersection forms of the vanishing homology on the generic Milnor fibers (lifted to an appropriate ramified double covering) coincide with those of the corresponding singularities of projections onto a line [18, 6].

REMARK 3.3. V. Arnold [5] classified the ratios of two holomorphic functions up to holomorphic diffeomorphisms of the source space, additions of germs of holomorphic functions, and multiplications by non-zero holomorphic functions, and obtained the same list of simple singularity classes, since not only the answer but the problem itself is actually the problem of flag-equivalence.

Now let $dim(X) + k = m$.

THEOREM 3.4. [15] *The simple F_2CS of i-germs (up to appropriate extensions and liftings) are as follows:*
   i $m = 3$, $k = 2$. *The set X is a smooth line in $\mathbb{C}^3$, with local coordinate x. The flag is $\mathbb{C}^3 \supset \mathbb{C}^2 (f = 0) \supset \mathbb{C}^1 (f = g = 0)$. The restrictions to $i(X)$ of f and g have the form*

$$B_{k,s} : f = x^k, \ g = x^s, \ k \geq s \geq 1.$$

   ii $m = 5$, $k = 3$. *The set X is a smooth surface in $\mathbb{C}^5$, with local coordinates x, y. The flag is $\mathbb{C}^3 (f_1 = f_2 = 0) \supset \mathbb{C}^2 (f_1 = f_2 = g = 0)$. The restrictions to $i(X)$ of f_1, f_2 and g have the form*

$$\begin{aligned}
I_{2,2}^4 &: f_1 = x^2, & f_2 = y^2, \ g = x + y; \\
I_{2,2}^5 &: f_1 = x^2, & f_2 = y^2, \ g = x; \\
I_{2,2}^6 &: f_1 = x^2, & f_2 = y^2, \ g = xy; \\
I_{2,2}^7 &: f_1 = x^2, & f_2 = y^2, \ g = 0;
\end{aligned}$$

$$\begin{aligned}
I_{2,3}^5 &: f_1 = x^2 + y^3, \ f_2 = xy, \ g = y; \\
I_{2,3}^6 &: f_1 = x^2 + y^3, \ f_2 = xy, \ g = x; \\
I_{2,3}^7 &: f_1 = x^2 + y^3, \ f_2 = xy, \ g = y^2; \\
I_{2,3}^8 &: f_1 = x^2 + y^3, \ f_2 = xy, \ g = y^3; \\
I_{2,3}^9 &: f_1 = x^2 + y^3, \ f_2 = xy, \ g = 0;
\end{aligned}$$

$$I_{2,k}^{k+2} : f_1 = x^2 + y^k, \ f_2 = xy, \ g = y, \ k \geq 4.$$

Due to homogeneity this list coincides with the list [18] of simple projections of fat points to a line.

THEOREM 3.5. [41] *The list of simple classes of contact of complete intersections X with a flag of three neighbouring dimensions, such that the generic intersection dimension of X with each element of the flag is non-negative, is as follows:*
   1. *The extensions of the simple classes from the previous theorems.*
   2. *The flag consists of the subspaces $\{w = 0\} \hookleftarrow \{w = x = 0\} \hookleftarrow \{w = x = y = 0\}$ in $\mathbb{C}^{n+1}$, where $x, y, w \in \mathbb{C}$, $z \in \mathbb{C}^{n-2}$. The set X is the regular hypersurface defined by $w = f(x, y, z)$, where f is one of the following: ($Q(z)$ denotes a non-degenerate quadratic form in the*

variables z):

$$(B_m, B_2) : \ f = Q(z) + x^m + y^2 \qquad\qquad m \geq 2$$
$$(C_m, B_m) : \ f = Q(z) + xy + y^m \qquad\qquad m \geq 3$$
$$(B_p, C_q) : \ f = x^p + yz_1 + z_1^q + Q(z_2 \ldots z_{n-2}) \quad q \geq 3 \ p \geq 2$$
$$(F_4, B_3) : \ f = x^2 + y^3 + Q(z)$$
$$(C_3, F_4) : \ f = xz_1 + y^2 + z_1^3 + Q(z_2 \ldots z_{n-2})$$
$$(F_4, F_4) : \ f = x^2 + y^2 + z_1^3 + Q(z_2 \ldots z_{n-2})$$

3. *The set X is the regular 3-manifold in $\mathbb{C}^5$ defined by the equations*
$w = f(x, y, z)$, $u = g(x, y, z)$, where $x, y, z, u, w \in \mathbb{C}$; and the flag
consists of the subspaces $\{w = 0\} \hookleftarrow \{w = x = 0\} \hookleftarrow \{w = x = u = 0\}$.
Here the pairs of functions (f,g) are the following :

$$\tilde{B}_{m,p,q} : \ f = x^m + y^2 + z^2, \ g = y^p + z^q \quad m \geq 2 \ \ p, q \geq 2$$
$$\tilde{C}_{3,m} : \ f = xy + y^3 + z^2, \ g = y^m \qquad m \geq 2$$
$$\tilde{C}_{3,m}^* : \ f = xy + y^3 + z^2, \ g = zy^m \qquad m \geq 1$$
$$\tilde{F}_{4,m} : \ f = x^2 + y^3 + z^2, \ g = y^m \qquad m \geq 2$$
$$\tilde{F}_{4,m}^* : \ f = x^2 + y^3 + z^2, \ g = zy^m \qquad m \geq 1$$

The adjacency diagram of the classes of theorem 3.5, part 2 is the following:

$$\cdots$$
$$\downarrow$$
$$(C_5, B_5)$$
$$\downarrow$$
$$(C_4, B_4)$$
$$\downarrow$$

$$(C_3, B_3) \leftarrow (C_3, F_4) \leftarrow (F_4, F_4)$$

(F_4, B_3)
$\nearrow$
$\quad (B_2, B_2) \leftarrow (B_2, C_3) \leftarrow (B_2, C_4) \leftarrow (B_2, C_5) \leftarrow \cdots$
$\searrow \qquad \uparrow \qquad\qquad \uparrow \qquad\qquad \uparrow \qquad\qquad \uparrow$
$\quad (B_3, B_2) \leftarrow (B_3, C_3) \leftarrow (B_3, C_4) \leftarrow (B_3, C_5) \leftarrow \cdots$
$\qquad \uparrow \qquad\qquad \uparrow \qquad\qquad \uparrow \qquad\qquad \uparrow$
$\quad (B_4, B_2) \leftarrow (B_4, C_3) \leftarrow (B_4, C_4) \leftarrow (B_4, C_5) \leftarrow \cdots$
$\qquad \uparrow \qquad\qquad \uparrow \qquad\qquad \uparrow \qquad\qquad \uparrow$
$\qquad \cdots \qquad\qquad \cdots \qquad\qquad \cdots \qquad\qquad \cdots$

Also, the class (F_4, F_4) is adjacent to (F_4, B_3).

This table is symmetric with respect to interchanging entries in each pair together with interchanging of B_k and C_k. This corresponds to the polar operation of reversing the flag.

Following the methods of [5] one can interpret the list of part 2 of Theorem 3.5 as the list of simple singularities of fractions of holomorphic germs on a space with complex boundary.

REMARK 3.6. Define a Milnor fibre of a flag contact singularity of a hypersurface embedding $i : X \to \mathbb{C}^m$ as the collection of intersections of Milnor fibers of $i(X)$ with the elements L_j of the flag. These fibers are local manifolds $\tilde{i}(X) \cup L_j$, where $\tilde{i}$ is a small perturbation of i. Each of them is homotopy equivalent to a wedge of spheres. The total number of these vanishing spheres (of different dimensions) will be called the *Milnor number* of the flag singularity.

Note that for each simple singularity in Theorem 3, its Milnor number coincides with the number of parameters in the corresponding versal deformation (Tjurina number).

REMARK 3.7. For two flag elements of neighbouring dimensions we get a boundary pair of respective Milnor fibers, as in the theory of boundary function singularities [6]. Using a ramified double covering over the boundary fiber we may define an intersection form on the relative vanishing homology group of the pair.

One can define an intersection form for the FCS of Theorem 3.5 in a similar way. For example, the substitution $x = \tilde{x}^2, y = \tilde{y}^2$ transforms each simple class of Theorem 3.5 part 2 into a $\mathbb{Z}_2 \times \mathbb{Z}_2$ symmetric function germ. Consider a distinguished basis B (which is symmetric) of the vanishing homology group of this function singularity. Let $B_0 \subset B$ be formed by the cycles corresponding to the critical points which lie in the subset $\{x \neq 0, y = 0\}$.

The symmetry group $\mathbb{Z}_2 \times \mathbb{Z}_2$ acts on the subspace of the vanishing homology group generated by $B \backslash B_0$. The restriction of the intersection form to the invariant part of this subspace will be called the intersection form of the original FCS.

The singularity classes of Theorem 3.5 part 2 have the following Dynkin diagrams, defining the corresponding intersection forms by standard rules:

$$(C_k, B_k): \overbrace{}^{k-1}$$

$$(B_k, C_l): \overbrace{}^{k-1} \qquad \overbrace{}^{l-1}$$

$$(C_3, F_4):$$

$$(F_4, B_3):$$

$$(F_4, F_4):$$

3.2. DIAGRAMS OF MAPPINGS

According to a classical theorem of J.Mather, right-left stability of a map-germ is equivalent to versality (with respect to the contact group) of a certain deformation of its graph map-germ.

Similarly, the stability of the row diagram of mappings

$$N \xrightarrow{G} \mathbf{R}^m \xrightarrow{\pi_1} \mathbf{R}_1^m \ldots \xrightarrow{\pi_k} \mathbf{R}_k^m,$$

where G is a smooth mapping and π_i are standard projections, is related to versality with respect to flag contact equivalence of a certain deformation of the mapping.

To avoid cumbersome notation, we describe this construction only for the case of diagrams corresponding to flags with two elements of neighbouring dimensions. The statements and proofs for the general case are analogous.

Given a pair (f, g), where $f : N \to \mathbf{R}$ and $g : N \to M = \mathbf{R}^m$, we form the row diagram

$$N \xrightarrow{(f,g)} \mathbf{R} \times \mathbf{R}^m \xrightarrow{\pi} \mathbf{R}^m. \quad .$$

The triples of diffeomorphisms $(\theta_1, \theta_2, \theta_3)$ giving rise to a commutative diagram:

$$\begin{array}{ccccc} N & \xrightarrow{(f_1,g_1)} & \mathbf{R} \times M & \xrightarrow{\pi} & M \\ \downarrow \theta_0 & & \downarrow \theta_1 & & \downarrow \theta_2 \\ N & \xrightarrow{(f_2,g_2)} & \mathbf{R} \times M & \xrightarrow{\pi} & M \end{array}$$

form a group, which we denote by D. This group acts on the space of pairs (f, g) (or, equivalently, on the space of row diagrams as above).

Denote by R^+ the subgroup of D consisting of triples for which θ_1 is simply a shift along each fiber of the projection π :

$$\theta_1 : (t, q) \to (t + \phi(q), Q(q)), \quad t \in \mathbb{R}, \ q \in M^m.$$

To a pair (f, g) we also associate the graph mapping

$$G_{f,g} : N \to K_0 := N \times \mathbb{R} \times \mathbb{R}^m, \quad G : x \mapsto (x, f(x), g(x)).$$

Consider the flag consisting of the two subspaces $K_1 \supset K_2$

$$K_1 = N \times R \times \{0\}; \quad K_2 = N \times \{0\} \times \{0\}$$

of K_0. The shift diffeomorphism

$$S : K_0 \to K_0, \quad S : (x, f, g) \mapsto (x, f - f(x_0), g - g(x_0))$$

takes the distinguished point $(x_0, f(x_0), g(x_0))$ of the graph to a point in K_2.

Two pairs will be called *flag-contact* equivalent if, after corresponding shifts, the graph of one can be sent to the graph of the other by a diffeomorphism Θ of K_0 which preserves the flag: $\Theta(K_1) = K_1, \Theta(K_2) = K_2$. Obviously D-equivalent pair-germs are flag-contact equivalent – take $\Theta = (\theta_0, \theta_1)$.

For a graph mapping $G_{f,g}$, define its *shift-deformation* $G_{f,g}(\lambda)$ with $m + 1$ additive parameters $\lambda_0, \lambda = (\lambda_1, \ldots, \lambda_m)$ as follows:

$$G_{f,g}(\lambda) : (x, \lambda) \mapsto (x, \ f(x) + \lambda_0, \ g(x) + \lambda).$$

THEOREM 3.8. *The D-simple and stable pairs are R^+-equivalent to the standard versal unfoldings of the simple pairs of theorems 3.1 and 3.4.*

The proof is based on the following propositions [44].

The natural notion of flag contact versality of a deformation is equivalent to the corresponding infinitesimal versality (since flag-contact equivalence is a geometrical subgroup of the group of right-left equivalences of graph mappings [12]). For shift-deformation this infinitesimal versality takes the following form.

Denote by $\mathcal{O}_N$ the ring of function-germs at the origin of N, and by I_g the ideal in $\mathcal{O}_N$ generated by the components of g. Denote by P the free $\mathcal{O}_N$-module on 2 generators, and by $K_{f,g}$ the submodule of P formed by pairs $(\tilde{\varphi}, \tilde{\psi})$ with $\tilde{\psi} \in I_g$ and $\tilde{\varphi} \in I_{f,g}$.

Let $W_{f,g}$ be the submodule of P of pairs of the form $i_v d(f, g)$, where v is the germ at the origin of a vector field on N. Finally write D_g for the submodule of $K_{f,g}$ of pairs with both components in I_g.

PROPOSITION 3.9. *The germ of the deformation $G_{f,g}(\lambda)$ is flag-contact versal if and only if the factor module $P/\{K_{f,g} + W_{f,g}\}$ is generated over $\mathbb{R}$ by constant mappings.*

A versal shift-deformation will be called *strictly* versal if the factor module $P/\{D_{f,g} + W_{f,g}\}$ is generated over $\mathbb{R}$ by constant mappings and pairs $f^j, 0$ where j belongs to a finite set. This is equivalent to the additional claim: some power of f belongs to the ideal I_g.

PROPOSITION 3.10. *If a pair-germ (multigerm) is D-stable, then its shift-deformation is versal (with respect to flag-contact equivalence).*
    Strict infinitesimal versality of the shift-deformation $G_{f,g}(\lambda)$ implies D-stability of the pair (f, g).

Let the mapping g have rank r at the origin. In linearly choosing adapted local coordinates (x, y) (with $x \in \mathbb{R}^{n-r}$, $y \in \mathbb{R}^r$) on $(N, 0)$ and (z, u) (with $z \in \mathbb{R}^{m-r}$, $u \in \mathbb{R}^r$) on $(M, 0)$, the mapping takes the form $g : (x, y) \mapsto (z(x, y), y)$, where the components $z_i(x, 0)$ belong to the square of the maximal ideal of the ring $\mathcal{O}_x$ of germs at the origin of functions in x.

The map-germ $z : \mathbb{R}^{n-r} \to \mathbb{R}^{m-r}$, $z : x \mapsto z(x, 0)$ is called the *genotype* of the map-germ g. Right-left stable map-germs are classified by the contact classes of their genotypes.

Now let (f, g) be a pair as above, so that f is a germ $M \to \mathbb{R}$. We define the *genotype* of (f, g) to be the pair $(w, z) : \mathbb{R}^{n-r} \to \mathbb{R} \times \mathbb{R}^{m-r}$, where z is the genotype of g and w the restriction $w(x) = f(x, y)|_{y=0}$.

Denote by $\xi_i(x)$ the derivative $\frac{\partial f}{\partial y_i}|_{y=0}$ and by $\chi_i(y)$ the derivative $\frac{\partial g}{\partial y_i}|_{y=0}$. The pair (f, g) with genotype (w, z) is called *geno-versal* if for any germ

$$(\varphi, \psi) : (\mathbb{R}^{n-r}, 0) \to \mathbb{R} \times \mathbb{R}^{m-r}$$

there exist:
   – a pair $(\tilde{\varphi}, \tilde{\psi})$ with $\tilde{\psi} \in I_z$, and $\tilde{\varphi} \in I_{w,z}$,
   – a germ of vector field v on $\mathbb{R}^{n-r}$ and a set of constants a_0, $a = (a_1, \ldots, a_m)$, b_j $(j = 1, \ldots, r)$ such that

$$\begin{cases} \varphi = \tilde{\varphi} + i_v df + a_0 + \xi_1 b_1 + \ldots + \xi_r b_r \\ \psi = \tilde{\psi} + i_v dg + a + \chi_1 b_1 + \ldots + \chi_r b. \end{cases}$$

A geno-versal pair germ will be called *strictly geno-versal* if some power of w belongs to the ideal $I_z + W_w$.

The Malgrange preparation theorem and the above infinitesimal stability criterion imply the following

PROPOSITION 3.11. *If a pair-germ is D-stable, then it is geno-versal.*

If (f, g) is strictly geno-versal, then it is D-stable.

Moreover, R^+-stability is equivalent to D-stability with the additional condition on the above decomposition: the function $\tilde{\varphi} \in I_z$.

If the pair germ (f, g) is weighted homogeneous, then $w \in J$ and geno-versality implies strict geno-versality and even R^+-geno-versality.

A D-stable pair-germ f, g with given genotype w, z is D-equivalent to a standard versal unfolding (F, G) of the genotype which is affine in the y-variables, and is defined as

$$F(x, y) = w(x) + \xi_1(x)y_1 + \ldots + \xi_r(x)y_r,$$

$$G(x, y) = (z(x) + \chi_1 y_1 + \ldots + \chi_r y_r, y),$$

where the pairs (ξ_i, χ_i) $(i = 1, \ldots, r)$, pairs of the form $(w^j, 0)$ and constant mappings form a system (not necessary minimal) of generators of the factor module $P_x / (W_{w,z} + D_z)$.

3.3. RELATIVE MINIMUM FUNCTION

The following elementary extremal problem with constraints arises as a model for various settings in parametric optimization theory.

Let $g : N^n \to M^m$ be a proper smooth mapping of smooth manifolds, and let $f : N^n \to \mathbb{R}$ be a smooth function on N.

The function

$$F(q) = \min \{f(p) \mid p \in N, \ g(p) = q\}, \ q \in g(N) \subset M,$$

defined on the image of g, is called *the relative minimum* function. It provides a solution of the extremal problem of minimising $f(p)$ under the constraint $g(p) = q$, where $q \in M$.

The relative minimum function is, in general, neither smooth nor continuous. Its singularities describe those of the boundary of domains of attainability (for a given time) of control systems, propagations of wave fronts, singularities of solutions of parameter-dependent extremal problems (for example, that of production of optimal mixtures) and so on.

When the constraint mapping is a submersion, the problem becomes that of the singularities of minima of functions depending on a parameter, which was studied in [10], [26] for example.

Two relative minimum function germs are called $\Gamma-$*equivalent* if their graphs can be mapped to each other by a germ of smooth diffeomorphism of the product of the parameter space by a line, which preserves the projection onto the parameter space. Generic singularities with respect to Γ-equivalence were described in [13, 14] (in the case $n \geq m$).

A property will be said to hold for a generic pair if it holds for all pairs (f, g) in a subset of the space of pairs, which is open and dense in the Whitney C^∞-fine topology.

The dimensions n and m are called *nice* if the germs of relative minimum function of generic pairs $(f, g) : N^n \to \mathbb{R} \times M^m$ are simple.

Recall the row diagram

$$N \xrightarrow{(f,g)} \mathbb{R} \times M \xrightarrow{\pi} M, \quad x \mapsto (f(x), g(x)) \mapsto g(x),$$

associated to the pair (f, g) (π is the natural projection).

Denote by Γ^+ the subgroup of the group D preserving the orientation of the fibers of π. Obviously the relative minimum functions of two Γ^+-equivalent diagrams are Γ-equivalent.

A genotype (w, z) (which is a germ at some point P) will be called *minimal* if the restriction of the w to the zero-level set of the constraint mapping z has a minimum at the point P.

Evidently, only collections of minimal genotypes form multigerms corresponding to singularities of the relative minimum function.

PROPOSITION 3.12. [15] *The dimensions $n = m$ are nice if and only if $m < 6$. The dimensions $n > m$ are nice if and only if $m < 4$.*

The inverse image under a generic pair-mapping of a point P on the graph of the relative minimum function consists of at most a finite number of points. Near P this graph coincides with the graph of the minimum of a finite number of *local* relative minimum functions.

The list of simple stable local singularities of relative minimum functions is the list of D-versal deformations of minimal pair genotypes.

If $m = 3$ the dimensions n and m are nice. If also $n = 3$ there are 34 different types of generic singularities. Only 24 of them remain when $n > 4$, and the list remains the same for all such n due to a certain stabilization property of singularities.

The lists of normal forms provide, of course, the diffeomorphism types of singularities of the domain where the relative minimum function is defined, as well as the sets of its discontinuity.

THEOREM 3.13. [14] *If $m = 3$, then the relative minimum function germ of a generic pair at an arbitrary point is stable, simple and is Γ-equivalent to the germ at zero of one of the functions in (x, y, z) in the second column of the following table. The codimension in $\mathbb{R}^3$ of the stratum of a given singularity is given in column 3 and the range of corresponding values of n is given in the last column.*

	Normal form	codim	n
1.	0	0	≥ 3
2.	$-\lvert x\rvert$	1	≥ 3
3.	$\min\{-\lvert x\rvert, y\}$	2	≥ 3
4.	$\min\{-\lvert x\rvert, y, z\}$	3	≥ 3
5.	$-\sqrt{x}$	1	≥ 3
6.	$\min\{-\sqrt{x}, 1\}$	1	≥ 3
7.	$\min\{-\sqrt{x}, -\lvert y\rvert + 1\}$	2	≥ 3
8.	$\min\{-\sqrt{x}, y\}$	2	≥ 3
9.	$\min\{-\sqrt{x}, -\lvert y\rvert + z\}$	3	≥ 3
10.	$\min\{-\sqrt{x}, \min\{-\lvert y\rvert, z\} + 1\}$	3	≥ 3
11.	$\min\{-\sqrt{x}, -\sqrt{y} + 1\}$	2	≥ 3
12.	$\min\{-\sqrt{x}, -\sqrt{y} + 1, 2\}$	2	≥ 3
13.	$\min\{-\sqrt{x}, -\sqrt{y} + z\}$	3	≥ 3
14.	$\min\{-\sqrt{x}, -\sqrt{y} + z, 1\}$	3	≥ 3
15.	$\min\{-\sqrt{x}, -\sqrt{y} + 1, -\lvert z\rvert + 2\}$	3	≥ 3
16.	$\min\{-\sqrt{x}, \min\{-\sqrt{y}, z\} + 1\}$	3	≥ 3
17.	$\min\{-\sqrt{x}, -\sqrt{y} + 1, -\sqrt{z} + 2\}$	3	≥ 3
18.	$\min\{-\sqrt{x}, -\sqrt{y} + 1, -\sqrt{z} + 2, 3\}$	3	≥ 3
19.	$\min\{w \mid w^3 + xw + y = 0\}$	2	≥ 3
20.	$\min\{\min\{w \mid w^3 + xw^2 + y = 0\}, z\}$	3	≥ 3
21.	$\min\{-\sqrt{z}, \min\{w \mid w^3 + xw + y = 0\} + 1\}$	3	≥ 3
22.	$\min\{w \mid w^4 + xw^2 + yw + z = 0\}$	3	≥ 3
23.	$\min\{\min\{w \mid w^4 + xw^2 + yw + z = 0\}, 1\}$	3	≥ 3
24.	$\min\{w^4 + xw^2 + yw \mid w \in R\}$	3	≥ 4
25.	$\min\{u^2 + w^2 + ux + wy \mid u^2 - w^2 = z\}$	3	$= 4$
26.	$\min\{uw + ux + wy \mid u^2 + w^2 = z\}$	3	$= 4$
27.	$\min\{\min\{uw + ux + yw \mid u^2 + w^2 = z\}, 1\}$	3	$= 4$
28.	$\min\{-\lvert y\rvert\sqrt{x}, z\}$	3	$= 3$
29.	$\min\{-\lvert y\rvert\sqrt{x}, -\sqrt{z} \pm 1\}$	3	$= 3$
30.	$\min\{-\lvert y\rvert\sqrt{x}, -\sqrt{z} \pm 1, 2\}$	3	$= 3$
31.	$-\lvert y\rvert\sqrt{x}$	2	$= 3$
32.	$\min\{\pm w^2 \mid w^3 + xw^2 + yw + z = 0\}$	3	$= 3$
33.	$\min\{-\lvert y\rvert\sqrt{x}, 1\}$	2	$= 3$
34.	$\min\{-\lvert y\rvert\sqrt{x}, -\lvert z\rvert + 1\}$	3	$= 3$

For classes 25 and 26, the complement of the singularity locus is not simply-connected, and a singularity of this type determines a winding number invariant for loops in the parameter space.

THEOREM 3.14. [15] *The germ of the relative minimum function of a generic pair depending on $m = 4$ parameters $q_1, \ldots, q_4$ is $\mathbb{R}^+$-equivalent* *at an arbitrary point to the germ at the origin of one of the following functions.*

A_3 $\min\{x^4 + q_1 x^2 + q_2 x \mid x \in R\}$
codim $= 2, n > 4$

A_5 $\min\{x^6 + q_1 x^4 + q_2 x^3 + q_3 x^2 + q_4 x \mid x \in R\}$
codim $= 4, n \geq 5$

$B_{k,1}$ $\min\{x \mid x^k + q_1 x^{k-2} + \ldots + q_{k-2} x + q_{k-1} = 0\}$
$1 \leq k \leq 5$ codim $= k - 1, n \geq 4$

$B_{k,2}$ $\min\{\pm x^2 \mid x^k + q_1 x^{k-1} + \ldots + q_{k-1} x + q_k = 0\}$
$2 \leq k \leq 4$ codim $= k, n = 4$

$B_{3,3}$ $\min\{x^3 + x q_4 \mid x^3 + q_1 x^2 + q_2 x + q_3 = 0\}$
codim $= 4, n = 4$

$X_4^{\pm}$ $\min\{x + 2y \mid xy + q_1 = x^2 \pm y^2 + x q_2 + y q_3 + q_4 = 0\}$
codim $= 4, n = 4$

$C_{2,2}^-$ $\min\{x^2 + y^2 + q_2 x + q_3 y \mid xy = q_1\}$
codim $= 4, n = 5$

$C_{2,2}^+$ $\min\{xy + q_2 x + q_3 y \mid x^2 + y^2 = q_1\}$
codim $= 4, n = 5$

F_5 $\min\{y^2 + x q_4 \mid x^2 + y^3 + q_1 y^2 + q_2 y + q_3 = 0\}$
codim $= 4, n = 5$

F_5^* $\min\{x + y^2 \lambda(q) \mid x^2 \pm y^4 + q_1 y^3 + q_2 y^2 + q_3 y + q_4 = 0\}$
codim $= 4, n \geq 5$

Z_6^e $\min\{\pm x^2 + \mu(q) y^2 + q_1 x + q_2 y + x_3 z \mid x^2 + y^2 + z^2 = q_4\}$
$\mu(0) \neq 0$, codim $= 4, n = 6$

Z_6^h $\min\{x^2 + \nu(q) y^2 + q_1 x + q_2 y + q_3 z \mid xy + z^2 = q_4\}$
$\nu(0) \neq 0$, codim $= 4, n = 6$,

where λ, μ, ν *are arbitrary smooth functions of* q_1, q_2, q_3 *(functional moduli).*

The relative minimum function germ of a generic pair at any point is Γ-equivalent to the minimum of a collection of the functions listed above with total codimension not exceeding n.

3.4. ENVELOPES OF WAVE FRONTS

Time-optimisation of a system with smooth strictly convex indicatrices gives rise to a special class of relative minimum singularities, related to the singularities of families of wave fronts [8, 42].

    Let $W_t \subset M$ be a time-dependent family of wave fronts on a manifold M. Taking a point $q \in M$ to the minimum time t such that $q \in W_t$ defines a relative minimum function. The singularities of these functions are special. Suppose the big front $\widetilde{W} = (t, W_t) \subset \mathbb{R} \times M$ in space-time is the projection of a smooth Legendrian submanifold $\mathcal{L}$ in $PT^*(\mathbb{R} \times M)$. Then the corresponding minimizing function is regular and the constraint mapping has corank at most one.

In particular, the classes of Theorems 3.13 and 3.14 satisfying the conditions mentioned above give all minimal singularities of singular loci [42] of families of wave-fronts in $\mathbb{R}^3$ and $\mathbb{R}^4$.

The critical value locus of the projection of $\widetilde{W}$ to the configuration space M along the time axis consists of three strata: the projection of the critical locus of $\widetilde{W}$ (caustic), the closure of the set of critical values of the projection of the regular part of $\widetilde{W}$ (envelope) and the Maxwell stratum (cut-locus), which is the projection of self-intersections of the big front.

THEOREM 3.15. [42] *The critical value surface of the projection of the wave front germ of a generic Legendre submanifold $\mathcal{L} \subset PT^*(\mathbb{R} \times \mathbb{R}^3)$ is diffeomorphic to the bifurcation set of a simple boundary singularity from the following list*

$$A_k \ (k = 1, 2, 3, 4); \ B_k \ (k = 2, 3, 4); \ C_k \ (k = 2, 3, 4); \ F_4; \ D_4.$$

3.5.   STABILITY OF A WAVE FRONT IN SUBRIEMANNIAN GEOMETRY

The following stability result (see [3, 4] for details) for a rather special flag contact singularity happens however to be the simplest generic example of a big wave front in subriemannian geometry [1, 2].

Consider a flag contact singularity determined by the germs at the origin of two coordinate functions $g_1(x, y)$, $g_2(x, y)$ on the plane

$$g_1 = x^2 + y^2 - \varepsilon_3, \quad g_2 = x\varepsilon_1 + y\varepsilon_2 + (x^2 + y^2)S(x, y, \varepsilon_1, \ldots, \varepsilon_4),$$

depending on parameters $\varepsilon_1, \ldots, \varepsilon_4$.

Suppose that the function S satisfies the following conditions: $S(0) = \frac{\partial S}{\partial x}|_0 = \frac{\partial S}{\partial y}|_0 = 0$, but $\frac{\partial S}{\partial \varepsilon_4}|_0 \neq 0$ and the second differential of $S|_{\varepsilon=0}$ is a non-umbilic (i.e. it is not proportional to the second differential of $x^2 + y^2$).

THEOREM 3.16. [4] *The bifurcation diagrams of all such FSC (g_1, g_2) are diffeomorphic.*

These bifurcation diagrams are essentially the germs of a modification of the big wave front W_E of the following mapping. Consider germs at the origin of a generic riemannian metric and a contact distribution in $\mathbb{R}^3$. Subriemannian geodesics (local length minimizers chozen among the curves which are tangent to the planes of the distribution) are projections to $\mathbb{R}^3$ of the trajectories of a certain Hamiltonian vector field [3] in $T^*\mathbb{R}^3$. The Hamiltonian flow determines a Legendre *exponential* mapping E from the product of the Hamiltonian level hypersurface H_0 in the cotangent space over the distinguished generic point in $\mathbb{R}^3$ by the time axis to the space-time.

In contrast to the Riemannian case, the hypersurface H_0 is not compact (it is isomorphic to a cylinder). Moreover the initial point belongs to the closure of the set Σ of conjugate points along the geodesics emanating from this point. This set coincides with the (caustic) image of the projection to the configuration space $\mathbb{R}^3$ of the singular point locus of the big wave front W_E.

The first, second, etc. conjugate points along the generic geodesic are distinct. They determine a decomposition of Σ into an infinite series of components.

The first component Σ_1 was described asymptotically by Agrachev [1].

A reparametrization of time (the coordinate ε_4 equals the time divided by the square root of the space coordinate, which vanishes on the distribution plane at the origin) permits us to separate these components.

Theorem 3.16 implies that for generic metrics and distributions the germs of the modified big wave fronts are diffeomorphic. However the corresponding diffeomorphisms of space-time do not preserve the projection along the time axis. The first caustic Σ_1 is not stable. It has functional moduli [4].

3.6. FINAL REMARKS

Constraints in optimization problems often occur in the form of inequalities. To find the critical values (in particular, extrema) of a function $h : \mathbb{R}^n \to \mathbb{R}$ subjected to equality constraints $g = 0$ (where $g : \mathbb{R}^n \to \mathbb{R}^k$, $k < n$) and an inequality $f \geq 0$ (where $f : \mathbb{R} \to \mathbb{R}$), one should distinguish the critical values of h on the total space $\mathbb{R}^n$, critical values of the restriction of h to the regular part of $g = 0$, values of h on the critical locus of the variety $g = 0$, critical values of h on the regular part of the variety $f = 0$, $g = 0$ and values of h on the critical locus of the restriction of f to the regular part of $g = 0$. If all the entries depend on parameters Λ then the critical values described above form, generally speaking, a reducible hypersurface Σ in the product $\Lambda \times \mathbb{R}$ of the parameter space and the set of values of h.

CLAIM 3.17. *Generically, when n and the dimension of Λ are small enough, Σ is an open subset of the bifurcation diagram of the flag* *contact singularity of the embedding*

$$\mathbb{R}^n \to \mathbb{R}^k \times \mathbb{R} \times \mathbb{R} \times \mathbb{R}^n, \quad (x) \mapsto (g(x), h(x), f(x), x))$$

and the flag

$$\{0\} \times \{0\} \times \{0\} \times \mathbb{R}^n \subset \{0\} \times \{0\} \times \{\mathbb{R}\} \times \mathbb{R}^n \subset$$

$$\{0\} \times \{\mathbb{R}\} \times \{\mathbb{R}\} \times \mathbb{R}^n \subset \{\mathbb{R}^k\} \times \{\mathbb{R}\} \times \{\mathbb{R}\} \times \mathbb{R}^n$$

with three elements.

Hence Theorem 3.5 provides the classification of all simple stable singularity classes of Σ in the complex case (the details in the real case have not been worked out). Of course certain classes are not minimal.

The geometry of submanifolds in Euclidean, hyperbolic, affine or projective spaces involves the following problem.

What are the singularities of contact of submanifolds with certain special flags or with a family of flags parametrized by the corresponding group G of orthogonal, projective etc. transformations of the ambient space?

For example, flattenings of curves are distinguished by the contact of the curve with the osculating complete flag [22]. Certain models of flattenings and inflections of a family of germs of curves determined as perturbations of a complete intersection are studied in [35, 17].

Even these simple cases require non trivial investigation of the bifurcation diagrams of flag contact singularities depending on parameter spaces which have a G-bundle structure.

References

1. A. A. Agrachev, Methods of control theory in nonholonomic geometry, *Proc. Intern. Congr. Math., Zürich, 1994*, (Birkhäuser, Basel, 1995) 12–19.
2. A. A. Agrachev, Exponential mappings for contact sub-Riemannian structures, Journal of dynamical and control systems **2**:3 (1996) 321–358.
3. A. A. Agrachev, G. Charlot, J.-P. Gauthier and V. Zakalyukin, On stability of generic subriemannian caustic in the three space, Comptes Rendus Acad. Sci. Paris, sér. I, **330** (2000) 465–470.
4. A. A. Agrachev, G. Charlot, J.-P. Gauthier and V. Zakalyukin, On subriemannian caustics and wave fronts for contact distributions in the three space, Journal of dynamical and control systems, **6**:3 (2000) 365–395.
5. V. I. Arnold, Singularities of fractions and behavior of polynomials at infinity, Proc. Steklov Math. Inst. **221** (1998) 48–68.
6. V. I. Arnold, V. V. Goryunov, O. V. Lyashko and V. A. Vassiliev, *Singularities II: classification and applications* (Dynamical systems VIII), Enc. Math. Sci. **39**, Springer-Verlag, Berlin a.o., 1993.

7. V. I. Arnold, A. N. Varchenko, and S. M. Gusein-Zade, *Singularities of differentiable mappings vol I*, Monographs in Math. **82**, Birkhäuser, Basel-Boston, 1985.

8. V. I. Arnold, *Singularities of caustics and wavefronts*, Acad. Math. and its Appl., Kluwer, 1990.

9. J. W. Bruce and J. M. West, Functions on cross-caps, Math. Proc. Camb. Phil. Soc. **123** (1998) 7–27.

10. L. N. Bryzgalova, On maximum function of family of functions, Funct. anal. appl. **12**:1 (1978) 66–67.

11. S. Chmutov and I. Scherback, Dynkin diagrams of dual boundary singularities are dual, J. Alg. Geom. **3** (1994) 449–462.

12. J. Damon, *The unfolding and determinacy theorems for subgroups of $\mathcal{A}$ and $\mathcal{K}$* Memoirs Amer. Math. Soc. **50 no. 306** (1984).

13. A. A. Davydov, Singularities of the maximum function over the preimage, *Geometry in nonlinear control and differential inclusions*, Banach Center Publ. **32**, 1995, pp 167–181.

14. A. A. Davydov and V. M. Zakalyukin, Point singularities of relative minimum on three dimensional manifold, Proc. Steklov Math. Inst. **220** (1998) 113–129.

15. A. A. Davydov and V. M. Zakalyukin, Classification of relative minima singularities, *Proc. Caustics-98*, Banach Center Publ. **50**, 1999, pp 74–90.

16. N. H. Duc, Involutive singularities, Kodai Math.J. **17** (1994) 627–635.

17. M. Garay, Vanishing flattening points and generalized Plücker formula, Comptes Rendus Acad. Sci. Paris, sér I, **323** (1996) 901–906.

18. V. V. Goryunov, Geometry of bifurcation diagrams of simple projections onto the line, Funct. Anal. Appl. **15**:2 (1981) 77–82.

19. V. V. Goryunov, Singularities of projections of complete intersections, Modern problems in mathematics, VINITI **22** (1983) 167–205.

20. M. Giusti, Classification des singularités isolées d'intersections complétes simples, Comptes Rendus Acad. Sci. Paris, sér I, **284** (1977) 167–170.

21. G. Ishikawa, Symplectic and Lagrange stabilities of open Whitney umbrellas, Invent. Math. **126** (1996) 215–234.

22. M. E. Kazarian, Flattenings of projective curves, singularities of Schubert stratifications of Grassmanians and flag varieties and bifurcations of Weierstrass points of algebraic curves, Russian Math. Surv. **46**:5 (1991) 91–136.

23. J. J. Koenderink, *Solid shape*, MIT press, 1993.

24. E. J. N. Looijenga, *Isolated singular points on complete intersections*, London Math. Soc. Lecture Notes **77**, Cambridge Univ. Press, 1984.

25. O. V. Lyashko, Classification of critical points of functions on a manifold with singular boundary, Funct. anal. appl. **17**:3 (1983) 28–36.

26. V. I. Matov, Topological classification of germs of maximum and minimax of generic families of functions, Russian Math. Surveys **37**:4 (1982) 167–168.

27. O. Myasnichenko and V. Zakalyukin, Lagrange singularities under symplectic reduction, Funct. anal. appl. **32**:1 (1998) 1–11.

28. M. Roberts and V. Zakalyukin, Symmetric wave fronts, caustics, and Coxeter groups, *Singularity theory*, Proc. College on Singularity Theory (Trieste 1991), (ed D. T. Lê, K. Saito and B. Teissier), (World Scientific Publ., 1994) pp 594–626.

29. B. Z. Shapiro, Normal forms of the Whitney umbrella with respect to a cone preserving contact group, Funct. Anal. Appl. **31**:2 (1997) 144–147.

30. I. Scherback, Singularities in the presence of symmetries, Amer. Math. Soc. Transl. (2) **180** (1997) 189–195.

31. I. Scherback and A. Szpirglas, Lagrange transformation and duality for corner and flag singularities, Israel J. Math. **111** (1999) 77–92.

32. O. Scherback, Projectively dual space curves and Legendre singularities, Proc. Univ. Tbilisi, **13-14** (1982) 280–336.

33. D. Siersma, Singularities of functions on boundaries, corners, etc., Quart. J. Math. Oxford ser.(2) **32** (1981) 119–127.

34. T. Tsukada, Reticular Lagrangian submanifolds, Asian J. Math. **1** (1997) 572–622.

35. R. Uribe, On the higher dimensional four vertex theorem, Comptes Rendus Acad. Sci. Paris, sér I, **321** (1997) 1353–1358.

36. K. Wirthmuller, Singularities determined by their discriminants, Math. Ann. **252** (1980) 237–245.

37. M. I. Zaharia, On the monodromy of a function germ defined on an arrangement of hyperplanes, Kodai Math. J. **22**:3 (1999) 438–445.

38. M. I. Zaharia, Function germ defined on an arrangement of coordinate hyperplanes, Stud. Cerc. Mat. **47**:3-4 (1995) 271– 282.

39. M. I. Zaharia, The homology of the Milnor fiber of a function germ defined on an arrangement of hyperplanes, Saitama Math. J. **15** (1997) 1–8.

40. V. M. Zakalyukin, Singularities of circles contact with surfaces and flags, Funct. Anal. Appl. **31**:2 (1997) 73–76.

41. V. M. Zakalyukin, Singularities of contact with three-flags, Funct. Anal. Appl. **33**:3 (1999) 67–69.

42. V. M. Zakalyukin, Envelopes of families of wave fronts and control theory, Proc. Steklov Math. Inst. **209** (1995) 133–142.

43. V. M. Zakalyukin, Generating ideals of Lagrangian varieties, *Theory of Singularities and its Applications* (ed V.I. Arnold), Advances in Soviet Math. **1** (1990) 201–210.

44. V. M. Zakalyukin, Flag contact singularities, *Real and Complex Singularities* (eds J.W. Bruce and F. Tari), Research Notes in Math. **412** (Chapman and Hall/CRC 2000) pp 134–146.

On Stokes Sets

Yuliy Baryshnikov

LAMA, Department of Mathematics, Université Versailles-Saint-Quentine
(yuliyb@math.uvsq.fr)

Introduction

The main goal of these notes is to describe the combinatorial structure
of the Stokes sets for polynomials in one variable, a certain bifurcation
diagram in the space of monic polynomials of given degree (the precise
definition is given in section 5). As it turns out, their structure is
intimately connected to other bifurcation diagrams (of quadratic differentials, or of Smale functions), and to various combinatorial structures,
most prominent among them being Stasheff polyhedra. These notes
are expository with proofs at best sketched. A detailed exposition will
appear elsewhere.

1. Stasheff polyhedra

1.1. GENERALITIES

We recall some interpretations of the *associahedra*, alias *Stasheff polyhedra*. Denote by B_n the set of all meaningful bracketings of $(n + 2)$
indeterminates written in a line (that is the set of ways to form products
in a nonassociative algebra).

Join a pair of bracketings by an edge if they are related by one
application of the associativity relation, $(a(bc)) = ((ab)c)$. It turns out
that the resulting graph is the 1-skeleton of a convex polyhedron whose
vertices are the elements of B_n, called the Stasheff polyhedron or the
associahedron, and denoted by K_n. The faces of K_n are again Stasheff
polyhedra and their products.

There are further interpretations of the Stasheff polyhedron (a general reference being, e.g. [15]): it is a geometric realization of the poset
of triangulations of a convex plane $(n+3)$-gon or of that of plane rooted
trees with $(n + 2)$ leaves. More precisely, an (incomplete) triangulation
of a convex polygon is just a collection of nonintersecting chords; ordered by inclusion, these triangulations form a poset dual to the face
poset of K_n. Similarly one defines a partial order on the set of plane

D. Siersma et al. (eds.), New Developments in Singularity Theory, 65–86.
© 2001 Kluwer Academic Publishers. Printed in the Netherlands.

rooted trees by saying that a tree obtained by contracting an edge is greater than the original one; this poset is isomorphic to the face poset of K_n.

There are several ways to construct convex polyhedra combinatorially equivalent to K_n. We will make use of the following explicit convex polyhedral realization of $K_n \subset \mathbb{R}^n$:

$$K_n(\epsilon) = \{(t_1, \ldots, t_n) : |t_i| \le 1, 1 \le i \le n; \sum_{i=j}^{k} t_i \le \epsilon, 1 \le j < k \le n\}.$$

(Here ϵ is a small enough positive number).

More symmetric convex realizations (having dihedral symmetry) can be obtained using the interpretation of the polyhedron K_n as the fiber polytope of the projection of an $(n+2)$ simplex onto the regular polygon with $(n+3)$ vertices [4].

1.2.   STASHEFF FAN

Recall that a fan in a real vector space is a finite stratification of $\mathbb{R}^n$ by convex polyhedral cones C_i such that the relative boundary of any of the cones is a union of cones of smaller dimensions and the intersection of any two cones is again a cone of the family.

The normal fan associated to a convex compact polyhedron P is defined as the collection of cones $\{C_f\}$ in the dual space, where f runs over faces of P: the cone C_f consists of the linear functionals attaining their maxima over P at f.

A *Stasheff fan* is a fan whose cones ordered by inclusion form a poset dual to that of the facets of K_n, for example, the normal fan to a Stasheff polyhedron (or rather to a convex realization of it, like $K(\epsilon)$, or the symmetric realization).

There is also an explicit construction of a Stasheff fan (related in a natural way to the triangulation interpretation) which we will use further:

Let $\mathbf{P}$ be a convex $(n+3)$-gon in the plane. We call a (real-valued) function on the set of vertices of $\mathbf{P}$ a *balanced weight* if its values add up to zero and the geometric center of mass is at the origin (so that the balanced weights form a real vector space of dimension n). A weight is called *degenerate* if there exists a linear function on the plane whose restriction to P majorizes the weight and coincides with it in at least four points.

The partition of $\mathbb{R}^n$ defined by degenerate balanced weights is a Stasheff fan $\Sigma_n \subset \mathbb{R}^n$. The natural stratification of $\mathbb{R}^n$ defined by the Stasheff fan Σ_n is simplicial (in other words, the Stasheff polyhedron is simple). The number of open simplicial cones in the Stasheff fan of dimension n is $\frac{1}{n+1}\binom{2n}{n}$, that is, the Catalan number c_n. The generating function for the Catalan number, $\Phi = \sum c_n t^n$ solves the functional equation $\Phi(t) = (1 + t\Phi)^2$ so that

$$\Phi = 1 + 2t + 5t^2 + 14t^3 + \ldots.$$

2. Quadratic differentials

2.1. BIFURCATION DIAGRAMS

Quadratic differentials (a general reference here is [13]), that is, differential elements $f(z)dz^2$ with holomorphic f, define a pair of orthogonal (singular) foliations on the Riemann surface V of z as follows. A vector $\xi \in T_z V$ is tangent to the *horizontal (vertical)* foliation iff $f(z)\xi^2 > 0$ ($f(z)\xi^2 < 0$, respectively). The foliations are well-defined everywhere outside the zeros of f; near a zero of order k they have $(k+2)$ rays emanating from the zero.

When the zeros of f are simple, the singularities of each foliation are "tripods" at each zero. For special parameter values special trajectories ("instantons") appear: some of the singular leaves can connect different zeros. This happens on a set of parameters of real codimension 1. The picture below shows a typical behavior of the singular trajectories of a quadratic differential along a 1-parameter transversal to the bifurcation diagram.

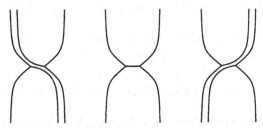

Figure 1. Homoclinic trajectory of a quadratic differential (λY-catastrophe).

DEFINITION 2.1. *Let $\mathcal{L}$ be the base of the versal deformation of a quadratic differential $f dz^2$. The bifurcation diagram $S \subset \mathcal{L}$ of the quadratic differential is the closure of the set of parameter values corresponding to quadratic differentials with a singular leaf (horizontal or vertical) joining two zeros of the differential.*

The bifurcation diagram naturally splits into the union of the *vertical* and *horizontal* ones, depending on which of the foliations has the nongeneric leaf: $\mathbf{S} = \mathbf{S}_v \cup \mathbf{S}_h$. The bifurcation diagram Δ of zeros of the quadratic differential, that is, the set of parameter values corresponding to differentials with multiple zeros, is a subset of $\mathbf{S}_v \cap \mathbf{S}_h$.

The study of the bifurcation diagrams of quadratic differentials was initiated by J. Bruce and D. O'Shea [5] in connection with their investigation of fields of principal directions on minimal surfaces in a 3-dimensional Euclidean space (such surfaces can be invariantly equipped with a conformal structure and a quadratic differential) It seems that the question about the local structure of these bifurcation diagrams has not been investigated earlier (see however [6]).

Below I describe, following [2], the combinatorial structure of the bifurcation diagram in the base of the versal deformation of 'A_n-singularity', that is the quadratic differential $z^{n+1} dz^2$.

The general theory of versal deformations of holomorphic forms $f dx^\lambda$ was developed by V. Kostov, S. Lando and others (a general reference is [11], see also [9]). In our case of quadratic differentials (which was already treated in [6]), the base of a versal deformation can be identified with the n-dimensional complex linear space of monic polynomials of degree $(n + 1)$ with vanishing sum of zeros.

In their preprint Bruce and O'Shea calculated the bifurcation diagram for the A_1 case. For the standard deformation $(z^2 - a) dz^2$ it consists of two straight lines in the a-plane intersecting at the origin.

2.2. COMBINATORICS OF BIFURCATION DIAGRAMS

Represent $\mathbb{C}^n$ as the product of two copies of $\mathbb{R}^n$ and consider two cylinders over Σ_n in each of the factors, $\bar{\Sigma}_v = \Sigma_n \times \mathbb{R}^n; \bar{\Sigma}_h = \mathbb{R}^n \times \Sigma_n$.

THEOREM 2.2. *The bifurcation diagram in the base of the versal deformation of the quadratic differential $z^{n+1} dz^2$ is homeomorphic to the union $\bar{\Sigma}_v \cup \bar{\Sigma}_h \subset \mathbb{R}^n \times \mathbb{R}^n$, that is, there exists a homeomorphism $h : (\bar{\Sigma}_v \cup \bar{\Sigma}_h, \mathbb{R}^m \times \mathbb{R}^m) \to (\mathbf{S}_v \cup \mathbf{S}_h, \mathcal{L})$ taking $\bar{\Sigma}_h$ to S_h and $\bar{\Sigma}_v$ to S_v .*

In other words, the bifurcation diagram $\mathbf{S}$ is homeomorphic to the fan dual to the *product* of two Stasheff polyhedra.

The number of connected components of the complement to the bifurcation diagram in the base of the versal deformation of A_n-singularity is therefore equal to c_n^2. For example, the cross of Bruce and O'Shea divides $\mathbb{R}^2$ into $4 = 2^2$ pieces.

Consider the *core* $h(\{0\} \times \mathbb{R}^n)$ of the vertical component of the bifurcation diagram (the subset of parameters for which $n - 1$ instantons exist). This is a topological submanifold. The intersections of the vertical component S_v with the germs of transversals to the core are combinatorially Stasheff fans, but their geometry varies strongly. Over a generic point it is diffeomorphic to the fan dual to the realization $K(\epsilon)$, as in the left part of Figure 2 below.

On the other hand, the intersection of the vertical component with the core of the horizontal component has the dihedral symmetry indicated on the right part of Figure 2.

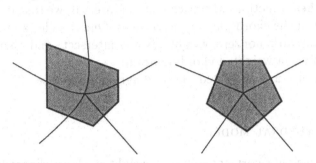

Figure 2. Possible geometries of the intersection of the vertical bifurcation diagram with transversals and their dual Stasheff polygons.

Both components of the bifurcation diagram, S_h and S_v, consist of a large number of pieces of hypersurfaces analytic outside of the discriminant. Quite surprisingly, these pieces glue together on analytic continuation.

THEOREM 2.3. *There exist two hypersurfaces H_v and H_h in $\mathcal{L} \cong \mathbb{R}^{2n}$ which are analytic and irreducible in the complement to the bifurcation diagram of zeros Δ, such that $S_v \subset H_v$ and $S_h \subset H_h$.*

The analytic hypersurface of Theorem 2.3 is non-algebraic, unlike the swallowtail. It has a logarithmic branching at Δ.

3. Weighted chord diagrams

3.1. DEFINITIONS

To construct the homeomorphism of Theorem 2.2 we will use the description of the Stasheff fan given in 1.2 in terms of balanced weights. The balanced weights give rise to weighted chord diagrams in the plane convex polygon $\mathbf{P}$ with $(n+3)$ vertices.

A *weighted chord diagram* in $\mathbf{P}$ is a set of non-intersecting chords (segments joining non-neighboring vertices of $\mathbf{P}$) with a positive number attached to each chord. This set will be called the *support* of the weighted chord diagram. If the number of chords is maximal (that is,

n) we will call the weighted chord diagram *complete*. Complete chord diagrams correspond to triangulations of **P**.

One can associate a weighted chord diagram to each balanced weight as follows. Fix a point p in the plane of **P** which is in general position with respect to the vertices of **P**. Let a balanced weight f be given. For any two non-neighboring vertices of **P** consider all linear functions coinciding with f at those vertices and majorizing f elsewhere. The values of these functions at p sweep out an interval; we take its length as the weight of the chord joining the vertices. One can check immediately that chords with nonzero weight do not intersect, and therefore the weight defines a weighted chord diagram.

Conversely, any weighted chord diagram gives rise to a balanced weight.

3.2. POLYHEDRAL MODEL

Balanced weights corresponding to weighted chord diagrams with a given support form a simplicial cone of dimension equal to the number of chords. Denote this cone by $C_d = \mathbb{R}^D$, where D is a set of nonintersecting chords (i.e. a chord diagram).

Recall that chord diagrams (without weights) are ordered by inclusion. For any ordered pair of chord diagrams $D_1 \subset D_2$ the cone C_{D_1} is embedded in C_{D_2}: just set the weights of the extra chords equal to 0. Gluing all these cones together along these mappings (or, equivalently, taking the inductive limit) one arrives at a "polyhedral model" of $\mathbb{R}^n$ built of simplicial cones corresponding to chord diagrams. The cones of the Stasheff fan of positive codimension correspond to the images of the cones C_d, $|D| < n$.

4. Quadratic differentials and weighted triangulations

4.1. FROM QUADRATIC DIFFERENTIALS TO WEIGHTED CHORD DIAGRAMS

According to [11], the standard affine deformation

$$f(z;a)dz^2 = (z^{n+1} + a_1 z^{n-1} + \ldots + a_n)dz^2$$

of the A_n singularity of the quadratic differential is versal. Now we associate a pair of weighted chord diagrams to each value of the parameter $a = (a_1, \ldots, a_n)$.

For each a one can choose $R > 0$ large enough so that outside of the circle $C_R = \{|z| \le R\}$ the field of directions defining either of the foliations is sufficiently close to the field of directions corresponding to the unperturbed differential $z^{n+1}dz^2$. In particular, the union of the leaves of (say) the horizontal foliation intersecting C_R consists of $(m+3)$ arms going to infinity. We identify once and for all these arms with the vertices of the convex polygon **P**.

Take R so large that the interior of C_R contains the set Z of zeros of f. We will call a nonsingular horizontal leaf *nontrivial* if its intersection with C_R represents a nontrivial element of $\pi_1(C_R - Z, \partial C_R)$. A connected component of the union of nontrivial leaves we will call a *stream*. Clearly a stream goes from infinity to infinity along two different arms. The integral of (a branch of) the 1-form $\sqrt{f}\,dz$ sends the stream to an infinite horizontal strip on the complex plane. The height of this strip we call the *weight of the stream*. Join the vertices of **P** corresponding to the arms containing the stream by the chord and attach to it the weight of the stream. What results is a weighted chord diagram. The same can be done with the vertical foliation.

Therefore one can associate a pair of weighted chord diagrams corresponding to the horizontal and vertical foliations to each (deformed) quadratic differential $f\,dz^2$ (see an example in Figure 3 below).

4.2. ...AND BACK

This correspondence can be inverted: for each pair of weighted chord diagrams, there exists a unique (given that the sum of its zeros vanishes) polynomial quadratic differential which generates exactly them.

For example, if both chord diagrams are empty (that is, the weights of all chords are zero) we arrive at the polynomial z^{n+1}.

To describe the inversion it will be convenient to draw the two polygons (with $(n+3)$ vertices each) in which the chord diagrams in question are given as inscribed with alternating vertices in a $2(n+3)$-gon **D**. We will call this configuration the *interlacing* polygons. This reflects the behavior of the arms of the horizontal and vertical foliations at infinity.

Let the chords with positive weights be represented by straight segments joining the vertices of the large polygon. They divide the interior of the polygon into several connected components. We pick one point in each of the components, and join these points by segments if the corresponding components share a segment of a chord on their boundaries. Further, we join the interior point in each component with the vertices of **D** which belong to the boundary of the component.

These segments divide the polygon **D** into convex polygons of the following types:

a)   triangles with one side being a side of **D**;

b) triangles with exactly one vertex being a vertex of **D**. Such a
    triangle has to contain a piece of exactly one chord.

c) 4-gons containing exactly one point of intersection of chords (one
    vertical and one horizontal). All the vertices of the 4-gon are
    interior points of **D**.

Replace each triangle of type a) by an infinite quadrant; each triangle
of type b) by a half-infinite rectangular strip of width equal the weight
of the chord piece which the rectangle contains; and each 4-gon by a
rectangle whose sides are equal to the weights of the chords intersecting
inside the component. One can identify the corresponding sides of the
boundaries of these new geometric pieces.

The resulting (topological) disk acquires a conformal structure, which
is flat everywhere outside the $(m + 1)$ internal points where more
than four right angles of the flat pieces come together. We will denote
these points by p_i. There exists a unique complex structure compatible
with the conformal one. As is easy to check, the disk is conformally
equivalent to the entire complex plane $\mathbb{C}$. Choose a coordinate z on
this plane so that the sum of the coordinates of the p_i's vanishes,
and the directions to the images of the vertices of **D** are positioned
correctly (remember that they are identified with the arms of a polynomial quadratic differential, which tend asymptotically to directions
$\exp(i\pi k/(n + 3))$, $k = 0, \ldots, 2n + 5$). This fixes the coordinate z. Let
the coordinates of the points p_i be z_i, $i = 1, \ldots, n + 1$ (counted
with multiplicities), and take f to be the monic polynomial of degree
$n + 1$ with roots z_i. Then the polynomial quadratic differential $f(z)dz^2$
generates the pair of weighted chord diagrams we started with. *This
proves Theorem 2.2.*

A pair of chord diagrams and the foliations defined by the corresponding polynomial quadratic differential of degree 4 are shown in
Figure 3. Here the solid diagonals in the left picture and the solid lines
in the right picture correspond to the vertical foliation; dashed diagonals and dashed lines to the horizontal one. The auxiliary segments
subdividing **D** are shown as dotted. The pieces corresponding to the
diagonal AB and the corresponding stream are filled.

The linear space of monic polynomials with vanishing zero sum can
be identified not only with the base space of the versal deformation of
the quadratic differential $z^{n+1}dz^2$ but also with the base space of the
versal deformation of the simple singularity A_n of a *plane curve* [1]:

$$a \mapsto P_a(z, w) = f(z, a) - w^2, \text{ where } f(z, a) = z^{n+1} + a_1 z^{n-1} + \ldots + a_n.$$

The discriminant Σ is the set of values of the parameters a for which
the level curve $X_a = \{(z, w) : P_a(z, w) = 0\}$ is singular (equivalently,

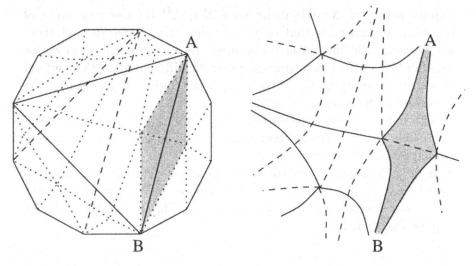

Figure 3.

for which $f(\cdot, a)$ has multiple roots). Let $\omega = wdz$. This holomorphic 1-form ω defines a section of the cohomological Milnor fibration over $\Lambda - \Delta$ (the fibers of this fibration are $H^1(X_a, \mathbb{C})$) defined by the restrictions of ω to the X_a. The integral of the form over a locally constant (with respect to the Gauss-Manin connection) section of the homological Milnor fibration defines a (multi-valued) holomorphic function on $\Lambda - \Delta$. For a cycle constructed from a vertical (horizontal) instanton the real (imaginary) part of this function vanishes. This shows that each of the two components of the bifurcation diagram, S_h and S_v, are real analytic hypersurfaces in $\Lambda \cong \mathbb{R}^{2m}$.

For example, for the standard deformation of the Morse singularity A_1,

$$f(z, a) = z^2 - a,$$

we have $h = a$ (up to a multiplicative constant), and the cross of Bruce and O'Shea is just the union of the lines $\Re(a) = 0$; $\Im(a) = 0$.

Consider the base space Λ as a real linear space and let $\Lambda^{\mathbf{D}}$ be its complexification (that is $\Lambda^{\mathbf{D}} \cong \Lambda \otimes_{\mathbb{R}} \mathbb{C}$). The function $\Re(h)$ locally defining the bifurcation diagram of the quadratic differential near a smooth point is locally holomorphic, and its zero locus is an analytic hypersurface. This hypersurface is singular and non-algebraic (the function h has logarithmic terms near Δ).

The discriminant Δ extends to $\Lambda^{\mathbf{D}}$ as a singular algebraic variety of (complex) codimension 2. In the example above the discriminant is just the origin in the plane. Both strata of the bifurcation diagram are smooth at the origin. Generally, both S_v and S_h are smooth at the

smooth points of Δ. Indeed, let $a_0 \in \Sigma \in \Lambda^{\mathbf{D}}$ be a generic point of
the discriminant, such that only two zeros of p_{a_0} coincide and there
are no nontrivial instanton trajectories. In a neighborhood of a_0 we
consider the set $\mathbf{S}_{v;a_0}$ consisting of quadratic differentials with a vertical
trajectory connecting the roots of p_a that collide at a. Clearly, the
hypersurface $\mathbf{S}_{v;a}$ is smooth at a.

The union of the hypersurfaces $\mathbf{S}_{v,\Delta} = \cap_{a \in \Delta} S_{-}v; a$ is a smooth,
irreducible (as Δ is irreducible) hypersurface in a tubular (of varying
radius) neighborhood of Δ in $\Lambda^{\mathbf{D}}$.

If a_0 is a generic point of the bifurcation diagram $\mathbf{S}_v$, the path
constructed at the beginning of this Section belongs to the smooth
part of the diagram and connects a_0 to $\mathbf{S}_{v,\Delta}$. The irreducibility of $\mathbf{S}_{v,\Delta}$
implies Theorem 2.3.

5.   Stokes sets

5.1. STOKES PHENOMENA

The motivation for the study of Stokes sets comes primarily from the
theory of asymptotic expansions of integrals using the steepest descent
method.

Consider the approximation of the integrals

$$I(k; l) = \int_C a(x; l) e^{k\Phi(x; l)} dx$$

for $k \to \infty$. Here the phase Φ is a function, analytic on $\mathbb{C}$, depending
analytically on some parameters l; and C is an infinite contour such
that $\Re\Phi \to -\infty$ at its ends (that is, C represents some element of
$H_1(\mathbb{C}, \{\Re\Phi \ll 0\})$). To approximate this integral one customarily de-
forms the contour to pass through the critical points of Φ and to go
along the trajectory of the steepest descent (that is the trajectory of the
gradient vector field of the real part of Φ). These trajectories are the
leaves of the foliation defined by the level curves of the imaginary part
of Φ. The neighborhoods of the critical points of the phase contribute
most to the integral.

The form of the asymptotic expansion thus obtained depends on the
number of critical points through which the steepest descent contour
passes. Typically, one has

$$I(k; l) = m_1 e^{k\Phi(x_1(l))} + m_2 e^{k\Phi(x_2(l))} + \dots,$$

where the x_i are the critical points of Φ on the deformed contour; and
the m_i are functions of k and l varying slowly (as compared to the

exponential function) with k. The main contribution comes, of course, from the critical values with the largest real part.

The Stokes phenomenon in this context is the discontinuous change of the coefficients m (Stokes' multipliers) as the parameters l change. The topological reason for this is the discontinuous behavior of the gradient trajectories at the parameter values when a segment of the trajectory connects two critical points. This is a nongeneric situation which happens in real codimension 1. A typical example is shown below.

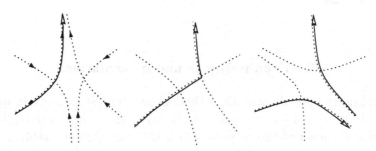

Figure 4.

Here the contour is shown as a solid line; dotted lines show the foliation by the level lines of the imaginary part of the phase.

Notice that at the bifurcation set the real parts of the critical values are strictly different: the Stokes phenomenon does *not* reduce to one critical value bypassing the other and becoming the leading exponent. The subset of parameters where this happens (antiStokes sets, see below) is also relevant in asymptotic analysis and in our constructions.

At the bifurcation set the imaginary parts of the critical values connected by the gradient trajectory coincide. However, the Maxwell stratum for the imaginary part of Φ is strictly larger than the set of parameters where the Stokes phenomenon occurs: the existence of the connecting gradient trajectory is necessary. The Maxwell strata for the imaginary parts of holomorphic germs were studied in [10].

5.2. STOKES SETS

The Stokes set associated with an asymptotic integral depends on the class of the integration contour; taking into account all possible contours, we get the following definition.

DEFINITION 5.1. *Let* $\Phi : U \times \mathcal{L} \to \mathbb{C}$, $U \subset \mathbb{C}$, $\mathcal{L} \cong \mathbb{C}^m$ *be a deformation of the function* $\phi(\cdot) = \Phi(\cdot; 0)$, $l \in \mathcal{L}$. *The Stokes set* $\mathbf{S}_v \subset \mathcal{L}$ *is the closure of the set of parameters* l *for which there exists a smooth component of a level set of the imaginary part of* $\Phi(\cdot, l)$ *with critical points at its ends.*

Similarly, the antiStokes set $S_h \subset \mathcal{L}$ *is defined in terms of the foliation by the level curves of the real part of* Φ.

The bifurcation diagram of functions $\mathbf{D}$ is clearly a stratum in $S_v \cap S_h$.

Example. The Stokes set for the Airy function (corresponding to $\Phi(x, l) = x^3/3 + xl, l \in \mathbb{C}$) is the union of three rays from the origin at angles of 120°; the antiStokes set is the centrally symmetric image of the Stokes set.

6.   Even polygons and quadrillages

For multiparameter deformations these elementary Stokes sets, in words of M. Berry, "coalesce or cross". He says further that "there ought to be a classification of the ways in which this can happen stably..." [3]. A description of the combinatorics of the Stokes and antiStokes sets follows.

6.1. DEFINITIONS

We will again use the polygons in a fashion similar to that of section 4.

All polygons we will consider here will be assumed "marked", meaning that their vertices are numbered from 1 through v counterclockwise. We will call a convex polygon with an even number of vertices simply an *even polygon*.

DEFINITION 6.1. *A quadrillage of an even polygon P is a set of non-intersecting chords which partition P into even polygons. The number of chords is called the size of a quadrillage. A quadrillage of maximal size (that is such that all polygons of the partition are 4-gons) is called complete.*

Complete quadrillages of marked even polygons correspond to rooted ternary trees just as triangulations correspond to rooted binary trees. The generating function $Q(t) = \sum_0 q_n t^n$, where q_n is the number of complete quadrillages of a $2n + 2$-gon, solves the functional equation

$$Q = (1 + tQ)^3$$

(the analogous equation for triangulations with squared, not cubed, term, leads to the Catalan numbers) and implies, via Lagrange inversion, $q_n = \frac{1}{2n+1}\binom{3n}{n}$;

$$\Phi = 1 + 3t + 12t^2 + 55t^3 + 273t^4 + \ldots.$$

For each even polygon we fix one of the two interlacing assignments of signs (+) and (−) to its vertices; an even polygon with such an assignment is called *signed*. For signed polygons one obtains an orientation of the chords of a quadrillage and of the sides of the polygon **P**: each of them is oriented from a (−) to a (+).

An (ordered) pair of *interlacing* polygons (having the same number of vertices v) is a convex polygon **D** with $2v$ vertices $1, 2, \ldots, 2v$ numbered counterclockwise; the even vertices thought of as the vertices of one polygon; the odd ones as the vertices of the other.

We consider two interlacing *even, signed* polygons $\mathbf{P}_v$ and $\mathbf{P}_h$ called respectively vertical and horizontal, and a quadrillage of each (also called vertical and horizontal). Fix the first vertex of **D** to be horizontal and the first two vertices to be (+)-signed. This sign assignment induces orientations (as described above) on the chords and sides in each of the interlacing even polygons.

DEFINITION 6.2. *Two quadrillages of a pair of interlacing even polygons P_h and P_v are called admissible if for any couple of intersecting chords c_h and c_v of the respective polygons, the sense of the orientation at the intersection point defined by the tangent vectors to c_h and c_v (in this order) is positive.*

Notice that the orientations at the intersections of chords and *sides* of the polygons are automatically positive, due to the chosen numbering/signing scheme.

For example, of the 3×3 pairs of complete quadrillages of interlacing 6-gons, 6 are admissible and 3 are not, see the figure below.

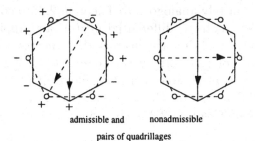

admissible and   nonadmissible

pairs of quadrillages

Figure 5. The dashed chords are horizontal; the solid ones — vertical.

6.2. COMBINATORICS OF STOKES/ANTISTOKES SETS

Similarly to 3.2, we associate to each admissible pair of quadrillages $\pi = (q_v, q_h)$ the simplicial cone C_π of the nonnegative weights on its chords. The dimension of this cone is clearly the sum of the sizes of the quadrillages.

Pairs of quadrillages are ordered by inclusion (as quadrillages themselves are). For any two ordered admissible pairs of quadrillages $\pi_1 = (q_v^1, q_h^1) \prec \pi^2$ we denote by $i_{\pi_1 \pi_2} : C_{\pi_1} \to C_{\pi_2}$ the embedding attaching zero weights to the missing chords.

Gluing the cones C_π together using these embeddings we arrive at a simplicial complex (or rather a cone over a simplicial complex) which we denote by Λ.

The union of the (images of) the cones with noncomplete vertical quadrillages will be denoted by Σ_v; the same for the horizontal quadrillages will be denoted Σ_h. Clearly, Σ_v and Σ_h are simplicial subcomplexes of Λ.

Consider also the admissible pairs of noncomplete quadrillages which have the property that they can be augmented by a pair of intersecting (one vertical, one horizontal) chords. The (image of the) union of the corresponding cones we denote by Σ.

The relevance of all these combinatorial constructions to our problem is explained by the following result.

THEOREM 6.3. *There exists a homeomorphism of quadruples*

$$w : (L; S_v, S_h; S) \to (\Lambda; \Sigma_v, \Sigma_h; \Sigma)$$

which is a diffeomorphism outside of S.

In particular, the space Λ is a cone over a sphere. The connectedness of the sphere implies, in particular, that any two complete quadrillages can be connected by a sequence of elementary flips (removing a chord and replacing it by another one in the resulting hexagon).

This result is entirely analogous to Theorem 2.2: to construct w one just considers the quadratic differential $(f_p' dz)^2$ and applies the gluing method. The only condition to check is that the resulting partitions of the interlacing polygons satisfy the conditions of the theorem and, conversely, that any admissible pair of weighted quadrillages gives rise to a quadratic differential on $\mathbb{C}$ with even zeros only.

The analogue of Theorem 2.3 is also valid:

THEOREM 6.4. *For $n \geq 3$ the union of the Stokes and antiStokes sets belongs to an irreducible (real) hypersurface in $\mathcal{L}$ (considered as a real affine space), analytic outside the discriminant.*

(The configuration of the Stokes and antiStokes sets for the Airy function shows that both of them belong to the same irreducible hypersurface).

The connected components of $\Lambda - \Sigma_v$ are numbered by complete quadrillages of the v-gon with ordered vertices.

PROPOSITION 6.5. *The connected components of* $\Lambda - \Sigma_v$ *are cells.*

7. Polyominos

Fix a quadrillage q subdividing the $2n + 2$-gon into n 4-gons. It will be convenient to think of the quadrillage as of a *polyomino* (e.g., a hexagon with 1 large diagonal corresponds to the domino).

An n-polyomino is defined as following. Consider a collection **S** of n copies of the unit square in the plane $\{|x| \leq 1/2, |y| \leq 1/2\}$ (thus equipped with an orientation). The sides of these squares are oriented by the condition that together with the outward normal they form a positive frame. A polyomino is the space resulting from the identifications of several sides of the squares in such a way that

a) the orientations of the identified sides are opposite if one of them is vertical and one is horizontal and the same otherwise;

b) any two squares have at most 1 side identified and

c) the resulting space is connected.

More formally, let $\mathbb{S}$ be the set of sides of the squares in **S**. Consider a partition of $\mathbb{S}$ into a family of subsets denoted by $\mathbb{E}$; the element to which the side α belongs is denoted by $e(\alpha)$. The partition of $\mathbb{S}$ into the sides belonging to the same square we will encode via the function $s : \mathbb{S} \to \mathbf{S}$; $s(\alpha)$ is therefore the square of which α is a side.

Further, take a family of isometries $\phi_\alpha : \alpha \to \mathbb{R}$, where α runs through $\mathbb{S}$, which *preserve orientations* when α is a horizontal edge and *reverse it* if it is a vertical one, and send the midpoints of the edges to zero. Now, we identify the points of the edge with the same $e(\alpha)$ having equal values of ϕ_α. The resulting cell complex is called a *polyomino* **p**.

We will assume (without loss of generality, to exclude some trivial cases) that each subset of the partition $\mathbb{E}$ with more than 1 element contains both vertical and horizontal sides.

The *backbone* of the polyomino is the graph whose vertices are the centers of the squares (these elements will be called s-vertices) and of their *identified* edges, that is of points numbered by elements of $\mathbb{E}$ (called e-vertices). The edges of the backbone are numbered by the elements of $\mathbb{S}$ and the edge α connects the s-vertex $s(\alpha)$ with the e-vertex $e(\alpha)$.

The backbone graph is clearly bipartite (into e- and s-vertices) and naturally oriented: an edge α points from an s-vertex towards an e-vertex if α is vertical and in the opposite direction if α is horizontal.

A polyomino is called *ordered* if with this orientation, the backbone graph has no oriented cycles.

We say that a polyomino is *simple* if any side of any square is identified with at most one side of some other square. Examples of nonsimple and unordered polyominos are given in the figure below.

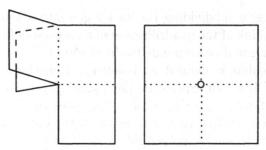

Figure 6. Nonsimple (left) and unordered (right) polyominos.

Denote by $V(\mathbf{p}) = \mathbb{R}^{S(\mathbf{p})}$ the real vector space of functions on the edges of the backbone. The projection sending an edge to its e-vertex embeds the space $E(\mathbf{p}) = \mathbb{R}^E$ of functions on e-vertices into $V(\mathbf{p})$.

To each polyomino $\mathbf{p}$ we associate a fan $S(\mathbf{p})$ in the real vector space $V(\mathbf{p})/E(\mathbf{p})$. To define a fan it is sufficient to define its intersection with a vicinity of the origin. An element of $V(\mathbf{p})/E(\mathbf{p})$ can be identified with a function w on S having the property that $\sum_{\alpha \in e} w(\alpha) = 0$ for any E-vertex e.

Provide each unit square of the polyomino with the piecewise linear foliation by the level sets of the function

$$|x - 1/2| - |y - 1/2|,$$

Now perturb the isometries ϕ_α defining the polyomino to

$$\phi_\alpha \mapsto \phi_\alpha + w(\alpha).$$

Gluing together the squares according to the modified isometries we obtain (for small w) again a topological space with an oriented foliation.

We say that the vector w is *degenerate* if there exists a trajectory (a piecewise linear immersion of the oriented segment into the polyomino tangent to the foliation at all points of linearity) of this foliation and connecting the centers of two squares.

The degenerate vectors form a germ of the associated fan in $V(\mathbf{p})/E(\mathbf{p})$ which we will denote as $S(\mathbf{p})$.

The relevance of the fans associated to polyominos is that they describe the local combinatorics of the Stokes sets.

PROPOSITION 7.1. *Let $C(\mathbf{p})$ be the cell in the complement to the antiStokes set corresponding to a polyomino $\mathbf{p}$. Then there exists a* *homeomorphism*

$$h : (C(\mathbf{p}), C(\mathbf{p}) \cap \mathbf{S}_v) \to (\mathbb{R}^n, \mathbf{S}(\mathbf{p})) \times \mathbb{R}^n.$$

8. Stokes polyhedra

It turns out that the fans $\mathbf{S}(\mathbf{p})$ for ordered polyominos are dual to certain finite convex polyhedra. We formulate the theorems in terms of polyominos, so that they are also valid for those not coming from quadrillages of even polygons.

THEOREM-PROPOSITION 8.1. *For every ordered polyomino* $\mathbf{p}$, *the fan* $\mathbf{S}(\mathbf{p})$ *is dual to a convex polyhedron* $St(\mathbf{p})$ *called the Stokes polyhedron of* $\mathbf{p}$.

The Stokes polyhedra form a family of polyhedra interpolating between the the Stasheff polyhedron and the cube. The Stokes polyhedra for polyominos with at most 4 squares are given in figure 7. More generally, call the *height* of a polyomino the length of the longest oriented chain in its backbone graph.

If the height of p is 2, then $St(\mathbf{p})$ is a cube. If the backbone graph itself is a chain, then $St(\mathbf{p})$ is a Stasheff polyhedron. More generally, one has the following result:

PROPOSITION 8.2. *Each Stokes polyhedron of an ordered polyomino is a Minkovsky sum of the Stasheff polyhedra corresponding to maximal chains in the backbone graph.*

The Minkovsky summands of some of the 3-dimensional Stokes polyhedra shown on Figure 7 are shaded.

PROPOSITION 8.3. *The polyomino is simple if and only if the Stokes polyhedron is simple.*

If an s-vertex σ of the backbone graph of $\mathbf{p}$ is *separating* (that is all adjacent edges are either all *in*- or all *out*-vertices), or, equivalently, the square is attached along only vertical or only horizontal sides, then $St(\mathbf{p})$ is the product of Stokes polyhedra corresponding to the polyominos obtained by removing the square σ and replacing it by its copy in each of the connected polyominos into which $\mathbf{p}$ splits upon this removal.

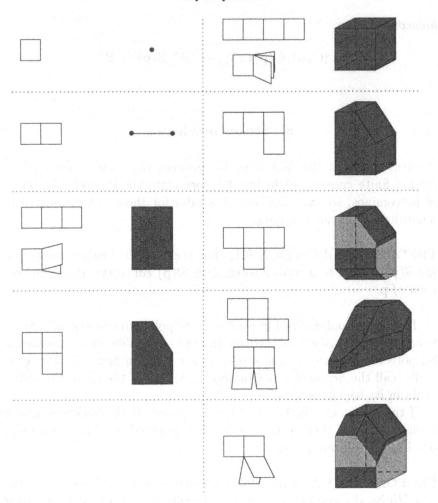

Figure 7.

9. Miscellaneous remarks

9.1. COMBINATORICS

The Stokes polyhedra have some obvious combinatorial interpretations in terms of enumerations of restricted triangulations (or bracketings). For example, the vertices of the Stokes polyhedron for the T-shaped tetromino enumerate the triangulations of a heptagon with exactly one of the chords $1 - 4$ and $1 - 5$ in it.

Enumerating these restricted triangulations allows one to quantify the claim made above that the Stokes polyhedra interpolate between the (most complicated) Stasheff polyhedron and (least complicated) cube.

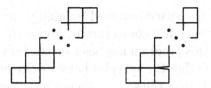

Figure 8.

The number of vertices of the Stokes polyhedron corresponding to
the A-type polyomino (left in Figure 8) is the Catalan number c_n ; the
number of vertices of the D-type polyomino (right) is $c_n - c_{n-2}$ (both
with $(n + 2)$ squares).

9.2. SINGULARITIES

There are some natural ramifications of the theory sketched above (to
be described elsewhere). Thus, (ordered) non-simply connected polyominos appear naturally in the description of the Stokes sets in versal
deformations of other singularities.

One can also study the real case (most interesting for the applications). All the constructions remain valid; the condition of reality
of the deformation just implies that the pairs of quadrillages and the
corresponding polyominos have an axis of symmetry.

9.3. K-THEORY

More surprisingly, Stokes polyhedra appear in algebraic K-theory.

Recall that the Steinberg group of order n over a ring R is the free
group with generators x_{ij}^a, $1 \leq i \neq j \leq n$, $a \in R$ satisfying the relations

$$x_{ij}^a x_{ij}^b = x_{ij}^{a+b}, \tag{A}$$

$$[x_{ij}^a, x_{kl}^b] = 0 \text{ if } i \neq l \text{ and} j \neq k, \tag{B}$$

and

$$[x_{ij}^a, x_{jk}^b] = x_{ik}^{ab}, \ i, j, k \text{ distinct.} \tag{C}$$

Kapranov and Saito undertook in [8] a geometric study of these
relations and syzygies between them, the first step of which is to represent the generators of a group as 1-simplices and the relations as
2-cells spanning them, so that the group itself is the fundamental group
of the resulting cell complex. The next steps then would consist in
finding the syzygies between the relations (corresponding to 3-cells),
syzygies between syzygies and so on. One can see immediately that the
Steinberg relations themselves can be represented as triangles (relation
A), squares (relation B) and pentagons (relation C).

The topology of the space obtained by gluing the cells representing the higher syzygies encodes, conjecturally, higher K-theory.

The first step in this program has been successfully realized: a family of polyhedra representing the syzygies between Steinberg relations was proposed in [8], such that the second homotopy group of the resulting CW-complex is equal to $K_3(R)$.

The surprising fact is that all these polyhedra (with the exception of those containing a triangle corresponding to the additive relation A) turned out to be the Stokes polyhedra listed in the right column of the figure 7.

An explanation of this comes from the Hatcher-Wagoner approach to K-theory [7].

Recall that a Morse-Smale function is a Morse function such that the stable and unstable manifolds $C^{\pm}(f, x)$ of the gradient flow corresponding to critical points x (that is unions of gradient trajectories of f approaching x as $t \to \pm\infty$) intersect transversally. (Of course this is a property of the pair [function, metric], or rather of the pair [function, gradient-like vector-field], not of the function per se.) The unstable manifold $S^-(f, x)$ for a Morse-Smale function is diffeomorphic to Euclidean space of dimension equal to the index of x.

Consider a smooth manifold M^m (with boundary) and a Morse-Smale function f having only Morse critical points of the same index $1 < i < m$ such that

a) all critical values are positive, and

b) the intersections of the unstable manifolds with $\{f = 0\}$ are spheres.

Then, provided with orientations, these unstable manifolds form a basis of $H_i(M, \{f \leq 0\}; \mathbb{Z})$.

Consider now the space $\mathcal{M}$ of Morse functions on M having r critical points and such that the two conditions above are satisfied. For a generic function, the stable and unstable manifolds of different critical points do not intersect, but for some of them this Smale condition is violated. Denote this bifurcation diagram Σ. A result of [7] states that under some dimensional conditions there exists an isomorphism of the Steinberg group with r generators over $R = \mathbb{Z}(\pi_1(M))$ onto $\pi_1(\mathcal{M}, \mathcal{M} - \Sigma)$.

A smooth point of the codimension 1 stratum of the bifurcation diagram corresponds to the transversal intersection of the stable and unstable manifolds $S^-(f, x)$ and $S^+(f, y)$ of some 2 critical points x and y along a gradient trajectory from y to x.

The element of this group represented by a generic 1-parameter deformation passing through a point on the codimension 1 stratum of Σ yields the change of the basis of H_i by an elementary operation.

The codimension 2 strata are either the self-intersections of the smooth parts of codimension 1 strata corresponding to pairs of disjoint trajectories connecting two pairs of critical points, or strata corresponding to chains of three critical points $x \to y \to z$ connected by the gradient trajectories. The former strata correspond to squares, or, algebraically, to the 4-term relation B; the latter to pentagons, that is, to the 5-term relation C.

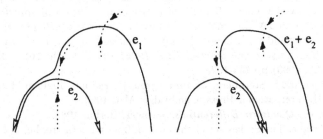

Figure 9. Changes of the homology basis associated with the crossing of the bifurcation diagram

More generally, for any partial order on the set of critical vertices, one can consider the strata consisting of a Morse function such that the ordered points correspond to the critical values are connected by a chain of gradient trajectories. One can show that at generic points the CW-complexes dual to this stratification again have the combinatorial type of certain convex polyhedra which are Minkovski sums of Stasheff polyhedra.

The bifurcation diagrams associated with polyominos described in Section 8 are obviously a special case of this construction. Moreover, in small codimensions (below 5) bifurcations coming from polyominos exhaust the combinatorial types of these polyhedra. This explains their appearance in our lists.

References

1. V. Arnold, A. Varchenko and S. Gusein-Zade, *Singularities of differentiable mappings, vol. 2*, Birkhäuser, Boston, 1988.
2. Yu. Baryshnikov, Bifurcation diagrams of quadratic differentials, Comptes Rendus Acad. Sci. Paris **325** (1997) 71–76.
3. M. Berry, Stokes' phenomenon: smoothing a Victorian discontinuity, Publ. Math. IHES **68** (1989) 211–221.
4. L. J. Billera and B. Sturmfels, Fiber polytopes, Ann. of Math. **135** (1992) 527–549.
5. J. W. Bruce and D. B. O'Shea, On binary differential equations and minimal surfaces, preprint, Liverpool, 1995.

6.   J. Hubbard and H. Masur, Quadratic differentials and foliations, Acta Math. **142** (1979).

7.   A. Hatcher and J. Wagoner, *Pseudo-isotopies of compact manifolds* Asterisque **6**, Soc. Math. France, Paris, 1973.

8.   M. Kapranov and M. Saito, Hidden Stasheff polytopes in algebraic K-theory and in the space of Morse functions, Contemp. Math. **227** (1998).

9.   V. P. Kostov, Versal deformations of differential forms of real degree on the real line, Math. USSR. Izv. **37**:3 (1991) 525–537.

10.   V. P. Kostov, On the stratification and singularities of the Stokes hypersurface of one- and two-parameter families of polynomials, *Theory of singularities and its applications* (ed V. I. Arnol'd), Advances in Soviet Math. **1** (1990) 251–271.

11.   V. P. Kostov and S. K. Lando, Versal deformations of powers of volume forms, *Computational algebraic geometry (Nice, 1992)*, Progr. Math. **109**, (Birkhäuser, Boston MA, 1993) pp 143-162.

12.   S. Shnider and S. Sternberg, *Quantum groups*, Graduate Texts in Mathematical Physics, II. International Press, Cambridge, MA, 1993.

13.   K. Strebel, *Quadratic Differentials*, Springer, Berlin, 1984.

14.   F. J. Wright, The Stokes set of the cusp diffraction catastrophe, J. Phys. **A13** (1980) 2913-2928.

15.   M. Ziegler, *Polytopes*, Springer, Berlin, 1995.

Resolutions of discriminants and topology of their complements

Victor Vassiliev
Steklov Mathematical Institute and Independent Moscow University
(vassil@vassil.mccme.rssi.ru)

The *discriminant* subsets of spaces of geometric objects are the sets of all objects with singularities of some chosen type. The important examples are: spaces of polynomials with multiple roots, *resultant* sets of polynomial systems having common roots, spaces of functions with degenerate singular points, of non-smooth algebraic varieties, of linear operators with zero or multiple eigenvalues, of smooth maps $S^1 \to M^n$ ($n \geq 3$) having singular or self-intersection points, of non-generic plane curves, and many others.

The discriminants are usually singular varieties, whose stratifications correspond to the classification of degenerations of the corresponding objects. E.g., the discriminant subset in the space of polynomials $x^3 + ax + b$ is the semicubical parabola $(a/3)^3 + (b/2)^2 = 0$: its regular points correspond to polynomials with a root of multiplicity exactly 2, and the vertex to the polynomial x^3. The discriminant in the space of polynomials $x^4 + ax^2 + bx + c$ is the *swallowtail*, i.e. the surface shown in the right-hand part of Fig. 1: its self-intersection curve consists of polynomials having two double roots, and the semicubical edges correspond to the polynomials with one triple root; the most singular point is the polynomial x^4 with a root of multiplicity 4. Similar stratifications hold for polynomials of all higher degrees: their strata are indexed by the multiplicities and orders in $\mathbb{R}^1$ of all corresponding multiple roots.

Usually one is interested in the space of non-singular objects which is the complement of the discriminant Σ, e.g. in the space of polynomials without multiple roots, of smooth varieties, of non-degenerate operators, or of knots, i.e. maps $S^1 \to \mathbb{R}^3$ having no self-intersection or singular points.

If the total space $\mathcal{F}$ of geometric objects is an N-dimensional vector space then the homology groups of these complementary spaces are related by the Alexander duality formula

$$H^i(\mathcal{F} \setminus \Sigma) \simeq \bar{H}_{N-i-1}(\Sigma), \qquad (1)$$

where $\bar{H}_*$ denotes the *Borel-Moore homology group*, i.e. the homology group of the one-point compactification relative to the added point. It was Arnold [3] who first used this reduction in the case of the space of

87

D. Siersma et al. (eds.), New Developments in Singularity Theory, 87–115.

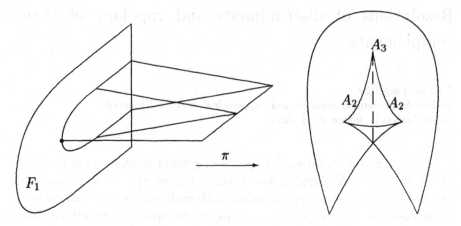

Figure 1. A swallowtail and its resolution

complex polynomials with(out) multiple roots, see also his works [4], [5], [7], [10] where this reduction is applied for some other discriminant spaces. This reduction is always very useful, because the space $\mathcal{F} \setminus \Sigma$ is an open manifold without any clear geometrical structure; on the other hand a lot of topological properties of the singular space Σ can be expressed in terms of its stratification.

The method of computing the groups $\bar{H}_*(\Sigma)$ invented in [57] is based on the notion of the simplicial resolution of the discriminant variety. One of its advantages consists in the fact that all its ingredients behave properly under stabilizations of $\mathcal{F}$ (e.g. if we consider the sequence of spaces of polynomials of increasing degree, or of operators of increasing order), and therefore allows us to calculate the left-hand groups in (1) even for infinite-dimensional spaces $\mathcal{F}$, when the right-hand part of (1) has, formally speaking, no sense.

Some first results of this method are described in [64], [55], [56]. Below we describe a more general version of it, based on the notions of *topological order complexes* and *conical resolutions*, and extending similar calculations to many new situations, especially related to *non-normal* discriminants. For a more extended description of this construction, see [73].

1. Order complexes of discrete posets and simplicial resolutions of subspace arrangements

In this section we demonstrate the method of simplicial resolutions in a simple 'discrete' case: that of *plane arrangements*, cf. [24], [28], [62],

[77]. The 'continuous' version of the method will be demonstrated in the next section.

DEFINITION 1.1. *Let $(A, \geq)$ be a discrete* poset *(=partially ordered set). The corresponding order complex $\Delta(A)$ is the simplicial complex whose vertices are the elements of A, and whose simplices span all strictly monotone finite sequences $\{a_1 < \ldots < a_m\}$, $a_i \in A$.*

Consider any *affine plane arrangement* $\mathcal{L}$, i.e. a finite collection of affine subspaces $L_1, \ldots, L_k$ of arbitrary dimensions in $\mathbb{R}^N$. Set $L = \cup L_i$, and, for any set of indices $I \subset \{1, \ldots, k\}$, $L_I \equiv \cap_{i \in I} L_i$. Then all possible nonempty planes L_I form a partially ordered set (by inclusion). Denote by $\Delta(\mathcal{L})$ the corresponding order complex. The simplicial resolution of the variety L can be constructed as a subset of the Cartesian product $\Delta(\mathcal{L}) \times \mathbb{R}^N$.

For any plane L_I the corresponding order subcomplex $\Delta(L_I) \subset \Delta(\mathcal{L})$ is defined as the union of the simplices all of whose vertices are subordinate to $\{L_I\}$, i.e. correspond to planes L_J *containing* L_I. This is a cone with vertex $\{L_I\}$. Denote by $\partial\Delta(L_I)$ its *link*, i.e. the union of all simplices in $\Delta(L_I)$ not containing the vertex $\{L_I\}$.

The resolution space $L' \subset \Delta(\mathcal{L}) \times \mathbb{R}^N$ is defined as the union of all spaces of the form $\Delta(L_I) \times L_I$ over all geometrically distinct planes L_I. The obvious projection $\Delta(\mathcal{L}) \times \mathbb{R}^N \to \mathbb{R}^N$ induces a map $\pi : L' \to L$. This map is proper, and all its fibers are contractible finite complexes of the form $\Delta(L_I)$. It follows easily that this map is a homotopy equivalence, and its extension to the map of one-point compactifications, $\bar{\pi} : \overline{L'} \to \bar{L}$, is also a homotopy equivalence.

EXAMPLE 1.2. Let L be the union of two intersecting lines a and b in $\mathbb{R}^2$, see the middle part of Fig. 2. The corresponding order complex $\Delta(\mathcal{L})$ consists of two segments (see the right-hand part of Fig. 2) joining the vertices (a) and (b) (corresponding to these lines) to the vertex (ab) (corresponding to the point of intersection). The resolution space L' consists of three complexes: the line $(a) \times a$, the line $(b) \times b$, and the complex $\Delta(\mathcal{L}) \times (a \cap b)$, see the left part of the picture.

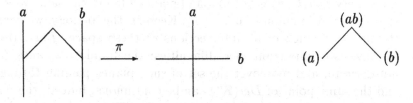

Figure 2. Resolution of a cross

In the general case, the resolution space L' has a natural increasing filtration $F_1 \subset F_2 \subset \cdots \subset F_{N-1} = L'$: the term F_m is the union of the spaces $\Delta(L_I) \times L_I$ over all planes L_I of *codimension* $\leq m$ in $\mathbb{R}^N$. The difference $F_m \setminus F_{m-1}$ is the disjoint union of the spaces $(\Delta(L_I) \setminus \partial\Delta(L_I)) \times L_I$ over all planes L_I of dimension exactly $N - m$. Also we get a filtration $\bar{F}_0 \subset \bar{F}_1 \subset \cdots \subset \bar{F}_{N-1} = \bar{L}'$ of the one-point compactification $\bar{L}'$ of the space L': the term $\bar{F}_0$ is the added pont, and each space $\bar{F}_i$, $i > 0$, is the closure of the corresponding subspace $F_i \subset L'$.

The results of [77] imply in particular that this filtration is homotopically split: there is a homotopy equivalence

$$\bar{L}' \sim \bar{F}_1 \vee (\bar{F}_2/\bar{F}_1) \vee \ldots \vee (\bar{F}_{N-1}/\bar{F}_{N-2}), \qquad (2)$$

where $\vee$ denotes the *wedge* ($\sim$ *bouquet*). An equivalent result was obtained simultaneously in [62].

This formula implies the Goresky–MacPherson formula for the co-homology of the complementary space $\mathbb{R}^N \setminus L$ (see [28]), and also the fact that the stable homotopy type of this space is determined by the dimensions of the planes L_I.

2. Conical resolutions of determinant sets

Let $\mathbb{K}$ be any of the fields $\mathbb{R}, \mathbb{C}$ or $\mathbb{H}$. The *determinant* variety $Det(\mathbb{K}^n) \subset End(\mathbb{K}^n)$ consists of all degenerate operators $\mathbb{K}^n \to \mathbb{K}^n$.

Its *tautological resolution* is defined by elimination of quantifiers. Namely, an operator A belongs to $Det(\mathbb{K}^n)$ if $\exists$ a point $x \in \mathbb{KP}^{n-1}$ such that $\{x\} \subset \ker A$. Define the resolution space $det_1(\mathbb{K}^n)$ as the space of all pairs $(x, A) \in \mathbb{KP}^{n-1} \times End(\mathbb{K}^n)$ such that $\{x\} \in \ker A$. This space admits the (tautological) structure of a $(n^2 - n)$-dimensional $\mathbb{K}$-vector bundle over $\mathbb{KP}^{n-1}$, whose fiber $L(x)$ consists of all A such that $x \in \ker A$. The obvious projection $\pi : det_1(\mathbb{K}^n) \to Det(\mathbb{K}^n)$ is regular over operators with 1-dimensional kernels, but the pre-image of an operator with $\dim \ker = l$ is isomorphic to $\mathbb{KP}^{l-1}$.

The situation is very similar to the one in the previous subsection: the variety $Det(\mathbb{K}, n)$ is the union of planes $L(x)$ in the same way as the space L was the union of planes L_i. Keeping the analogy, we construct the order complex of all intersections of these spaces $L(x)$. However we have two important new difficulties: the family of planes $L(x)$ is not discrete, and moreover the set of such planes passing through one and the same point of $Det(\mathbb{K}^n)$ can be continuous. Indeed, the possible intersections of several planes $L(x_j) \subset End(\mathbb{K}^n)$ are just the planes of the form $L(X)$ where X is a subspace of a certain dimension in $\mathbb{K}^n$ (i.e.

a point of a certain Grassmannian manifold $G_i(\mathbb{K}^n)$, $i \in [1, n]$), and $L(X)$ consists of all operators whose kernels contain X.

Thus the set of parameters indexing the planes in our poset is the *disjoint union of all Grassmann manifolds* $G_1(\mathbb{K}^n), \ldots, G_{n-1}(\mathbb{K}^n), G_n(\mathbb{K}^n)$. The *continuous order complex* of all these Grassmannians is defined as follows. Consider the join $G_1(\mathbb{K}^n) * \ldots * G_n(\mathbb{K}^n)$, i.e., roughly speaking, the union of all simplices whose vertices correspond to points of different Grassmannians. Such a simplex is *coherent* if the planes corresponding to its vertices form a flag. The desired order complex $\Theta(\mathbb{K}^n)$ is the union of all coherent simplices, with topology induced from that of the join. This is a cone with vertex $\{\mathbb{K}^n\} \in G_n(\mathbb{K}^n)$. Its *link* $\partial\Theta(\mathbb{K}^n)$ is the union of the coherent simplices not containing this vertex $\{\mathbb{K}^n\}$.

This link $\partial\Theta(\mathbb{K}^n)$ is homeomorphic to the sphere S^M, $M = \frac{1}{2}n(n-1)(\dim_{\mathbb{R}}\mathbb{K}) + n - 2$. (Probably this fact is assumed in the remark 1.4 of [20], see also [61], [56].) Hence $\Theta(\mathbb{K}^n)$ is homeomorphic to a ball.

The conical resolution of $Det(\mathbb{K}^n)$ is constructed as a subset of the direct product $\Theta(\mathbb{K}^n) \times Det(\mathbb{K}^n)$. For example, let $\mathbb{K} = \mathbb{R}$, $n = 2$. The space $End(\mathbb{R}^2)$ of all operators $\mathbb{R}^2 \to \mathbb{R}^2$ is 4-dimensional, and $Det(\mathbb{R}^2)$ is a 3-dimensional conical subvariety in it. There is a single point in $Det(\mathbb{R}^2)$, over which the tautological resolution is not a homeomorphism: the zero operator. Its preimage is the line $\mathbb{R}P^1$. In order to get from this resolution a space homotopy equivalent to $Det(\mathbb{R}^2)$, we need to insert a disc whose boundary coincides with this preimage. It is useful to take this disc as the space $\Theta(\mathbb{R}^2)$.

We proceed in a similar way for any n and $\mathbb{K}$. To any plane $X \subset \mathbb{K}^n$ there corresponds a subspace $\Theta(X) \subset \Theta(\mathbb{K}^n)$, namely, the union of all coherent simplices whose vertices correspond to planes lying in X. This is a cone with vertex $\{X\}$, and is homeomorphic to a closed ball. Define the conical resolution $\delta(\mathbb{K}^n) \subset \Theta(\mathbb{K}^n) \times Det(\mathbb{K}^n)$ as the union of the products $\Theta(X) \times L(X)$ over all planes X of dimensions 1, $\ldots$, n. It is easy to see that the obvious projection $\delta(\mathbb{K}^n) \to Det(\mathbb{K}^n)$ induces a homotopy equivalence of one-point compactifications of these spaces (indeed, this projection is proper and semialgebraic, and all its fibers are contractible cones of the form $\Theta(X)$). On the other hand, the space $\delta(\mathbb{K}^n)$ has a nice filtration: its term F_i is the union of products $\Theta(X) \times L(X)$ over planes X of dimensions $\leq i$. The term $F_i \setminus F_{i-1}$ of this filtration is the total space of a fibre bundle over $G_i(\mathbb{K}^n)$. Its fiber over a point $\{X\}$ is the space $(\Theta(X) \setminus \partial\Theta(X)) \times L(X)$, and is homeomorphic to an Euclidean space. Thus the Borel–Moore homology group of this term can be reduced to that of the base. The spectral sequence, generated by this filtration and converging to the Borel–

Moore homology group of $Det(\mathbb{K}^n)$ (or, equivalently, to the cohomology group of the complementary space $GL(\mathbb{K}^n)$), degenerates at the first term and gives, in particular, the homological *Miller splitting*

$$H_m(GL(\mathbb{C}^n)) = \bigoplus_{k=0}^{n} H_{m-k^2}(G_k(\mathbb{C}^n)) \tag{3}$$

and similar splittings over $\mathbb{R}$ and $\mathbb{H}$.

3. Some more examples and difficulties

All the other discriminant spaces can be resolved in a similar way. However, in some examples we meet two further difficulties: 1) the families of spaces L_I forming our posets may not be closed, and 2) the function space can be infinite-dimensional, so that the Alexander duality (1) formally does not work in it. We discuss these difficulties in the next two subsections and show in the easiest examples how to overcome them.

3.1. ALL THE FAMILIES OF PLANES SHOULD BE CLOSED

Following [3], [7], consider the space $\mathcal{F}_d$ of polynomials $x^d + a_1 x^{d-1} + \ldots + a_d$ over $\mathbb{K} = \mathbb{R}$ or $\mathbb{C}$, and the discriminant space Σ_k consisting of polynomials having at least one root of multiplicity $\geq k$. It can be swept out by a family of planes $L(x)$, $x \in \mathbb{K}^1$, of codimension k in $\mathcal{F}_d$: any such plane consists of polynomials with a k-fold root at x. Nonempty intersections of such planes are parametrized by the points of configuration spaces $B(\mathbb{K}^1, i)$, i.e. by unordered collections $X = (x_1, \ldots, x_i)$ of i distinct points in $\mathbb{K}^1$, $i = 1, \ldots, [d/k]$. Unfortunately, if $i > 1$ then the set of all such planes $L(X)$ is not closed in the manifold of all planes of codimension ik in $\mathcal{F}_d$. For instance, if all the i points of a configuration X (depending on a parameter) tend smoothly to one and the same point $x \in \mathbb{K}^1$, then the corresponding planes $L(X)$ tend to some limit position, which is not of the form $L(X')$ for any $X' \in B(\mathbb{K}^1, i)$ (but lies in the plane $L(x)$; this limit position depends on the relative velocities with which our points tend to x). Therefore, if we formally apply the construction of §§1, 2 to the poset formed by spaces $B(\mathbb{K}^1, i)$, then we get a non-closed resolution space, whose projection onto Σ_k is not proper and does not preserve its local homotopy type. The space $B(\mathbb{K}^1, i)$ of such planes needs to be completed. In our 1-dimensional case such a completion $\overline{B(\mathbb{K}^1, i)}$ is obvious: it is the space $S^i(\mathbb{K}^1)$ of all collections of i not necessarily distinct points (or, equivalently, the space of all ideals of codimension i in $\mathbb{K}[t]$). In more general cases, when we resolve discriminants in $C^\infty(M^n, \mathbb{R})$, we can take any reasonable completion of the configuration space, e.g. just its closure in the space of all affine planes of codimension k in $C^\infty(M^n, \mathbb{R})$ with respect to the natural Grassmannian topology.

The disjoint union of these closures forms a poset (by the inclusions of corresponding ideals), and we can construct its continuous order complex and the resolution of the discriminant exactly as previously; for subtleties see [55], [56], [73], [68].

This situation is shown in the left part of Fig. 1. The right-hand segment in it symbolizes the space $\overline{B(\mathbb{R}^1, 2)}$. Its interior points are regular configurations (x, y), $x \neq y$; they are connected by two coherent segments with the points $\{x\}$ and $\{y\}$ of the space $B(\mathbb{R}^1, 1) \sim \mathbb{R}^1$ shown by the parabola on the left. The endpoint of the segment corresponds to a degenerate configuration (x, x) and is joined with only one point $\{x\} \in B(\mathbb{R}^1, 1)$.

3.1.1. *Geometrization*

This construction can be slightly simplified: all the 'nongeometrical' coherent simplices arising in the construction of the resolution (i.e. the simplices containing 'nongeometrical' vertices corresponding to the added boundary points of configuration spaces) can be contracted onto their maximal 'geometrical' faces: this contraction does not change the homotopy type of the one-point compactification of the resolution space. For instance, in Fig. 1 we can contract the segment $[(x, x), \{x\}]$ into a point. (For a direct construction of such a 'geometrical' resolution in many cases see [55], [56].) The quotient space obtained is in obvious set-theoretical bijection with the subset of the resolution space consisting of coherent simplices all whose vertices are 'geometric', i.e. belong to spaces $B(\mathbb{R}^1, i)$ and not to their boundaries. However as topological spaces they are different.

In our case of the discriminant $\Sigma_k \subset \mathcal{F}_d$ the spectral sequence again degenerates at the first term, $E^1 \equiv E^\infty$ (see [55], [58]). Moreover, in this case we also have homotopy splittings of discriminants similar to (2). This is especially obvious if $\mathbb{K} = \mathbb{R}$: indeed, in this case any term $F_i \setminus F_{i-1}$, $i \leq [d/k]$, is fibered over the trivial space $B(\mathbb{R}^1, i) \sim \mathbb{R}^i$ with fiber equal to the product of the affine space of dimension $d - ik$ and an open $(i-1)$-dimensional simplex. So these terms are open cells of decreasing dimensions, and the summands of the wedge will be the spheres of the same dimensions.

3.2. STABILIZATION AND RESOLUTION OF INFINITE DIMENSIONAL DISCRIMINANTS

The power of the above-described construction of resolutions is shown by its perfect functoriality under embeddings of functional spaces and their discriminants. For instance, let us consider the space $\mathcal{F}_d$ from the previous section, and a polynomial $f \in \Sigma_k \subset \mathcal{F}_d$ having exactly one root of multiplicity $c \in [k, d]$ and no other multiple roots. In a neighborhood of f, the variety $\Sigma_k \subset \mathcal{F}_d$ is ambient diffeomorphic to the direct product of the variety $\Sigma_k \subset \mathcal{F}_c$ and the space $\mathbb{K}^{d-c}$. (E.g. the strata $\{A_2\}$ of the swallowtail in Fig. 1 are locally direct products of $\mathbb{R}^1$ and the semicubical parabola.) In the restriction to this neighborhood, our resolution of the discriminant $\Sigma \subset \mathcal{F}_d$ coincides with the resolution of $\Sigma_k \subset \mathcal{F}_c$ multiplied by $\mathbb{R}^{d-c}$. Therefore we get a morphism of spectral sequences converging to the Borel–Moore homology groups of these discriminants: it maps any cell $E^r_{p,q}(d)$ to $E^r_{p,q-(d-c)}(c)$. Further, we can formally replace these homological sequences by the cohomological ones (converging to the Alexander dual cohomology groups of complements of discriminants),

$$E^{p,q}_r(d) \equiv E^r_{-p,D-q-1}(d) \tag{4}$$

(where D is the dimension of the functional space and in our particular case is equal to d). These 'inverted' spectral sequences lie in the second quadrant, $\{p \leq 0, q \geq 0\}$. The induced morphisms of them preserve both gradings p and q; moreover their final action on the groups E_∞ is compatible with the cohomology map induced by the corresponding embedding $\mathcal{F}_c \setminus \Sigma_k \subset \mathcal{F}_d \setminus \Sigma_k$.

This allows us to define a *stable* spectral sequence $E^{p,q}_r(\infty)$, converging to the cohomology of some limit space: '$\mathcal{F}_d \setminus \Sigma_k$ with infinitely large d'. For $\mathbb{K} = \mathbb{R}$, the last space can be realized as the space of all smooth functions $\mathbb{R}^1 \to \mathbb{R}^1$ with some standard behavior at infinity (say, equal identically to 1 outside some compact) and having no zeros of multiplicity k. This is the simplest manifestation of a general method of computing cohomology groups of complements of discriminants in infinite-dimensional functional spaces: we consider an increasing sequence of finite-dimensional approximations, consider resolutions of their intersections with the discriminant set, and then prove a stabilization theorem for the corresponding spectral sequences converging to the cohomology groups of the complements of these intersections. These theorems can be of different strength in different situations: we will discuss some of them in items G (convergence) and I (stabilization) in the next section.

4. List of examples

4.1. WHAT CAN BE SAID ON A DISCRIMINANT AND ITS RESOLUTION

In this section we outline in a uniform way resolutions of different discriminant spaces and results on the topology of their complements. The description of each case consists of the following items (some of which may be omitted):

A. FUNCTIONAL SPACE.

B. THE DISCRIMINANT.

C. THE TAUTOLOGICAL RESOLUTION (the set of maximal planes sweeping out the discriminant.)

D. DESCRIPTION OF THE POSET.

E. TOPOLOGY OF THE CORRESPONDING CONTINUOUS ORDER COMPLEX (or of its link if the poset has a unique maximal element).

F. THE SUPPORT OF THE COHOMOLOGICAL SPECTRAL SEQUENCE.

G. CONVERGENCE OF THE SPECTRAL SEQUENCE: does it converge (in some sense) to the whole cohomology group of the complement of Σ? Or perhaps to an important subgroup of it? Of course this question is trivial if the function space is finite-dimensional, or if on any line $\{p + q = const\}$ we have only a finite number of nonzero groups $E_1^{p,q}$ and all groups are finitely generated.

H. DEGENERATION OF THE SPECTRAL SEQUENCE: what is the least r for which we have $E_\infty \equiv E_r$? If $r = 1$, then maybe even a homotopical splitting (2) of discriminants holds ?

I. STABILIZATION OF SPECTRAL SEQUENCES. Often our problems form a directed family: e.g. the discriminants $\Sigma_k \subset \mathcal{F}_d$ with different d, or determinant varieties in $End(\mathbb{K}^n)$ with different n. Then their resolutions and the corresponding spectral sequences can stabilize to interesting limit objects.

J. COMPARISON THEOREMS AND SMALE–HIRSCH PRINCIPLE. Often we have isomorphic spectral sequences for related but different discriminant spaces. This proves the isomorphism of the cohomology groups of their complements, and often reflects the fact that these complements are homotopy equivalent (at least stably). Such comparison theorems often provide versions of the generalized *Smale–Hirsch principle* relating the space of smooth maps without singularities of certain types and the space of continuous sections of the jet bundle not intersecting the corresponding singular set. Example: the space of functions $S^1 \to \mathbb{R}^1$ without zeros of multiplicity k is homotopy equivalent to the space of all maps of S^1 into $\mathbb{R}^k \setminus 0$ (or, equivalently, to S^{k-1}): this equivalence is induced by the jet extension map sending any function f to the collection of k functions $f, f', \ldots, f^{k-1}$.

K. EXPLICIT FORMULAS expressing cohomology classes obtained from the spectral sequence (i.e. in terms of the Alexander dual cycles) in terms intrinsic to the complement of the discriminant.

L. MULTIPLICATION IN THE SPECTRAL SEQUENCE.

4.2. MONIC POLYNOMIALS WITHOUT k-FOLD ROOTS

A. THE SPACE $\mathcal{F}_d$ of all polynomials $x^d + a_1 x^{d-1} + \ldots + a_d$, $a_i \in \mathbb{K} = \mathbb{C}$ or $\mathbb{R}$.

B. THE DISCRIMINANT is the set Σ_k of polynomials having at least one root of multiplicity k or more in $\mathbb{K}^1$. For $\mathbb{K} = \mathbb{C}$ and $k = 2$, the space $\mathcal{F}_d \setminus \Sigma_k$ is the classifying space for the group of braids with d strings.

C. THE TAUTOLOGICAL RESOLUTION is fibered over $\mathbb{K}^1$ with fiber $\mathbb{K}^{d-k}$, see §3.1.

D. THE POSET consists of $[d/k]$ terms $\overline{B(\mathbb{K}^1, i)}$, $i = 1, \ldots, [d/k]$.

E. THE ORDER COMPLEX is contractible, as $\mathbb{K}^1$ is.

F. THE SUPPORT. For $\mathbb{K} = \mathbb{R}$, it consists of $[d/k] + 1$ nonzero terms $E_1^{p,q} \sim \mathbb{Z}$ with $p = 0, -1, \ldots, -[d/k]$ and $q + (k-1)p = 0$. For $\mathbb{K} = \mathbb{C}$ it lies in the domain $\{(p,q) | p = 0, -1, \ldots, -[d/k], q \in [-p(2k - 3), -p(2k - 2) - 1]\}$.

G. CONVERGENCE. Yes, as $\dim \mathcal{F}_d$ is finite.

H. DEGENERATION. Yes, $E_\infty \equiv E_1$, and for $\mathbb{K} = \mathbb{R}$ even the homotopy splitting (2) holds: probably also for $\mathbb{K} = \mathbb{C}$.

I. STABILIZATION. For $\mathbb{K} = \mathbb{R}$ the spectral sequences stabilize to one calculating the cohomology groups of the space $\mathcal{F}_\infty \setminus \Sigma_k$ of functions $\mathbb{R}^1 \to \mathbb{R}^1$ with fixed behavior at infinity and without k-fold zeros, and also the cohomology group of the loop space $\Omega(\mathbb{R}^k \setminus 0) \sim \Omega S^{k-1}$, see [58], [55], [56]. For $\mathbb{K} = \mathbb{C}$ and $k = 2$ the stabilized spectral sequence calculates the cohomology group of the stable braid group (with infinitely many strings). For any k it also calculates the cohomology group of the double loop space $\Omega^2(\mathbb{C}^k \setminus 0) \sim \Omega^2 S^{2k-1}$.

J. COMPARISON AND SMALE–HIRSCH PRINCIPLE. For $\mathbb{K} = \mathbb{R}$, the limit space $\mathcal{F}_\infty \setminus \Sigma_k$ is homotopy equivalent (via the k-jet extension map) to the loop space $\Omega(\mathbb{R}^k \setminus 0)$. Any particular term $\mathcal{F}_d \setminus \Sigma_k$ is homotopy equivalent (via the same embedding) to the space of loops of length $< 2\pi([d/k] + 1)$ in the unit sphere S^{k-1}.

For $\mathbb{K} = \mathbb{C}$, the comparison of stable spectral sequences allows us to establish a stable homotopy equivalence between the limit space $\lim_{d \to \infty} \mathcal{F}_d \setminus \Sigma_k$ and $\Omega^2 S^{2k-1}$. In particular, these spaces have the same cohomology groups. For $k = 2$ this stable homotopy equivalence was first proved by J.-P. May and G. Segal, see [39], [47].

K. EXPLICIT FORMULAS. These formulas can be induced via the same embedding from the known expressions for the space ΩS^{k-1}. For $\mathbb{K} =$

$\mathbb{R}, k = 3$, the generator of the group $H_1(\mathcal{F}_d \setminus \Sigma_3) \simeq \pi_1(\Omega S^2) \simeq \pi_2(S^2)$ (with index equal to 1) is identified in [3], the generator of $\pi_2(\mathcal{F}_d \setminus \Sigma_3) \simeq_{d \geq 9} \pi_3(S^2)$ (with Hopf invariant 1) in [58], [55], [56].

L. MULTIPLICATION. In this case, the spectral sequence coincides with the Adams-Eilenberg-Moore-Anderson spectral sequence for loop spaces (see e.g. [1]), whose 'de Rhamization' is also known as the theory of iterated path integrals, and admits a natural multiplicative structure compatible with the multiplication in the limit cohomology group.

For $\mathbb{K} = \mathbb{C}$ these spaces were studied by V. Arnold from 1968: in his work [3] he invented the seminal reduction (1) and essentially started the topological study of discriminant sets. For $k = 2$, the complete calculation of cohomology rings was then obtained (by different methods) by D. B. Fuchs (for $\mathbb{Z}_2$-coefficients) and F. R. Cohen (over $\mathbb{Z}$). Similar problems for $\mathbb{K} = \mathbb{R}$ were also considered by Arnold in [7], where in particular the cohomology groups of $\mathcal{F}_d \setminus \Sigma_k$ were calculated. The Smale-Hirsch principle for stabilizations of these spaces was found in [58] in answering Arnold's question on the multiplicative structure in these groups.

4.3. RESULTANTS

A. THE SPACE $Syst(k,m)$ of all systems of k polynomials of the form $x^m + b_1 x^{m-1} + \ldots + b_m$ over $\mathbb{K} = \mathbb{C}$ or $\mathbb{R}$.

B. THE DISCRIMINANT is the resultant set $Res(k,m)$ of all systems having a repeated root in $\mathbb{K}^1$.

C. THE TAUTOLOGICAL RESOLUTION is fibered over $\mathbb{K}^1$ with fiber over $x \in \mathbb{K}^1$ equal to the space of all systems having a repeated root at x.

D, E, F, G, H, K, L are exactly the same as in the previous subsection with the same k and $d = km$.

The unique difference for items I, J is that for $\mathbb{K} = \mathbb{C}$ the limit of spaces $Syst(k,m) \setminus Res(k,m)$ as $m \to \infty$ is homotopy equivalent to $\Omega^2 S^{2k-1}$ (and not just stably homotopy equivalent as in the previous case); this is a theorem of G. Segal [48].

Moreover, a comparison of resolved discriminants and resultants allows us to prove that for any finite k and m the space $Syst(k,m) \setminus Res(k,m)$ is stably homotopy equivalent to $\mathcal{F}_{km} \setminus \Sigma_k$. For $k = 2$ this fact was proved in [23], for arbitrary k in [55]. As usual, it is easier here to prove stable homotopy equivalence of one-point compactifications of resultants and discriminants, so that the stable homotopy equivalence of their complements follows by *Spanier–Whitehead duality*, see [55].

In [34] an even stronger comparison theorem for *real* discriminants and resultants was proved: the resultant variety in the space of pairs of two monic polynomials of degree m is homeomorphic to the discriminant subset of the space of such polynomials of degree $2m$.

4.4. Homogeneous polynomials $\mathbb{R}^2 \to \mathbb{R}$ without multiple roots

A. The space $\mathcal{H}_d$ of all homogeneous polynomials $\mathbb{R}^2 \to \mathbb{R}$ of degree d.

B. The discriminant Σ_k consists of polynomials divisible by the k-th power of some linear function.

C. The tautological resolution is fibered over $\mathbb{RP}^1$ with fiber equal to the space of polynomials having a zero of multiplicity k along the corresponding line.

D. The poset is the disjoint union of all compactified configuration spaces $\overline{B(\mathbb{RP}^1, i)}$, $i = 1, \ldots, [d/k]$, *plus the one-point set* $\{\mathbb{RP}^1\}$ (corresponding to the identically zero polynomial).

E. The link is homotopy equivalent to $S^{2[d/k]-1}$ (theorem of C. Caratheodory).

F. The support belongs to the union $2[d/k] + 3$ cells (p, q): for any $p = 0, -1, \ldots, -[d/k]$, the number q can be equal to either $-p(k-1)$ or $-p(k-1) + 1$, and there is one cell more, $(p, q) = (-[d/k] - 1, d - [d/k] + 1)$. For a description of the corresponding groups $E_r^{p,q}$ see [67], [56].

G. Convergence. Yes, as $\dim \mathcal{H}_d < \infty$.

H. Degeneration. $E_\infty \equiv E_1$ *except for the case when k is odd and d is a multiple of k*: in this case $E_\infty \equiv E_2 \neq E_1$.

I. Stabilization. For any fixed $k > 2$ and $d \to \infty$ the corresponding spectral sequences stabilize to one calculating the cohomology group of one of two homotopy equivalent spaces of smooth functions $S^1 \to \mathbb{R}^1$, those even or odd under the involution of S^1, and having no zeros of multiplicity k.

For $k = 2$ the stable spectral sequence is also well-defined but does not give the entire cohomology group of the space of functions without double roots. Indeed, there is an obvious 0-dimensional cohomology class (i.e. an invariant of such functions): the number of their simple zeros. It turns out that all other invariants arising from our spectral sequence depend polynomially on this one. In particular they cannot distinguish the identically positive and negative functions. These cohomology classes are exact analogues of finite-type knot invariants, and we get a picture showing *how the space of such invariants may not be complete*.

J. Comparison and Smale–Hirsch principle. The spectral sequence allows us to prove the homotopy equivalence of these spaces with the free loop space $\Omega_f S^{k-1} \equiv \{S^1 \to S^{k-1}\}$ (or, in the case of odd functions, with the other component of the space of free loops $\Omega_f(\mathbb{RP}^{k-1})$ which consists of loops not liftable to closed loops in S^{n-1}).

K. EXPLICIT FORMULAS. Again, can be inferred from these for $H^*(\Omega_f(\mathbb{RP}^{k-1}))$.

L. MULTIPLICATION. Again, this is induced from the Eilenberg-Moore spectral sequence for the loop space.

4.5. SPACES OF SMOOTH FUNCTIONS $M^m \to \mathbb{R}^n$ WITHOUT COMPLICATED SINGULARITIES

A. THE SPACE of all smooth functions $M^m \to \mathbb{R}^n$ (with some fixed behavior close to ∂M^m, if the latter is nonempty).

B. THE DISCRIMINANT is the space of functions having singular points of some class of codimension $\sigma \geq 2$ in the function space (i.e. defined by any $\mathrm{Diff}(\mathbb{R}^m)$-invariant closed subvariety of codimension $m + \sigma \geq m + 2$ in the jet space $J^T(\mathbb{R}^m, \mathbb{R}^n)$ for some T).

C. THE TAUTOLOGICAL RESOLUTION is fibered over the space of pairs $(x \in M^m; \varphi \in J_x^T(M^m, \mathbb{R}^n))$ such that φ belongs to our singularity class, with the fiber equal to the space of functions $M^m \to \mathbb{R}^n$ with this T-jet at the point x.

D. THE POSET is the disjoint union of appropriate compactifications of configuration spaces $B(M^m, i)$ over all natural numbers i.

F. THE SUPPORT is in the wedge $\{(p, q) : p \leq 0, q + \sigma p \geq 0\}$.

G. CONVERGENCE. Yes, because there are only finitely many nonzero terms $E_1^{p,q}$ on any line $\{p + q = \mathrm{const}\}$.

H. DEGENERATION. The entire spectral sequences (especially their *higher differentials*) for different singularity classes seem to be strong invariants of smooth manifolds. To what extent can they be derived from topological invariants?

J. COMPARISON AND SMALE–HIRSCH PRINCIPLE. The spectral sequence coincides with one calculating the cohomology group of the corresponding space of all continuous sections of the jet bundle not intersecting our singular subset. This allows us to prove the *homological* Smale-Hirsch principle stating the homology (and even stable homotopy) equivalence of these spaces (and hence, if $\sigma \geq 3$, even the usual homotopy equivalence), see [58], [55], [56]. Moreover, in the most classical case of codimension 2, when $n = 1$ and the forbidden singularity class consists of all germs more complicated than Morse and A_2, ordinary homotopy equivalence also holds: it was proved up to dimension $m - 1$ in [32] and in all dimensions in [25].

4.6. SPACES OF CONTINUOUS MAPS OF m-DIMENSIONAL TOPOLOGICAL SPACES INTO $(m-1)$-CONNECTED ONES

This is essentially the special case of the calculation given in the previous subsection when the forbidden singularity class is defined in the terms of the 0-jets of maps. In this case the source space does not need to be a smooth manifold.

Any finite $(m-1)$-connected cell complex Y is homotopy equivalent to the complement of a closed subset $\Lambda(Y)$ of codimension $\geq m+1$ in some space $\mathbb{R}^N$. If the topological space X is $\leq m$-dimensional, then the maps $X \to \mathbb{R}^N$ not intersecting $\Lambda(Y)$ are dense in the space of all maps. Therefore we get the following way to study the homotopical properties of the space of maps $X \to Y$ (maybe fixed on some subcomplex $Z \subset X$). (If X is a manifold, then the spectral sequence obtained in this way coincides with one constructed by D. Anderson, see [1], but our approach allows us to remove this restriction.)

A. THE SPACE of all continuous maps $X \to \mathbb{R}^N$ (maybe coinciding with a fixed map $X \to \mathbb{R}^N \setminus \Lambda(Y)$ on some subcomplex $Z \subset X$.)

B. THE DISCRIMINANT consists of maps whose images meet $\Lambda(Y)$.

C. THE TAUTOLOGICAL RESOLUTION is fibered over the space of pairs $(x \in X \setminus Z, y \in \Lambda(Y))$ with fiber equal to the space of maps sending x to y.

D. THE POSET is the disjoint union of suitably completed configuration spaces $B(X \setminus Z, i)$ with all natural numbers i.

F. THE SUPPORT is in the wedge $\{(p,q) : p \leq 0, q + \sigma p \geq 0\}$, where $\sigma = n - \dim X - \dim \Lambda(Y)$.

G. CONVERGENCE. If $\sigma \geq 2$ and both X and Y are finite cell complexes (or, more generally, finite type cell complexes, i.e. have finitely many cells in any given dimension) then the spectral sequence obviously converges to the cohomology group of the space of continuous maps $X \to Y$.

H. DEGENERATION. Often we have $E_\infty \equiv E_1$. For instance this holds if $X = S^m, Y = S^n, m < n$ and $Z =$(one point), so that the space of maps considered is the iterated loop space $\Omega^m S^n$; in that case the degeneration theorem provides the *Snaith splitting formula* (more precisely, its homological version)

$$H_t(\Omega^m S^n) \simeq \bigoplus_{i=0}^{\infty} H_{t-i(n-m)}(B(\mathbb{R}^m, i), \pm \mathbb{Z}^{\otimes(n-m)}) \qquad (5)$$

for the homology group of this space of maps, where $\pm \mathbb{Z}$ is the 'sign' local system over the configuration space, cf. [62], [56]. Many other cases when such a decomposition holds were found by C.-F. Bödigheimer, F. Cohen, L. Taylor and others, see the references in [56].

J. COMPARISON AND SMALE–HIRSCH PRINCIPLE. Conversely, the spaces of such maps are the ultimate objects to which one tries to reduce function spaces defined in terms of jet extensions, see e.g. all the previous items of this subsection.

K. EXPLICIT FORMULAS. For instance, the splitting (5) can be realized as follows. Let us fix any standard $(n - m)$-spheroid Ξ generating the group $\pi_{n-m}(\Omega^m S^n) \sim \pi_m(S^m) \simeq \mathbb{Z}$ as a family of maps $\mathbb{R}^m \to S^n$ equal to the constant map outside a ball of small radius ρ centered at 0 and depending on a parameter running over the sphere S^{n-m}. For any i-configuration $X = (x_1, \ldots, x_i) \in B(\mathbb{R}^m, i)$ such that the ρ-neighborhoods of all its points are disjoint, we can consider the $i(n - m)$-parameter family Ξ^X of maps $\mathbb{R}^m \to S^n$ constant outside these neighborhoods and in the neighborhood of any point $x_j \in X$ coinciding up to a parallel translation $\{0 \to x_j\}$ with maps of the family Ξ. Any homology class in $B(\mathbb{R}^m, i)$ can be realized by a compact cycle in the set of configurations X all whose points are 2ρ-separated. Associating to any point of such a cycle the corresponding cycle Ξ^X we sweep out a cycle in $\Omega^m S^n$. Homology splittings for other spaces of maps $X \to Y$ as above usually can be realized in a similar way.

L. MULTIPLICATION. If X is a manifold, then this spectral sequence coincides with that of [1], and also admits a natural multiplicative structure. How can this be extended to the most general situation?

4.7. DETERMINANTS

A. THE SPACE is $End(\mathbb{K}^n)$, $\mathbb{K} = \mathbb{R}, \mathbb{C}$ or $\mathbb{H}$.

B. THE DISCRIMINANT is the set $Det(\mathbb{K}^n)$ of degenerate operators.

C. THE TAUTOLOGICAL RESOLUTION is fibered over $\mathbb{KP}^{n-1}$ with fiber over x equal to the space of operators whose kernel contains the line $\{x\}$.

D. THE POSET is the disjoint union of Grassmannians $G_i(\mathbb{K}^n)$, $i = 1, \ldots, n - 1, n$.

E. THE LINK is PL-homeomorphic to S^M, $M = \dim_\mathbb{R} \mathbb{K} \cdot n(n - 1)/2 + n - 2$.

F. THE SUPPORT is in the wedge

$$\{(p, q) : p \in [-n, 0], \, q \in [\dim_\mathbb{R} \mathbb{K}(p(p-1)/2)+p, \dim_\mathbb{R} \mathbb{K}(-np-p(p+1)/2)+p]\}.$$

G. CONVERGENCE. Yes, as the function space is finite-dimensional.

H. DEGENERATION. $E_\infty \equiv E_1$. See the end of §2.

I. STABILIZATION. For $n \to \infty$, the spectral sequences stabilize to one calculating the cohomology group of the stable group $GL(\mathbb{K}, \infty)$.

K. EXPLICIT FORMULAS. The Miller splitting (3) for $H_*(GL(\mathbb{C}^n)) \sim H_*(U(n))$ has the following realization (see [61], [56]). For any $i =$

$1, \ldots, n$ and any point $L \in G_i(\mathbb{C}^n)$ we imbed a copy of the group $U(i)$ into $U(n)$ as the set of all unitary operators acting trivially on the orthogonal complement of the i-plane $\{L\}$. When L runs over a cycle in the Grassmann manifold, such copies of $U(i)$ sweep out a cycle in $U(n)$. For $\mathbb{K} = \mathbb{R}$ or $\mathbb{H}$ realizations of homology groups of $O(n)$ and $Sp(n)$ are exactly the same (although in the real case we need to take care of orientations and use homology with twisted coefficients).

L. MULTIPLICATION. The ring $H^*(U(n))$ is the exterior algebra with canonical generators $\alpha_1, \alpha_3, \ldots, \alpha_{2n-1}$ of corresponding dimensions. For any $i = 1, \ldots, n$, the term F_i of its filtration induced by the above spectral sequence (i.e. consisting of linking numbers with cycles in the i-th term of the filtration of the resolved determinant) is spanned by all monomials with $\leq i$ factors α_{2j-1}.

4.8. KNOTS AND LINKS IN $\mathbb{R}^n$, $n \geq 3$, AND IN OTHER MANIFOLDS

The 0-dimensional cohomology classes of the space of knots in $\mathbb{R}^3$ are exactly the numerical knot invariants. The invariants arising from the resolutions of discriminants are exactly the *finite-type* knot invariants, see [14]. However the study of the whole cohomology rings of spaces of knots is a more natural problem, leading to equally beautiful algebraic structures, of which the algebraic theory of invariants can be obtained by easy factorization, see [52].

A. THE SPACE of all smooth maps $S^1 \to \mathbb{R}^n$, $n \geq 3$. (Variants: all smooth maps of S^1 to any manifold M^n, all smooth maps of a finite collection of circles to $\mathbb{R}^n$ or M^n; all smooth maps $\mathbb{R}^1 \to \mathbb{R}^n$ coinciding with a standard embedding outside a compact subset in $\mathbb{R}^1$ ('long knots').

B. THE DISCRIMINANT is the set of maps $S^1 \to M^n$ (or $\mathbb{R}^1 \to M^n$ etc.) which are not smooth embeddings, i.e. have either self-intersections or singular points with vanishing derivative.

C. THE TAUTOLOGICAL RESOLUTION is fibered over the completed configuration space $\overline{B(S^1, 2)}$ with fiber over the configuration $(x, y) \subset S^1$ equal to the space of maps $f : S^1 \to M^n$ such that $f(x) = f(y)$ if $x \neq y$ or $f'(x) = 0$ if $x = y$. If $M^n \neq \mathbb{R}^n$ then this space can be not a plane.

D. THE POSET is a model example of a *poset of multi-configurations* in the same way as the ones in subsections 3.1, 4.4 were the typical examples of posets of (mono)configurations. Namely, consider a multi-index $A = (a_1 \geq \ldots \geq a_k)$, where all a_i are natural numbers greater than 1. Given a topological space N (say, $N = S^1$), a *multi-configuration of type A in N* is a collection of $a_1 + \cdots + a_k$ distinct points in N divided into groups of cardinalities $a_1, \ldots, a_k$. Denote by $V(N, A)$ the set of all A-configurations in N. It is convenient to consider any such configuration as a subspace (even a subring) in the space of continuous (or smooth if N is a manifold) functions $N \to \mathbb{R}^1$: namely, as the space of all functions taking equal values at the points of any group. The codimension of this subspace is equal to $\sum_{i=1}^{k}(a_i - 1)$, therefore this number is called the *complexity* of the multi-index A and of any multi-configuration of type A. Let $\overline{V(N, A)}$ be the closure of $V(N, A)$ in the corresponding Grassmannian topology. For example, if $N = S^1$, $k = 1$ and $a_1 = 2$, then the space $V(N, A)$ is the configuration space $B(S^1, 2)$, i.e. an open Möbius band, and $\overline{V(N, A)}$ is the space $\overline{B(S^1, 2)}$, i.e. a closed Möbius band.

For any natural number s, we consider the union $\rho(N, s)$ of spaces $\overline{V(N, A)}$ over all possible multi-indices A of complexity s. The disjoint union of such unions with different s is a poset under the natural subordination of multi-configurations (this subordination can be interpreted as the inverse inclusion of the corresponding functional subspaces).

E. ORDER COMPLEXES. For any natural d consider the topological order complex $\Omega(S^1, d)$ of all spaces $\rho(N, s)$ with $s \le d$. The homological study of these order complexes $\Omega(S^1, d)$ is known as the theory of finite-type knot invariants (and other cohomology classes of the space of knots) and is very complicated. Indeed, the homology group of the quotient space $\Omega(S^1, d)/\Omega(S^1, d - 1)$ is the first (and, accordingly to M. Kontsevich, in the case of rational coefficients also the last) step in the calculation of all such invariants and classes of order d modulo similar classes of order $d - 1$.

F. THE SUPPORT. The wedge $\{(p, q) : p \le 0, q + (n - 2)p \ge 0\}$. If $M^n = \mathbb{R}^n$ then we can indicate also the upper boundary of this support: $q \le np + [-p/2] + 1$, see [60], [55], [56].

G. CONVERGENCE. If $n > 3$, then the spectral sequence converges to the entire cohomology group of the space of knots in M^n, and there are only finitely many nonzero terms $E_1^{p,q}$ on any line $\{p + q = const\}$. If additionally $M^n = \mathbb{R}^n$ then all terms $E_1^{p,q}$ are finitely generated (and explicitly described in [55], [56], [69]). On the other hand, for $n = 3$ already the problem of the convergence on the line $\{p + q = 0\}$ (responsible for the 0-dimensional cohomology classes, i.e. the knot invariants) is unsolved (and is now one of the main problems of knot theory).

H. DEGENERATION. The well-known Kontsevich integral [36] proves that for $M^n = \mathbb{R}^3$ the spectral sequence (with complex coefficients) degenerates on the main diagonal $\{p + q = 0\}$. The same construction proves degeneration for any $\mathbb{R}^n$ on the main diagonal $\{q + (n - 2)p = 0\}$. Moreover, Kontsevich knows (at least since 1994) a similar proof of degeneracy of the entire spectral sequence for any $\mathbb{R}^n$: $E_\infty/\mathbb{C} \equiv$

$E_1/\mathbb{C}$ (still unpublished). My guess is that a) for *long knots* $\mathbb{R}^1 \to \mathbb{R}^n$ even a homotopy splitting of the discriminant like (2) holds (in some exact 'stable' sense, although the discriminant and all terms of its filtration are infinite dimensional, see Problem 5.1 in [72]), but b) for standard knots $S^1 \to \mathbb{R}^n$ such a splitting (and moreover even the integer homology splitting) does not hold because of torsion terms arising from the topological nontriviality of the source manifold S^1. On the other hand, there are easy counterexamples to the degeneracy property in the case of manifolds $M^n \neq \mathbb{R}^n$, see e.g. [66].

I. STABILIZATION. Our spectral sequences calculating the cohomology groups of spaces of knots in vector spaces $\mathbb{R}^n, \mathbb{R}^m$ are very similar if n and m are of the same parity. Namely, in this case their initial terms E_1 coincide up to shifts: $E_1^{p,q-pn}(n) \simeq E_1^{p,q-pm}(m)$. In the case of spectral sequences with coefficients in $\mathbb{Z}_2$ such isomorphisms hold independently of the parity of $m - n$. All this follows immediately from the cellular structure of the resolved discriminant, see [60], [55].

J. COMPARISON AND SMALE–HIRSCH PRINCIPLE. Any smooth function $f : S^1 \to \mathbb{R}^1$ defines a curve in $\mathbb{R}^n$ given by its $(n-1)$-jet extension $(f, f', \ldots, f^{n-1}) : S^1 \to \mathbb{R}^n$. Such a curve is called a *holonomic knot* if this map is a smooth embedding. If $n > 3$ then the space of holonomic knots (or links) in $\mathbb{R}^n$ is homotopy equivalent to the space of standard knots (respectively, links with the same number of components). If $n = 3$ then any isotopy class of links can be represented by a holonomic link (see [65]); conversely, any two holonomic links isotopic in the space of all links are isotopic also in the space of holonomic links (see [16]).

K. EXPLICIT FORMULAS for all cohomology classes can be found (as well as in all other problems discussed here) by the direct calculation of the spectral sequence. In the case of knots in $\mathbb{R}^n$, any element γ of the term $E_1^{p,q}$ can be encoded as a linear combination of certain graphs (known as *chord diagrams* for $q+p(n-2) = 0$ and slightly more general for arbitrary p and q) satisfying certain homological condition (ensuring that the corresponding chain in the term $F_i \setminus F_{i-1}$ of the filtration, $i = -p$, consisting of cells encoded by these graphs, actually is a cycle modulo F_{i-1}). The conscientious calculation of the spectral sequence is the following process: we find the boundary of the chain γ in the next term $F_{i-1} \setminus F_{i-2}$ and span it by some chain in this term (it does no matter that this chain will be of infinite dimension). Then find the boundary of this spanning chain in $F_{i-2} \setminus F_{i-3}$, etc. By Kontsevich's theorem, all these steps can actually be performed, i.e. such spanning chains always exist (at least in the case of complex coefficients). At the last step we already have a cycle in all of Σ and span it by a relative chain in the space of all maps $N^1 \to \mathbb{R}^n$ mod Σ. The corresponding explicit formula works as follows: to any generic cycle in the space of knots $\mathcal{F} \setminus \Sigma$ it associates the number of its intersection points with this relative chain (counted with appropriate signs).

For invariants (i.e. 0-dimensional cohomology classes) of knots in $\mathbb{R}^3$, some combinatorial expressions were obtained by J. Lannes, M. Polyak and O. Viro, P. Cartier, S. Piunikhin, S. Tyurina, a.o., see [37], [45], [53], [54]. It was then proved by M. Goussarov [31] that expressions of Polyak–Viro type exist for any invariants of finite filtration for long knots $\mathbb{R}^1 \to \mathbb{R}^3$.

These expressions (and their extensions to the case of $n > 3$) arise naturally in the above-described algorithm if we choose the spanning chains in some natural way (a semialgebraic chain in $\mathbb{R}^N$, distinguished by several equations, the last of which is $f(x) = g(x)$, should be spanned by the chain given by all the same conditions with the last one replaced by $f(x) \geq g(x)$, etc.).

Several examples of cohomology classes of spaces of knots other than the knot invariants (and not related to them by the stabilization mentioned in item I above) are known, see [66], [69], [56]. Namely, for 'compact' knots $S^1 \to \mathbb{R}^n$ there are two linearly independent cohomology classes of filtration 1 (of dimensions $n - 2$ and $n - 1$) and two cohomology classes of filtration 2 (one of which is the well-known knot invariant or its stabilization mentioned in I and has dimension $2(n-3)$, and the second is of dimension $2n - 3$). For 'long' knots $\mathbb{R}^1 \to \mathbb{R}^n$ there are no cohomology classes of filtration 1 or 2 other than the knot invariant or its stabilization, and in filtration 3 for any n there is exactly one more independent cohomology class having dimension $3n - 8$: it was found by D. Teiblum and V. Turchin in the case of odd n and in [69], [56] for even n. Combinatorial formulas for all these classes will be given in [74].

L. MULTIPLICATION. The multiplication formula for chord diagrams expressing the multiplication of corresponding knot invariants was found by Kontsevich. For a similar formula for higher cohomology classes, see [52].

There exists a huge theory of finite type knot invariants (i.e. zero-dimensional cohomology classes), see [14]. In this case the natural filtration on the space of such classes has an elementary characterization in terms of finite differences (see e.g. §0.2 in [60]). However the direct translation of this elementary definition turns out to be very misleading if one tries to apply it to different problems such as the study of generic plane curves or the calculation of higher-dimensional cohomology classes of spaces of knots. The families of classes and invariants arising from such a direct translation are usually unnatural and only weakly related to more classical ones, and the algebraic structures describing them are non aesthetic[1]. The reason for this is that in these cases *the singularities of discriminant spaces essential for the calculation of these classes and invariants are more complicated than just normal crossings.*

4.9. GENERIC PLANE CURVES

There are dozens of problems of this kind. We shall consider four of them and denote them by (i), (si), (d) and (o) (for 'immersions', 'immersions/strangeness', 'doodles' and 'ornaments', respectively).

A. THE SPACE consists for (i) and (si) of all smooth *immersions* of a circle (or a collection of circles) to $\mathbb{R}^2$, for (d) of all smooth maps of a circle to $\mathbb{R}^2$, and for (o) of all smooth maps of the union N^1 of $s \geq 3$ circles to $\mathbb{R}^2$.

B. THE DISCRIMINANT consists for (i) of non-generic immersions (i.e. having self-tangencies or triple points of the image); for (si) of immersions with triple points; for (d) of maps with triple points or their degenerations (i.e. double points at one of which $f' = 0$ or single points at which $f' = f'' = 0$); and for (o) of maps such that images of some 3 *different* components meet at some point in $\mathbb{R}^2$.

The study of the complement of the discriminant in case (i) (and its subproblem (si)) was initiated in [8], [9]; for (d) a similar problem was formulated in [62], [63] and studied in various versions in [35], [41], [42], [70] under the name of the *theory of doodles*. The problem (o) in the general multidimensional situation goes back at least to Kronecker, see also [26], [27], [34]. The homological problems related to the study of the the corresponding resolved discriminants were formulated in [62], [63] and studied, in particular, in [18], [63], [40].

C. THE TAUTOLOGICAL RESOLUTION for (si) is fibered over the configuration space $B(S^1, 3)$. The fiber over a triple $(x, y, z) \subset S^1$ consists of all *immersions* f such that $f(x) = f(y) = f(z)$. For (i) the resolution space also includes the set fibered over $B(S^1, 2)$ whose fiber over $(x, y) \subset S^1$ is the space of immersions f such that $f(x) = f(y)$ and $f'(x)$ is collinear with $f'(y)$. For (d) the resolution space is fibered over the completed configuration space $\overline{B(S^1, 3)} = S^3(S^1)$ (or $\overline{B(N^1, 3)}$ if we consider many-component curves). The fiber consists of maps gluing together the points of the configuration, in particular it is an affine subspace of codimension 4 in the space of all maps. For (o) all is the same, but only

---

[1] A nice counterexample is provided by the theory of finite type invariants of 3-manifolds started by T. Ohtsuki and extended by S. Garoufalidis, M. Goussarov and others. Unfortunately I cannot include this wonderful theory in the general framework of discriminant theory

the configurations whose points belong to different components of the curve are considered.

We mention below only the problems (si), (d) and (o). The remaining theory (i)\(si) includes, among others, Arnold's basic invariants J^+ and J^- (dual to the sets of immersed curves with codirected and counter-directed self-tangencies), and the theory of Legendrian knots, see [9], [11].

D. THE POSET. See the description of the poset of multi-singularities in §4.8, with the sole difference that all indices a_i of a multi-index A should be greater than 2, and not only greater than 1. Additionally, in the case (si) we do not consider completions of spaces of multi-configurations, but only their regular points, and in the case (o) we consider only multi-configurations such that in any of the groups of cardinalities $a_1, \ldots, a_k$ there are present points of at least three different components of the manifold N^1.

E. TOPOLOGY OF THE ORDER COMPLEX is a good open problem.

F. THE SUPPORT. For problems (d), (o) the wedge $\{(p,q) : p \leq 0, p + q \geq 0\}$. For the analogous problem on plane curves without self-intersections of multiplicity k, the wedge $\{(p,q) : p \leq 0, p + q \geq k - 3\}$.

G. CONVERGENCE. In the cases (is) and (d), the terms $E^{p,q}$ with $p + q = 0$ converge (weakly) to the group of invariants of the corresponding objects: all nonequivalent doodles or generic immersions can be distinguished by finite-type invariants, see [42].

H. DEGENERATION. For cases (o) and (d), the spectral sequences conjecturally degenerate at the first term on the diagonal $\{p + q = 0\}$, i.e. $E_\infty^{p,q} = E_1^{p,q}$ for such p, q. This is an experimental fact proved (by A. B. Merkov) up to filtration 6, as well as for many other important cases, but in general it is a conjecture.

K. EXPLICIT FORMULAS. The first explicit formulas for non-obvious finite-type invariants of generic curves in problem (o) were found in [63]: we can count all intersection points of some two components of a curve with their signs (i.e. mutual orientations of these components at these points) and weights (which are polynomial functions of the indices of these points with respect to different components of the curve), see [63]. Strong generalizations of these *index-type invariants* were found in [40], see also [46]. Similar expressions were found in [50], [51] for the simplest *strangeness* invariant in problem (si) of [8], [9] and some of its generalizations; combinatorial expressions for the invariants J^+, J^- were constructed in [76].

4.10. NONSINGULAR HYPERSURFACES OF DEGREE d IN $\mathbb{CP}^n$

A. THE SPACE $\mathcal{H}(d, n)$ of all homogeneous polynomials $\mathbb{C}^{n+1} \to \mathbb{C}$ of degree d.

B. THE DISCRIMINANT consists of polynomials whose zeros form a non-smooth hypersurface in $\mathbb{CP}^n$.

C. THE TAUTOLOGICAL RESOLUTION is fibered over $\mathbb{CP}^n$ with fiber over x equal to the space of polynomials with zero differential at the corresponding line in $\mathbb{C}^{n+1}$.

D. THE POSET. We take the space of all possible singular sets defined by such polynomials in $\mathbb{CP}^n$ and supply it with the topology induced from the Grassmann manifolds $G_i(\mathcal{H}(d, n))$ by the map sending any such set into the space of all polynomials whose singular set contains this one. Finally, we take the closures of these spaces in this topology.

E. THE ORDER COMPLEX. For $d = 2$, this order complex coincides with that considered in §4.7 for the space $End(\mathbb{C}^{n+1})$, and hence its link is homeomorphic to the sphere of dimension $(n+1)^2 - 2$. For $d = 3$ and $n = 2, 3$ the *rational* homology groups of these links vanish in all positive dimensions. For $(d, n) = (4, 2)$ its rational Poincaré polynomial is equal to $t^{14}(1 + t^3)(1 + t^5)$, see [68].

G. CONVERGENCE. Yes, as the dimension is finite.

H. DEGENERATION. For all cases calculated with $d > 2$, i.e. for $(d, n) = (3, 2), (3, 3)$ or $(4, 2)$ the *rational* spectral sequence converges at the first term: $E_\infty/\mathbb{Q} \equiv E_1/\mathbb{Q}$. However, this is just an experimental fact, and not a part of a general theorem. Also, in the most 'rigid' case $d = 2$ the spectral sequence does not degenerate at E_1, see [68].

I. STABILIZATION. For any fixed n and $d \to \infty$ the corresponding spectral sequences stabilize to one calculating the cohomology of the space of non-vanishing sections of an $(n+1)$-dimensional vector bundle over $\mathbb{CP}^n$.

J. COMPARISON AND SMALE–HIRSCH PRINCIPLE. The gradient mapping $\mathcal{H}(3, 2) \to \mathcal{H}(2, 2)^3$, sending any polynomial of degree 3 in $\mathbb{C}^3$ into the triple of its partial derivatives, induces an isomorphism between the *rational* cohomology groups of the space $\mathcal{H}(3, 2) \setminus \Sigma$ of nonsingular polynomials and the space of quadratic vector fields in $\mathbb{C}^3$ with unique singular point at 0. This isomorphism follows by comparison of the corresponding spectral sequences, see [68].

K. EXPLICIT FORMULAS. For $d = 2$, the spaces $\mathcal{H}(d, n) \setminus \Sigma$ are homotopy equivalent to the corresponding Lagrange Grassmannian manifolds $U(n + 1)/O(n + 1)$, whose homology groups are well known together with their various realizations. For any d, n the projective linear group $PGL(n+1, \mathbb{C})$ acts on the space of nonsingular hypersurfaces in $\mathbb{C}^{n+1}$. In particular any orbit is the image of a map of this group into this space. If $(d, n) = (3, 2)$ or $(3, 3)$ then any such map defines an isomorphism of rational cohomology groups. Thus the cohomology classes of our spaces can be expressed in terms of generators of the cohomology of the group. On the other hand, the Poincaré polynomial of the rational cohomology group of the space of nonsingular *quartics* in $\mathbb{C}P^2$ is equal to $(1 + t^3)(1 + t^5)(1 + t^6)$ (see [68]), i.e., we have a new generator of dimension 6. As J. Steenbrink explained to me, this generator is induced from a cohomology class of the moduli space of curves of genus 3: the cohomology groups of this space were calculated by E. Looijenga in [38].

The problem of calculating topological invariants of spaces of non-singular plane algebraic curves was posed by V. Arnold, see problems 1970-13 and 1981-13 in [11]. V. Kharlamov [33] used the topology of the *real* discriminant in the parallel theory of rigid isotopy classification of real algebraic plane curves.

4.11. HERMITIAN MATRICES WITH SIMPLE SPECTRA

A. THE SPACE of all Hermitian operators in $\mathbb{C}^n$.

B. THE DISCRIMINANT is the set of operators having at least two equal eigenvalues. This set (as well as its complement) was studied in Arnold's papers [4], [10]. For some related physical motivations see also [44]. The cohomology ring of the complementary space is well known at least since [19] (as it coincides with the cohomology ring of the space of complete flags in $\mathbb{C}^n$), nevertheless the induced 'stable' structures in it arising from the resolutions seem to be interesting.

C. THE TAUTOLOGICAL RESOLUTION is fibered over the Grassmannian manifold $G_2(\mathbb{C}^n)$; the fiber over any point is the space of all operators whose restrictions to the corresponding 2-plane are scalar.

D. THE POSET is indexed by the same multi-indices $A = (a_1, \ldots, a_k)$, $a_i \geq 2$, as in §4.8 (with the additional condition $\sum a_i \leq n$).

E. THE LINK. The explicit formula for the ranks of its homology groups in the case of general n is unknown to me. It was calculated in [71] for $n = 3, 4$ and 5: the Poincaré polynomials of such rational groups (reduced modulo a point) are equal in these cases to $t^2(1 + t^2)$, $t^3(1 + t^4)(1 + t^2 + t^4)$ and $t^4(1 + t^2 + t^4 + t^6)(1 + t^2 + t^4 + t^6 + t^8 + t^{10})$ respectively. For any n, such rational homology groups are trivial in all dimensions of the same parity as n.

G. CONVERGENCE. Yes, as the problem is finite dimensional.

H. DEGENERATION. $E_\infty \equiv E_1$ in the case of rational coefficients. However, homotopy splitting surely does not hold since the groups $E_1^{p,q}$ with integer coefficients can have torsion (due to coinciding indices a_i). See [71].

I. STABILIZATION. There is a natural stabilization of our spectral sequences as n increases. The stabilized spectral sequence converges to the cohomology group of the space of infinite Hermitian matrices with simple spectra (with the topology of the direct limit) and provides a natural filtration on this group. All cells $E_1^{p,q}$ of this stable spectral sequence are finitely generated, although the limit cohomology group certainly is not; thus we get the notion of *finite type cohomology classes* of the space of infinite Hermitian matrices with simple spectra, see [71].

K. EXPLICIT FORMULAS. A few are given in [71]. Finding the others (i.e. the expression of our filtration in the terms of the Chern classes of tautological bundles) seems to be an interesting problem.

L. MULTIPLICATION. A conjectural multiplication formula was given in [71] but is not yet proved.

4.12. (STABILIZED) COHOMOLOGY GROUPS OF COMPLEMENTS OF BIFURCATION DIAGRAMS OF ZEROS OF COMPLEX FUNCTION SINGULARITIES

A. THE SPACE is the parameter space of a deformation (say, a versal deformation) of a complex function germ $f : (\mathbb{C}^n, 0) \to (\mathbb{C}, 0)$ with isolated singularity. E.g., the space of complex polynomials as in §4.2 considered as a deformation of the function x^d.

B. THE DISCRIMINANT is (the germ at the origin of) the set of parameter values, for which the corresponding perturbation of f has a critical point close to the origin in $\mathbb{C}^n$ with critical value 0.

C. THE TAUTOLOGICAL RESOLUTION is fibered over (a neighborhood of the origin in) $\mathbb{C}^n$, the fiber over x consists of all parameter values for which the corresponding perturbed function has x as a critical point.

D. THE POSET. If the singularity is sufficiently complicated with respect to the number d, and the deformation is versal, then its elements corresponding to planes of complex codimension $\leq d(n+1)$ are the (completed) configuration spaces $B(\mathbb{C}^n, i)$, $i = 1, 2, \ldots, d$ (see [57]).

F. THE SUPPORT is in the wedge $\{(p,q) : p \leq 0, \ q + 2p \geq 0\}$.

G. CONVERGENCE. Yes, as the problem is finite dimensional (for any particular f and its deformation).

H. DEGENERATION. If the singularity is sufficiently complicated with respect to d (as in D) then all its differentials d^r, $r \geq 1$, act trivially on the groups $E_r^{p,q}$ with $p \geq -d$. Since by F all nontrivial groups $E_r^{p,q}$ with smaller d satisfy the inequality $p+q > d$, we get the degeneration $E_\infty^{p,q} \equiv E_1^{p,q}$ for pairs (p,q) in the domain $\{p+q < d\}$, see [57]. Moreover, for such d we have a homotopy splitting $\bar{F}_d \sim \bar{F}_1 \vee (\bar{F}_2/\bar{F}_1) \vee \ldots \vee (\bar{F}_d/\bar{F}_{d-1})$, see [55], [56].

I. STABILIZATION. If a singularity f is 'more complicated' than g (i.e. in any versal deformation of f we have singularities equivalent to g), then the parameter space of a versal deformation of g can be embedded into that of f in such a way that the discriminant goes to the discriminant, see problem 17 in [5] and also problems 1975-19, 1976-28 and 1980-15 in [11]. These embeddings induce morphisms of the corresponding spectral sequences, which stabilize to a limit spectral sequence calculating the limit cohomology group (which can be thought of as that of the complement of the discriminant of an immensely complicated isolated singularity).

J. COMPARISON AND SMALE–HIRSCH PRINCIPLE. The stable spectral sequence coincides with that (described in §4.6) calculating the cohomology ring of the space $\Omega^{2n}(\mathbb{R}^{2n+2} \setminus 0)$ of all continuous maps $\mathbb{R}^{2n} \to \mathbb{R}^{2n+2}$ with fixed behavior at infinity and avoiding 0. For any isolated complex function singularity and its deformation, the 1-jet extensions of functions $[\varphi \mapsto (\varphi, \partial\varphi/\partial x_1, \ldots, \partial\varphi/\partial x_n)]$ define an embedding of the complement of the corresponding discriminant into this iterated loop space. Our comparison theorem proves that for sufficiently complicated functions this map induces a homology (and even stable homotopy) equivalence up to some high dimension.

K. EXPLICIT FORMULAS. Follow from those for iterated loop spaces.

The above stabilization map defined by adjacency of functions was proposed by Arnold about 1975 (see [5]) together with the problem of computing the corresponding stable cohomology rings (and of proving that they are well defined). Stating this problem was one of the main steps in all the theory described in this section, since it forced one to find 'stable' structures of discriminants and 'stable' methods of computing their homology groups. The corresponding resolution and stable spectral sequence were constructed in April 1985, see [57]. This was the first of the series of calculations listed in this paper.

References

1. D. W. Anderson, A generalization of the Eilenberg–Moore spectral sequence, Bull. Amer. Math. Soc. **78** (1972) 784–788.
2. V. I. Arnold, On a characteristic class entering into conditions of quantization, Funct. Anal. Appl. 1:1 (1967) 1–13.
3. V. I. Arnold, On some topological invariants of algebraic functions, Transact. (Trudy) Moscow Mat. Soc. **21** (1970) 27–46.
4. V. I. Arnold, Modes and quasimodes, Funct. Anal. Appl. 6:2 (1972) 94–101.
5. V. I. Arnold, Some unsolved problems of the singularity theory, *Proc. of the S. L. Sobolev Seminar* (ed S. V. Uspenskii), (Math. Inst. of the Siberian branch of Russian Ac. Sci., Novosibirsk, 1976) pp 5–15. In Russian; extended Engl. transl. in [6].

6. V. I. Arnold, Some unsolved problems in the theory of singularities, *Singularities* (ed Peter Orlik), Proc. Symp. Pure Math. **40**:2, (Amer. Math. Soc., 1983) pp 57–69.

7. V. I. Arnold, Spaces of functions with mild singularities, Funct. Anal. Appl. **23**:3 (1989) 1–10.

8. V. I. Arnold, Plane curves, their invariants, perestroikas and classifications, *Singularities and Bifurcations* (ed V. I. Arnold), Adv. in Soviet Math. **21**, (Amer. Math. Soc., Providence, RI, 1994) pp 33–91.

9. V. I. Arnold, Invariants and perestroikas of fronts in the plane, Proc. Steklov Math. Inst. **209** (1995) 14–64.

10. V. I. Arnold, Remarks on eigenvalues and eigenvectors of Hermitian matrices, Berry phases, adiabatic connections and quantum Hall effect, Selecta Math. (N. S.) **1**:1 (1995) 1–19.

11. V. I. Arnold, *Arnold's Problems*, Phasis, Moscow, 2000 (in Russian).

12. V. I. Arnold, V. V. Goryunov, O. V. Lyashko and V. A. Vassiliev, *Singularities I* (Dynamical systems VI), Enc. Math. Sci. **6**, Springer-Verlag, Berlin a.o., 1993.

13. D. Bar-Natan, On the Vassiliev knot invariants, Topology **34** (1995) 423–472.

14. D. Bar-Natan, D. Bibliography of Vassiliev Invariants, 1994-.
    http://www.ma.huji.ac.il/ drorbn/VasBib/VasBib.html

15. J. S. Birman and X.-S. Lin, Knot polynomials and Vassiliev's invariants, Invent. Math. **111** (1993) 225–270.

16. J. S. Birman and N. Wrinkle, Holonomic and Legendrian parametrizations of knots, J. of Knot Theory and its Ramifications, **9**:3 (1999) 293–309.

17. A. Björner, Continuous partition lattice, Proc. Nat. Acad. Sci. USA **84** (1987) 6327–6329.

18. A. Björner and V. Welker, The homology of '*k*-equal' manifolds and related partition lattices, Adv. Math. **110** (1995) 277–313.

19. A. Borel, Sur la cohomologie des espaces fibrés principaux et des espaces homogènes de groupes de Lie compacts, Ann. of Math. **57** (1957) 115–207.

20. A. Borel and J.-P. Serre, Cohomologie d'immeubles et de groupes *S*-arithmetiques, Topology **15** (1976) 211–231.

21. R. Bott and C. Taubes, On self-linking of knots, J. Math. Phys. **35** (1994) 5247–5287.

22. P. Cartier, Construction combinatoire des invariants de Vassiliev, Comptes Rendus Acad. Sci. Paris, Sér. I **316** (1993) 1205–1210.

23. F. R. Cohen, R. L. Cohen, B. M. Mann and R. J. Milgram, The topology of rational functions and divisors of surfaces, Acta Math. **166** (1991) 163–221.

24. P. Deligne, Immeubles des groupes de tresses généraisés, Invent. Math. **17** (1972) 273–302.

25. Ya. M. Eliashberg and N. M. Mishachev, Wrinkling of smooth mappings and its applications, Invent. Math. **130** (1997) 345–369.

26. R. Fenn and P. Taylor, Introducing doodles, *Topology of low-dimensional manifolds* (ed R. Fenn), Springer Lecture Notes in Math. **722** (1977) 37–43.

27. V. V. Fock, N. A. Nekrasov, A. A. Rosly and K. G. Selivanov, What we think about the higher-dimensional Chern-Simons theories, *Sakharov Memorial Lectures in Physics*, (Nova Science Publishers, Commack, NY, 1992) pp 465–471.

28. M. Goresky and R. MacPherson, *Stratified Morse Theory*, Springer, Berlin a.o., 1988.

29. V. V. Goryunov, Monodromy of the image of the mapping $\mathbb{C}^2 \rightarrow \mathbb{C}^3$, Funct. Anal. Appl. **25**:3 (1991) 174–180.

30. V. Goryunov and D. Mond, Vanishing cohomology of singularities of mappings, Compositio Math. **89** (1993) 45–80.

31. M. Goussarov, M. Polyak, and O. Viro, Finite type invariants of classical and virtual knots, Topology **39** (2000) 1045–1068.
math.GT/9810073 v2

32. K. Igusa, Higher singularities of smooth functions are unnecessary, Ann. of Math. (2) **119** (1984) 1–58.

33. V. M. Kharlamov, Rigid isotopy classification of real plane curves of degree 5, Funct. Anal. Appl. **15**:1 (1981) 88–89.

34. B. A. Khesin, Ergodic interpretation of integral hydrodynamic invariants, J. Geom. Phys. **9**:1 (1992) 101–110.

35. M. Khovanov, Doodle groups, Trans. Amer. Math. Soc. **349** (1996) 2297–2315.

36. M. Kontsevich, Vassiliev's knot invariants, Adv. in Sov. Math. **16**:2 (1993) 137–150.

37. J. Lannes, Sur les invariants de Vassiliev de degré inferieur ou égal à 3, L'Enseignement Math. **39** (1993) 295–316.

38. E. J. N. Looijenga, Cohomology of $\mathcal{M}_3$ and $\mathcal{M}_3^1$, *Mapping class groups and moduli spaces of Riemann surfaces (Göttingen, 1991/Seattle, WA, 1991)*, Contemp. Math. **150**, (Amer. Math. Soc., Providence RI, 1993) pp 205–228.

39. J. P. May, *The Geometry of Iterated Loop Spaces*, Springer Lecture Notes in Math. **271** (1972).

40. A. B. Merkov, Finite order invariants of ornaments, J. Math. Sci. **90**:4 (1998) 2215–2273.

41. A. B. Merkov, Vassiliev invariants classify flat braids, *Differential and symplectic topology of knots and curves* (ed S. L. Tabachnikov), AMS Translations Ser. 2 **190**, (Amer. Math. Soc., Providence RI, 1999) pp 83–102.

42. A. B. Merkov, Vassiliev invariants classify plane curves and doodles, preprint 1998.
http://www.botik.ru/ duzhin/as-papers/finv-dvi.zip

43. A. B. Merkov, Segment-arrow diagrams and invariants of ornaments, Mat. Sbornik, to appear.

44. S. P. Novikov, Two-dimensional Schroedinger operator in the periodic field, Current Problems in Math. (VINITI, Moscow) **23** (1983) 3–22.

45. M. Polyak and O. Viro, Gauss diagram formulas for Vassiliev invariants, Internat. Math. Res. Notes **11** (1994) 445–453.

46. M. Polyak, Invariants of plane curves and fronts via Gauss diagrams, Topology **37** (1998) 989–1010.

47. G. B. Segal, Configuration spaces and iterated loop spaces, Invent. Math. **21** (1973) 213–221.

48. G. B. Segal, Topology of spaces of rational functions, Acta Math. **143** (1979) 39–72.

49. B. Z. Shapiro and B. A. Khesin, Swallowtails and Whitney umbrellas are homeomorphic, J. Alg. Geom. **1** (1992) 549–560.

50. A. Shumakovich, Explicit formulas for strangeness of plane curves, St. Petersburg Math. J. **7**:3 (1996) 445–472.

51. A. Shumakovich, Strangeness and invariants of finite degree, Ph. D. thesis, Uppsala Univ., 1996.

52. V. Tourtchine, Sur l'homologie des espaces des nœuds non-compacts, preprint. arXiv:math.QA/0010017

53. S. D. Tyurina, Diagrammatic formulas of the Viro-Polyak type for finite degree invariants, Russian Math. Surv. (Uspekhi) **54**:3 (1999).

54. S. D. Tyurina, On the Lannes and Viro-Polyak type formulas for finite type invariants, Matem. Zametki (Math. Notes) **65**:6 (1999).

55. V. A. Vassiliev, *Complements of discriminants of smooth maps: topology and applications, Revised ed.*, Translations of Math. Monographs **98**, Amer. Math. Soc., Providence RI, 1994.

56. V. A. Vassiliev, *Topology of complements of discriminants* (in Russian), Phasis, Moscow, 1997.

57. V. A. Vassiliev, Stable cohomology of complements of discriminant manifolds of singularities of holomorphic functions, Russian Math. Surveys **42**:2 (1987) 307–308 and J. Soviet Math. **52**:4 3217–3230.

58. V. A. Vassiliev, Topology of spaces of functions without complicated singularities, Funct. Anal. Appl. **23**:4 (1989).

59. V. A. Vassiliev, Topology of complements to discriminants and loop spaces, *Theory of Singularities and its Applications* (ed V. I. Arnold), Advances in Soviet Math. **1** (1990) 9–21.

60. V. A. Vassiliev, Cohomology of knot spaces, *Theory of Singularities and its Applications* (ed V. I. Arnold), Advances in Soviet Math. **1** (1990) 23–69.

61. V. A. Vassiliev, A geometric realization of the homology of classical Lie groups, and complexes, S-dual to the flag manifolds, Algebra i Analiz **3**:4 (1991) 113–120. English transl. St. Petersburg Math. J. **3**:4 809–815.

62. V. A. Vassiliev, Complexes of connected graphs, *The I. M. Gel'fand mathematical seminars 1990–1992* (eds L. Corvin, I. Gel'fand and J. Lepovsky), (Birkhäuser, Basel, 1993) pp 223–235.

63. V. A. Vassiliev, Invariants of ornaments, *Singularities and Bifurcations* (ed V.I. Arnold), Advances in Soviet Math. **21** (1994) 225–262.

64. V. A. Vassiliev, Topology of discriminants and their complements, *Proc. Intern. Congr. Math., Zürich, 1994*, (Birkhäuser, Basel, 1995) pp 209–226.

65. V. A. Vassiliev, Holonomic links and Smale principles for multisingularities, J. of Knot Theory and its Ramifications **6**:1 (1997) 115–123.

66. V. A. Vassiliev, On invariants and homology of spaces of knots in arbitrary manifolds, *Topics in quantum groups and finite-type invariants* (eds B. Feigin and V. Vassiliev), AMS Translations. Ser. 2 **185** (Amer. Math. Soc., Providence RI, 1998) pp 155–182.

67. V. A. Vassiliev, Homology of spaces of homogeneous polynomials without multiple roots in $\mathbf{R}^2$, Proc. Steklov Math. Inst. **221** (1998) 143–148.

68. V. A. Vassiliev, How to calculate homology groups of spaces of nonsingular algebraic projective hypersurfaces, Proc. Steklov Math. Inst. **225** (1999) 121–140.

69. V. A. Vassiliev, Topology of two-connected graphs and homology of spaces of knots, *Differential and symplectic topology of knots and curves* (ed S. L. Tabachnikov), AMS Transl. Ser. 2 **190**, (Amer. Math. Soc., Providence RI, 1999) pp 253–286.

70. V. A. Vassiliev, On finite-order invariants of triple points free plane curves, *D. B. Fuchs Anniversary Volume* (eds A. Astashkevich and S. Tabachnikov) AMS Transl. Ser. 2 **194**, (Amer. Math. Soc., Providence RI, 1999) pp 275–300.

71. V. A. Vassiliev, Spaces of Hermitian operators with simple spectra and their finite-order cohomology, *Arnold's Mathematical Seminar: algebraic and topological problems of singularity theory*, AMS Transl., (Amer. Math. Soc., Providence RI., 2001), to appear.

72. V. I. Vassiliev, Homology of i-connected graphs and invariants of knots, plane arrangements, etc., *Proc. of the Arnoldfest conference*, Fields Inst. Communications **24**, (Amer. Math. Soc., Providence RI, 1999) pp 451–469.

73. V. A. Vassiliev, Topological order complexes and resolutions of discriminant sets, Publ. l'Inst. Math. Belgrade, Nouvelle sér. **66(80)** (1999) 165–185.

74. V. A. Vassiliev, On combinatorial formulas for cohomology of spaces of knots, Moscow Math. J. (2001), to appear.

75. V. A. Vassiliev and V. V. Serganova, On the number of real and complex moduli of singularities of smooth functions and realizations of matroids, Matem. Zametki (Math. Notes) **49**:1 (1991) 19–27.

76. O. Viro, First degree invariants of generic curves on surfaces, Report 1994:21, Dept. of Math., Uppsala Univ., (1994).

77. G. M. Ziegler and R. T. Živaljević, Homotopy type of arrangements via diagrams of spaces, Math. Ann. **295** (1993) 527–548.

78. R. T. Živaljević, Combinatorics of topological posets, MSRI Preprint 1998-009, 1998.

Classifying Spaces of Singularities and Thom Polynomials

Maxim Kazarian*
Steklov Mathematical Institute and Independent Moscow University
(kazarian@mccme.ru)

1. Thom polynomials

Theorems of global singularity theory express global topological invariants (of manifolds, bundles, etc.) in terms of the geometry of singularities of some differential geometry structures. A classical example is the Hopf theorem expressing the Euler characteristic of a manifold via singular points of a vector field on it. Another example is the Maslov class of a Lagrange submanifold in the cotangent bundle defined as the cohomology class Poincaré dual to the (properly co-oriented) critical set of the projection to the base of the bundle, see e.g. [1]. Many results in this theory are formulated as theorems on existence and computation of so called *Thom polynomials*. In these notes we explain the definition of these polynomials based on the notion of the classifying space of singularities. This approach makes the 'existence theorem' trivial and also gives some ideas on computing these polynomials.

1.1. THEOREM ON THE EXISTENCE OF THOM POLYNOMIAL

Many classification problems in singularity theory can be formulated as the problem of classification of orbits for some Lie group action. In the case of singularities of maps consider the space

$$V = J_0^K(\mathbb{R}^m, \mathbb{R}^n) \tag{1}$$

of K-jets at the origin of map germs $(\mathbb{R}^m, 0) \to (\mathbb{R}^n, 0)$. We consider this space together with the action of the Lie group

$$G = J_0^K \mathrm{Diff}(\mathbb{R}^m) \times J_0^K \mathrm{Diff}(\mathbb{R}^n) \tag{2}$$

of K-jets of left-right changes at the origin.

DEFINITION 1.1. *A singularity class is any G-invariant subset $\Sigma \subset V$.*

* Partially supported by the grants RFBR 99-01-01109 and NWO-047.008.005

117

D. Siersma et al. (eds.), New Developments in Singularity Theory, 117–134.

If $f : M^m \to N^n$ is a generic smooth map then the singularity locus $\Sigma(f) \subset M$ consisting of points with the given local singularity type Σ has the same codimension as $\Sigma \subset V$. Moreover if the singularity class $\Sigma \subset V$ is an algebraic subvariety then the singularity locus $\Sigma(f)$ carries a fundamental $\mathbb{Z}_2$-homology class and its Poincaré dual cohomology class is well defined.

THEOREM 1.2 (Thom). *For any algebraic singularity class $\Sigma \subset V$ there exists a universal polynomial P_Σ (called Thom polynomial) in Stiefel-Whitney classes $\omega_1(M), \ldots, \omega_m(M), f^*\omega_1(N), \ldots, f^*\omega_n(N)$, such that for any generic map $f : M \to N$ the cohomology class Poincaré dual to the singularity locus $\Sigma(f)$ is given by this polynomial,*

$$[\Sigma(f)] = P_\Sigma(\omega(M), f^*\omega(N)) \in H^*(M, \mathbb{Z}_2).$$

1.2. GENERICITY CONDITION

The genericity condition in the theorem above is formulated as follows. Consider the smooth fiber bundle $E \to M$ whose fiber E_x over a point x is isomorphic to V and consists of all K-jets of map germs $(M, x) \to (N, f(x))$. The structure group of this bundle can be reduced to G, and up to an isomorphism of G-bundles, it is independent of the representative f in the same homotopy class of smooth maps $M \to N$. The singularity class Σ gives rise to a well-defined subvariety $\Sigma(E) \subset E$ in the jet bundle E. If Σ is locally algebraic then so $\Sigma(E)$ is and the codimension of Σ in V is equal to that of $\Sigma(E)$ in E. Therefore its dual cohomology class $[\Sigma(E)] \in H^*(E, \mathbb{Z}_2)$ is always well defined. The map f defines a natural section s_f of the bundle E whose value at a point $x \in M$ is the K-jet of the map f itself at this point. A reformulation of Theorem 1.2 claims that for any (generic or not) map f there is an equality

$$P_\Sigma(\omega(M), f^*\omega(N)) = s_f^*[\Sigma(E)] \in H^*(M, \mathbb{Z}_2).$$

If Σ is algebraic then the space E admits a Whitney stratification such that $\Sigma(E)$ is the union of several strata.

DEFINITION 1.3. *The map $f : M \to N$ is called* transversal *if the jet extension section $s_f : M \to E$ is transversal to each stratum of the stratification of E.*

The generic maps of Theorem 1.2 are those which are transversal. By the transversality theorem, any generic section $s : M \to E$ is transversal. Moreover by the strong version of this theorem due to Thom the transversality condition for the section s_f can be achieved by some small perturbation of the map f, i.e. within the class of 'integrable' sections.

1.3. COMPLEX VERSION

There is a complex version of Theorem 1.2 where smooth maps of smooth manifolds are replaced by holomorphic maps of complex analytic manifolds, $\mathbb{Z}_2$-cohomology by integer cohomology, Stiefel-Whitney classes by the corresponding Chern classes, etc. In the complex case the transversality condition is open but not necessarily dense. For any map f the characteristic class represented by the corresponding Thom polynomial can be defined as

$$P_\Sigma(c(M), f^*c(N)) = s_f^*[\Sigma(E)]. \tag{3}$$

It can be interpreted as follows.

- If the section s_f is transversal then $P_\Sigma(c(M), f^*c(N)) = [\Sigma(f)]$.

- The equality (3) can be applied if the singularity locus $\Sigma(f)$ has the 'expected codimension'. In this case the components of $\Sigma(f)$ should be taken with multiplicities prescribed by the scheme structure of $\Sigma(f) = s_f^{-1}(\Sigma(E))$.

- If the codimension of $\Sigma(f)$ is less than that expected, then the Poincaré dual of $P_\Sigma(c(M), f^*c(N))$ can be represented by some closed singular chain in $\Sigma(f)$. It follows, in particular, that $P_\Sigma(c(M), f^*c(N)) \neq 0$ implies $\Sigma(f) \neq \varnothing$.

- In any case one may neglect the holomorphic structure on M and consider a generic C^∞-perturbation $s : M \to E$ of the section s_f. Then $P_\Sigma(c(M), f^*c(N))$ is Poincaré dual to the singularity locus $\Sigma(s)$ (which is a real locally analytic co-oriented subvariety in M).

In many problems there is a correspondence between the classifications of real and complex singularities: every complex singularity has a real representative and the real codimension of a real singularity class is equal to the complex codimension of its complexification. There is no *à priori* proof of this statement. Moreover there are counterexamples that show that this is not always the case. All known counterexamples are very degenerate and have a very large codimension, see e.g. [22]. So we can formulate a general *complexification principle* ([4]) which should be proved in each particular case independently: *the Thom polynomial of a real singularity can be obtained from the Thom polynomial of the corresponding complex singularity by replacing the Chern classes by the corresponding Stiefel-Whitney classes and reducing all coefficients modulo 2.*

1.4. CLASSIFYING SPACE OF SINGULARITIES AND DETERMINATION OF THOM POLYNOMIALS

The characteristic classes dual to singularity loci of a smooth map $f : M \to N$ are defined using an auxiliary (V, G)-bundle $E \to M$, see (1), (2). This bundle can be induced from the universal classifying (V, G)-bundle $\mathbf{BV} \to \mathbf{B}G$. The construction of the classifying space $\mathbf{BV}$ presented below is used also in Borel's definition of equivariant cohomology for the G-space V.

Consider the classifying principal G-bundle $\mathbf{E}G \to \mathbf{B}G$, i.e. a contractible space $\mathbf{E}G$ with a free action of the group G. This action extends to the diagonal action on the product space $V \times \mathbf{E}G$.

DEFINITION 1.4 ([10, 11]). *The classifying space of singularities* $\mathbf{BV}$ *is the total space of the* (V, G)*-bundle associated with the classifying principal bundle* $\mathbf{E}G \to \mathbf{B}G$,

$$\mathbf{BV} = V \times_G \mathbf{E}G = (V \times \mathbf{E}G)/G.$$

The projection to the second factor $\mathbf{BV} \to \mathbf{E}G/G = \mathbf{B}G$ is a bundle with fiber isomorphic to V and structure group G. Since the space V is contractible the projection $\mathbf{BV} \to \mathbf{B}G$ induces an isomorphism of (co)homology groups,

$$H^*(\mathbf{BV}, \mathbb{Z}_2) \cong H^*(\mathbf{B}G, \mathbb{Z}_2).$$

On the other hand each singularity class $\Sigma \subset V$ defines a subspace $\mathbf{B}\Sigma = \Sigma \times_G \mathbf{E}G \subset \mathbf{BV}$. If Σ is an algebraic subvariety then $\operatorname{codim}_{\mathbf{BV}} \mathbf{B}\Sigma = \operatorname{codim}_V \Sigma$ and the cohomology class dual to $\mathbf{B}\Sigma$ is well defined.

DEFINITION 1.5. *The* Thom polynomial *of the singularity class* Σ *is the cohomology class* $P_\Sigma \in H^*(\mathbf{BV}, \mathbb{Z}_2) = H^*(\mathbf{B}G, \mathbb{Z}_2)$ *dual to the locus* $\mathbf{B}\Sigma \subset \mathbf{BV}$.

The Lie group G of jets of left-right changes is contractible to its subgroup $\mathrm{GL}(m) \times \mathrm{GL}(n)$ of linear changes and hence to its maximal compact subgroup $\mathrm{O}(m) \times \mathrm{O}(n)$. Therefore the Thom polynomial is an element of the ring

$$H^*(\mathbf{B}G, \mathbb{Z}_2) \cong H^*(\mathbf{BO}(m) \times \mathbf{BO}(n), \mathbb{Z}_2) \cong \mathbb{Z}_2[\omega_1, \dots, \omega_m, \omega'_1, \dots, \omega'_n]$$

of polynomials in Stiefel-Whitney classes.

REMARK 1.6.   The classifying spaces $\mathbf{B}G$, $\mathbf{B}V$, etc. have infinite dimensions and thus the definition above should be clarified. The simplest way to overcome this difficulty is to replace the classifying principal bundle $\mathbf{E}G \to \mathbf{B}G$ by a finite dimensional smooth principal bundle $\mathbf{E}G_N \to \mathbf{B}G_N$ with N-connected total space $\mathbf{E}G_N$. Then we get isomorphisms $H^p(\mathbf{B}G_N, \mathbb{Z}_2) \cong H^p(\mathbf{B}G, \mathbb{Z}_2)$ for all $p < N$ and can set

$$P_\Sigma = [\Sigma \times_G \mathbf{E}G_N] \in H^c(V \times_G \mathbf{E}G_N, \mathbb{Z}_2) \cong H^c(\mathbf{B}G_N, \mathbb{Z}_2) \cong$$
$$\cong H^c(\mathbf{B}G, \mathbb{Z}_2),$$

where $c = \operatorname{codim} \Sigma$. This cohomology class is independent of the choice of the finite dimensional approximation $\mathbf{E}G_N$ provided that $N > \operatorname{codim} \Sigma$.

REMARK 1.7.   In [18] Szücs and Rimányi used an alternative approach to the definition of the classifying space of singularities based on Szücs's idea of gluing the classifying spaces of symmetry groups of singularities. They considered only simple singularities, and the very clear topology of the classifying space does not follow from their construction. It should be noticed nevertheless that their construction works as well for the case of multisingularities, see [16, 17, 19] for some applications. It is an interesting problem to find an *à priori* construction for the classifying space of multisingularities and to describe its topology (the work [18] implies that it should be related to cobordism theory).

1.5. PROOF OF THEOREM 1.2

Let $f : M \to N$ be a smooth map and $E \to M$ be the associated (V, G)-bundle whose fiber over a point $x \in M$ consists of all K-jets of map germs $(M, x) \to (N, f(x))$. This bundle, like any other G-bundle, can be induced from the classifying space $\mathbf{B}G$ by some continuous map $\kappa : M \to \mathbf{B}G$. This map extends to the map $\tilde{\kappa} : E \to \mathbf{B}V$ of total spaces of (V, G)-bundles. Thus we get the diagram of maps

$$M \xrightarrow{s_f} E \xrightarrow{\tilde{\kappa}} \mathbf{B}V.$$

The maps in this diagram induce both characteristic classes and partitions by singularity classes. Hence the induced homomorphism of cohomology groups

$$H^*(\mathbf{B}G, \mathbb{Z}_2) \xrightarrow{\tilde{\kappa}^*} H^*(E, \mathbb{Z}_2) \xrightarrow{s_f^*} H^*(M, \mathbb{Z}_2)$$

sends the Thom polynomial of Σ to the corresponding polynomial in Stiefel-Whitney classes of M and N, and the cohomology class dual to the singularity locus $B\Sigma$ to the cohomology class dual to the singularity locus $\Sigma(f)$.   □

1.6. STABILIZATION

One of the most important invariants of a map germ $y = f(x)$, $f :$ $(\mathbb{R}^m, 0) \to (\mathbb{R}^{m+k}, 0)$ is the *local algebra* $Q_f = \mathfrak{m}_x / f^* \mathfrak{m}_y$. This is the quotient algebra of the algebra of function germs $(\mathbb{R}^m, 0) \to (\mathbb{R}, 0)$ over the ideal generated by the components f_i of the germ f.

DEFINITION 1.8. *Two map germs* $f : (\mathbb{R}^m, 0) \to (\mathbb{R}^{m+k}, 0)$ *and* $f' :$ $(\mathbb{R}^{m'}, 0) \to (\mathbb{R}^{m'+k}, 0)$ *(with the same k and possibly different m, m') are called* stably equivalent *if they have isomorphic local algebras, $Q_f \cong Q_{f'}$. A class Σ of singularities is called* stable *if it contains together with each map germ f any map germ stably equivalent to it. (Do not confuse with the notion of a stable singularity!)*

   The stabilization allows us to compare singularities of map germs of manifolds of different dimensions. The codimension of a stable singularity class of map germs $(\mathbb{R}^m, 0) \to (\mathbb{R}^{m+k}, 0)$ given by some collection of local algebras does not depend on m (but it *does* depend on k). Therefore for each k we get an independent problem of stable classification of map germs $(\mathbb{R}^m, 0) \to (\mathbb{R}^{m+k}, 0)$ for all m.

THEOREM 1.9 (cf. [5]). *The Thom polynomial of any stable class of singularities can be expressed as a polynomial in the relative Stiefel-Whitney classes $\omega_i(f^*TN - TM)$ defined as the homogeneous components of the expression*

$$\omega(f^*TN - TM) = \frac{1 + f^*\omega_1(N) + f^*\omega_2(N) + \dots}{1 + \omega_1(M) + \omega_2(M) + \dots}.$$

Proof. First note that the proof of Theorem 1.2 implies that the statement can be extended to a generic section of any (V, G)-bundle over M, not necessary related to a map $M \to N$. A (V, G)-bundle and a section may be given by a family of germs of manifolds E_x, F_x, and of map germs $E_x \to F_x$ depending on the point $x \in M$. In particular, to any map $f : M \to N$ we associate the family of map germs $f_x : (M, x) \to (N, f(x))$, $x \in M$. The Stiefel-Whitney classes in this extended version of Theorem 1.2 are those of the vector bundles $\bigcup_x T_0 E_x$ and $\bigcup_x T_0 F_x$ respectively.

   Now a map germ $x \mapsto f(x)$ is stably equivalent to the map germ $(x, z) \mapsto (f(x), z)$ where $z = (z_1, \dots, z_l)$ is any number of additional variables. This observation may be globalized as follows. The jet extension of a map $f : M \to N$ may be considered as the family of map germs $f_x : (M, x) \to (N, f(x))$ depending on the point $x \in M$. Consider an arbitrary vector bundle $U \to M$. Then we may construct a new family of map germs

$$f_x \times \mathrm{id} : (M \times U_x, x \times 0) \to (N \times U_x, f(x) \times 0).$$

The map germs of the new family are stably equivalent to those of the original one. Therefore the singularity loci and their cohomology classes coincide for both families. By the extended version of Theorem 1.2 mentioned above the cohomology class dual to the locus of singularity Σ is given by $P_\Sigma(\omega(TM \oplus U), \omega(f^*TN \oplus U))$. We may chose the bundle U arbitrary. For example, we can chose it in such a way that the bundle $TM \oplus U$ is trivial. Then $\omega(f^*TN \oplus U) = \omega(f^*TN - TM)$ and we get

$$P_\Sigma(\omega(TM), \omega(f^*TN)) = P_\Sigma(1, \omega(f^*TN - TM)). \qquad \square$$

Theorem 1.9 implies that for computing Thom polynomials of a stable class of singularities it is sufficient to consider the case when the target space is a fixed Euclidean space $N = \mathbb{R}^n$.

REMARK 1.10. The formal definition of the classifying space given above admits a very simple geometrical realization of this space. We present a construction for a version of the classifying space which takes into account the stabilization used in Theorem 1.9. For fixed k, K choose large integers $m, N \gg 0$ and consider the Euclidean space $\mathbb{R}^{N+n} = \mathbb{R}^N \times \mathbb{R}^n$, $n = m + k$. Denote by $\mathcal{G}_m(\mathbb{R}^{N+n})$ *the manifold of all K-jets of germs at 0 of m-dimensional submanifolds in $(\mathbb{R}^{N+n}, 0)$.* This manifold is homotopy equivalent to the Grassmannian $G_m(\mathbb{R}^{N+n})$ (since the space of germs of submanifolds with a fixed tangent plane is contractible) and the $\mathbb{Z}_2$-cohomology ring of this space is generated by Stiefel-Whitney classes. The points of this space are classified according to the singularities of the projection to the coordinate plane $\mathbb{R}^n \subset \mathbb{R}^N \times \mathbb{R}^n$. Denote by $\widetilde{\Sigma} \subset \mathcal{G}_m(\mathbb{R}^{N+n})$ the collection of points for which this projection belongs to the given stable class Σ of singularities. The cohomology class dual to this cycle

$$[\widetilde{\Sigma}] \in H^*(\mathcal{G}_m(\mathbb{R}^{N+n}), \mathbb{Z}_2)$$

(or, more exactly, the expression of this class in terms of the multiplicative generators of the cohomology ring of the Grassmannian) may be taken as an independent definition of the Thom polynomial. This construction reduces the problem of finding Thom polynomials to the study of the geometry of Grassmannians.

REMARK 1.11. Similar stabilizations allow one to compare singularities in the spaces of different dimensions exist for other problems of singularity theory. For instance, the stable classification of critical points of functions is related to Lagrange singularities and leads to Lagrange characteristic classes, see [20, 13] and Section 3.4 below.

2. Computing Thom polynomials

A large number of Thom polynomials for various kinds of singularities were found by different authors, see [2, 16] and references therein. As an example of computation of Thom polynomials, we present several proofs of the classical formulas for the Thom-Porteous classes. These proofs illustrate different methods which can be used for other classes. In this section by cohomology we mean cohomology with $\mathbb{Z}_2$-coefficients.

2.1. THOM-PORTEOUS CLASSES

Let E, F be two vector bundles of ranks $m, n = m+k$ respectively over a smooth manifold M and $f : E \to F$ be a generic morphism of vector bundles. Denote by $\Sigma_d \subset M$ the set of points $x \in M$ where the rank of the linear map $f_x : E_x \to F_x$ is at most $m - d$ (i.e. the dimension of the kernel of f_x is at least d; we assume that $d \geq \max(0, -k)$). The classes dual to the loci Σ_d are called *Thom-Porteous classes*.

THEOREM 2.1 ([15]). *Generically Σ_d is a subvariety of codimension $d(d+k)$ and its Thom polynomial is given by $[\Sigma_d] = \Delta_{d,d+k}(\omega(F - E))$, where for a formal series $a = 1 + a_1 + a_2 + \dots$ we denote*

$$\Delta_{p,q}(a) = \begin{vmatrix} a_q & a_{q+1} & \cdots & a_{q+p-1} \\ a_{q-1} & a_q & \cdots & a_{q+p-2} \\ \vdots & \vdots & \ddots & \vdots \\ a_{q-p+1} & a_{q-p+2} & \cdots & a_q \end{vmatrix}.$$

In particular, let $f : M \to N$ be a generic smooth map. Set $E = TM$, $F = f^*TN$. Then the formula of Theorem 2.1 expresses the Thom polynomial for the locus $\Sigma_d(f)$ consisting of points $x \in M$ where the rank of the differential $df_x : T_xM \to T_{f(x)}N$ is at most $m - d$.

In the case of Thom-Porteous classes the singularity type is determined by 1-jet of the map and it is sufficient to set $K = 1$. The classifying space is the Grassmann manifold $G_m(\mathbb{R}^{N+n})$, where $N, m \gg 0$, $n = m + k$. The cycle $\tilde{\Sigma}_d$ is formed by the m-planes whose projections to the fixed subspace $\mathbb{R}^n$ have rank at most $m - d$. Equivalently, it is formed by planes whose intersections with the fixed subspace $\mathbb{R}^N$ are at least d-dimensional. Theorem 2.1 is equivalent therefore to the equality

$$[\tilde{\Sigma}_d] = \Delta_{d,d+k}(\omega(-E)) \in H^*(G_m(\mathbb{R}^{N+n})), \qquad (4)$$

where E denotes the m-dimensional tautological vector bundle on the Grassmannian.

2.2. SCHUBERT CALCULUS

The cohomology group of the Grassmannian has two natural bases. The first is given by monomials in Stiefel-Whitney classes and the second by classes of Schubert cells. The Schubert basis is more geometric and in many cases it is possible to express the cohomology classes given by particular cycles on the Grassmannian via Schubert cells. The passage between the two bases is described by the Giambelli formula ([8]).

In the case of Theorem 2.1, the cycle $\tilde{\Sigma}_d \subset G_m(\mathbb{R}^{N+n})$ is the closure of the Schubert cell denoted by $(d + k, \ldots, d + k)$ (d entries) in the Schubert calculus. The formula (4) is a particular case of the Giambelli formula. $\qquad\qquad\square$

2.3. RESOLUTIONS OF SINGULARITIES

Another direct method of proving rem 2.1 2.1 uses resolutions of singularities. This method may be used for proving Giambelli formulas as well as for finding Thom polynomials for other singularities. Consider the product space $G_d(\mathbb{R}^N) \times G_m(\mathbb{R}^{N+n})$, where we identify $\mathbb{R}^N$ with a fixed N-subspace in $\mathbb{R}^{N+n}$. Consider the submanifold $Z \subset G_d(\mathbb{R}^N) \times G_m(\mathbb{R}^{N+n})$ formed by pairs $(K, L) \in G_d(\mathbb{R}^N) \times G_m(\mathbb{R}^{N+n})$ such that $K \subset L$. Then $\tilde{\Sigma}_d$ is the image of Z under the projection

$$\pi : G_d(\mathbb{R}^N) \times G_m(\mathbb{R}^{N+n}) \to G_m(\mathbb{R}^{N+n})$$

to the second factor. Moreover the restriction $\pi|_Z : Z \to \Sigma_d$ is one-to-one over an open dense set, so that $[\tilde{\Sigma}_d] = \pi_*[Z]$, where $\pi_* : H^*(G_d(\mathbb{R}^N) \times G_m(\mathbb{R}^{N+n})) \to H^*(G_m(\mathbb{R}^{N+n}))$ is the Gysin homomorphism. So the problem is split into two: computing the class $[Z]$ and computing the homomorphism π_*.

For the first problem we note that the cycle Z may be identified with the zero section of the bundle $\mathrm{Hom}(K, \mathbb{R}^{N+n}/L)$, where we denote by K, L the tautological bundles on the two Grassmannians. Therefore $[Z]$ is the top Stiefel-Whitney class of this bundle,

$$[Z] = \omega_{d(N+k)}(\mathrm{Hom}(K, \mathbb{R}^{N+n}/L)) = \Delta_{d,N+k}(\omega(-K - L)).$$

The last equality can by proved purely algebraically using, for example, the splitting principle.

Now we compute the homomorphism π_*. Denote by $Q = \mathbb{R}^N/K$ the universal $(N - d)$-dimensional quotient bundle on $G_d(\mathbb{R}^N)$. Then

$$\omega_s(-K - L) = \omega_s(Q - L) = \sum_{i=0}^{N-d} \omega_{N-d-i}(Q)\omega_{s-(N-d)+i}(-L).$$

Substituting this in the determinant $\Delta_{d,N+k}$ we obtain

$$[Z] = \Delta_{d,N+k}(\omega(Q - L)) = (\omega_{N-d}(Q))^d \Delta_{d,d+k}(\omega(-L)) + \ldots,$$

where the dots denote terms whose degree in $\omega_i(Q)$ is strictly less than $d(N - d) = \dim G_d(\mathbb{R}^N)$. The Gysin homomorphism vanishes on these terms for dimensional reasons and thus

$$[\tilde{\Sigma}_d] = \pi_*[Z] = \pi_*((\omega_{N-d}(Q))^d)\,\Delta_{d,d+k}(\omega(-L)).$$

It remains to note that the equality $\pi_*((\omega_{N-d}(Q))^d) = 1$ reflects the fact that given d generic lines in $\mathbb{R}^N$, there exists a unique d-plane containing them. $\square$

2.4. Symmetries of singularities

Recently R. Rimányi [16] invented a new method for finding Thom polynomials. His method reduces this problem to the linear algebra problem of inverting a large matrix. In general it requires less computations for computing particular Thom polynomials, though it usually does not give closed formulas for series of singularity classes. The main idea is very simple. Since the Thom polynomials are universal, every example where we may compute both the Stiefel-Whitney classes and the class dual to the singularity locus gives linear relations on the coefficients of the Thom polynomial. If the number of examples is large enough then these relations should be sufficient to determine the polynomial completely.

Many examples may be produced in the following way. Let Σ be an orbit of the action of the equivalence group G on the jet space V. Then as a test manifold we can take a tubular neighborhood U of the submanifold $\mathbf{B}\Sigma$ in the classifying space $\mathbf{B}V$ (it can be identified with the total space of a normal bundle of $\mathbf{B}\Sigma$). The test manifold in this case is homotopy equivalent to $\mathbf{B}\Sigma = (\Sigma \times \mathbf{E}G)/G = (\mathrm{pt} \times \mathbf{E}G)/G_\Sigma \cong \mathbf{B}G_\Sigma$, where G_Σ is the 'symmetry group' of the singularity Σ, the stationary group of any point $\mathrm{pt} \in \Sigma$ (or a maximal compact subgroup in it). Moreover the normal bundle of $\mathbf{B}\Sigma$ may be identified with the space of the universal bundle over $\mathbf{B}G_\Sigma$ associated with the action of G_Σ on any G_Σ-invariant transversal slice to Σ. The locus of the singularity Σ for this test manifold is the zero section of the normal bundle and hence its dual coincides with the Euler class of the bundle. It is usually not difficult to describe explicitly the homomorphism $H^*(\mathbf{B}G) \to H^*(\mathbf{B}G_\Sigma)$ and to compute the corresponding Euler class.

In the particular case of Thom-Porteous singularities the arguments above can be reduced to the following. Consider vector bundles K, L of ranks d, $d+k$ respectively with Stiefel-Whitney classes $a_i = \omega_i(K)$, $b_j = \omega_j(L)$ over some smooth base B. Let the test manifold M be the total space of the bundle $\mathrm{Hom}(K, L)$ over B. The singularity locus $\Sigma_d(M)$ in this case is the zero section of the bundle $M \to B$. Therefore

$$[\Sigma_d(M)] = \omega_{d(d+k)}(\mathrm{Hom}(K, L))$$
$$= \Delta_{d,d+k}(\omega(L - K)) \in H^*(B) \cong H^*(M).$$

(The last equality is an algebraic exercise on the application of the splitting principle.) The base B can be chosen arbitrary. For example we can chose B to be (a finite dimensional approximation of) the product space $\mathbf{BO}(d) \times \mathbf{BO}(d + k)$ of the classifying spaces for d- and $(d + k)$-dimensional vector bundles respectively, and K, L to be the corresponding canonical bundles. We see that the Thom polynomial $P_{\Sigma_d}(\omega_1, \omega_2, \ldots)$ has the following property: *after the substitution* $\omega = \frac{1+a_1+\ldots+a_d}{1+b_1+\ldots+b_{d+k}}$ *it coincides with* $\Delta_{d,d+k}(\omega)$. One can verify that the homomorphism $H^*(\mathbf{BO}) \to H^*(\mathbf{BO}(d) \times \mathbf{BO}(d + k))$ given by $1 + \omega_1 + \ldots \mapsto \frac{1+a_1+\ldots+a_d}{1+b_1+\ldots+b_{d+k}}$ is injective up to degree $d(d + k)$ so the relation above determines the polynomial completely. □

REMARK 2.2. There are many results in global singularity theory that involve classes dual to cycles of multisingularities (see, for example, [2] and references therein). The method of Rimányi may be effectively applied to this kind of problem as well, see [16, 17].

3. Universal complex of singularity classes and characteristic spectral sequence

The singularity classes in *real* classification problems usually form semi-algebraic rather than algebraic subvarieties in the jet space. In order for the cohomology class dual to some union of singularity classes to be well defined, the (formal) boundary of this union must vanish. A similar problem appears when one tries to define an integer characteristic class dual to some combination of singularity classes. In this case all singularity classes of this combination must be co-oriented and the (formal) co-oriented boundary of the combination must vanish. These observations are formalized in the notions of the universal complex of singularity classes [20] and the characteristic spectral sequence [10, 11, 14].

3.1. CLASSIFICATIONS

Consider a classification problem of singularity theory formulated as the classification of orbits of an equivalence Lie group G acting on a contractible jet space V. A *finite G-classification* ([20]) is a finite G-invariant Whitney stratification of V. If the group G is not connected then its elements may permute some strata. Unions of strata containing points of one orbit are called *classes*.

Some classes may consist of only one orbit. For other classes orbits may form families (modules). If, for each class, the moduli space of orbits is smooth and contractible then the G-classification is called *cellular*. In this case, a maximal compact subgroup in the stationary group is independent (up to an isomorphism) of a point of the given class Σ. This group is called the *symmetry group* of the class. The existence of cellular G-classifications for any algebraic action is proved in [20].

A singularity class $\Sigma \subset V$ is called *co-orientable* if it admits a G-invariant co-orientation in V. For cellular classifications this is equivalent to the condition that the symmetry group preserves the orientation of the normal space to the class.

3.2. CHARACTERISTIC SPECTRAL SEQUENCE

For a given G-classification on V consider the filtration formed by *open* subspaces

$$F_0(V) \subset F_1(V) \subset \cdots \subset V,$$

where F_i is the union of classes of codimension less than or equal to i. This filtration defines an *equivariant* spectral sequence $E_*^{*,*}$ called the *characteristic spectral sequence*. This sequence converges to the *equivariant* cohomology $H_G^*(V)$ of V. Since V is contractible, one has $H_G^*(V) \cong H_G^*(\mathrm{pt}) \cong H^*(BG)$.

The reformulation of this definition in the language of classical cohomology groups is as follows. The filtration on V induces a corresponding filtration on the classifying space $\mathbf{B}V = V \times_G \mathbf{E}G$,

$$F_0(\mathbf{B}V) \subset F_1(\mathbf{B}V) \subset \cdots \subset \mathbf{B}V, \qquad F_p(\mathbf{B}V) = F_p(V) \times_G \mathbf{E}G.$$

The characteristic spectral sequence $E_*^{*,*}$ defined by this filtration converges to $H^*(\mathbf{B}V) \cong H^*(\mathbf{B}G)$.

This spectral sequence contains all cohomological information on adjacencies of singularities and their symmetry groups. Its initial term $E_1^{p,*} \cong H_G^*(F_p(V), F_{p-1}(V)) \cong H^*(F_p(\mathbf{B}V), F_{p-1}(\mathbf{B}V))$ is isomorphic to the cohomology group of the Thom space of the normal bundle of the codimension p smooth manifold $F_p(\mathbf{B}V) \setminus F_{p-1}(\mathbf{B}V)$. It is the direct sum of the cohomology groups of the corresponding Thom spaces over all classes of codimension p. In the case of cellular classifications both the submanifold $\mathbf{B}\Sigma \cong \mathbf{B}G_\Sigma$ and its normal bundle (the universal bundle over $\mathbf{B}G_\Sigma$ corresponding to the action of G_Σ on a G_Σ-invariant transversal slice to Σ) may be determined intrinsically in terms of the singularity class Σ and its symmetry group G_Σ. The first differential δ_1 is given by adjacencies of singularities of neighboring codimensions; the higher differentials δ_r correspond to adjacencies of singularities whose codimensions differ by r; for details see [10, 11, 14] and the recent preprint [7].

3.3. UNIVERSAL COMPLEX OF SINGULARITY CLASSES

The cohomology classes corresponding to the fundamental cycles of singularity loci are described by the row $E_*^{*,0}$ of the spectral sequence.

DEFINITION 3.1. *The row* $(E_1^{*,0}, \delta_1)$ *is called the* universal complex of singularity classes.

The cohomology classes of this complex give rise to well defined characteristic classes via the canonical homomorphism

$$E_2^{*,0} \to E_\infty^{*,0} \subset H^*(\mathbf{B}G).$$

Among these are, for example, the cohomology classes dual to algebraic singularity loci. Below we give an abstract geometric-algebraic definition of this complex as it appeared in [20].

In the case of cohomology with $\mathbf{Z}_2$-coefficients, the free generators of this complex in degree p correspond to the singularity classes of codimension p. The differential is given by

$$\delta\Sigma = \sum_{\text{codim}\,\Omega = \text{codim}\,\Sigma + 1} [\Sigma, \Omega]\,\Omega,$$

where the *incidence coefficients* $[\Sigma, \Omega] \in \mathbf{Z}_2$ are defined as follows. Consider a germ of some (codim Ω)-dimensional transversal T to the class $\Omega \subset V$. The points of singularity type Σ form a collection of curves in T going out of the origin. The coefficient $[\Sigma, \Omega]$ is equal to the parity of the number of these curves.

In the case of integer coefficients, the term of degree p in the universal complex is freely generated by *co-orientable* classes (with some fixed choice of the co-orientations). The coboundary operator is defined in a similar way, but now the incidence coefficient $[\Sigma, \Omega]$ is an integer. It is defined as the algebraic number of curves (together with their signs) of singularity type Σ in the transversal T to the singularity class Ω. The *sign* of every such curve (positive or negative) is defined as follows. Consider a small sphere in T centered at the origin. This sphere is oriented as the boundary of a small ball oriented by the chosen co-orientation of Ω. In a neighborhood of the intersection point with a curve of singularity Σ the sphere has an additional orientation as the germ of a transversal to the singularity class Σ. The sign is positive (negative) if the two orientations on the sphere coincide (respectively, are opposite).

3.4. EXAMPLE: FIBER SINGULARITIES OF FUNCTIONS

A number of applications of the notions introduced in this section to different problems of singularity theory are considered in [20], [21] and [11]–[14]. In this section we discuss characteristic classes related to the classification of critical points of functions. Consider the following diagram of holomorphic maps of complex analytic manifolds:

$$
\begin{array}{ccc}
W & \xrightarrow{f} & C \\
\downarrow{\scriptstyle \pi} & & \\
B & &
\end{array}
\tag{5}
$$

We assume that the differential of π is surjective at every point, so the fibers of π form locally a smooth fibration; C is a complex curve. (The case when π is the trivial bundle and $C = \mathbb{C}P^1$ is already interesting enough.) We study singularities of the restrictions of f to the fibers. Let $M \subset W$ be the subset of all critical points of such restrictions. Generically M is smooth and has codimension $n = \dim W - \dim B$. It can be identified with the zero locus of the section $df|_V$ of the bundle $\mathrm{Hom}(V, I)$, where $V \subset TW$ is the subbundle of vectors tangent to the fibers of π and I is the complex line bundle $I = f^*TC$.

Let Ω be any class of singularities of functions (an algebraic subvariety in some jet space of function germs $\mathbb{C}^n, 0 \to \mathbb{C}, 0$ which is invariant with respect to the group of left-right changes of co-ordinates). We shall use the same letter Ω to denote the class of function germs $\mathbb{C}^{n'}, 0 \to \mathbb{C}, 0$, $n' \neq n$, stably equivalent to the functions from Ω. Recall that two germs of functions on spaces of possibly different dimensions are called *stably equivalent* if after adding suitable non-degenerate quadratic forms in new variables they can be reduced to each other by a left-right change of variables.

Define $\Omega(f) \subset M$ as the locus of points at which the restriction of f to the fiber belongs to the given singularity class Ω. According to the general principle of Thom the cohomology class Poincaré dual to the locus $\Omega(f)$ is independent of f (provided that genericity conditions for f analogous to those of Section 1.2 are satisfied) and can be expressed as a universal polynomial in Chern classes of W, B, C. We claim that this polynomial can be expressed in terms of some particular combinations of these classes. Namely, denote $u = c_1(I) = f^* c_1(TC)$, $c_i = c_i(V) = c_i(TW - \pi^* TB)$, and define the classes $a_i = c_i(V^* \otimes I - V)$ as the homogeneous components in the expansion of

$$1 + a_1 + a_2 + \ldots = \frac{(1+u)^n - (1+u)^{n-1}c_1 + (1+u)^{n-2}c_2 - \ldots \pm c_n}{1 + c_1 + c_2 + \ldots + c_n}. \quad (6)$$

These classes satisfy the relations

$$(1 + a_1 + a_2 + \ldots)\left(1 - \frac{a_1}{1+u} + \frac{a_2}{(1+u)^2} - \ldots\right) = 1, \quad (7)$$

following from the identity $U + U^* \otimes I = 0$, where U is the formal difference $U = V^* \otimes I - V$. These relations allow us to expand the squares of the classes a_i, and hence any polynomial in $u, a_1, a_2, \ldots$ can be expressed as a linear combination of monomials $u^{i_0} a_1^{i_1} a_2^{i_2} \ldots$, $i_0 \geq 0$, $i_k \in \{0, 1\}$ $(k > 0)$.

THEOREM 3.2 ([13]). *For any singularity class Ω, the class in $H^*(M)$ Poincaré dual to the locus $\Omega(f)$ can be expressed as a universal polynomial P_Ω in $u, a_1, a_2, \ldots$. This polynomial (called the* Thom polynomial*) is independent of n.*

The Poincaré dual of the locus $\Omega(f)$ considered as a locus in W is equal to $(u^n - u^{n-1}c_1 + \ldots \pm c_n) P_\Omega(u, a_1, a_2, \ldots) \in H^*(W)$.

For the singularity classes of codimension not greater than 6 the Thom polynomials are given in Table I.

This theorem can be formally applied to the case when the manifolds W, B, C are real and the function f is smooth (with Chern classes replaced by the corresponding Stiefel-Whitney classes). But in the real case the $\mathbb{Z}_2$-classes u, a_i vanish and so all cohomology classes of singularities are trivial. Moreover ([11, 14]), *for any locally trivial fiber bundle with compact fibers there exists a real-valued function on the total space whose restrictions to the fibers have no singularities more complicated than A_2.*

Table I. Thom polynomials of singularities of functions of codim ≤ 6

$$A_2 = a_1$$

$$A_3 = 3a_2 + ua_1$$

$$A_4 = 3a_1a_2 + 6a_3 + 4ua_2 + u^2a_1$$

$$D_4 = a_1a_2 - 2a_3 - ua_2$$

$$A_5 = 27a_1a_3 + 6a_4 + u(16a_1a_2 - 12a_3) - 4u^2a_2 + u^3a_1$$

$$D_5 = 6a_1a_3 - 12a_4 + u(4a_1a_2 - 14a_43) - 4u^2a_2$$

$$A_6 = 87a_2a_3 + 54a_1a_4 + 78a_5 +$$
$$u(127a_1a_3 - 53a_4) + u^2(59a_1a_2 - 126a_3) - 41u^3a_2 + u^4a_1$$

$$D_6 = 12a_2a_3 - 24a_5 + u(14a_1a_3 - 40a_4) + u^2(8a_1a_2 - 30a_3) - 8u^3a_2$$

$$E_6 = 9a_2a_3 - 12a_1a_4 + 6a_5 + 3ua_4 + u^2(3a_1a_2 - 6a_3) - 3u^3a_2$$

$$A_7 = 135a_1a_2a_3 + 465a_2a_4 + 264a_1a_5 + 522a_6 + u(516a_2a_3 - 16a_1a_4 + 485a_5) +$$
$$u^2(305a_1a_3 - 70a_4) + u^3(190a_1a_2 - 440a_3) - 165u^4a_2 + u^5a_1$$

$$D_7 = 24a_1a_2a_3 - 24a_2a_4 + 48a_1a_5 - 144a_6 + u(8a_2a_3 + 44a_1a_4 - 224a_5) +$$
$$u^2(48a_1a_3 - 172a_4) + u^3(20a_1a_2 - 88a_3) - 20u^4a_2$$

$$E_7 = 9a_1a_2a_3 + 6a_2a_4 - 42a_1a_5 + 36a_6 + u(21a_2a_3 - 61a_1a_4 + 80a_5) +$$
$$u^2(43a_4 - 6a_1a_3) + u^3(7a_1a_2 - 8a_3) - 7u^4a_2$$

$$P_8 = a_1a_2a_3 - 6a_2a_4 + 6a_1a_5 - 4a_6 +$$
$$u(7a_1a_4 - 4a_2a_3 - 10a_5) + u^2(2a_1a_3 - 8a_4) - 2u^3a_3$$

The $\mathbf{Z}_2$-reductions of the polynomials listed in Table I are non-trivial when they are considered in the context of the *theory of Lagrange and Legendre singularities*, see [20, 21, 10, 13]. Non-trivial classes appear also if we consider the *global* singularities of the restriction of f to the fibers. Assume that in diagram (5) π is a smooth locally trivial bundle with oriented fibers diffeomorphic to S^1. Then the Chern-Euler class $e = c_1(\pi) \in H^2(B)$ of the bundle π can be interpreted as follows in terms of the fiber singularities of a generic smooth function $f : W \to \mathbf{R}$ on the total space of the bundle.

We study the global minima of the restrictions $f_b : W_b \cong S^1 \to \mathbf{R}$, $b \in B$. Denote by $(a_1, \ldots, a_l) = (a_1, \ldots, a_l)_f \subset B$ the locus of points $b \in B$ such that the function f_b attains its global minimum at l consecutive points $x_1, \ldots, x_l$ on the circle W_b, and has a critical point of multiplicity a_i at x_i (i.e. f_b is equivalent to $(x - x_i)^{a_i+1}$ near x_i). The numbers a_i are odd positive integers; their order is defined up to a cyclic permutation. Generically the locus $(a_1, \ldots, a_l)$ is smooth and has codimension $(\sum a_i) - 1$.

THEOREM 3.3 ([12]). *Every singularity class* $(a_1, \ldots, a_l) \subset B$ *of even codimension has a natural co-orientation. For any integer* $r > 0$ *there is a universal (independent of f) linear combination with rational co-efficients of the classes* $(a_1, \ldots, a_l)$ *of codimension $2r$ such that the cohomology class dual to this combination is well defined and equals the* *characteristic class e^r of the bundle π. For $r \leq 4$ these combinations are given in Table II.*

Table II. Characteristic classes of codim$_\mathbb{R} \leq 8$ singularities of the global minimum.

$$-2e = (1^3) - (3),$$
$$12e^2 = (1^5) - (3, 1^2) + 2(5),$$
$$-120e^3 = (1^7) - (3, 1^4) + (3^2, 1) + 2(5, 1^2) - 5(7),$$
$$1680e^4 = (1^9) - (3, 1^6) + 2(5, 1^4) + \tfrac{31}{15}(3^2, 1^3) - \tfrac{1}{15}(3, 1^2, 3, 1) -$$
$$\tfrac{21}{5}(3^3) - \tfrac{14}{15}(5, 3, 1) - \tfrac{14}{15}(5, 1, 3) - \tfrac{91}{15}(7, 1^2) + 14(9).$$

In Table II (m^l) stands for $(m, \ldots, m)$ (l times). The universal complex of singularity classes responsible for this problem is closely related to cyclic homology theory. This relation is studied in [12].

References

1. V. I. Arnold, On a characteristic class entering into conditions of quantization, Funct. Anal. Appl. **1**:1 (1967) 1–13.
2. V. I. Arnold, V. V. Goryunov, O. V. Lyashko and V. A. Vassiliev, *Singularities I*, Enc. Math. Sci. **6** (Dynamical systems VI), Springer-Verlag, Berlin a.o., 1993.
3. A. Borel, Sur la cohomologie des espaces fibrés principaux et des espaces homogènes de groupes de Lie compacts, Ann of Math. **57** (1957) 115–207.
4. A. Borel and A. Haefliger, La classe d'homologie fondamentale d'une espace analytique, Bull. Soc. Math. France **89** (1961) 461–513.
5. J. Damon, Thom polynomials for contact class singularities, Ph.D. thesis, Harvard University, 1972.
6. Ya. M. Eliashberg and N. M. Mishachev, Wrinkling of smooth mappings and its applications, Invent. Math. **130** (1997) 345–369.
7. L. Fehér and R. Rimányi, Calculation of Thom polynomials for group actions, preprint.
8. P. Griffiths, and J. Harris, *Principles of algebraic geometry*, Wiley, NY, 1978.
9. K. Igusa, Higher singularities of smooth functions are unnecessary, Ann. of Math. **119** (1984) 1–58.
10. M. Kazarian, Characteristic classes of Lagrange and Legendre singularities, (Russian) Uspekhi Mat. Nauk **50(304)** (1995) 45–70.
11. M. Kazarian, Characteristic classes of singularity theory, *Arnold-Gelfand Mathematical Seminars*, (Birkhäuser, Basel, 1997) pp 325–340.
12. M. Kazarian, Relative Morse theory of circle bundles and cyclic homology, Func. Anal. Appl. **31**:1 (1997) 20–31.
13. M. Kazarian, Thom polynomials for Lagrange, Legendre and isolated hypersurface singularities, preprint.
14. M. Kazarian, Characteristic spectral sequence of singularity classes, Appendix in [21], 243–310.
15. I. R. Porteous, Simple singularities of maps, *Proceedings of Liverpool Singularities Symposium I*, (ed C. T. C. Wall) Springer Lecture Notes in Math. **192** (1970) 286–307.

16. R. Rimányi, Thom polynomials, symmetries and incidences of singularities, preprint.

17. R. Rimányi, Multiple point formulas—a new point of view, Pacific J. Math, to appear.

18. R. Rimányi and A. Szücs, Generalized Pontrjagin-Thom construction for maps with singularities, Topology **37** (1998) 1177–1191.

19. A. Szücs, Multiple points of singular maps, Math. Proc. Camb. Phil. Soc., **100** (1986), 331–346.

20. V. A. Vassiliev, *Lagrange and Legendre Characteristic Classes*, 2nd edition, Gordon and Breach, New York a.o., 1993. See also [21].

21. V. A. Vassiliev, *Lagrange and Legendre Characteristic Classes*, MCCME, Moscow, 2000. (Extended Russian translation of [20].)

22. V. A. Vassiliev and V. V. Serganova, On the number of real and complex moduli of singularities of smooth functions and of realizations of matroids (Russian), Mat. Zametki **49**:1 (1991) 19–27, 159; translation in Math. Notes **49**:1-2, (1991) 15–20.

Singularities and Noncommutative Geometry

Jean-Paul Brasselet
Institut de Mathématiques de Luminy - CNRS, Marseille
(jpb@iml.univ-mrs.fr)

Introduction

During the last decades, two important discoveries have provided fundamental progress in the areas of geometry and topology: Noncommutative Geometry and Intersection Homology Theory.

The purpose of noncommutative geometry, following Connes, is to extend the correspondence between geometric spaces and commutative algebras to the noncommutative case, in the framework of real analysis. Noncommutative geometry has been developed in particular for spaces such as the space of leaves of a foliation, which corresponds to a noncommutative algebra. Also, the theory provides new points of view for classical spaces. The example of the Gauss-Bonnet formula (see [14, Introduction]) is very instructive. Differential calculus and topology extend through new algebraic tools such as cyclic homology. While having its origins in physics, the theory has many applications in geometry and topology: the Atiyah-Singer index theorem, Atiyah-Hirzebruch topological K-theory, signature of manifolds, the Novikov conjecture, etc...

The Atiyah-Singer index theorem asserts an equality between the analytic index and the topological index of an elliptic differential operator on a smooth compact manifold. The general aim of this paper is to present and to give a contribution to the programme: prove an index theorem for singular varieties. To fulfil such a programme, it is necessary to define and to give sense to the different ingredients which appear in the formulæ of analytic and topological indices, in the singular case. Many authors already proposed generalizations of some of the ingredients and, for the moment, they appear as pieces of a puzzle to be fitted together in order to prove the result. It is impossible to give an exhaustive list: many authors introduced classes of (pseudo)differential operators on manifolds with edges, corners or manifolds with conical singularities, in particular: Brüning-Seeley [7], [8], [9], Lesch [21], [22], Melrose [25], Monthubert [26], Schrohe [27], Schülze, [28], [29], [30].

On another hand definitions of Chern character for singular varieties have been given by Baum-Fulton-MacPherson [1] and Schwartz [31], see

D. Siersma et al. (eds.), New Developments in Singularity Theory, 135–155.

also Kwieciński [20] and Yokura [34]. Recently Suwa [32] gave a nice definition satisfying the suitable properties.

On another hand, the extension of the Chern character to singular varieties has been studied by Baum-Fulton-MacPherson [1], Schwartz [31], Kwieciński [20] and Yokura [34].

The Chern character is in fact one of the crucial ingredients in the definition of the topological index. As indicated by Connes, cyclic homology appears to be the good receptacle for the Chern character (in K-homology). For a compact manifold M, Connes expressed (periodic) cyclic cohomology of the algebra of smooth functions on M in terms of the classical de Rham cohomology of M. The precise aim of this paper is to prove an equivalent of Connes' result in the case of a singular variety defining and using a suitable cyclic homology and a suitable substitute for de Rham cohomology.

This is realized in the framework of intersection homology. Intersection homology defined by Goresky and MacPherson, allows one to recover, for singular varieties, many of the properties which are no longer true using ordinary homology: Poincaré duality, the de Rham theorem, Hodge theory, Morse theory etc... In particular, Cheeger [10], [11] and Cheeger-Goresky-MacPherson [12], then many authors proved in particular cases an isomorphism between $\mathcal{L}^2$-cohomology of differential forms on the regular part of the variety and intersection homology. The shadow forms give an explicit and geometric point of view for such a 'singular' de Rham theorem. Moreover, they provide good intuition for the construction of an algebra of 'controlled functions' and cyclic homology for singular varieties, i.e. considering smooth functions defined on the regular part which have suitable behaviour in the neighbourhood of the singular strata.

In the first section of these notes, we recall briefly the ingredients used in the definition of analytic and topological indices. Section 2 is devoted to the definition of intersection homology, due to Mark Goresky and Robert MacPherson, then we introduce the shadow forms, common work with these two authors [2]. Controlled functions and controlled differential forms on a singular variety are defined using the idea of shadow forms. Restricting ourselves to the case of a cone on a manifold, we obtain a 'mixed complex' whose cyclic homology is related to intersection homology. This part comes from work with André Legrand ([3],[4],[5]). The author thanks IMPA, Rio de Janeiro for hospitality during the preparation of the manuscript.

1. The index theorem.

In this section, we recall the definitions of analytic and topological indices in the case of a smooth manifold, in particular the different ingredients used in their definitions. General references for this section are Hirzebruch [18] and Teleman [33].

Let M be a (compact) differentiable n-dimensional manifold with a Riemannian metric and let E and F be differentiable complex vector bundles over M. We denote by $\Gamma(E)$ and $\Gamma(F)$ the vector spaces of global differentiable sections. A differential operator D of order m is a linear operator $D : \Gamma(E) \to \Gamma(F)$ such that in local coordinates (frames) ε_i and η_j in E and F respectively, one has

$$D(f\varepsilon_i) = \sum_j D_i^j(f)\eta_j \quad \text{with} \quad D_i^j(f)(x) = \sum_{|\alpha| \le m} A_{i\alpha}^j \frac{\partial^{|\alpha|} f}{\partial x^\alpha}(x)$$

for $f \in C^\infty(M)$ and $\alpha = (\alpha_1, \ldots, \alpha_r)$ a multi-index of length $|\alpha| = \alpha_1 + \cdots + \alpha_r$.

The completion $\Gamma_k(E)$ of $\Gamma(E)$ with metric

$$\|\lambda\|_k^2 = \sum_{|\alpha| \le k} \left\| \frac{\partial^{|\alpha|} f}{\partial x^\alpha}(x) \right\|^2$$

is a Hilbert space (the Sobolev space of order k of sections of the bundle E) and the differential operator D of order m induces a bounded operator

$$\Gamma_k(E) \to \Gamma_{k-m}(F)$$

Recall that a (bounded) operator between Hilbert spaces is called Fredholm if $\dim_\mathbb{C} Ker D < \infty$, the image $Im\, D$ of D is closed in $\Gamma(F)$ and $\dim_\mathbb{C} Coker D < \infty$.

DEFINITION 1.1.   *The analytic index of a Fredholm operator D is defined as*

$$i_a(D) = \dim_\mathbb{C} Ker D - \dim_\mathbb{C} Coker D$$

EXAMPLE 1.2.   Consider the classical operators $*$ (Hodge), d and $\delta = - * d*$, then for $D = d + \delta$, we have $i_a(D) = 0$.

The symbol of the operator D corresponds to the highest order terms of the operator. More precisely, denote by T^*M the cotangent bundle of M and by π the projection $\pi : T^*M \to M$. Then the symbol of D, denoted by $\sigma_D : \pi^*E \to \pi^*F$, is the morphism of vector bundles defined by

$$\sigma_D(x,\xi) : [\pi^*E]_{(x,\xi)} \to [\pi^*F]_{(x,\xi)} \qquad \sigma_D(x,\xi)(\omega) = D(f^m \tilde{\omega})(x)$$

where $x \in M$, $\xi \in T_x^*(M)$, $\omega \in E_x$, f is a C^∞ function on M such that $f(x) = 0$ and $(df)_x = \xi$ and $\tilde{\omega}$ is a section of E such that $\tilde{\omega}(x) = \omega$.

Remark. σ_D is a morphism of vector bundles satisfying

$$\sigma_D(x, t\xi) = t^m \sigma_D(x, \xi)$$

EXAMPLE 1.3. $\sigma_d(x, \xi)(\omega) = \xi \wedge \omega$, $\sigma_\delta(x, \xi)(\omega) = -i_\xi(\omega)$ and for the Laplace operator $\Delta = d\delta + \delta d$, $\sigma_\Delta(x, \xi)(\omega) = -\|\xi\|^2(\omega)$.

DEFINITION 1.4.   *The differential operator D is said to be elliptic if $\sigma_D(x, \xi)$ is an isomorphism for any $\xi \neq 0$.*

Denote by $(T^*M)_0$ the complement of the zero section. The first ingredient for the topological index is given by the following proposition.

PROPOSITION 1.5.   *If D is elliptic; it defines a class*

$$[\sigma_D] \in K^0(T^*M, (T^*M)_0) = K_c^0(T^*M)$$

(K_c denotes K-theory with compact supports).

The two other ingredients are the Chern character and the Todd class, which are defined by means of a formal factorisation of the Chern class (see [18], §10). Let E be a complex vector bundle on a manifold X, with Chern class $\sum_{j=0}^q c_j(E)$. In terms of the formal factorisation $\sum_{j=0}^q c_j(E)t^j = \prod_{i=1}^q (1 + \gamma_i t)$, the Chern character is defined by $ch(E) = \sum_{i=1}^q e^{\gamma_i}$ and the Todd class by $\tau(E) = \prod_{i=1}^q \frac{\gamma_i}{1 - e^{-\gamma_i}}$.

On one hand, the Chern character defines a map $ch : K(T^*M) \otimes \mathbb{Q} \to H^*(T^*M; \mathbb{Q})$. On the other hand, denoting by $T_\mathbb{C}^*M$ the complexified bundle of the cotangent bundle, the Todd class of M is defined by $\tau(M) = \tau(\pi^* T_\mathbb{C}^* M)$.

DEFINITION 1.6.   *Denote by $[T^*M] \in H_{2n}(T^*M)$ the fundamental class of the $2n$-dimensional manifold T^*M. The topological index of an elliptic operator D is defined as the evaluation*

$$i_t(D) = (-1)^n \left(ch[\sigma_D] \cup \tau(M) \right) [T^*M].$$

THEOREM 1.7.   (Index theorem) *Let D be an elliptic differential operator. Then the analytic and topological indices coincide : $i_a(D) = i_t(D)$.*

To prove an index theorem in the singular case, it is necessary to define in that context the different ingredients which appear in the formulae of analytic and topological indices. As mentioned in introduction, some of these ingredients have been defined by various authors in particular cases of singular varieties. The aim of the present paper is to provide a suitable general setup in which the formula can be established. This is made in the context of Hochschild homology, following Connes' ideas and using intersection homology.

2. Intersection homology.

We consider a singular variety X which is a pseudovariety. In particular, the singular part has codimension at least 2, and we suppose that X is endowed with a Thom-Mather C^∞ stratification, i.e. a filtration of X by closed subsets

$$X = X_n \supset X_{n-1} \supset X_{n-2} \supset \cdots \supset X_1 \supset X_0 \supset X_{-1} = \emptyset$$

such that the connected components of $X_k - X_{k-1}$ are k-dimensional C^∞-manifolds, denoted by S_i^k and called the strata. For each of them, there exist: an open neighbourhood T_i of S_i^k in X, a continuous retraction π_i of T_i on S_i^k, and a continuous function $\rho_i : T_i \to [0,1[$, such that $S_i^k = \{x \in T_i | \rho_i(x) = 0\}$ and the (T_i, π_i, ρ_i) satisfy Mather's axioms (see [24]).

According to Thom's first isotopy lemma, these data imply the following local triviality condition: every point x in S_i^k admits an open neighbourhood $U_x \subset X$ such that there is a homeomorphism

$$\psi_x : U_x \to B^k \times c(L_x)$$

where B^k is the standard k-dimensional open ball and $c(L_x)$ is the (open) cone over the link L_x. (By definition $c(\emptyset) = \{point\}$). The link is assumed to be stratified and independent of the point $x \in S_i^k$. We stratify the cone $c(L_x)$ with the vertex as the 0-dimensional stratum V_0 and the (open) generators through a stratum of L_x forming a stratum of $c(L_x)$. There is a filtration

$$L_x = L_{n-k-1} \supset L_{n-k-2} \supset \cdots \supset L_0 \supset L_{-1} = \emptyset,$$

and ψ_x carries the stratification of U_x induced by that of X to the product stratification of $B^k \times c(L_x)$.

The parameter of the cone corresponds to the Mather distance function ρ_i. For a complete definition see for instance [16].

We now recall the definition of intersection homology due to M. Goresky and R. MacPherson [15]. Given a stratified singular variety, the idea of intersection homology is to consider only chains and cycles whose intersections with the strata are 'not too big'. The allowed chains and cycles meet the strata with a controlled and defect of transversality. This defect is given by a sequence of integers $\bar{p} = (p_0, p_1, \ldots, p_\alpha, \ldots, p_n)$, called the *perversity*, which satisfies:

$$p_0 = p_1 = p_2 = 0 \qquad \text{and} \qquad p_\alpha \leq p_{\alpha+1} \leq p_\alpha + 1 \quad \text{for } \alpha \geq 2.$$

Let $C_i(X)$ be any 'classical' chain complex on X with integer coefficients, and define the complex of $\bar{p}$-*allowable* chains:

$$IC_i^{\bar{p}}(X) = \{\xi \in C_i(X) \ : \ \dim(|\xi| \cap X_{n-\alpha}) \leq i - \alpha + p_\alpha$$
$$\text{and } \dim(|\partial \xi| \cap X_{n-\alpha}) \leq i - 1 - \alpha + p_\alpha\}.$$

The *intersection homology groups* (with $\mathbb{Z}$ coefficients) $IH_i^{\bar{p}}(X)$ are the homology groups of this complex.

Note that if ξ and $X_{n-\alpha}$ are transverse, then their intersection dimension is precisely $i - \alpha$, so the perversity corresponds to an allowed defect of transversality.

Intersection homology does not satisfy all properties of classical homology. For instance, intersection homology is not a homotopy invariant. The main property satisfied by intersection homology is Poincaré duality. If $\bar{p}$, $\bar{q}$ and $\bar{r}$ are perversities such that $\bar{p} + \bar{q} \leq \bar{r}$, there is an unique pairing

$$IH_i^{\bar{p}}(X) \times IH_j^{\bar{q}}(X) \to IH_{i+j-n}^{\bar{r}}(X)$$

realized by intersection of cycles.

THEOREM 2.1. (Poincaré duality [15]) *Let X be a compact stratified pseudovariety, let $\bar{p}$ and $\bar{q}$ be complementary perversities (i.e. $p_\alpha + q_\alpha = t_\alpha = \alpha - 2$ for $\alpha \geq 2$), then the augmented pairing*

$$IH_i^{\bar{p}}(X; \mathbb{Q}) \times IH_{n-i}^{\bar{q}}(X; \mathbb{Q}) \to IH_0^{\bar{t}}(X; \mathbb{Q}) \overset{\varepsilon}{\to} \mathbb{Q}$$

is nondegenerate.

Intersection homology satisfies the relative exact sequence and the Mayer-Vietoris exact sequence but the Künneth formula does not hold in general. It holds, for example, if one of the varieties involved in the product is smooth (Goresky).

In section 4, we will define complexes of sheaves on the pseudovariety X and will prove that their hypercohomology is equal to the intersection homology of X. This is a consequence of the Theorem 2.2: if a complex of sheaves on X satisfies the so-called *perverse sheaves* axioms (see [16]), then the hypercohomology of (X with value in) this perverse sheaf is the intersection homology of X. In fact, the main axioms of perverse sheaves originate from the following local property: Let x be a point in an α-codimensional stratum, and $U_x \cong B^{n-\alpha} \times c(L_x)$ a distinguished neighbourhood of x. Let $\bar{p}$ be any perversity, then:

$$IH_i^{\bar{p}}(U_x) \cong \begin{cases} IH_i^{\bar{p}}(L_x) & i < \alpha - p_\alpha - 1 \\ 0 & i \geq \alpha - p_\alpha - 1. \end{cases}$$

Let us give a sketch of the easy and very instructive proof of this local computation: The Künneth formula (true in this case) implies

$$IH_i^{\bar{p}}(U_x) \cong IH_i^{\bar{p}}(B^{n-\alpha} \times c(L_x)) \cong IH_i^{\bar{p}}(c(L_x))$$

where $c(L_x)$ is an α-dimensional pseudovariety V, with the stratification induced from that of L_x.

(i) If $i < \alpha - p_\alpha - 1$ and ξ is an allowed i-cycle in $c(L_x)$, then

$$\dim(|\xi| \cap V_0) \leq i - \alpha + p_\alpha < -1$$

so ξ does not contain the vertex of the cone. We can deform ξ along the generatrices of the cone to an allowed i-cycle of the base L_x. We obtain in this way a cycle in $IC_i^{\bar{p}}(L_x)$ homologous to ξ by an allowed homology.

(ii) If $i \geq \alpha - p_\alpha - 1$ and ξ is an allowed i-cycle in $c(L_x)$, then the cone on ξ with vertex $\{x\}$, denoted by $c(\xi)$, is an allowed $i+1$-chain. We have, in particular

$$0 = \dim(|c(\xi)| \cap V_0) \leq (i+1) - \alpha + p_\alpha.$$

The chain $\xi = \partial(c(\xi))$ is an allowed boundary in $IC_i^{\bar{p}}(c(L_x))$ and is homologous to 0. $\qquad\qquad\square$

Let us now denote, in general $U_j = X - X_{n-j}$ $\quad (2 \leq j \leq n+1)$ and by i_j the natural inclusion $i_j : U_j \hookrightarrow U_{j+1}$. For every closed subset F of X, let i be the inclusion $i : X - F \hookrightarrow X$. If $\mathcal{F}^\bullet$ is a complex of sheaves on X, the composition :

$$\mathcal{F}^\bullet \longrightarrow i_* i^* \mathcal{F}^\bullet \overset{\mu}{\longrightarrow} Ri_* i^* \mathcal{F}^\bullet$$

is called the *attaching map*. It is a quasi-isomorphism at all $x \in X - F$. If $\mathcal{F}^\bullet$ is a complex of soft sheaves, μ is a quasi-isomorphism for all $x \in X$.

THEOREM 2.2. [16] *Let $\mathcal{F}^\bullet$ be a complex of fine (or soft) sheaves on X satisfying the following axioms:*

1. $\mathcal{F}^\bullet$ is a bounded complex, $\mathcal{F}^k = 0$ for $k < 0$, and $\mathcal{F}^\bullet_{U_2} = \mathbb{C}$,

2. for all $x \in X_{n-j} - X_{n-j-1}$, $H^k(\mathcal{F}^\bullet_x) = 0$, for $k > q_j$, $(j = 2, \ldots, n)$,

3. the attaching map $\mathcal{F}^k|_{U_{j+1}} \to Ri_{j*}\mathcal{F}^k|_{U_j}$ is a quasi-isomorphism for $k \leq q_j$,

then the hypercohomology groups (with compact support) $\mathbb{H}^k_c(X; \mathcal{F}^\bullet)$ are isomorphic to $Hom(IH^{\bar{p}}_k(X); \mathbb{R})$, where $\bar{p}$ is the complementary perversity.

3. The de Rham theorem.

The Whitney forms, defined by Whitney in the smooth case in order to prove the de Rham theorem, are the main tool used in the definition of shadow forms, providing a de Rham theorem in the singular case.

3.1. THE SMOOTH CASE: WHITNEY FORMS

In the case of a compact manifold M, the integration map:

$$\Omega^r(M) \to C^r(M) \qquad \omega \mapsto \left(c \to \int_c \omega\right)$$

from differential forms to singular cochains (with respect to smooth chains c in $C_r(M)$) gives rise to the de Rham isomorphism in cohomology. The proof given by Whitney [35] uses the following Whitney map, providing a homotopy inverse to the integration map.

Let K be a triangulation of M and σ an r-simplex of K, with vertices v^i, $1 \leq i \leq r+1$. Then barycentric coordinates on σ are functions x_i such that for every point x in σ, $x = \sum_{i=1}^{r+1} x_i v^i$ with $\sum_{i=1}^{r+1} x_i = 1$ and $0 \leq x_i \leq 1$.

In terms of such coordinates, the Whitney form associated to σ is defined by

$$W(\sigma) = r! \sum_{j=1}^{r+1}(-1)^{j+1} x_j dx_1 \wedge \cdots \wedge \widehat{dx_j} \wedge \cdots \wedge dx_{r+1}.$$

In fact, the Whitney construction uses smoothing of the barycentric coordinate functions as these are piecewise linear. The Whitney map associates to the elementary r-cochain whose value is 1 on the r-simplex σ and 0 on other simplices, the Whitney form $W(\sigma)$.

Composed with Poincaré duality, the de Rham isomorphism gives isomorphisms

$$H_{\mathrm{dR}}^{n-k}(M) \cong H^{n-k}(M) \cong H_k(M) \qquad \text{(we put } r = n - k\text{)}.$$

3.2. THE SINGULAR CASE: SHADOW FORMS

In the remainder of this section, X will be a n-pseudomanifold triangulated by a simplicial complex K linearly embedded in a euclidean space.

The idea of shadow forms is to extend the Whitney map to the singular case, providing an isomorphism between the (de Rham) cohomology of a suitable complex of differential forms defined on the regular part of X and the intersection homology of X (there is no longer a Poincaré isomorphism as X is singular).

Denote by Σ the singular part of X. The construction of shadow forms associates explicitly a differential $(n - k)$-form $\omega(\xi)$ on $X - \Sigma$ to any k-simplex ξ of the barycentric subdivision K' of K, transverse to the $(n - 1)$-skeleton of K. There is a natural stratification of X corresponding to the filtration of K by skeleta of successive dimensions. The shadow forms satisfy:

     i) if ξ is transverse to a stratum of K, then $\omega(\xi)$ is defined and regular on this stratum,

     ii) if ξ meets an α-codimensional stratum S with a failure of transversality p_α, then the shadow form $\omega(\xi)$ has a pole of maximum order p_α on the stratum S.

We will recall briefly the construction of [2] on the standard n-simplex and we will make precise the behaviour of shadow forms in the neighbourhood of singular strata.

Let $\Delta = \Delta^n$ be the standard n-simplex. The shadow forms are defined for k-simplices ξ of the barycentric subdivision Δ' which do not lie in the boundary of Δ. Such a barycentric subdivision can be realized for each point p in the interior of Δ, requiring that p is the barycenter of Δ and that for each pair $F' < F$ of faces of Δ, the barycenters of F, F' and of the face opposite to F' in F are collinear. The corresponding barycentric subdivision of Δ will be denoted by $\Delta'(p)$. Every k-simplex ξ of the (abstract) barycentric subdivision Δ' admits a geometrical realisation $\xi(p)$ in $\Delta'(p)$. For c a singular $(n - k)$-chain with support in the interior of Δ, the shadow $S_\xi(c)$ cast by c with respect to ξ is the set of all points p such that $\xi(p)$ intersects c.

THEOREM 3.1. [2] *The current defined by $c \mapsto$ volume $S_\xi(c)$ is realizable by a differential form $\omega(\xi)$ (the shadow form), i.e. $\omega(\xi)$ satisfies*

$$\int_c \omega(\xi) = \text{ volume } S_\xi(c).$$

We give an explicit formula for $\omega(\xi)$, allowing one to determine explicitly the order of its poles. The Whitney form associated to a face of Δ with vertices $v^{i_1}, \ldots, v^{i_{r+1}}$ is the differential form defined on Δ by

$$W_\Delta(x_{i_1}, \ldots, x_{i_{r+1}}) = r! \sum_{j=1}^{r+1} (-1)^{j+1} x_{i_j} dx_{i_1} \wedge \cdots \wedge \widehat{dx_{i_j}} \wedge \cdots \wedge dx_{i_{r+1}}.$$

We will denote it also by $W_\Delta(J)$, J being the set of indices $\{i_1, \ldots, i_{r+1}\}$.

For example, in Δ of any dimension, $W_\Delta(x_1, x_2) = x_1 dx_2 - x_2 dx_1$.

Let ξ be a k-simplex of Δ', not lying in the boundary of Δ, the vertices of ξ are barycenters of incident faces of Δ:

$$F_1 < F_2 < \cdots < F_{k+1} = \Delta.$$

We denote the sets of indices of vertices as follows. J_1 is the set of indices of vertices in F_1 and, in general, J_ℓ is the set of indices of vertices in F_ℓ, which are not in $F_{\ell-1}$. The set of indices $J_1, \ldots, J_{k+1}$ defines a partition of $\{1, \ldots, n+1\}$. We write ξ as

$$\xi = (\{x_i\}_{i \in J_1})(\{x_i\}_{i \in J_2}) \cdots (\{x_i\}_{i \in J_{k+1}}) \qquad (3.1)$$

and we denote $s_j = \text{card}(J_j)$ and $\lambda_j = \sum_{i \in J_j} x_i$. Also, we define the $(k+1)$-tuples $s = (s_1, \ldots, s_{k+1})$ and $\lambda = (\lambda_1, \ldots, \lambda_{k+1})$.

Define the function $\phi_s(u_1, \ldots, u_k)$ by

$$\phi_s(u_1, \ldots, u_k) = (u_1)^{s_1-1} \cdots (u_k)^{s_k-1}(1 - u_1 - \cdots - u_k)^{s_{k+1}-1}$$

and the domains

$$R = \{(u_1, \ldots, u_k) \in \mathbf{R}^k : u_1 \geq 0, \ldots, u_k \geq 0, \sum_{i=1}^{k} u_i \leq 1\},$$

$$R(\lambda) = \{(u_1, \ldots, u_k) \in R : 0 \leq \frac{u_1}{\lambda_1} \leq \cdots \frac{u_k}{\lambda_k} \leq \frac{1 - u_1 - \cdots - u_{k-1}}{1 - \lambda_1 - \cdots - \lambda_{k-1}}.$$

The following function is defined in terms of Dirichlet type integrals (in fact a function of the λ_j's):

$$\Psi(x_1, \ldots, x_{n+1}) = \frac{\int_{R(\lambda)} \phi_s(u_1, \ldots, u_k) du_k \cdots du_1}{\int_R \phi_s(u_1, \ldots, u_k) du_k \cdots du_1}.$$

We can now give the explicit expression of the shadow forms:

THEOREM 3.2. [2] *The shadow form $\omega(\xi)$ associated to the k-simplex (3.1) is given by*

$$\omega(\xi) = \varepsilon(s) \frac{W_\Delta(J_1)}{(\lambda_1)^{s_1}} \wedge \cdots \wedge \frac{W_\Delta(J_{k+1})}{(\lambda_{k+1})^{s_{k+1}}} \Psi(x_1, \ldots, x_{n+1})$$

where $\varepsilon(s) = (-1)^{s_2 + 2s_3 + \cdots k s_{k+1}}$.

For example, in Δ^2:

$$\xi_1 = (x_1, x_2)(x_3) \qquad \omega(\xi_1) = -W_\Delta(x_1, x_2),$$

$$\xi_2 = (x_1)(x_2, x_3) \qquad \omega(\xi_2) = W_\Delta(x_2, x_3) \frac{x_1(2 - x_1)}{(1 - x_1)^2},$$

$$\xi_3 = (x_1)(x_2)(x_3) \qquad \omega(\xi_3) = \frac{x_1 x_2}{1 - x_1}.$$

Remark that the simplices ξ_1 and ξ_3 are transverse to all faces of Δ, and the shadow forms $\omega(\xi_1)$ and $\omega(\xi_3)$ are smooth. The simplex ξ_2 is not transverse to the vertex $\{v^1\}$ and the shadow from $\omega(\xi_2)$ has a pole of order 1 in this point.

The explicit formula for a shadow form provides its behaviour in the neighbourhood of the singular strata, with respect to the position of ξ relative to these strata. The position of ξ relative to the faces of Δ is given by its profile. The profile p_ξ is a function graph defined as follows:

1) We mark successive points ($\bullet$) on the real axis, 0, then s_{k+1}, $s_{k+1} + s_k, \cdots, s_{k+1} + s_k + \cdots + s_2$,

2) The starting point of the graph is the origin of the real axis,

3) Each time we reach a marked point ($\bullet$) of the real axis, the graph goes on the right for one unit, then grows along a parallel to the first diagonal until reaching the next marked point ($\bullet$).

The profile gives the bound of the intersection dimension of the k-simplex ξ and faces of Δ. For a face F with codimension α, we have:

$$\dim(\xi \cap F) \leq (k - \alpha + p_\xi(\alpha)). \tag{3.2}$$

We remark that, if ξ and F are transverse and intersect, the dimension of $\xi \cap F$ is $k - \alpha$ and there are faces which realize the equality.

As ξ is transverse to the $(n-1)$-skeleton of Δ, the profile satisfies all conditions to be a perversity. In fact, the profile is the smallest perversity for which ξ is an allowable chain. On a face F of codimension α such that (3.2) is an equality, the shadow form $\omega(\xi)$ admits a pole of order $p_\xi(\alpha)$ and that is the maximum order possible for poles of $\omega(\xi)$.

THEOREM 3.3. ([2] Corollary 9.3) *Let X be a triangulated pseudovariety, let $1 \leq q \leq \infty$ and write $\bar{p}_q$ for the largest perversity such that $\bar{p}(\alpha) < \alpha/q$ for all α. The shadow forms corresponding to simplices ξ such that $p_\xi \leq \bar{p}_q$ generate a complex whose cohomology is the intersection homology $IH_*^{\bar{p}_q}(X)$.*

Thus shadow forms provide a good intuition for what can substitute for the de Rham complex in regard to Hochschild homology.

4. Controlled differential forms

We return to the situation of a singular pseudovariety endowed with a Thom-Mather stratification. Intersection differential forms will be 'controlled' differential forms with poles of maximum fixed order along every stratum of the singular part (the order depends only on the stratum). The complex generated by these forms is an intersection complex whose perversity is related to the order of poles. The construction (Brasselet-Legrand [4] [5]) is mainly inspired by the properties of shadow forms.

4.1. THE CASE OF THE CONE

Remarking that the cone $c(*)$ over a point is $[0, 1[$, we define the algebra

$$A = \{a \in C^\infty(]0, 1[) \; : \; \forall \, k \; \sup_{r \in]0,1[} |(r\partial_r)^k a(r)| < +\infty\}$$

and, for $\beta \in \mathbb{R}$, the module of differential forms

$$B_\beta^* = r^{-\beta} \left(A \oplus A \frac{dr}{r} \right) \subset \Omega^*(]0, 1[) .$$

Here, we have to be cautious: in order to show that controlled geometry is naturally associated to an algebra of differential functions, we need topological versions of the standard algebraic constructions of homological algebra. In particular, for the natural Fréchet topology, the set of C^∞ functions with compact support in $]0, 1[$ is not dense in A. This is true for the locally convex topology and that is the reason why we use locally convex spaces, injective and projective tensor products (see [17]). This is carefully done in [4] where the reader will find suitable justifications.

For each graded A-module $\mathcal{D}^*$ of locally convex spaces we define the complex

$$\left(B_\beta^* \, \hat{\otimes}_A \, \mathcal{D}^* \right)^k = r^{-\beta} \left(\mathcal{D}^k \oplus \frac{dr}{r} \wedge \mathcal{D}^{k-1} \right)$$

with the adapted locally convex topology and injective tensor product.

Let L be a smooth $(n-1)$-dimensional compact Riemannian manifold. We can now define an intersection complex of differential forms on the cone $c(L) = L \times [0,1]/L \times \{0\}$ on L.

DEFINITION 4.1. [5] *The algebra of intersection functions on the cone cL on a smooth manifold L is the 'algebraic cone' of $C^\infty L$:*

$$IC^\infty(cL) = rA \,\widehat{\otimes}\, C^\infty L + A.$$

In the de Rham subalgebra of $\Omega^*(]0,1[\times L)$ generated by elements $f \in IC^\infty(cL)$ and dg where $g \in IC^\infty(cL)$, we consider the subalgebra $\widetilde{\Omega}^*(cL)$ of differential forms which do not contain terms with element dr. The associated de Rham complex of forms with poles of order β is defined by

$$I\Omega_\beta^*(cL) = B_\beta^* \,\widehat{\otimes}_A\, \widetilde{\Omega}^*(cL).$$

PROPOSITION 4.2. [5] *(Poincaré lemma for $I\Omega_\beta^*(cL)$). Let $\beta > 0$, $\beta \notin \mathbb{N}$, then*

$$H^k(I\Omega_\beta^*(cL)) = \begin{cases} H^k(\Omega^*L) & \text{if } k \leq [\beta] - 1 \\ 0 & \text{if } k > [\beta] - 1. \end{cases}$$

There are (at least) two ways to prove such a result, the first one geometric, inspired by classical proofs of the Poincaré lemma (see [10]); the second algebraic, using spectral sequences (cf [5]). We give a sketch of the first. Decompose ω as $\omega = \eta + dr \wedge \phi$, and define, for $r_0 > 0$, a homotopy operator

$$\mathcal{K}_{r_0}(\omega) = \int_{r_0}^r \phi \, dr.$$

It is easy to see that

$$(\mathcal{K}_{r_0}d + d\mathcal{K}_{r_0})(\omega) = \omega - \omega(r_0)$$

If $k \leq \beta - 1$, then $\mathcal{K}_{r_0}(\omega) \in I\Omega_\beta^{k-1}(cL)$ and $d\mathcal{K}_{r_0}(\omega) \in I\Omega_\beta^k(cL)$.

If $k > \beta - 1$, then, for r_0 going to 0, the integral $\mathcal{K}_0(\omega)$ converges and $\mathcal{K}_{r_0}d(\omega)$ converges to $\mathcal{K}_0 d(\omega)$ in $I\Omega_\beta^*(cL)$. As $\omega(r_0)$ converges to 0 (the coefficients are bounded by $Cr_0^{k-\beta}$), we obtain $(\mathcal{K}_0 d + d\mathcal{K}_0)(\omega) = \omega$ and the Proposition.

4.2. A MODEL OF STRATIFIED SPACE.

The elementary example of stratified variety satisfying the Thom-Mather conditions which is our model for the atlas charts of singular varieties is the ℓ-th iterated cone $V = M_0 \times c\,(M_1 \times c\,(M_2 \times \cdots \times c(M_\ell)\cdots))$ on a family of smooth manifolds $(M_i)_{0 \leq i \leq \ell}$.

The regular stratum is

$$V_{\text{reg}} = M_0 \times]0,1[\times M_1 \times]0,1[\times M_2 \times \cdots \times]0,1[\times M_\ell.$$

The singular strata are $S^{k_i} = M_0 \times]0,1[\times M_1 \times]0,1[\times M_2 \times \cdots \times]0,1[\times M_i$ where $0 \leq i < \ell$ and the corresponding link is the iterated cone

$$N_i = M_{i+1} \times c\,(M_{i+2} \times \cdots \times c(M_\ell)\cdots) = M_{i+1} \times c(N_{i+1})$$

for $0 \leq i < \ell - 1$ and $N_{\ell-1} = M_\ell$.

We now extend the definition of the algebra $IC^\infty(cL)$ of intersection functions on a cone cL (Definition 4.1) to an iterated cone $V = M_0 \times c(M_1) \times \ldots \times c(M_\ell)\ldots$.

DEFINITION 4.3. [5] *The intersection function algebra on the iterated cone V, denoted by $IC^\infty(V)$, is the locally convex algebra of differentiable functions defined iteratively by*

$$IC^\infty(N_{\ell-1}) = C^\infty(M_\ell), \quad \text{and}$$

$$IC^\infty(N_{i-1}) = C^\infty(M_i) \,\hat{\otimes}\, c(IC^\infty(N_i)).$$

To define the de Rham complex of differential forms with polar coefficients, we use the following definition of the cone over a positively graded A-module of locally convex spaces C^*:

$$(cC)^k = r^{k+1} A \,\hat{\otimes}\, C^k + r^k A \,\hat{\otimes}\, dC^{k-1}.$$

DEFINITION 4.4. [5] *For any sequence $\bar{\gamma} = (\gamma_1, \gamma_2, \ldots, \gamma_\ell)$ of real numbers, the intersection de Rham differential module $I\Omega^*_{\bar{\gamma}}(V)$ of forms on $IC^\infty(V)$, with poles of order less than $\bar{\gamma}$, is defined iteratively by:*

$$\Omega^*_\emptyset(N_{\ell-1}) = \Omega^*(M_\ell) \quad \text{and}$$

$$I\Omega^*_{(\gamma_{i+1},\ldots,\gamma_\ell)}(N_{i-1}) = \Omega^*(M_i) \,\hat{\otimes}\, \left(B^*_{\gamma_{i+1}} \,\hat{\otimes}_{A_{i+1}}\, c\,[I\Omega^*_{(\gamma_{i+2},\ldots,\gamma_\ell)}(N_i)] \right).$$

4.3. THE CASE OF STRATIFIED SPACES.

In order to generalize the previous constructions to stratified varieties, we will use atlases whose charts are iterated cones. Let x be in a stratum $S_{n-\alpha_1}$ of the stratified space X and $\psi_x : U_x \overset{\cong}{\to} B^{n-\alpha_1} \times cL_x$ a distinguished open neighbourhood of x. The link L_x is itself a singular variety covered by distinguished open sets of the same type. By iteration, we obtain a system of *iterated* cone charts of the type:

$$W_{\bar{\alpha}} = \mathbf{B}^{n-\alpha_1} \times c\left(\mathbf{B}^{\alpha_1-1-\alpha_2} \times c\left(\mathbf{B}^{\alpha_2-1-\alpha_3} \times \cdots \times c(L^{\alpha_\ell-1})\cdots\right)\right)$$

where $\bar{\alpha} = \{n+1 = \alpha_0 > \alpha_1 > \alpha_2 > \cdots > \alpha_\ell > \alpha_{\ell+1} = 0\}$ denotes a decreasing sequence of integers, $\mathbf{B}^{\alpha_t-1-\alpha_{t+1}}$ is an open ball in $\mathbb{R}^{\alpha_t-1-\alpha_{t+1}}$ (possibly a point) and $L^{\alpha_\ell-1}$ is a finite disjoint union of $(\alpha_\ell - 1)$-dimensional balls. Via the homeomorphism ψ_x, this chart corresponds to the following chain of elements of the filtration of X:

$$\emptyset = X_{n-\alpha_0} \subset X_{n-\alpha_1} \subset X_{n-\alpha_2} \subset \cdots \subset X_{n-\alpha_\ell} \subset X = X_n = X_{n-\alpha_{\ell+1}} \ .$$

We will denote in the same way the iterated cone and its image in X. We obtain in this way a covering of U_x by charts which are iterated cones and correspond to different sequences $\bar{\alpha}$. Each chart $W_{\bar{\alpha}}$ is identified with a model iterated cone $V = M_0 \times c\,(M_1 \times \cdots \times c(M_\ell)\cdots)$.

Let us fix a sequence of real numbers $\bar{\beta} = (\beta_2, \beta_3, \ldots, \beta_n)$, each β_α being associated with the stratum $S_{n-\alpha}$. Given an iterated cone chart $W_{\bar{\alpha}}$, hence a sequence $\bar{\alpha}$ as above, we associate the sequence $\bar{\beta}_{\bar{\alpha}} = (\beta_{\alpha_1}, \beta_{\alpha_2}, \ldots, \beta_{\alpha_\ell})$, i.e. such that $(\bar{\beta}_{\bar{\alpha}})_i = \beta_{\alpha_i}$.

DEFINITION 4.5. *Let X be a stratified singular variety endowed with an atlas $(W_{\bar{\alpha}})$ of iterated cones. We denote by $\mathcal{I}\Omega^k_{\bar{\beta}}$ the sheaf on X whose sections over every open set U of X are C^∞-differential forms ω on $U \cap X_{\text{reg}}$ such that, for each iterated cone $W_{\bar{\alpha}}$, $\omega \in I\Omega^*_{\bar{\beta}_{\bar{\alpha}}}(W_{\bar{\alpha}} \cap U)$.*

The sheaf $\mathcal{I}\Omega^k_{\bar{\beta}}$ satisfies the following Poincaré lemma [4].

PROPOSITION 4.6. *Let cX be the cone on a stratified $(n-1)$-variety X. Let $\bar{\beta}$ be a sequence of positive numbers as previously defined. Assume that no β_α is an integer. Then, there is an isomorphism*

$$H^k(\mathcal{I}\Omega^*_{\bar{\beta}_{\bar{\alpha}}}(cX)) = \begin{cases} H^k(\mathcal{I}\Omega^*_{\bar{\beta}_{\bar{\alpha}}}(X)) & \text{if } k \leq [\beta_n] - 1 \\ 0 & \text{if } k > [\beta_n] - 1. \end{cases}$$

As a consequence, we obtain the following de Rham theorem (see [3] and [5]):

THEOREM 4.7. *Let X be a stratified singular variety, endowed with an atlas of iterated cones. Let $\bar{\beta}$ be a sequence of positive real numbers, satisfying the perversity condition:*

$$[\beta_\alpha] \leq [\beta_{\alpha+1}] \leq [\beta_\alpha] + 1 \ ,$$

for all $2 \leq \alpha \leq n$ and such that no β_α is an integer. Then,

1. $\mathcal{I}\Omega^*_{\bar{\beta}}$ is a perverse sheaf for the perversity $\bar{q}$ defined by $q_\alpha = [\beta_\alpha] - 1$,

2. there is an isomorphism

$$H^k(\mathcal{I}\Omega^*_{\bar{\beta}}(X)) \cong Hom(IH^{\bar{p}}_k(X, \mathbb{R}); \mathbb{R}),$$

where $\bar{p}$ is the complementary perversity of $\bar{q}$.

5. Hochschild and cyclic homology.

In this section, we recall the definitions of Hochschild and cyclic homology: the basic references are [14] and [23]. Let Λ be a ring and $\mathcal{A}$ be a Λ-algebra with unit, so $\Lambda \subset \mathcal{A}$. For any $\mathcal{A}$-bimodule N, the Hochschild complex $(C_*(\mathcal{A}, N), b)$ with coefficients in N is defined by $C_k(\mathcal{A}, N) = N \otimes_\Lambda \mathcal{A}^{(\otimes^k_\Lambda)}$, with boundary map

$$b(n \otimes a_1 \otimes \cdots \otimes a_k) = na_1 \otimes a_2 \otimes \cdots \otimes a_k +$$

$$\sum_{j=1}^{k-1} (-1)^j n \otimes a_1 \otimes \cdots \otimes a_j a_{j+1} \otimes \cdots \otimes a_k$$

$$+ (-1)^k a_k n \otimes a_1 \otimes \cdots \otimes a_{k-1}. \quad (5.3)$$

Its homology, called Hochschild homology, is denoted by $HH_*(\mathcal{A}, N)$.

When $N = \mathcal{A}$, the Hochschild complex is denoted by $C_*(\mathcal{A})$ and Hochschild homology by $HH_*(\mathcal{A})$. In that case, $C_*(\mathcal{A})$ is a cyclic module, i.e. it admits a cyclic action

$$\tau(a_0 \otimes \cdots \otimes a_k) = (-1)^k a_k \otimes a_0 \otimes a_1 \otimes \cdots \otimes a_{k-1}.$$

We have $\tau^{k+1} = id$, so, for each k, τ defines an action of $\mathbb{Z}/(k+1)\mathbb{Z}$ on $C_k(\mathcal{A})$.

If Λ is a field of characteristic 0, the Connes cyclic homology of $\mathcal{A}$, denoted by $HC_*(\mathcal{A})$, is the homology of the complex $(C_*(\mathcal{A})/(1-\tau), b)$ where b is induced by the Hochschild boundary. The relation between Hochschild and cyclic homology is given by the Connes exact sequence

$$\cdots \to HH_k(\mathcal{A}) \overset{I}{\to} HC_k(\mathcal{A}) \overset{S}{\to} HC_{k-2}(\mathcal{A}) \overset{B}{\to} HH_{k-1}(\mathcal{A}) \to \cdots$$

Using the so called *periodicity operator* S, we define the periodic cyclic homology (odd or even)

$$PHC_{o/e}(\mathcal{A}) = \lim_k \left[HC_k(\mathcal{A}) \overset{S}{\to} HC_{k-2}(\mathcal{A}) \right].$$

The importance of the above definitions is shown by the following result. Let M be a compact C^∞ manifold, $\mathcal{A} = C^\infty(M)$ the Fréchet algebra of differentiable functions on M, and $\Omega^*(M)$ the associated de Rham algebra. Replace everywhere $\otimes$ by the projective tensor product $\hat{\otimes}$ (Grothendieck [17]). Then the map $\pi : C_k(C^\infty(M)) \to \Omega^k(M)$ defined by $\pi(f_0 \otimes \cdots \otimes f_k) = f_0 df_1 \wedge \cdots \wedge df_k$ induces the following isomorphisms [13]:

$$HH_*(C^\infty(M)) \cong \Omega^*(M); \quad PHC_{o/e}(C^\infty(M)) \cong \oplus_{o/e} H^*_{dR}(M). \quad (5.4)$$

From this result, Connes recovers the de Rham complex of the manifold X without using the commutativity of the algebra $\mathcal{A}$, which is needed in the construction of the exterior algebra.

If Λ is a ring, as it will be the case in the singular framework, the definition of cyclic homology needs a more general setting, the notion of mixed complex that we briefly describe now (see [19]).

A *mixed complex* is a triple (E_*, b, B), where E_* is a graded module and b, resp. B, is a differential on E_* of degree -1, resp. $+1$, such that $bB + Bb = 0$. It defines a bicomplex $E_*[u]$ with differentials $b(xu^k) = (bx)u^k$, $B(xu^k) = (Bx)u^{k-1}$ where $\text{degree}(u) = 2$. We define the *Hochschild homology* of (E_*, b, B) as $HH_*(E_*) = H_*(E_*, b)$ and the *cyclic homology* as $HC_*(E_*) = H_*(E_*[u], b + B)$. There is again a Connes exact sequence and we can define the periodic cyclic homology $PHC(E_*)$.

The Hochschild complex $C_*(\mathcal{A})$ admits a mixed complex structure where b is the Hochschild boundary and B is the operator defined by

$$B(a_0 \otimes \cdots \otimes a_k) = \sum_{j=0}^{k-1} (-1)^{kj} 1 \otimes a_j \otimes \cdots \otimes a_k \otimes a_0 \otimes \cdots \otimes a_{j-1}$$

$$- (-1)^{k(j-1)} a_{j-1} \otimes 1 \otimes a_j \otimes \cdots \otimes a_k \otimes a_0 \otimes \cdots \otimes a_{j-2}. \quad (5.5)$$

Let us remark that, when Λ is a field of characteristic 0, we recover the previous definition of cyclic homology of $\mathcal{A}$ in the following way. Replacing the quotient $C_k(\mathcal{A})/(1 - \tau)$ by a $(\mathbb{Z}/(k+1)\mathbb{Z})$-free resolution of $C_k(\mathcal{A})$, we obtain a bicomplex which is quasi-isomorphic to the bicomplex associated to the mixed complex structure of $C_*(\mathcal{A})$. Then the two definitions of cyclic homology agree.

6. The singular case.

The aim of this section is to prove a version of Connes' result (5.4). This is performed in the case of a cone on a smooth manifold. The general result requires more sophisticated techniques (see [5]).

In the case of the cone, we use a slight generalization of the previous definition, taking $\Lambda = A$, $\mathcal{A} = IC^\infty(cL)$ and we define a mixed complex structure on the Hochschild complex $C_*^A(IC^\infty(cL), B_\beta^*)$ of $IC^\infty(cL)$ with coefficients in the differential module B_β^* (see [4]).

More precisely, consider the A-Hochschild complex of the Fréchet algebra $IC^\infty(cL)$,

$$C_k^A(IC^\infty(cL)) = IC^\infty(cL)\hat{\otimes}_A \cdots \hat{\otimes}_A IC^\infty(cL) \qquad (k+1)\text{-terms}$$

and denote by b_A its differential. We define the *Hochschild intersection complex* by

$$IC_k^\beta(cL) = C_k^A(IC^\infty(cL), B_\beta^*) = B_\beta^* \hat{\otimes}_A C_{k-*}^A(IC^\infty(cL)).$$

The elements of degree k are sums of terms

$$r^{-\beta} f_0 \otimes f_1 \otimes \cdots \otimes f_k + r^{-\beta}\frac{dr}{r} g_0 \otimes g_1 \otimes \cdots \otimes g_{k-1}$$

such that $f_i, g_j \in IC^\infty(cL)$. The differential b is given by

$$
\begin{array}{ccccc}
IC_k^\beta(cL) & = & r^{-\beta} C_k^A(IC^\infty(cL)) & \oplus & r^{-\beta}A\frac{dr}{r}\hat{\otimes}_A C_{k-1}^A(IC^\infty(cL)) \\
\uparrow b & & b_A^{k+1} \uparrow\downarrow B_A^k & \searrow d_r & b_A^k \uparrow\downarrow B_A^{k-1} \\
IC_{k+1}^\beta(cL) & = & r^{-\beta} C_{k+1}^A(IC^\infty(cL)) & \oplus & r^{-\beta}A\frac{dr}{r}\hat{\otimes}_A C_k^A(IC^\infty(cL))
\end{array}
$$

where the operators b_A and B_A are defined in (5.3) and (5.5), with cyclic action defined by

$$\tau(r^{-\beta} g_0 \otimes \cdots \otimes g_k) = (-1)^k r^{-\beta} g_k \otimes g_0 \otimes \cdots \otimes g_{k-1}$$

and d_r is the following derivation

$$d_r(r^{-\beta} f_0 \otimes f_1 \otimes \cdots \otimes f_k) =$$

$$(-1)^k r^{-\beta}\frac{dr}{r} \wedge \left[(-\beta) f_0 \otimes \cdots \otimes f_k + \sum_{i=0}^k f_0 \otimes \cdots \otimes \partial_r f_i \otimes \cdots \otimes f_k \right].$$

So $(b_A^* \oplus b_A^{*-1})$ has degree -1 and $(B_A^* \oplus B_A^{*-1} + d_r)$ has degree $+1$.

LEMMA 6.1. *The following triple is a mixed complex:*

$$(IC_*^\beta(cL), \ b = b_A^* \oplus b_A^{*-1}, \ B = B_A^* \oplus B_A^{*-1} + d_r)$$

Denote by $\pi : IC_*^\beta(cL) \to I\Omega_\beta^*(cL)$ the morphism defined by

$$\pi(r^{-\beta} f_0 \otimes f_1 \otimes \cdots \otimes f_k + r^{-\beta}\frac{dr}{r} g_0 \otimes g_1 \otimes \cdots \otimes g_{k-1})$$

$$= r^{-\beta} f_0 d_L f_1 \wedge \cdots \wedge d_L f_k + r^{-\beta}\frac{dr}{r} \wedge g_0 d_L g_1 \wedge \cdots \wedge d_L g_{k-1}$$

and consider the kernel (K_*, b, B) of π equipped with the induced mixed complex structure.

Denote by C_* the Hochschild complex of $C^\infty(L)$ and by $s : C_{k-1} \to C_k$ the homotopy operator $s(f_1, \ldots, f_k) = (1, f_1, \ldots, f_k)$.

LEMMA 6.2. [4] *The Hochschild homology of* K_* *satisfies the following exact sequence*

$$0 \to \frac{r^{k+1}A \,\widehat{\otimes}\, bC_{k+1}/bsC_k}{r^{k+2}A \,\widehat{\otimes}\, bC_{k+1}/bsC_k} \to HH_k(K_*) \to \frac{r^k A \,\widehat{\otimes}\, bC_{k+1} \cap sC_{k-1}}{r^{k+1}A \,\widehat{\otimes}\, bC_{k+1} \cap sC_{k-1}} \to 0$$

There is also a similar formula for

$$PHC_{o/e}(K_*) = \lim_k \left[HC_k(K_*) \xrightarrow{S} HC_{k-2}(K_*) \right].$$

THEOREM 6.3. [4] *The Hochschild homology and the cyclic homology of the mixed complex* $(IC_*^\beta(cL), b, B)$ *enter in the following exact sequences*

$$0 \to HH_k(K_*) \to HH_k(IC_*^\beta(cL)) \xrightarrow{\pi} I\Omega_\beta^k(cL) \to 0$$

$$0 \to PHC_{o/e}(K_*) \to PHC_{o/e}(IC_*^\beta(cL)) \to \bigoplus_{o/e} IH_{\bar{q}}^*(cL) \to 0$$

where the perversity $\bar{q}$ *satisfies* $q_n = [\beta]$.

The result can be generalized to the case of a stratified variety, using the results of [5]. This will appear in a forthcoming publication.

References

1. P. Baum, W. Fulton and R. MacPherson, Riemann-Roch and topological K theory for singular varieties, Acta Math. **143** (1979) 155–192.
2. J.-P. Brasselet, M. Goresky and R. MacPherson, Simplicial differential forms with poles, Amer. Jour. Math. **113** (1991), 1019–1052.
3. J.-P. Brasselet et A. Legrand, Un complexe de formes différentielles à croissance bornée sur une variété stratifiée, Ann. Scuola Normale Sup. Pisa, Ser. IV, **XXI**:2 (1994) 213–234

4. J.-P. Brasselet and A. Legrand, Differential forms on singular varieties and cyclic homology, *Singularity Theory* (eds Bill Bruce and David Mond), London Math. Soc. Lecture Notes **263** (Cambridge Univ. Press, 1999) pp 175–188.

5. J.-P. Brasselet and A. Legrand, Algebras of differential functions on singular varieties, preprint no 118, Université Paul Sabatier, Toulouse, 1998.

6. J.-P. Brasselet, A. Legrand and N. Teleman, Hochschild homology of controlled algebras and piecewise linear functions, preprint, Toulouse, 1997.

7. J. Brüning, The signature theorem for manifolds with metric horns, preprint, Saint Jean des Monts, 1996.

8. J. Brüning and R. Seeley, The resolvent expansion for second order regular singular operators, J. Funct. Anal. **73** (1987) 369–429.

9. J. Brüning and R. Seeley, The expansion of the resolvent near a singular stratum of conical type, J. Funct. Anal. **95** (1991), 255–290.

10. J. Cheeger, On the spectral geometry of spaces with cone-like singularities, Proc. Nat. Acad. Sci. U.S.A. **76** (1979), 2103–2106.

11. J. Cheeger, On the Hodge theory of Riemannian pseudomanifolds, Proc. Symp. in Pure Math. **36**, (Amer. Math. Soc., Providence, R.I., 1980) pp 91–146.

12. J. Cheeger, M. Goresky and R. MacPherson, $\mathcal{L}^2$-cohomology and intersection cohomology for singular varieties, *Seminar on Differential Geometry* (ed S.-T. Yau), Ann. of Math. Studies **102**, (Princeton University Press, Princeton N.J., 1982) pp 303–340.

13. A. Connes, Noncommutative differential geometry, Publ. Math. I.H.E.S. **62** (1986) 257–360.

14. A. Connes, *Noncommutative Geometry*, Academic Press, 1994.

15. M. Goresky and R. MacPherson, Intersection homology theory, Topology **19** (1980) 135–162.

16. M. Goresky and R. MacPherson, Intersection homology theory II, Invent. Math. **71** (1983) 77–129.

17. A. Grothendieck, *Produits tensoriels topologiques et espaces nucléaires*, Mem. Amer. Math. Soc. **16**, 1955.

18. F. Hirzebruch, *Topological Methods in Algebraic Geometry*, Springer-Verlag, 1956.

19. C. Kassel, Cyclic homology, comodules and mixed complexes, J. of Algebra **107** (1987) 195–216.

20. M. Kwieciński, Sur le transformé de Nash et la construction du graphe de MacPherson, thèse, Marseille, 1994.

21. M. Lesch, *Operators of Fuchs type, conical singularities, and asymptotic methods*, Teubner Texte zur Math. **136**, B.G. Teubner, Leipzig, 1997.

22. M. Lesch and N. Peyerimhoff, On the index formula for manifolds with metric horns, Comm. Partial Diff. Equations **23**:3-4 (1998) 649–684.

23. J.-L. Loday, *Cyclic homology*, Grund. der math. Wiss. **301**, Springer Verlag, 1992.

24. J. Mather, Notes on topological stability, preprint, Harvard University, 1970.

25. R. B. Melrose, *The Atiyah-Patodi-Singer index theorem*, Research Notes in Math. 4, A K Peters, Wellesley MA, 1993.

26. B. Monthubert, Pseudodifferential calculus on manifolds with corners and groupoids, Proc. Amer. Math. Soc. **127** (1999) 2871–2881.

27. E. Schrohe, Noncommutative residues and manifolds with conical singularities, J. Funk. Anal. **150** (1997) 146–176.

28. B. V. Schülze, *Pseudo-differential operators on manifolds with singularities*, North-Holland, Amsterdam, 1991.

29. B. V. Schülze, *Pseudo-differential boundary value problems, conical singularities, and asymptotics*, Akademie-Verlag, Berlin, 1994.

30. B. V. Schülze, B. Yu Sternin and V. E. Shatalov, On the index of differential operators on manifolds with conical singularities, Ann. Global Anal. Geom. **16** (1998)141–172.

31. M.-H. Schwartz, Classes et caractères de Chern-Mather des espaces linéaires, Comptes Rendus Acad. Sci. Paris Sér. I **295**:5 (1982) 399–402.

32. T. Suwa, Characteristic classes of coherent sheaves on singular varieties, *Singularities, Sapporo 1998*, Advanced Studies in Pure Mathematics **29** (2000), 279-297.

33. N. Teleman, From index theorem to non-commutative geometry, *Symétries quantiques (Les Houches 1995)* (North Holland, Amsterdam, 1998) pp 787–844.

34. S. Yokura, An extension of Baum-Fulton-MacPherson's Riemann-Roch theorem for singular varieties, Publ. Res. Inst. Math. Sci. **29** (1993) 997–1020.

35. H. Whitney, *Geometric integration theory*, Princeton University Press, Princeton, N.J., 1957.

Part B: Singular complex varietes

The Geometry of Families of Singular Curves

Gert-Martin Greuel
Fachbereich Mathematik, Universität Kaiserslautern
(greuel@mathematik.uni-kl.de)

Christoph Lossen
Fachbereich Mathematik, Universität Kaiserslautern
(lossen@mathematik.uni-kl.de)

Introduction

It is a classical and interesting problem, which is still in the centre of theoretical research, to study the variety V of (reduced, irreducible) curves $C \subset \mathbb{P}^2 = \mathbb{P}^2(\mathbb{C})$ of degree d having exactly r singularities of prescribed (topological or analytic) types $S_1, \ldots, S_r$. Among the most important questions are:

— Is $V \neq \emptyset$ (*existence problem*)?

— Is V irreducible (*irreducibility problem*)?

— Is V smooth and of expected dimension (*T-smoothness problem*)?

The first case under consideration was the case of nodal curves. In 1920, Severi [49] showed that the variety $V_d^{irr}(r \cdot A_1)$ of irreducible plane curves of degree d having r nodes as only singularities is non-empty and T-smooth if and only if $r \leq (d-1)(d-2)/2$. The necessity of this bound is easy since $(d-1)(d-2)/2 - r$ is just the geometric genus which must be positive. Severi claimed also the irreducibility, but his proof contained a gap. Finally, in 1985 the irreducibility problem ("Severi's conjecture") was settled by Harris [24], who proved that each non-empty $V_d^{irr}(r \cdot A_1)$ is irreducible. Note that there is no assumption about the position of the points.

So far, this is the only case where such a complete answer could be found. Even for cuspidal curves there is no sufficient *and* necessary answer to any of the above questions. The known results so far suggest that one can even hardly expect such an answer to exist.

In this survey article we should like to report on recent progress with respect to the above three questions. The emphasis lies on *sufficient conditions* for existence, T-smoothness and irreducibility which are, asymptotically with respect to d, as close as possible to the known necessary conditions (cf. Section 4).

D. Siersma et al. (eds.), *New Developments in Singularity Theory*, 159–192.
© 2001 Kluwer Academic Publishers. Printed in the Netherlands.

We try to present the material in a self-contained form, starting with the deformation theoretical background (in a slightly more general context than needed for the study of the varieties V). In Section 2 we recall the definition of the varieties V as equisingularity strata in the Hilbert scheme and show how to reduce the problems under consideration to H^1-vanishing problems for ideal sheaves of zero-dimensional schemes. Here, we are not only restricted to the case of curves in $\mathbb{P}^2$, but consider the more general situation of curves in a smooth complex projective surface Σ. Section 3 is devoted to the sketch of four quite different approaches leading to appropriate H^1-vanishing criteria which we apply in Section 4, and, finally, in Section 5 we discuss open problems and conjectures.

1. Deformation Theory

Throughout the following let Σ be a smooth, connected, complex projective surface and $C \subset \Sigma$ a reduced curve. Then a *deformation of C/Σ* (or an *embedded deformation* of $C \subset \Sigma$) over the pointed complex space T, $t_0 \in T$, is a triple $(\mathscr{C}, i, j)$ of a complex space and closed embeddings defining a Cartesian diagram

$$
\begin{array}{ccc}
C & \overset{i}{\hookrightarrow} & \mathscr{C} \\
\cap & & \cap\, j \\
\Sigma & \hookrightarrow & \Sigma \times T \\
\downarrow & & \downarrow \pi \\
t_0 & \in & T
\end{array}
\quad\right] \Phi \text{ flat}
$$

such that the composition $\Phi = \pi \circ j$ is flat ($\Sigma \hookrightarrow \Sigma \times T$ denotes the canonical embedding with image $\Sigma \times \{t_0\}$ and π the projection). By abuse of notation, we shall frequently write $\mathscr{C}$ instead of the triple $(\mathscr{C}, i, j)$. A *deformation with sections* is a triple $(\mathscr{C}, i, j)$, as before, together with a set of morphisms $\sigma = \{\sigma_1, \ldots, \sigma_r\}$, $\sigma_i : T \to \mathscr{C}$, such that $\Phi \circ \sigma_i = \mathrm{id}_T$. It is called a *deformation with trivial sections* if the compositions $j \circ \sigma_i$ coincide with the trivial sections $t \mapsto (\sigma_i(t_0), t)$ in $\Sigma \times T$.

Two deformations $(\mathscr{C}, i, j)$, $(\mathscr{C}', i', j')$ (resp. with sections σ, σ') of C/Σ over T are *isomorphic* if there exists an isomorphism $\psi : \mathscr{C} \overset{\cong}{\longrightarrow} \mathscr{C}'$ such that the obvious diagram with the identity on $\Sigma \times T$ (resp. including the sections) commutes, that is, $j = j' \circ \psi$, $\psi \circ i = i'$ (resp. $\psi \circ \sigma_i = \sigma_i'$, $i = 1, \ldots, r$). Note that this implies that the subspaces $j(\mathscr{C})$ and $j(\mathscr{C}')$ of $\Sigma \times T$ coincide.

We introduce the deformation functors $\mathcal{D}ef_{C/\Sigma}$, $\mathcal{D}ef^\sigma_{C/\Sigma}$ from pointed complex spaces to sets

$$\mathcal{D}ef_{C/\Sigma}(T) := \{\text{isomorphism classes of deformations of } C/\Sigma \text{ over } T\},$$

$$\mathcal{D}ef^\sigma_{C/\Sigma}(T) := \left\{ \begin{array}{c} \text{isomorphism classes of deformations with} \\ \text{trivial sections of } C/\Sigma \text{ over } T \end{array} \right\}.$$

There are natural forgetful morphisms $\mathcal{D}ef_{C/\Sigma} \to \mathcal{D}ef_C$, $\mathcal{D}ef^\sigma_{C/\Sigma} \to \mathcal{D}ef^\sigma_C$, with $\mathcal{D}ef_C$, $\mathcal{D}ef^\sigma_C$ the functors of isomorphism classes of deformations (with section) of C, that is, forgetting the embedding.

Furthermore, for each point $z \in C$, any deformation of C induces a deformation of the germ (C, z). Hence, we obtain natural transformations $\mathcal{D}ef_C \to \mathcal{D}ef_{C,z}$, $\mathcal{D}ef^\sigma_C \to \mathcal{D}ef^\sigma_{C,z}$, where $\mathcal{D}ef_{C,z}$ (respectively $\mathcal{D}ef^\sigma_{C,z}$) denotes the functor of isomorphism classes of deformations (with section) of the analytic germ (C, z). Deformations of germs are defined as above, by using representatives.

Note that each deformation of the analytic germ $(C, z) \subset (\Sigma, z)$ can be embedded, that is, extended to a deformation of $(C, z)/(\Sigma, z)$. Moreover, we can choose the embedding in such a way that a given section σ maps to the trivial section in $(\Sigma, z) \times T$. The proof given in [4] for affine varieties works for germs, too.

This is, however, in general not true in the global situation, that is, for $C \subset \Sigma$ a compact curve the morphism $\mathcal{D}ef_{C/\Sigma} \to \mathcal{D}ef_C$ is in general not surjective.

We are interested in subfunctors of $\mathcal{D}ef_{C/\Sigma}$, respectively $\mathcal{D}ef^\sigma_{C/\Sigma}$, such that specified invariants of the germs $(C, \sigma_i(t))$, for instance the (embedded) topological type or the analytic type, are constant (along some, respectively trivial, sections σ_i) during the deformation. In order to be able to apply cohomological methods, we have to develop the infinitesimal theory of such deformations, that is, deformations with base space $T_\varepsilon := \mathrm{Spec}\,(\mathbb{C}[\varepsilon]/\varepsilon^2)$, the simplest fat point (a point with tangent direction).

Our approach works in all cases where the considered set of isomorphism classes of deformations over T_ε is actually an ideal. Therefore, we consider subfunctors $\mathcal{D}ef'_{C,z}$ of $\mathcal{D}ef_{C,z}$, respectively of $\mathcal{D}ef^\sigma_{C,z}$, such that

$$(T^1)' := \mathcal{D}ef'_{C,z}(T_\varepsilon) \subset \left\{ \begin{array}{l} \mathcal{D}ef_{C,z}(T_\varepsilon) \cong \mathcal{O}_{\Sigma,z}/I^{ea}(C,z)\,, \text{ resp.} \\ \mathcal{D}ef^\sigma_{C,z}(T_\varepsilon) \cong \mathfrak{m}_z/I^{ea}_{fix}(C,z)\,, \end{array} \right.$$

is an ideal.

Here $I^{ea}(C, z) \subset \mathcal{O}_{\Sigma,z}$ denotes the Tjurina ideal $\langle f, f_u, f_v \rangle$ where f is a local equation for $(C, z) \subset (\Sigma, z)$ and f_u, f_v are the partial derivatives, and $I^{\text{ea}}_{\text{fix}}(C,z) := \langle f \rangle + \mathfrak{m}_z \cdot I^{\text{ea}}(C,z) \subset \mathcal{O}_{\Sigma,z}$, where $\mathfrak{m}_z \subset \mathcal{O}_{\Sigma,z}$ denotes the maximal ideal. Note that $(T^1)'$ is the Zariski tangent space to the base space of the semiuniversal deformation of $\mathcal{D}ef'_{C,z}$, if it exists. We define $\tau'(C) := \sum_{z \in C} \tau'(C,z)$ where

$$\tau'(C,z) := \dim_{\mathbb{C}} \left(\mathcal{D}ef_{C,z}(T_\varepsilon)/\mathcal{D}ef'_{C,z}(T_\varepsilon) \right).$$

Having defined the "local" deformation functors $\mathcal{D}ef'_{C,z}$, $z \in \operatorname{Sing} C$, we define a "global" subfunctor $\mathcal{D}ef'_{C/\Sigma}$ of $\mathcal{D}ef_{C/\Sigma}$ (resp. of $\mathcal{D}ef^\sigma_{C/\Sigma}$), as the sets $\mathcal{D}ef'_{C/\Sigma}(T)$ consisting exactly of all those classes in $\mathcal{D}ef_{C/\Sigma}(T)$ which are mapped to $\mathcal{D}ef'_{C,z}(T)$ for all z in (a subset of) $\operatorname{Sing} C$. Let us consider some interesting subfunctors $\mathcal{D}ef'_{C/\Sigma}$:

Examples. (a) $\mathcal{D}ef^{\text{ea}}_{C/\Sigma} \subset \mathcal{D}ef_{C/\Sigma}$, the subfunctor of isomorphism classes of *equianalytic deformations*, that is, of those deformations whose induced deformations of the germs (C,z), $z \in \operatorname{Sing} C$, are isomorphic to trivial deformations. Here, $\mathcal{D}ef^{\text{ea}}_{C,z}(T_\varepsilon) = 0 \subset \mathcal{O}_{\Sigma,z}/I^{\text{ea}}(C,z)$ and $\tau^{\text{ea}}(C,z) = \tau(C,z) = \dim_{\mathbb{C}} \mathcal{O}_{\Sigma,z}/I^{\text{ea}}(C,z)$ is the Tjurina number.

(b) $\mathcal{D}ef^{\sigma,\text{ea}}_{C/\Sigma} \subset \mathcal{D}ef^\sigma_{C/\Sigma}$, the subfunctor of classes of *equianalytic deformations with fixed position* (of the singularities), that is, the induced local deformations are supposed to be isomorphic, as deformations with sections, to trivial deformations. The tangent space is $\mathcal{D}ef^{\sigma,\text{ea}}_{C,z}(T_\varepsilon) = 0 \subset \mathfrak{m}_z/I^{\text{ea}}_{\text{fix}}(C,z)$ and $\tau^{\text{ea}}_{\text{fix}}(C,z) = \dim_{\mathbb{C}} \mathcal{O}_{\Sigma,z}/I^{\text{ea}}_{\text{fix}}(C,z) = \tau(C,z)+2$.

(c) $\mathcal{D}ef^{\text{es}}_{C/\Sigma} \subset \mathcal{D}ef_{C/\Sigma}$, the subfunctor of classes of *equisingular deformations*, that is, of those deformations whose induced local deformations are equisingular (cf. [64]). Here $\mathcal{D}ef^{\text{es}}_{C,z}(T_\varepsilon) \cong I^{\text{es}}(C,z)/I^{\text{ea}}(C,z)$ with $I^{\text{es}}(C,z) := \{g \in \mathcal{O}_{\Sigma,z} \mid f+\varepsilon g \text{ equisingular over } T_\varepsilon\}$, the equisingularity ideal (in the sense of Wahl).

(d) $\mathcal{D}ef^{\sigma,\text{es}}_{C/\Sigma} \subset \mathcal{D}ef^\sigma_{C/\Sigma}$, the subfunctor of classes of *equisingular deformations with fixed position*. $\mathcal{D}ef^{\sigma,\text{es}}_{C,z}(T_\varepsilon) \cong I^{\text{es}}_{\text{fix}}(C,z)/I^{\text{ea}}_{\text{fix}}(C,z)$, where $I^{\text{es}}_{\text{fix}}(C,z) := \{g \in \mathcal{O}_{\Sigma,z} \mid f+\varepsilon g \text{ equisingular along the trivial section}\}$ is an ideal in $\mathcal{O}_{\Sigma,z}$.

(e) Further examples are the equimultiple, equigeneric and equiclassical deformation functors (cf. [12]), and many more.

We should like to comment on (c): Over a reduced base T "equisingular" is equivalent to "μ-constant", where $\mu = \mu(f) = \mathbb{C}\{u,v\}/\langle f_u, f_v \rangle$ is the Milnor number of (C,z), respectively of f, a local equation for (C,z). Since in the semiuniversal deformation of (C,z) the μ-constant stratum coincides with the equisingularity stratum which is smooth, by Wahl [64], $\tau^{\text{es}}(C,z)$ equals the dimension of the μ-constant stratum which coincides with $\mu(C,z) - \operatorname{mod}(C,z)$ (cf. [14]). Here, mod denotes the modality with respect to right equivalence (cf. [3]).

We denote by $Z'(C, z) \subset (\Sigma, z)$ the zero-dimensional scheme defined by the preimage of $(T^1)'$ under the projection $\mathcal{O}_{\Sigma,z} \to \mathcal{O}_{\Sigma,z}/I^{\mathrm{ea}}(C, z)$, respectively $\mathcal{O}_{\Sigma,z} \to \mathcal{O}_{\Sigma,z}/I^{\mathrm{ea}}_{\mathrm{fix}}(C, z)$, and set

$$Z'(C) := \bigcup_{z \in \mathrm{Sing}\, C} Z'(C, z) \subset \Sigma.$$

Note that $Z'(C)$ is also a subscheme of C, and we write $\mathcal{J}_{Z'(C)/\Sigma} \subset \mathcal{O}_\Sigma$, respectively $\mathcal{J}_{Z'(C)/C} = \mathcal{J}_{Z'(C)/\Sigma} \otimes \mathcal{O}_C \subset \mathcal{O}_C$, for the corresponding ideal sheaves in $\mathcal{O}_\Sigma$, resp. $\mathcal{O}_C$. We set $\deg Z'(C, z) = \dim_{\mathbb{C}} \mathcal{O}_{Z'(C,z)}$ and

$$\deg Z'(C) = \sum_{z \in \mathrm{Sing}\, C} \deg Z'(C, z) = \dim_{\mathbb{C}} H^0(\mathcal{O}_{Z'(C)}).$$

LEMMA 1.1. *There is a canonical isomorphism of complex vector spaces* $\Phi : \mathcal{D}ef'_{C/\Sigma}(T_\varepsilon) \xrightarrow{\cong} H^0(\mathcal{J}_{Z'(C)/C}(C))$.

In particular, taking $Z'(C)$ the empty scheme, we obtain the well-known isomorphism $\mathcal{D}ef_{C/\Sigma}(T_\varepsilon) \xrightarrow{\cong} H^0(\mathcal{O}_C(C))$.

Proof. Each representative of an element in $\mathcal{D}ef'_{C/\Sigma}(T_\varepsilon)$ is given by local equations $(f_i + \varepsilon g_i = 0)_{i \in I}$, $f_i, g_i \in \Gamma(U_i, \mathcal{O}_\Sigma)$ for an open covering $(U_i)_{i \in I}$ of Σ, which satisfy

(1) $(f_i = 0)_{i \in I}$ are local equations for $C \subset \Sigma$,

(2) $f_i + \varepsilon g_i = (a_{ij} + \varepsilon b_{ij}) \cdot (f_j + \varepsilon g_j)$ on $U_i \cap U_j =: U_{ij}$ with a_{ij} a unit in $\Gamma(U_{ij}, \mathcal{O}_\Sigma)$ and $b_{ij} \in \Gamma(U_{ij}, \mathcal{O}_\Sigma)$,

(3) the germ of g_i at each $z \in \mathrm{Sing}\, C \cap U_i$ maps to a class in $(T^1)' = \mathcal{D}ef'_{C,z}(T_\varepsilon)$.

For the induced sections $g_i/f_i \in \Gamma(U_i, \mathcal{O}_\Sigma(C))$ it follows immediately

$$\frac{g_i}{f_i} - \frac{g_j}{f_j} = \frac{a_{ij}g_j + b_{ij}f_j}{a_{ij}f_j} - \frac{g_j}{f_j} = \frac{b_{ij}}{a_{ij}} \equiv 0 \in \Gamma(U_{ij}, \mathcal{O}_C(C)).$$

Hence, $(g_i/f_i)_{i \in I}$ defines a global section in $\mathcal{O}_C(C)$ which, due to property (3), corresponds to an element of $H^0(\mathcal{J}_{Z'(C)/C}(C))$.

It is easy to check that in this way we get an isomorphism of vector spaces. $\square$

In the following, let $\mathcal{D}ef'_{C/\Sigma} \subset \mathcal{D}ef_{C/\Sigma}$ (respectively $\mathcal{D}ef'_{C/\Sigma} \subset \mathcal{D}ef^\sigma_{C/\Sigma}$) be a subfunctor as above (that is, with $(T^1)'$ an ideal) and assume additionally that the corresponding local subfunctors (restricted to the category of fat points) satisfy:

- $Def'_{C,z}$ has a *very good deformation theory*, that is, satisfies Schlessinger's conditions (H1), (H2), (H4) (cf. [46]), and

- $Def'_{C,z}$ is *unobstructed* (smooth).

REMARK 1.2. The functors $Def^{ea}_{C,z}, Def^{\sigma,ea}_{C,z}$ are well-known to be smooth and have a very good deformation theory. The results of [64] show that the same holds true for $Def^{es}_{C,z}, Def^{\sigma,es}_{C,z}$.

We should like to make precise what we mean when speaking about "fat points". If A is a local Artinian C-algebra, we denote by $\operatorname{Spec} A$ the complex space consisting of one point with structure sheaf A. The set of all such $\operatorname{Spec} A$ together with the morphisms (induced by C-algebra homomorphisms) forms a category, the category of "fat points".

PROPOSITION 1.3. *If* $H^1(\mathcal{J}_{Z'(C)/\Sigma}(C)) \to H^1(\mathcal{J}_{Z'(C)/C}(C))$ *is the zero morphism then the functor* $Def'_{C/\Sigma}$ *is unobstructed on the category of fat points, that is,* $Def'_{C/\Sigma}(\operatorname{Spec} A') \to Def'_{C/\Sigma}(\operatorname{Spec} A)$ *is surjective for any surjection* $A' \to A$ *of local Artinian* C-*algebras.*

In particular, if $Z'(C)$ is the empty scheme, the obstructions against lifting a deformation of $C \subset \Sigma$ to higher order lie in the image of $H^1(\mathcal{O}_\Sigma(C)) \to H^1(\mathcal{O}_C(C))$.

Being unobstructed is an important property of deformations. It implies that the base space of the semiuniversal deformation is smooth if it exists.

Proof. Since each surjection of local Artinian C-algebras factors through small extensions, it suffices to consider the special case of a small extension $A \longrightarrow A/(\eta) =: \overline{A}$, $\eta \cdot \mathfrak{m}_A = 0$.

Let $\overline{\mathscr{C}}$ represent an isomorphism class in $Def'_{C/\Sigma}(\operatorname{Spec} \overline{A})$. We have to show that (up to isomorphism) it lifts to an embedded deformation $\mathscr{C}$ of C/Σ over $\operatorname{Spec} A$ representing a class in $Def'_{C/\Sigma}(\operatorname{Spec} A)$.

The deformation $\overline{\mathscr{C}}$ of C/Σ is given by local equations $\overline{F}_i \in \Gamma(U_i, \mathcal{O}_\Sigma \otimes \overline{A})$, where $(U_i)_{i \in I}$ is an open covering of Σ, such that

(1) on $U_{ij} := U_i \cap U_j$, $\overline{F}_i = \overline{G}_{ij} \cdot \overline{F}_j$ with a unit $\overline{G}_{ij}$,

(2) the image $F_i^{(0)}$ of $\overline{F}_i$ in $\Gamma(U_i, \mathcal{O}_\Sigma \otimes A/\mathfrak{m}_A) = \Gamma(U_i, \mathcal{O}_\Sigma \otimes C)$ is a local equation for $C \subset \Sigma$,

(3) the germs of $\overline{F}_i$ at each point $z \in \operatorname{Sing} C \cap U_i$ map to classes in $Def'_{C,z}(\operatorname{Spec} \overline{A})$.

Without loss of generality, we may assume that each U_i contains, at most, one singular point z_i, $i \in I$.

Recall that the local functors $\mathcal{D}ef'_{C,z_i}$ are supposed to be smooth and to have a very good deformation theory. Using the results of Schlessinger ([46], Remark 2.15), this guarantees in particular

- the existence of a lifting $F_i \in \Gamma(U_i, \mathcal{O}_\Sigma \otimes A)$ of $\overline{F}_i$ mapping to an element of $\mathcal{D}ef'_{C,z_i}(\operatorname{Spec} A)$.

- for any lifting $G_{ij} \in \Gamma(U_{ij}, \mathcal{O}_\Sigma \otimes A)$ of $\overline{G}_{ij}$ the existence of $h_{ij} \in \Gamma(U_{ij}, \mathcal{O}_\Sigma)$ such that $F_i = G_{ij} \cdot F_j + \eta \cdot h_{ij}$ and the germ of h_{ij} at $z \in \operatorname{Sing} C \cap U_{ij}$ maps to a class in $\mathcal{D}ef'_{C,z}(T_\varepsilon)$.

To be able to glue the local liftings, we have to modify the F_i and G_{ij} in a suitable way, such that the h_{ij} become 0. We know

$$
\begin{aligned}
\eta \cdot h_{ij} + \eta \cdot G_{ij} \cdot h_{jk} &= F_i - G_{ij} \cdot F_j + G_{ij} \cdot (F_j - G_{jk} \cdot F_k) \\
&= F_i + G_{ij} \cdot G_{jk} \cdot F_k \\
&= \eta \cdot h_{ik} + (G_{ik} - G_{ij} \cdot G_{jk}) \cdot F_k \,,
\end{aligned}
$$

and $\eta \cdot \mathfrak{m}_A = 0$ implies that $(G_{ik} - G_{ij} \cdot G_{jk}) \in \Gamma(U_{ijk}, \mathcal{O}_\Sigma \otimes (\eta))$. As sections in $\mathcal{O}_\Sigma \otimes A/\mathfrak{m}_A \cong \mathcal{O}_\Sigma$ we obtain

$$
h_{ij} + G_{ij}^{(0)} \cdot h_{jk} = h_{ik} + \left[\frac{1 - G_{ij} \cdot G_{jk} \cdot G_{ik}^{-1}}{\eta} \right] \cdot G_{ik}^{(0)} \cdot F_k^{(0)} .
$$

Furthermore, $G_{ij}^{(0)} = F_i^{(0)} / F_j^{(0)}$, which implies in $\Gamma(U_{ijk}, \mathcal{O}_C(C))$ the cocycle condition

$$
\frac{h_{ij}}{F_i^{(0)}} + \frac{h_{jk}}{F_j^{(0)}} = \frac{h_{ik}}{F_i^{(0)}} .
$$

From the definition of the h_{ij} it follows that $\left(h_{ij}/F_i^{(0)} \right)_{i,j \in I}$, represents an element in $H^1(\mathcal{J}_{Z'(C)/\Sigma}(C))$. Since its image in $H^1(\mathcal{J}_{Z'(C)/C}(C))$ is zero, by assumption, the cocycle is a coboundary and there exist $f_i \in \Gamma(U_i, \mathcal{J}_{Z'(C)/\Sigma})$ such that

$$
\frac{h_{ij}}{F_i^{(0)}} \equiv \frac{f_j}{F_j^{(0)}} - \frac{f_i}{F_i^{(0)}}
$$

as sections in $\mathcal{J}_{Z'(C)/C}(C)$. Hence, $h_{ij} + f_i - f_j \cdot G_{ij}^{(0)} \in \Gamma(U_{ij}, \mathcal{J}_C)$ and the germs of f_i at z_i map to classes in $\mathcal{D}ef'_{C,z_i}(T_\varepsilon)$. Defining

$$
g_{ij} := \frac{h_{ij} + f_i - f_j \cdot G_{ij}^{(0)}}{F_j^{(0)}}, \quad \tilde{F}_i := F_i + \eta \cdot f_i, \quad \tilde{G}_{ij} := G_{ij} + \eta \cdot g_{ij}
$$

we can replace F_i by $\tilde{F}_i$ which glue together to a lifted deformation over $\operatorname{Spec} A$.   □

2. Equianalytic and equisingular families

We generalize the notion of a deformation slightly to the notion of a family. The difference is basically that in a family we do not fix a special fibre and our base spaces are just complex spaces rather than germs. The main results of deformation theory can be applied to families to derive properties which are local with respect to the base space.

Let T be a complex space, then by a *family of* reduced (irreducible) *curves on* Σ *over* T we mean a commutative diagram

$$\mathscr{C} \overset{j}{\hookrightarrow} \Sigma \times T$$
$$\varphi \searrow \quad \swarrow \mathrm{pr}$$
$$T$$

where φ is a proper and flat morphism such that for all $t \in T$ the fibre $\mathscr{C}_t := \varphi^{-1}(t)$ is a reduced (irreducible) curve on Σ, $j : \mathscr{C} \hookrightarrow \Sigma \times T$ is a closed embedding and pr denotes the natural projection.

A *family with sections* is a diagram as above, together with sections $\sigma_1, \ldots, \sigma_r : T \to \mathscr{C}$ of φ.

To a family of reduced plane curves and a point $t_0 \in T$ we can associate, in a functorial way, the deformation

$$\coprod_i (\mathscr{C}, z_i) \to (T, t_0)$$

of the multigerm $(C, \mathrm{Sing}\, C) = \coprod_i (C, z_i)$ over the germ (T, t_0), $C = \mathscr{C}_{t_0}$ being the fibre over t_0. Having a family with sections $\sigma_1, \ldots, \sigma_r$, $\sigma_i(t_0) = z_i$, we obtain in the same way a deformation of $\coprod_i (C, z_i)$ over (T, t_0) with sections.

A family $\mathscr{C} \hookrightarrow \Sigma \times T \to T$ of reduced curves (with sections) is called *equianalytic* (along the sections) if, for each $t \in T$, the induced deformation of the multigerm $(\mathscr{C}_t, \mathrm{Sing}\, \mathscr{C}_t)$ is isomorphic (isomorphic as deformation with section) to the trivial deformation (along the trivial section). It is called *equisingular* (along the sections) if, for each $t \in T$, the induced deformation of the multigerm $(\mathscr{C}_t, \mathrm{Sing}\, \mathscr{C}_t)$ is isomorphic (isomorphic as deformation with section) to an equisingular deformation along the trivial section (cf. [64]).

The *Hilbert functor* $\mathcal{H}ilb_\Sigma$ on the category of complex spaces defined by

$$\mathcal{H}ilb_\Sigma(T) := \{\mathscr{C} \hookrightarrow \Sigma \times T \to T, \text{ family of reduced curves over } T\}$$

is known to be representable by a complex space Hilb_Σ (cf. [6]). More-over, given any $\mathscr{Y} \subset \mathbb{P}^N \times T$, flat over T, the *Hilbert polynomial* h_t of the fibre $\mathscr{Y}_t$, $t \in T$, defined by $h_t(n) = \chi(\mathcal{O}_{\mathscr{Y}_t}(n))$, is constant on each connected component of T. In particular, the universal family of reduced curves on Σ "breaks up" into strata with fixed Hilbert polynomials, more precisely, $\text{Hilb}_\Sigma = \coprod_{h \in \mathcal{C}[z]} \text{Hilb}_\Sigma^h$ where the Hilb_Σ^h are (unions of) connected components of Hilb_Σ. Note that Hilb_Σ^h represents the functor $\mathcal{H}ilb_\Sigma^h$ defined by

$$\mathcal{H}ilb_\Sigma^h(T) := \{\mathscr{C} \hookrightarrow \Sigma \times T \to T \text{ family of reduced curves with } h_t = h\}.$$

We have the following classical result.

LEMMA 2.1. $H^0(\mathcal{O}_C(C))$ *is the Zariski tangent space to* Hilb_Σ *at* C.
   Moreover, Hilb_Σ *is smooth at* C *of dimension* $h^0(\mathcal{O}_C(C))$ *if and only if the natural map* $H^1(\mathcal{O}_\Sigma(C)) \to H^1(\mathcal{O}_C(C))$ *is zero.*

Proof. It follows from the universal property of Hilb_Σ and from the definition of $\mathcal{D}ef_{C/\Sigma}$ that the germ (Hilb_Σ, C) is the base space of the semi-universal deformation of C/Σ, the tangent space of which is $H^0(\mathcal{O}_C(C))$, by Lemma 1.1.
   If $H^1(\mathcal{O}_\Sigma(C)) \to H^1(\mathcal{O}_C(C))$ is the zero map then the completion of the local ring of the base space of the semi-universal deformation of C/Σ at C is a formal power series ring. This and the converse follows from Proposition 1.3, using, e.g., the smoothness criterion of [39], Lecture 22. Since an analytic local ring is a power series ring iff its completion is, the result follows.   □

We are interested in subfamilies of curves on Σ having fixed analytic, respectively topological, types of singularities. To be specific, let $S_1, \ldots, S_r$ be analytic, respectively topological, types.

DEFINITION 2.2. Let $V_h(S_1, \ldots, S_r) \subset \text{Hilb}_\Sigma^h$ denote the *"equisingular stratum"* of reduced curves having precisely r singularities which are of types $S_1, \ldots, S_r$, and let $V_h^{\text{irr}}(S_1, \ldots, S_r) \subset V_h(S_1, \ldots, S_r)$ denote the open subspace parametrizing irreducible curves.

Note that $V_h(S_1, \ldots, S_r)$ is a locally closed subspace of Hilb_Σ^h and represents the functor of equianalytic, respectively equisingular families of given types $S_1, \ldots, S_r$ (cf. [17], Proposition 2.1).

Notation. If $S_1, \ldots, S_r$ are *ordinary multiple points* (that is, all branches are smooth and intersect transversally) of multiplicities $m_1, \ldots, m_r$, respectively, then we write $V_h(m_1, \ldots, m_r)$ instead of $V_h(S_1, \ldots, S_r)$, respectively $V_h^{\text{irr}}(m_1, \ldots, m_r)$, instead of $V_h^{\text{irr}}(S_1, \ldots, S_r)$.

Classically, instead of Hilb_Σ^h, a complete linear system on Σ, $|C| = \mathbb{P}(H^0(\mathcal{O}_\Sigma(C)))$, respectively the open subset $U \subset |C|$ corresponding to reduced curves, is considered. The two notions are related as follows: fixing the Hilbert polynomial of a curve $C \subset \Sigma$ is equivalent to fixing the arithmetic genus $p_a(C)$ and either the degree of the embedding or the self-intersection C^2 or the intersection with the canonical divisor $C.K_\Sigma$. In particular, the Hilbert polynomial $h = h(C')$ is constant for all (reduced) curves $C' \in |C|$, and, since $|C|$ is reduced, it follows that the incidence variety $\{(x, C') \in \Sigma \times U \mid x \in C'\} \subset \Sigma \times U$ is flat over U.

Hence, there exists a unique (injective) morphism $U \to \mathrm{Hilb}_\Sigma^h$, which on the tangent level corresponds to

$$H^0(\mathcal{O}_\Sigma(C))/H^0(\mathcal{O}_\Sigma) \hookrightarrow H^0(\mathcal{O}_C(C)),$$

and we may consider U as a subscheme of Hilb_Σ^h. In particular, for a regular surface (that is, a surface satisfying $H^1(\mathcal{O}_\Sigma) = 0$) the above injection is an isomorphism and U is an open subscheme of Hilb_Σ^h (cf. also Remark 2.7).

DEFINITION 2.3. $V_{|C|}(S_1, \ldots, S_r) \subset |C|$, resp. $V_{|C|}^{\mathrm{irr}}(S_1, \ldots, S_r) \subset |C|$, denotes the locally closed subspace of reduced, resp. irreducible, curves $C' \subset \Sigma$ having precisely r singularities which are of types $S_1, \ldots, S_r$.

In the following, we give a geometric interpretation of the zero-th and first cohomology of the ideal sheaves of certain zero-dimensional schemes. We write $C \in V_h(S_1, \ldots, S_r)$ to denote either the point in $V_h(S_1, \ldots, S_r)$ or the curve corresponding to the point, that is, the corresponding fibre of the universal family. Consider the map

$$\Phi_h : V_h(S_1, \ldots, S_r) \longrightarrow \mathrm{Sym}^r\Sigma, \quad C \longmapsto (z_1 + \ldots + z_r), \qquad (2.1)$$

where $\mathrm{Sym}^r\Sigma$ is the r-fold symmetric product of Σ and $(z_1 + \ldots + z_r)$ the non-ordered tuple of the singularities of C. Since any equisingular, in particular any equianalytic, deformation of a germ admits a unique singular section (cf. [58]), the universal family

$$\mathcal{U}_h(S_1, \ldots, S_r) \hookrightarrow \Sigma \times V_h(S_1, \ldots, S_r) \to V_h(S_1, \ldots, S_r)$$

admits, locally at C, r singular sections. Composing these sections with the projections to Σ gives a local description of the map Φ_h and shows in particular that Φ_h is a well-defined morphism, even if $V_h(S_1, \ldots, S_r)$ is not reduced.

DEFINITION 2.4. $V_{h,\mathrm{fix}}(S_1, \ldots, S_r)$, denotes the complex space consisting of the disjoint union of the fibres of Φ_h.

REMARK 2.5. It follows from the universal property of $V_h(S_1, \ldots, S_r)$ and from the above construction that $V_{h,\text{fix}}(S_1, \ldots, S_r)$, together with the induced universal family on each fibre, represents the functor of equianalytic, respectively equisingular, families of given types $S_1, \ldots, S_r$ along trivial sections.

T-SMOOTHNESS PROBLEM

To shorten notation, we write in the following proposition $Z'(C)$ instead of $Z^{\text{ea}}(C)$, $Z^{\text{ea}}_{\text{fix}}(C)$, $Z^{\text{es}}(C)$, $Z^{\text{es}}_{\text{fix}}(C)$, respectively, if the statement holds in all four cases (cf. Section 1 for the definitions). Moreover, we write V_h to denote $V_h(S_1, \ldots, S_r)$, respectively $V_{h,\text{fix}}(S_1, \ldots, S_r)$, and $V_{|C|}$ to denote $V_{|C|}(S_1, \ldots, S_r)$, respectively $V_{|C|,\text{fix}}(S_1, \ldots, S_r)$.

PROPOSITION 2.6. *Let $C \subset \Sigma$ be a reduced curve with Hilbert polynomial h having precisely r singularities $z_1, \ldots, z_r$ of analytic or topological types $S_1, \ldots, S_r$.*

(a) *The Zariski tangent space of V_h at C is $H^0(\mathcal{J}_{Z'(C)/C}(C))$, while the Zariski tangent space of $V_{|C|}$ at C is $H^0(\mathcal{J}_{Z'(C)/\Sigma}(C))/H^0(\mathcal{O}_\Sigma)$.*

(b) $h^0(\mathcal{J}_{Z'(C)/C}(C)) - h^1(\mathcal{J}_{Z'(C)/C}(C)) \leq \dim(V_h, C)$
$$\leq h^0(\mathcal{J}_{Z'(C)/C}(C)).$$

(c1) *If $H^1(\mathcal{J}_{Z'(C)/C}(C)) = 0$ then V_h is T-smooth at C, that is, smooth of the expected dimension $h^0(\mathcal{O}_C(C)) - \deg Z'(C) = C^2 + 1 - p_a(C) - \deg Z'(C)$.*

(c2) *If $H^1(\mathcal{J}_{Z'(C)/\Sigma}(C)) = 0$ then $V_{|C|}$ is T-smooth at C, that is, smooth of the expected dimension $h^0(\mathcal{O}_\Sigma(C)) - 1 - \deg Z'(C)$.*

(d) *If $H^1(\mathcal{J}_{Z^{\text{ea}}(C)/C}(C)) = 0$ then the natural morphism of germs*

$$(\text{Hilb}^h_\Sigma, C) \longrightarrow \prod_{i=1}^{r} \text{Def}(C, z_i)$$

is smooth of fibre dimension $h^0(\mathcal{J}_{Z^{\text{ea}}(C)/C}(C))$.

Here, $\prod_{i=1}^{r} \text{Def}(C, z_i)$ is the cartesian product of the base spaces of the semi-universal deformations of the germs (C, z_i).

(e) *Let $Z_{\text{fix}}(C)$ denote $Z^{\text{ea}}_{\text{fix}}(C)$, respectively $Z^{\text{es}}_{\text{fix}}(C)$. Then the vanishing of $H^1(\mathcal{J}_{Z_{\text{fix}}(C)/C}(C))$ implies that the morphism of germs*

$$\Phi_h : (V_h(S_1, \ldots, S_r), C) \to (\text{Sym}^r\Sigma, (z_1 + \ldots + z_r))$$

is smooth of fibre dimension $h^0(\mathcal{J}_{Z_{\text{fix}}(C)/C}(C))$.

Recall that a morphism of germs is smooth if it is locally isomorphic to the projection from the product of the target space with a smooth factor to the target space.

Proof. The Zariski tangent space to V_h is isomorphic to $\mathcal{D}ef'_{C/\Sigma}(T_\varepsilon)$. Hence, (a) follows from Lemma 1.1 and the above interpretation of $V_{|C|}$ as a subscheme of V_h. In particular, the second estimate in (b) holds true. The first is due to the fact that (V_h, C) is the fibre over the origin of a (non-linear) obstruction map $H^0(\mathcal{J}_{Z'(C)/C}(C)) \to H^1(\mathcal{J}_{Z'(C)/C}(C))$ (cf. [32], Theorem 4.2.4).

(c1) is a consequence of Proposition 1.3, applying the smoothness criterion of [39], Lecture 22. To obtain (c2), one uses in addition that the vanishing of $H^1(\mathcal{J}_{Z'(C)/\Sigma}(C))$ implies

$$h^0(\mathcal{J}_{Z'(C)/\Sigma}(C)) - h^0(\mathcal{O}_\Sigma) = h^0(\mathcal{O}_\Sigma(C)) - \deg Z'(C) - 1,$$

whence the T-smoothness of $V_{|C|} = |C| \cap V_h$.

(d) was proved in [16], Theorem 6.1. To see (e), we apply (c1) to $Z_{fix}(C)$ and notice that this implies that Φ_h has a smooth fibre through C of the claimed dimension. Moreover, $\mathcal{J}_{Z_{fix}(C)/C}(C)$ is a subsheaf of $\mathcal{J}_{Z'(C)/C}(C)$, of (finite) codimension $2r$. In particular, the vanishing of $H^1(\mathcal{J}_{Z_{fix}(C)/C}(C))$ implies that $H^1(\mathcal{J}_{Z'(C)/C}(C)) = 0$ and, by (c1), $V_h(S_1, \ldots, S_r)$ is smooth at C, the fibre having codimension $2r$. It follows that Φ_h is flat with smooth fibre, hence smooth.   □

REMARK 2.7. Let Σ be a regular surface, that is, $h^1(\mathcal{O}_\Sigma) = 0$. Then the long exact cohomology sequence to

$$0 \longrightarrow \underbrace{\mathcal{J}_{C/\Sigma}(C)}_{\cong \mathcal{O}_\Sigma} \longrightarrow \mathcal{J}_{Z'(C)/\Sigma}(C) \longrightarrow \underbrace{\mathcal{J}_{Z'(C)/\Sigma}(C) \otimes \mathcal{O}_C}_{= \mathcal{J}_{Z'(C)/C}(C)} \longrightarrow 0,$$

implies $H^0(\mathcal{J}_{Z'(C)/C}(C)) \cong H^0(\mathcal{J}_{Z'(C)/\Sigma}(C))/H^0(\mathcal{O}_\Sigma)$, that is. the Zariski tangent spaces to V_h and $V_{|C|}$ are isomorphic, and T-smoothness of $V_{|C|}$ implies T-smoothness of V_h.

If, additionally, $h^2(\mathcal{O}_\Sigma) = 0$, as, for instance, for rational surfaces Σ, then also $H^1(\mathcal{J}_{Z'(C)/C}(C))$ and $H^1(\mathcal{J}_{Z'(C)/\Sigma}(C))$ are isomorphic.

Let's reformulate and strengthen Proposition 2.6 in the case of plane curves, that is, $\Sigma = \mathbb{P}^2$, which is of special interest. Of course, since $h^1(\mathcal{O}_{\mathbb{P}^2}) = h^2(\mathcal{O}_{\mathbb{P}^2}) = 0$, there is no difference whether we consider the curves in the linear system $|dH|$, H the hyperplane section, or curves with fixed Hilbert polynomial $h(z) = dz - (d^2 - 3d)/2$. We denote the corresponding varieties by $V_d = V_d(S_1, \ldots, S_r)$, respectively $V_{d,fix}(S_1, \ldots, S_r)$. Using the above notation, we obtain:

PROPOSITION 2.8. *Let $C \subset \mathbb{P}^2$ be a reduced curve of degree d with precisely r singularities $z_1, \ldots, z_r$ of analytic, or topological, types $S_1, \ldots, S_r$.*

(a) *$H^0(\mathcal{J}_{Z'(C)/\mathbb{P}^2}(d))/H^0(\mathcal{O}_{\mathbb{P}^2})$ is isomorphic to the Zariski tangent space of V_d at C.*

(b) *$h^0(\mathcal{J}_{Z'(C)/\mathbb{P}^2}(d)) - h^1(\mathcal{J}_{Z'(C)/\mathbb{P}^2}(d)) - 1 \leq \dim(V_d, C)$*
    $\leq h^0(\mathcal{J}_{Z'(C)/\mathbb{P}^2}(d)) - 1.$

(c) *$H^1(\mathcal{J}_{Z'(C)/\mathbb{P}^2}(d)) = 0$ if and only if V_d is T-smooth at C, that is, smooth of the expected dimension $d(d+3)/2 - \deg Z'(C)$.*

(d) *$H^1(\mathcal{J}_{Z^{ea}(C)/\mathbb{P}^2}(d)) = 0$ if and only if the natural morphism of germs $(\mathbb{P}(H^0(\mathcal{O}_{\mathbb{P}^2}(d))), C) \to \prod_{i=1}^{r} Def(C, z_i)$ is smooth (hence surjective) of fibre dimension $h^0(\mathcal{J}_{Z^{ea}(C)/\mathbb{P}^2}(d)) - 1.$*

(e) *Let $Z_{fix}(C)$ denote $Z_{fix}^{ea}(C)$, resp. $Z_{fix}^{es}(C)$. Then $H^1(\mathcal{J}_{Z_{fix}(C)/\mathbb{P}^2}(d))$ vanishes if and only if the morphism of germs*

$$\Phi_d : (V_d(S_1, \ldots, S_r), C) \to (\operatorname{Sym}^r \mathbb{P}^2, (z_1 + \ldots + z_r))$$

is smooth of fibre dimension $h^0(\mathcal{J}_{Z_{fix}(C)/\mathbb{P}^2}(d)) - 1$. In particular, the vanishing of $H^1(\mathcal{J}_{Z_{fix}(C)/\mathbb{P}^2}(d))$ implies that arbitrarily close to C there are curves in $V_d(S_1, \ldots, S_r)$ having their singularities in general position in $\mathbb{P}^2$.

Proof. Most parts of this proposition follow immediately from Proposition 2.6 and Remark 2.7. We need to show only the "if" parts in the statements (c)–(e): for (d), (e) cf. [16], Corollary 6.3; since the expected dimension of V_d is $d(d+3)/2 - \deg Z'(C) = h^0(\mathcal{O}_{\mathbb{P}^2}(d)) - h^0(\mathcal{O}_{Z'(C)}(d))$, (c) follows from the exact sequence

$$0 \longrightarrow H^0(\mathcal{J}_{Z'(C)/\mathbb{P}^2}(d)) \longrightarrow H^0(\mathcal{O}_{\mathbb{P}^2}(d)) \longrightarrow H^0(\mathcal{O}_{Z'(C)}(d))$$
$$\longrightarrow H^1(\mathcal{J}_{Z'(C)/\mathbb{P}^2}(d)) \longrightarrow 0. \qquad \square$$

The statements of Proposition 2.6 (c1), (c2), respectively 2.8 (c), reduce the T-smoothness problem to an H^1-vanishing problem for ideal sheaves of zero-dimensional schemes. In the following we indicate how to do the same for the existence and irreducibility problem:

EXISTENCE PROBLEM

The key point is to introduce for any $f \in \mathcal{O}_{\Sigma, z} \cong \mathbb{C}\{u, v\}$ a zero-dimensional ideal $I(f)$ such that

(1) $f \in I(f)$,

(2) each generic element $g \in I(f)$ defines a singularity of the same type as f,

(3) $I(f)$ depends only on the ν-jet of f for some $\nu > 1$,

(4) $I(\psi^*(uf)) = \psi^* I(f)$ for any unit u, and ψ a local analytic isomorphism.

As an example, we should like to mention the case of *topological types*, where one associates

- to an ordinary singularity f the ideal $I^s(f) := \mathfrak{m}_z^m$, $m = \mathrm{mt}(f)$;

- to an arbitrary f the ideal $I^s(f)$ defined via the *cluster* associated to f (cf. [19]).

In the following, we restrict ourselves to curves with *only ordinary singularities*. The general case is much more involved (cf. [19]).

DEFINITION 2.9. Let $\underline{m} = (m_1, \ldots, m_r)$ be an r-tuple of positive integers and $\underline{z} = (z_1, \ldots, z_r)$ be an r-tuple of distinct points in Σ. Then we denote by $Z(\underline{m}, \underline{z})$, the zero-dimensional scheme with support $\{z_1, \ldots, z_r\}$ and locally at z_i being defined by $\mathfrak{m}_{z_i}^{m_i}$, $i = 1, \ldots, r$.

We call the zero-dimensional schemes $Z(\underline{m}, \underline{z})$ (ordinary) fat point schemes.

PROPOSITION 2.10. *If $H^1(\mathcal{J}_{Z(\underline{m},\underline{z})/\mathbf{P}^2}(d-1)) = 0$ then there exists a reduced, irreducible plane curve of degree d with r ordinary singular points of multiplicities $\underline{m} = (m_1, \ldots, m_r)$ at $\underline{z} = (z_1, \ldots, z_r)$ as only singularities.*

Sketch of proof. Let $Z^{(j)} := Z(\underline{m}^{(j)}, \underline{z})$ denote the ordinary fat point scheme with multiplicities

$$m_i^{(j)} := \begin{cases} m_i + 1 & \text{if } i = j, \\ m_i & \text{otherwise.} \end{cases}$$

Then $Z^{(j)} \supset Z := Z(\underline{m}, \underline{z})$ and the corresponding inclusion of ideal sheaves induces an exact sequence

$$0 \longrightarrow H^0(\mathcal{J}_{Z^{(j)}/\mathbf{P}^2}(d)) \longrightarrow H^0(\mathcal{J}_{Z/\mathbf{P}^2}(d)) \longrightarrow \overbrace{H^0(\mathcal{J}_{Z/\mathbf{P}^2}/\mathcal{J}_{Z^{(j)}/\mathbf{P}^2})}^{\neq 0}$$
$$\longrightarrow H^1(\mathcal{J}_{Z^{(j)}/\mathbf{P}^2}(d)).$$

On the other hand, it is not difficult to show that the vanishing of $H^1(\mathcal{J}_{Z(\underline{m},\underline{z})/\mathbf{P}^2}(d-1))$ implies $H^1(\mathcal{J}_{Z^{(j)}/\mathbf{P}^2}(d)) = 0$, and we can conclude the existence of a curve $D_j \in H^0(\mathcal{J}_{Z/\mathbf{P}^2}(d)) \setminus H^0(\mathcal{J}_{Z^{(j)}/\mathbf{P}^2}(d))$, in particular, with multiplicity exactly m_j at z_j. A similar reasoning shows that we can even prescribe the tangent directions at z_j, and Bertini type arguments allow to conclude the statement.   □

For a detailed proof we refer to [35], Proposition 2.1.

IRREDUCIBILITY PROBLEM

Let $V = V_{|C|}(S_1, \ldots, S_r)$. Roughly speaking, the procedure is as follows: first, we need to show the existence of an *irreducible* Hilbert scheme of zero-dimensional schemes, that is, a variety $\mathcal{M}$ parametrizing the schemes $Z(C)$ (concentrated in $\mathrm{Sing}\, C$ and locally defined by the ideals $I(f)$ introduced for the existence problem) when C varies in V (cf. [21]). $\mathcal{M}$ should represent the functor of such families of zero-dimensional schemes. Then, by the universal property, there is a morphism $\Psi : V \to \mathcal{M}$ which can be thought of as associating to a curve C the corresponding zero-dimensional scheme $Z(C)$. The restriction Ψ' of Ψ to

$$V_{\mathrm{reg}} := \{C \in V \mid H^1(\mathcal{J}_{Z(C)/\Sigma}(C)) = 0\}$$

has smooth and equidimensional fibres. Moreover, an easy dimension count shows that Ψ' is dominant on $\mathcal{M}$, implying the irreducibility of V_{reg}. It remains to show that V_{reg} is *dense* in V, that is, there is no component of V whose *generic* element C satisfies $H^1(\mathcal{J}_{Z(C)/\Sigma}(C)) \neq 0$.

In [21], we give a more refined reasoning, by considering additionally the morphism Φ_d, which allows to restrict ourselves to curves with singular points in general position (cf. Proposition 2.6 (e)). We obtain the density of V_{reg} in V by applying an H^1-vanishing theorem for generic fat point schemes (of local multiplicities given by the determinacy bounds) and by showing that there is no component of V whose *generic* element C satisfies $H^1(\mathcal{J}_{Z'_{\mathrm{fix}}(C)/\Sigma}(C)) \neq 0$.

3. H^1-vanishing criteria for zero-dimensional schemes

In the previous section, we have shown, that H^1-vanishing for ideal sheaves of some zero-dimensional schemes has immediate geometric consequences. In this section, we describe different approaches to prove H^1-vanishing.

When looking for appropriate H^1-vanishing theorems for the problems under consideration, one has to be aware that the needed types of those are quite different.

- *Existence problem:* to show that $V \neq \emptyset$ it suffices to have an H^1-vanishing theorem for certain *generic fat point*, respectively *cluster*, schemes. This means that the support of the fat point scheme, resp. together with some of its infinitely near points, is in general position.

- *T-smoothness problem:* to prove the T-smoothness of V we need H^1-vanishing for the zero-dimensional schemes $Z'(C)$ (described above) associated with *any curve* $C \in V$.

- *Irreducibility problem:* to have irreducibility it suffices to show the vanishing of $H^1(\mathcal{J}_{Z(C)/\Sigma}(C))$ for *generic curves* $C \in V$.

A large variety of approaches has been used in connection with the problems stated. In the following, we describe four of these,

1. the classical approach,

2. Hirschowitz' approach by the so-called "Horace method",

3. Xu's approach based on Kodaira vanishing,

4. Barkats' approach based on properties of the Castelnuovo function.

We present the obtained H^1-vanishing criteria in some detail. All these approaches have been used and improved in [17, 19, 20, 21]. It should be mentioned that Hirschowitz' and Xu's approaches are best suited for the existence problem, while the remaining two approaches can be used to obtain appropriate sufficient conditions for T-smoothness and irreducibility.

3.1. The classical approach

The classical idea, already applied by Severi, Segre and Zariski, is to consider the zero-dimensional scheme as a subscheme of the curve $C \in V$ itself and to apply vanishing theorems for coherent $\mathcal{O}_C$-modules (e.g., based on the Riemann-Roch theorem for curves).

PROPOSITION 3.1. *Let Σ be a smooth complex projective surface, $C \subset \Sigma$ a reduced curve with irreducible components $C_1, \ldots, C_s$ and $Z \subset C$ a zero-dimensional scheme. Then $H^1(\mathcal{J}_{Z/C}(C))$ vanishes if for any $i = 1, \ldots, s$*

$$\deg(Z \cap C_i) - isod_{C_i}(\mathcal{J}_{Z/C}, \mathcal{O}_C) < -K_\Sigma \cdot C_i, \qquad (3.2)$$

where K_Σ denotes the canonical divisor on Σ. $isod_{C_i}$ is the total (local) isomorphism defect on C_i, that is,

$$isod_{C_i}(\mathcal{J}_{Z/C}, \mathcal{O}_C) := \sum_{x \in C_i} \min_\varphi \dim_{\mathbb{C}} Coker(\varphi_{C_i} : \mathcal{J}_{Z \cap C_i/C_i, x} \to \mathcal{O}_{C_i, x}),$$

the minimum being taken over all φ_{C_i} with $\varphi : \mathcal{J}_{Z/C,x} \to \mathcal{O}_{C,x}$ an injective local homomorphism.

Proof ([16]). By Serre duality, $H^1(\mathcal{J}_{Z/C}(C)) \cong \mathrm{Hom}(\mathcal{J}_{Z/C}(C), \omega_C)^*$, ω_C the dualizing sheaf of C. Hence, the non-vanishing of $H^1(\mathcal{J}_{Z/C}(C))$ implies the existence of an $\mathcal{O}_C$-linear map $0 \neq \varphi : \mathcal{J}_{Z/C}(C) \to \omega_C$. Since $\omega_C, \mathcal{J}_{Z/C}(C)$ are torsion free and have rank 1 on each irreducible component C_i of C, there is at least one C_i such that the restriction

$$\varphi_i : \mathcal{J}_{Z\cap C_i/C_i} \cong \mathcal{J}_{Z/C}(C) \otimes_{\mathcal{O}_C} \mathcal{O}_{C_i} \longrightarrow \omega_C \otimes_{\mathcal{O}_C} \mathcal{O}_{C_i}$$

is injective and has image of rank 1. In particular, $\mathrm{Coker}\,(\varphi_i)$ is a torsion sheaf. Since C is Gorenstein, $\omega_{C,x} \cong \mathcal{O}_{C,x}$, and the additivity of the Euler characteristic implies

$$\chi(\mathcal{J}_{Z\cap C_i/C_i}) \leq \chi(\omega_C \otimes_{\mathcal{O}_C} \mathcal{O}_{C_i}) - \mathrm{isod}_{C_i}(\mathcal{J}_{Z/C}, \mathcal{O}_C).$$

It remains to compute the two Euler characteristics:

$$\begin{aligned}
\chi(\mathcal{J}_{Z\cap C_i/C_i}(C)) &= \chi(\mathcal{O}_{C_i}(C)) - \chi(\mathcal{O}_{Z\cap C_i}) \\
&= \chi(\mathcal{O}_{C_i}) + C \cdot C_i - \deg(Z \cap C_i),
\end{aligned}$$

and, by Riemann-Roch and the adjunction formula,

$$\chi(\omega_C \otimes_{\mathcal{O}_C} \mathcal{O}_{C_i}) = \chi(\mathcal{O}_{C_i}) + K_\Sigma \cdot C_i + C \cdot C_i. \qquad \square$$

In many cases, already a slight modification of this approach (replacing C by some auxiliary curve C') leads to better results: consider any curve $C' \subset \Sigma$ *containing the zero-dimensional scheme* Z. There is an obvious commutative diagram with exact horizontal arrows

$$\begin{array}{c}
\quad \xrightarrow{\Phi} H^0(\mathcal{O}_C(C)) \to H^0(\mathcal{O}_Z) \to H^1(\mathcal{J}_{Z/C}(C)) \to H^1(\mathcal{O}_C(C)) \\
H^0(\mathcal{O}_\Sigma(C)) \qquad\qquad\qquad \| \\
\quad \xrightarrow{\Phi'} H^0(\mathcal{O}_{C'}(C)) \to H^0(\mathcal{O}_Z) \to H^1(\mathcal{J}_{Z/C'}(C)).
\end{array}$$

Suppose Φ' is surjective and $H^1(\mathcal{O}_C(C)) = 0$. Then the vanishing of $H^1(\mathcal{J}_{Z/C'}(C))$ implies, obviously, $H^1(\mathcal{J}_{Z/C}(C)) = 0$.

In particular, if $-K_\Sigma$ is ample then this reasoning, together with the proof of Prop. 3.1, shows that $H^1(\mathcal{J}_{Z/C}(C))$ vanishes if there exists an *irreducible* curve C' containing Z, such that $H^1(\mathcal{O}_\Sigma(C - C')) = 0$ and

$$\deg(Z) < (C - C' - K_\Sigma) \cdot C'. \qquad (3.3)$$

3.2. THE HORACE METHOD

Hirschowitz [27, 28] initiated a new approach by the so-called *Horace method*, applicable mainly to the existence problem. The Horace method, as well as its recent modifications and improvements [1, 2, 38], is mainly based on two procedures: reduction (by a curve) and specialization (with respect to the curve).

Reduction: Let Σ be a smooth projective surface, $H \subset \Sigma$ a curve and $Z \subset \Sigma$ a zero-dimensional scheme. Then for any divisor D on Σ there is an obvious *"residual"* exact sequence

$$0 \longrightarrow \mathcal{J}_{Z:H/\Sigma}(D-H) \longrightarrow \mathcal{J}_{Z/\Sigma}(D) \longrightarrow \mathcal{J}_{Z\cap H/H}(D) \longrightarrow 0\,,$$

$Z:H$ denoting the *residue* (given by the ideal quotient $I_Z : I_H$) and $Z \cap H$ the *trace* of Z w.r.t. H. The corresponding long exact cohomology sequence implies that $H^1(\mathcal{J}_{Z/\Sigma}(D))$ vanishes if

$$H^1(\mathcal{J}_{Z:H/\Sigma}(D-H))= 0 \quad \text{and} \quad H^1(\mathcal{J}_{Z\cap H/H}(D))= 0\,, \qquad (3.4)$$

the first being a condition of *"reduced degree"* (of the zero-dimensional scheme and of the divisor), the second of *"reduced dimension"* (of the ambient space).

For instance, in the case $\Sigma = \mathbb{P}^2$ and $H \subset \mathbb{P}^2$ a reduced, irreducible curve of degree h, the vanishing of $H^1(\mathcal{J}_{Z/\mathbb{P}^2}(d))$ is implied by

$$H^1(\mathcal{J}_{Z:H/\mathbb{P}^2}(d-h))= 0, \quad \deg(Z \cap H) < h(d-h+3)\,, \qquad (3.5)$$

(cf. Section 3.1). This allows, obviously, to proceed by induction on d.

Specialization: The crucial point for the success of such an inductive procedure is that in each step the residue scheme should become as small as possible, that is, $\deg(Z \cap H)$ has to be "made as big as possible", but still satisfying the upper bound in (3.5). To do so, one "specializes" in each inductive step a certain number of points of the support of Z on H. Finally, due to semi-continuity, the h^1-vanishing for the specialized scheme implies h^1-vanishing for the scheme under consideration.

Throughout the following, we use the notation $Z(\underline{m})$ for any fat point scheme $Z(\underline{m}, \underline{z}) \subset \mathbb{P}^2$ with $z_1, \ldots, z_r$ in generic position.

PROPOSITION 3.2 (Hirschowitz, 1989). *Let d, r, $m_1 \geq \ldots \geq m_r$, be positive integers such that*

$$\sum_{i=1}^{r} \frac{m_i(m_i+1)}{2} < \left\lfloor \frac{(d+3)^2}{4} \right\rfloor\,. \qquad (3.6)$$

Then $H^1(\mathcal{J}_{Z(\underline{m})/\mathbb{P}^2}(d))$ vanishes.

Proof. We proceed by induction on d to prove the following, more general, statement: *let $L \subset \mathbb{P}^2$ be a fixed straight line, $0 \leq r' \leq r$, and let $(z_1 + \ldots + z_r)$ be generic in $Sym^{r'}(L) + Sym^{r-r'}(\mathbb{P}^2) \subset Sym^r(\mathbb{P}^2)$. If (3.6) holds and*

$$\sum_{z_i \in L} m_i \leq d+1 \qquad (3.7)$$

then $H^1(\mathcal{J}_{Z(\underline{m};\underline{z})/\mathbb{P}^2}(d))$ vanishes. For $d = 0, 1$ the statement is obviously satisfied. Let $d \geq 2$ and $Z = Z(\underline{m}, \underline{z})$. One has to consider the following three cases:

Case 1. $\dfrac{d+2}{2} \leq \sum_{z_i \in L} m_i \leq d+1.$

In this case, *reduction by L*, implies the statement:

$$\deg(Z : L) = \deg Z - \deg(Z \cap L) \leq \left\lfloor \frac{(d+3)^2}{4} \right\rfloor - \left\lceil \frac{d+2}{2} \right\rceil = \left\lfloor \frac{(d+2)^2}{4} \right\rfloor,$$

whence, by the induction assumption, (3.5) is satisfied.

Case 2. $\displaystyle\sum_{z_i \in L} m_i \leq \dfrac{d+1}{2}$ and $m_1 \geq \dfrac{d+2}{2}.$

Here, *reduction by L'*, a generic line through z_1, gives the wanted result (cf. Case 1, and recall that $m_1 \leq d+1$, by assumption (3.6)).

Case 3. $\displaystyle\sum_{z_i \in L} m_i \leq \dfrac{d+1}{2}$ and $m_1 \leq \dfrac{d+1}{2}.$

This is the situation where we use a *specialization* argument. Successively specializing points on L (starting with z_1), we end up either with a situation satisfying the conditions of Case 1, respectively 2, or with $r = r'$ and $\sum_{i=1}^{r} m_i \leq (d+1)/2$. Note that the latter implies

$$\deg(Z : L) = \sum_{i=1}^{r} \frac{(m_i - 1) m_i}{2} \leq \frac{d^2 - 1}{4} < \left\lfloor \frac{(d+2)^2}{4} \right\rfloor.$$

In any case, reduction by L (respectively some other line L', cf. Case 2) allows to conclude H^1-vanishing for the "specialized" position of points. Finally, the semicontinuity theorem ([25], III, 12.8) gives the statement for the original, more general, position.   □

REMARK 3.3. The right-hand side of (3.6) cannot be improved without adding an extra condition, as the example $r = 2$, $m_1 = \lfloor d/2 \rfloor + 1$ and $m_2 = d+1 - \lfloor d/2 \rfloor$ shows. In this case, the right- and left-hand side of (3.6) are equal, and h^1 is positive, since the trace of the fat point scheme on the line through the two points of its support has degree $d + 2$.

On the other hand, it seems that such phenomena occur only for a small number of points r, the reason being the existence of a curve of small degree, passing through the points. After introducing a suitable extra condition which restricts the largest occuring multiplicities, one expects to have H^1-vanishing whenever the expected dimension is non-negative (cf. [23, 28]):

CONJECTURE 3.4 (Harbourne/Hirschowitz). *If $m_1 \geq \ldots \geq m_r$ are non-negative integers satisfying $m_1 + m_2 + m_3 \leq d$ and*

$$\sum_{i=1}^{r} \frac{m_i(m_i+1)}{2} \leq \frac{(d+1)(d+2)}{2}$$

then $H^1(\mathcal{J}_{Z(\underline{m})/\mathbf{P}^2}(d))$ vanishes.

Even if there have been lots of contributions, the conjecture is still unproven. In the following, we present the most promising results for the general case (cf. Propositions 3.5, 3.7). For special cases, we refer to the literature, e.g., [10, 37, 9].

There is one main disadvantage of the "basic" Horace method, as applied in the proof of Proposition 3.2. One has only the choice to specialize the position of a point on H or not to do so, but not, as one would like to do, to specialize part of the structure of the fat point. For instance, each ordinary multiple point specialized to H increases the degree of the trace $Z \cap H$ by the multiplicity m_i of the point. For curves in $\mathbf{P}^2$ this implies, that one can, at most, guarantee

$$h(d - h + 3) - \min_i\{m_i\} \leq \deg(Z \cap H) < h(d - h + 3).$$

while, of course, the optimal choice in each inductive step would be

$$\deg(Z \cap H) = h(d - h + 3) - 1.$$

In [1, 2], Alexander and Hirschowitz developed a new, *"differential"*. *Horace method* to overcome this technical obstacle. On the one hand side, this method allowed to solve some low-degree and low-multiplicity cases, on the other hand, it was central for the proof of

PROPOSITION 3.5 (Alexander/Hirschowitz, 2000). *Let Σ be a smooth projective surface, $D, L \in Pic(\Sigma)$ with L ample, and let m be a positive integer.*

*Then there is an integer $d_0 = d_0(m)$ such that $H^1(\mathcal{J}_{Z(\underline{m})/\Sigma}(D+dL))$
vanishes for any $d \geq d_0$ and any $m \geq m_1 \geq \ldots \geq m_r$ satisfying*

$$\sum_{i=1}^{r} \frac{m_i(m_i+1)}{2} \leq h^0(\mathcal{O}_\Sigma(D+dL)). \qquad (3.8)$$

The importance of this result is that the bound d_0 is independent on
r, the number of points (always assumed to be in general position).

Finally, we should like to point out the main problems of generalizing
the above approach to other classes $\mathcal{C}$ of zero-dimensional schemes:
firstly, it has to be *closed* under the reduction operation. Secondly,
in general, it is not sufficient to be able to specialize the position of
(possibly infinitely near) points, belonging to the support of the scheme
Z, on the curve H. Hence, one has to think about another kind of
possible "specializations" for the schemes under consideration. Finally,
there is another big difference to the case of generic fat point schemes.
For the latter, we always have

$$\deg((Z:H) \cap H) = \deg(Z \cap H) - \#(Z \cap H) \leq \deg(Z \cap H) - 1.$$

In particular, when reducing by a line $L \subset \mathbb{P}^2$ we can choose a condition of type (3.7) to run an inductive procedure. For other classes of
schemes it may happen that $\deg((Z:H) \cap H) = \deg(Z \cap H)$, whence
one needs another type of extra condition.

In [19], these problems were solved for the class of *"cluster schemes"*
(or, *"generalized singularity schemes"*). The obtained H^1-vanishing
theorem lead to the first asymptotically proper sufficient existence
condition in the case of topological types (cf. Section 4.1, below).

3.3. KODAIRA VANISHING

Xu [66] used the well-known Kodaira vanishing theorem to derive a
new H^1-vanishing theorem for *generic fat point schemes* in $\mathbb{P}^2$.

THEOREM 3.6 (Kodaira, 1953). *Let Y be a smooth complex projective
variety, K_Y the canonical divisor and $D \in Pic(Y)$ ample. Then*

$$H^i(\mathcal{O}_Y(D + K_Y)) = 0 \text{ for any } i > 0.$$

The key point of Xu's approach is to consider the blow-up $\pi : \Sigma \to \mathbb{P}^2$
of $\mathbb{P}^2$ in r (generic) points $z_1, \ldots, z_r$. Obviously,

$$H^1(\mathcal{J}_{Z(\underline{m},\underline{z})/\mathbb{P}^2}(d)) = 0 \Longleftrightarrow H^1(\mathcal{O}_\Sigma(dH - m_1 E_1 - \ldots - m_r E_r)) = 0,$$

E_i being the exceptional divisor corresponding to the blown-up point
z_i, $i = 1, \ldots, r$. Applying the Nakai-Moishezon criterion for ampleness,

he obtained sufficient conditions for the divisor $dH - \sum_{i=1}^{r} m_i E_i - K_\Sigma$ to be ample and, as a corollary,

PROPOSITION 3.7 (Geng Xu, 1995). *Let $m_1 \geq \ldots \geq m_r$ and d be positive integers satisfying $m_1 + m_2 \leq d$, $m_1 + \ldots + m_5 \leq 2d$ and*

$$\sum_{i=1}^{r} \frac{(m_i+1)^2}{2} < \frac{9(d+3)^2}{20}.$$

Then $H^1(\mathcal{J}_{Z(\underline{m})/\mathbb{P}^2}(d))$ vanishes.

3.4. CASTELNUOVO FUNCTION

The *Castelnuovo function* of a 0-dimensional scheme $Z \subset \mathbb{P}^2$,

$$C_Z : \mathbb{Z}_{\geq 0} \longrightarrow \mathbb{Z}_{\geq 0}, \quad d \longmapsto h^1(\mathcal{J}_{Z/\mathbb{P}^2}(d-1)) - h^1(\mathcal{J}_{Z/\mathbb{P}^2}(d)),$$

satisfies the following more or less obvious properties

1. $C_Y(d) \leq C_Z(d)$ for any subscheme $Y \subset Z$.
2. $C_Z(0) + \ldots + C_Z(d) = \deg Z - h^1(\mathcal{J}_{Z/\mathbb{P}^2}(d))$.
3. There exist integers $a(Z) \leq t(Z)$ such that

$$C_Z(d) \begin{cases} = d+1 & \text{if } d < a(Z), \\ \leq C_Z(d-1) & \text{if } a(Z) \leq d \leq t(Z), \\ = 0 & \text{if } d > t(Z). \end{cases}$$

Moreover, $a(Z) = \min\{d \in \mathbb{Z} \mid h^0(\mathcal{J}_{Z/\mathbb{P}^2}(d)) > 0\}$, and the minimal possible choice for t is $t(Z) := \min\{d \in \mathbb{Z} \mid h^1(\mathcal{J}_{Z/\mathbb{P}^2}(d)) = 0\}$.

Examples. (a) Let $Z \subset \mathbb{P}^2$ be an *ordinary fat point* of multiplicity m, that is, Z is supported at a unique point z and $\mathcal{O}_{Z,z} \cong \mathcal{O}_{\mathbb{P}^2,z}/\mathfrak{m}_z^m$. Then

$$C_Z(d) = \begin{cases} d+1 & \text{if } d \leq m-1, \\ 0 & \text{if } d \geq m. \end{cases}$$

(b) Let $Z \subset \mathbb{P}^2$ be a *complete intersection*, that is, $Z = C_m \cap C_k$, where C_m, C_k are plane curves of degree m, respectively k ($k \leq m$), without common components. Then Bézout's Theorem implies that

$$C_Z(d) = \begin{cases} d+1 & \text{if } d \leq k-1, \\ k & \text{if } k-1 \leq d \leq m-1, \\ k+m-d-1 & \text{if } m-1 \leq d \leq m+k-1, \\ 0 & \text{if } d \geq m+k-1. \end{cases}$$

(c) Let $Z \subset \mathbb{P}^2$ be the union of 9 "simple" points in the following position:

- generic   • 6 on a line + 3 generic   • 7 on a line + 2 generic.

Then the graph of the corresponding Castelnuovo function looks

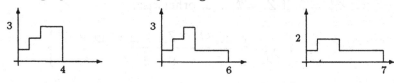

The latter example suggests that the presence of "long stairs" is closely related to a "special position" of the scheme (w.r.t. a curve D). Indeed, one has the following

LEMMA 3.8 (Davis, 1986). *Let* $Z \subset \mathbb{P}^2$ *be a zero-dimensional scheme,* $d_0 \geq a(Z)$ *such that* $C_Z(d_0) = C_Z(d_0+1)$. *Then there exists a fixed curve* D *of degree* $C_Z(d_0)$ *in the complete linear system* $|H^0(\mathcal{J}_{Z/\mathbb{P}^2}(d_0))|$ *with the additional property that for each* $d \geq 0$ *we have*

$$C_{Z \cap D}(d) = \min \{C_Z(d), C_Z(d_0)\}.$$

We call a zero-dimensional scheme $Z \subset \mathbb{P}^2$ *decomposable* if the graph of C_Z has a "long stair", that is, there is a positive integer d_0 such that $C_Z(d_0-1) > C_Z(d_0) = C_Z(d_0+1) > 0$. Barkats [5] showed how to apply the above properties of the Castelnuovo function for the computation of h^1 in relation to the smoothness, resp. irreducibility, problem for families of curves in $\mathbb{P}^2$. His fundamental lemma is

LEMMA 3.9 (Barkats, 1993). *Let* $C_d \subset \mathbb{P}^2$ *be an irreducible curve of degree* $d > 0$, *and* $Z \subset C_d$ *a zero-dimensional scheme satisfying* $h^1(\mathcal{J}_{Z/\mathbb{P}^2}(d)) > 0$. *Suppose, moreover,* $d > a(Z)$. *Then there is a curve* C_k *of degree* $k \geq 3$ *such that the scheme* $Y = C_k \cap Z$ *is non-decomposable and satisfies*

(a) $h^1(\mathcal{J}_{Y/\mathbb{P}^2}(d)) = h^1(\mathcal{J}_{Z/\mathbb{P}^2}(d))$,

(b) $\deg Y \geq k \cdot (d+3-k)$ *if* $k \leq \frac{d+3}{2}$, *and* $\deg Y > \frac{(d+3)^2}{4}$ *if* $k > \frac{d+3}{2}$.

Combining Barkats' approach with new ideas, and introducing a new invariant γ associated to a reduced curve singularity and a zero-dimensional scheme on it, it is not difficult to derive the following H^1-vanishing theorem (cf. [21], Proposition 4.1):

PROPOSITION 3.10. *Let* $C \subset \mathbb{P}^2$ *be an irreducible curve of degree* $d \geq 6$ *with* r *singular points* $z_1, \ldots, z_r$, $Z = Z_1 \cup \ldots \cup Z_r$ *a zero-dimensional scheme such that* $Z_i \subset Z^{ea}(C, z_i)$, $i = 1, \ldots, r$.

Then $H^1(\mathcal{J}_{Z/\mathbf{P}^2}(d))$ *vanishes if*

$$\sum_{i=1}^{r} \gamma(C; Z_i) \leq (d+3)^2. \qquad (3.9)$$

Here $\gamma(C; Z_i) := 0$, if $Z_i = \emptyset$, and, otherwise,

$$\gamma(C; Z_i) := \max_{(D,z)} \left\{ \frac{(\deg(D \cap Z_i) + \Delta(C, D; Z_i))^2}{\Delta(C, D; Z_i)} \right\},$$

where $\Delta(C, D; Z_i) := \min\{i(C, D; z) - \deg(D \cap Z_i), \deg(D \cap Z_i)\}$, and the maximum is taken over all curve germs $(D, z) \subset (\mathbf{P}^2, z)$ having no component in common with (C, z) ($i(C, D; z)$ denotes the local intersection multiplicity of C and D at z).

4. Results

Throughout the following, we consider both, equianalytic and equisingular families of plane curves. To save space, we use the superscript $'$ to denote either ^{ea} (in the equianalytic case) or ^{es} (in the equisingular case).

4.1. EXISTENCE PROBLEM

The statements of Prop. 2.10 and 3.7 imply the non-emptiness of $V_d^{irr}(m_1, \ldots, m_r)$ whenever $m_1 + m_2 \leq d-1$, $m_1 + \ldots + m_5 \leq 2d-2$, and

$$\sum_{i=1}^{r} \frac{(m_i + 1)^2}{2} < \frac{9(d+2)^2}{20},$$

which is (at the time of writing) the best known general existence theorem for curves with ordinary singularities. Moreover, by applying H^1-vanishing theorems for cluster schemes (resp. by giving explicit equations in the case of simple singularities), we proved the nonemptiness of $V_d^{irr}(S)$, S an arbitrary topological type of plane curve singularities, provided that $d \geq \sqrt{29\mu(S)} + \frac{9}{2}$ ([34]).

To construct curves with many singularities, we apply *Viro's glueing method* as modified and generalized by Shustin [50, 52, 53]. This method appears to be a strong tool for constructing controllable deformations. While the original Viro method ([60, 61, 62]) was invented for the construction of non-singular real algebraic hypersurfaces with prescribed topological properties, Shustin's modification allows to glue *singular* algebraic curves, resp. hypersurfaces, with adjacent Newton

polyhedra. Unlike the global topological results, which are based on
the use of toric varieties, the possibility to control the singularities
which may appear in the deformation of a degenerate singular object
depends on the T-smoothness of certain equisingularity strata.

In particular, if the degenerate object is a plane curve with ordi-
nary singularities, $C \in V_d^{irr}(m_1, \ldots, m_r)$, such that $V_d^{irr}(m_1, \ldots, m_r)$
is T-smooth at C, then the modified Viro method gives the exis-
tence of a deformation to a curve $C_t \in V_d^{irr}(S_1, \ldots, S_r)$ provided that
$V_{m_i-1}(S_i) \neq \emptyset$ has a (generically) T-smooth component. As a result,
we obtain

PROPOSITION 4.1 ([19, 35]). *Let d, r be positive integers, $S_1, \ldots, S_r$*
topological types of plane curve singularities with Milnor numbers
$\mu(S_1) \geq \cdots \geq \mu(S_r)$. Suppose that

$$\sqrt{\mu(S_1)} + \sqrt{\mu(S_2)} \leq \frac{d-12}{\sqrt{29}}, \quad \sqrt{\mu(S_1)} + \ldots + \sqrt{\mu(S_5)} \leq \frac{4d-59}{2\sqrt{29}}.$$

Then $V_d^{irr}(S_1, \ldots, S_r) \neq \emptyset$ provided that

$$\sum_{i=1}^{r} \mu(S_i) \leq \frac{(d+2)^2}{46}. \tag{4.10}$$

Note that the intersection of two generic polar curves of a curve in
$V_d(S_1, \ldots, S_r)$ leads to the necessary condition

$$\sum_{i=1}^{r} \mu(S_i) \leq (d-1)^2 \tag{4.11}$$

for the non-emptiness of $V_d(S_1, \ldots, S_r)$. Hence, the sufficient condition (4.10) is *asymptotically proper*, that is, has the same asymptotics
(with respect to d) as a necessary condition up to a constant factor
$\alpha = 46 > 1$.

REMARK 4.2.   In special cases there are much sharper bounds than
(4.10). Recall, for instance, that for nodal curves there is even a complete answer. For cuspidal curves one has a complete answer only for
small degrees ($d \leq 10$). For general d, we know at least that there
exist irreducible cuspidal curves of degree d having k cusps when-
ever $k \leq \frac{1}{4}d^2 - \frac{3}{4}d + 1$ ([50]). On the other hand, the existence of such
a curve implies $k \leq \frac{5}{16}d^2 - \frac{3}{8}d$ ([30, 45]), that is, asymptotically the
"uncertain region" is bounded by the lines $y = x/8$ and $y = 5x/32$.

In general, there are sharper upper bounds for the existence obtained
by applying the log-Miyaoka inequality ([45]), resp. the semi-continuity

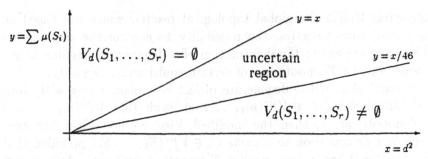

Figure 1. Necessary and sufficient conditions for the existence of curves with arbitrary topological types of singularities.

of the spectrum ([59]). On the other hand, taking into account special constructions and results for simple (ADE) and ordinary singularities, we can refine the sufficient condition (4.10) to (cf. [34])

$$(d+2)^2 \geq 10 \cdot \#\{\text{nodes}\} + \sum_{2 \leq \mu(S_i) \leq 8} 14 \cdot \mu(S_i) + \sum_{\substack{S_i \text{ ADE} \\ \mu(S_i) \geq 9}} 10 \cdot \mu(S_i)$$

$$+ \sum_{\substack{S_i \text{ ordinary} \\ \mu(S_i) \geq 9}} 4 \cdot \mu(S_i) + \sum_{\text{other } S_i} 46 \cdot \mu(S_i).$$

4.2. T-SMOOTHNESS PROBLEM

The classical approach to the T-smoothness problem, as applied by Severi, Segre, Zariski (cf. Section 3.1), leads to

COROLLARY 4.3 ([16, 17]). *Let* $S_1, \ldots, S_r$ *be topological (respectively analytic) types. Then*

(a) $V_d^{irr}(S_1, \ldots, S_r)$ *is T-smooth (or empty) if*

$$\sum_{i=1}^{r} (\tau'(S_i) - 1) < 3d. \tag{4.12}$$

(b) $V_d(S_1, \ldots, S_r)$ *is T-smooth (or empty) if*

$$\sum_{i=1}^{r} \tau'(S_i) < 4d - 4. \tag{4.13}$$

The latter being implied by (3.3), choosing C' to be a generic polar curve of C (which is irreducible and contains the scheme $Z^{ea}(C)$).

In the case of ordinary singularities $S_1, \ldots, S_r$, a more refined view on the isomorphism defect appearing in condition (3.2) gives:

COROLLARY 4.4 ([15, 17]). *Let $C \subset \mathbb{P}^2$ be a reduced curve with ordinary singularities of multiplicities $m_1, \ldots, m_r$ as only singularities and $C = C_1 \cup \ldots \cup C_s$ its decomposition into irreducible components, $\deg(C_i) = d_i$. Then $V_d(m_1, \ldots, m_r)$ is T-smooth at C if for $i = 1, \ldots, s$*

$$\sum_{x \in C_i \cap Sing\, C;\, mt(C,x) > 2} mt\,(C_i, x) < 3d_i. \tag{4.14}$$

In particular, $V_d^{irr}(m_1, \ldots, m_r)$ is T-smooth (or empty) if

$$\sum_{m_i > 2} m_i < 3d. \tag{4.15}$$

Note that all these, more or less classical, conditions are far from being asymptotically proper. However, the $3d$-formulas (4.12), (4.14) and (4.15) have the advantage that nodes do not count. In particular, (4.14) provides the fact that the Severi variety of nodal curves is T-smooth.

A qualitatively new result is obtained, when combining Proposition 3.10 with the approach of Chiantini and Sernesi (cf. [8, 20]), using Bogomolov's theory of unstable rank two vector bundles on surfaces.

PROPOSITION 4.5. *Let $C \subset \mathbb{P}^2$ be an irreducible curve of degree $d \geq 6$ having r singularities $z_1, \ldots, z_r$ of topological (respectively analytic) types $S_1, \ldots, S_r$ as its only singularities. Then $V_d^{irr}(S_1, \ldots, S_r)$ is T-smooth at C if*

$$\sum_{i=1}^{r} \gamma'(C, z_i) \leq (d+3)^2. \tag{4.16}$$

The invariants on the left-hand side being defined as

$$\gamma'(C, z_i) := \max\{\gamma\,(C; Z) \mid Z \subset Z'(C, z_i) \text{ a complete intersection}\}.$$

Note that $\gamma'(C, z_i) \leq (\tau'(C, z_i) + 1)^2$, cf. [21].

Proof. Assume that $V_d^{irr}(S_1, \ldots, S_r)$ is not T-smooth at C, that is, $h^1(\mathcal{J}_{Z'(C)/\mathbb{P}^2}(d))$ is positive. Let $\emptyset \neq Z_0 \subset Z'(C)$ be minimal with

$$h^1(\mathcal{J}_{Z_0/\mathbb{P}^2}(d)) \neq 0.$$

Then, by Serre-Grothendieck duality, $\text{Ext}^1(\mathcal{J}_{Z_0/\mathbb{P}^2}(d), \mathcal{O}_{\mathbb{P}^2}(-3)) \neq 0$, and a generic element defines an extension

$$0 \longrightarrow \mathcal{O}_{\mathbb{P}^2} \longrightarrow E \longrightarrow \mathcal{J}_{Z_0/\mathbb{P}^2}(d+3) \longrightarrow 0,$$

where E is a rank 2 vector bundle on $\mathbb{P}^2$ (cf., e.g., [33]). In particular, Z_0 is a (locally) complete intersection scheme.

Finally, the statement follows, by applying Proposition 3.10 to the zero-dimensional scheme $Z_0 \subset Z^{ea}(C)$. $\qquad\qquad\qquad\qquad\qquad\qquad$ □

Examples (cf. [36]) Let $(C,z) \subset (I\!\!P^2, z)$ be a *simple singularity*. Then $\gamma' = \gamma^{es} = \gamma^{ea}$ and

$$\gamma'(C,z) = \begin{cases} (k+1)^2 & \text{if } (C,z) \text{ is an } A_k\text{-singularity}, \\ (k-1)^2 & \text{if } (C,z) \text{ is a } D_k\text{-singularity}, \; k \geq 9, \\ \frac{1}{2}(k+2)^2 & \text{if } (C,z) \text{ is a } D_k\text{- or } E_k\text{-singularity}, \; k \leq 8. \end{cases}$$

Let $(C,z) \subset (I\!\!P^2, z)$ be an *ordinary singularity* of multiplicity $m \geq 2$. Then

$$\gamma^{es}(C,z) = \begin{cases} 4 & \text{if } m = 2, \\ 2m^2 & \text{if } m \geq 3. \end{cases}$$

The latter gives

COROLLARY 4.6. *$V_d^{irr}(m_1, \ldots, m_r)$ is empty or T-smooth if*

$$4 \cdot \#(nodes) + \sum_{m_i \geq 3} 2 \cdot m_i^2 \leq (d+3)^2. \qquad (4.17)$$

4.3. IRREDUCIBILITY PROBLEM

The method discussed at the end of Section 2 leads to the following proposition, which contains the best known general sufficient condition for the irreducibility of V (cf. [21], Theorem 2, with a slightly stronger formulation).

PROPOSITION 4.7. *Let $S_1, \ldots, S_r$ be topological (respectively analytic) types of plane curve singularities. If $\max_i \tau'(S_i) \leq (2/5)d - 1$ and*

$$\frac{25}{2} \cdot \#(nodes) + 18 \cdot \#(cusps) + \frac{10}{9} \cdot \sum_{\tau'(S_i) \geq 3} (\tau'(S_i)+2)^2 < d^2 \quad (4.18)$$

then the family $V_d^{irr}(S_1, \ldots, S_r)$ is irreducible.

5. Open problems and perspectives

While some of the results collected in Section 4 are sharp, others seem to be still far from a final form. In the following, we should like to discuss conjectures and perspectives.

5.1. EXISTENCE PROBLEM

A first natural question is: *How to improve the constant coefficient* 1/46 *in the general sufficient condition (4.10)?* Concerning our method, which uses H^1-vanishing theorems for generic fat point schemes, a proof of the Harbourne-Hirschowitz conjecture (the best possible H^1-vanishing criterion) would give such an improvement.

A probably more interesting question is the following. We provide asymptotically proper sufficient existence conditions only for plane curves with given *topological types* of singularities. We conjecture that the same asymptotics hold for analytic types. By our method of proof this conjecture actually reduces to

CONJECTURE 5.1.   *There is an absolute constant* $A > 0$ *such that, for any* $\mu > 0$ *and any singular point with Milnor number* μ, *there exists a curve of degree* $\leq A\sqrt{\mu}$ *having a singular point analytically equivalent to the given one.*

Other open questions concern the "uncertain region" in Figure 4.1. For instance: *Let* $k' < k$. *Is it possible that there exist cuspidal curves of degree* d *with* k *cusps, while no curve of degree* d *has exactly* k' *cusps?* Possible candidates could be Hirano's [26] examples of cuspidal curves with $d = 2 \cdot 3^r$, $k = 9(9^r-1)/8$, $r \geq 1$.

5.2. T-SMOOTHNESS PROBLEM

In the case of topological types of singularities, the following conjecture on the asymptotical properness of the sufficient condition (4.16) for T-smoothness seems to be realistic:

CONJECTURE 5.2.   *There exists an absolute constant* $B > 0$ *such that for any topological singularity* S *there are infinitely many pairs* $(r, d) \in \mathbb{N}^2$ *such that* $V_d^{irr}(r \cdot S)$ *is empty or not smooth or has dimension greater than the expected one and* $r \cdot \gamma^{es}(S) \leq B \cdot d^2$.

We know that the exponent 2 of d in the right-hand side of (4.16) cannot be raised in any reasonable sufficient criterion for T-property with the left-hand side being the sum of local singularity invariants. Hence, for an asymptotically proper sufficient criterion for T-property the right-hand side is correct. On the other hand, for the left-hand side of such a sufficient criterion different invariants can be used. What we conjecture is that the new invariant γ^{es} is the "correct" one for an asymptotically proper bound in the case of topological singularities.

Conjecture 5.2 is true for ordinary singularities, since the inequality $\sum_{i=1}^{r} m_i(m_i - 1) \leq (d-1)(d-2)$ is necessary for the existence of an irreducible curve with ordinary singularities of multiplicities $m_1, \ldots, m_r$. The results in [51, 20], respectively [36], show that Conjecture 5.2 does also hold for infinite series of singularities of types A and D.

We propose a similar conjecture in the case of analytic types, though it is confirmed only for simple singularities (in which case it coincides with Conjecture 5.2).

In connection with these conjectures another question concerns the invariants γ^{es}, γ^{ea}: *Find an explicit formula or algorithm to compute these invariants for any plane curve singularity (C, z). Is γ^{es} a topological invariant ?*

5.3. IRREDUCIBILITY PROBLEM

Our sufficient conditions for irreducibility seem to be far from being asymptotically proper. Nevertheless, we state the following problem: *Find asymptotically proper sufficient conditions for irreducibility, or show that the conditions in section 4.3 are asymptotically proper.*

To reach such an asymptotically proper sufficient irreducibility condition one should try to improve the results obtained, by reducing the singularity invariants in the left-hand side of the inequalities, or find examples of reducible ESF with asymptotics of the singularity invariants being as close as possible to that in the sufficient conditions.

At this point, we should like to comment on *Zariski pairs*, that is, pairs of curves of the same degree and with the same collection of singularities, but which have different fundamental groups of the complement in the plane. Note that a Zariski pair gives rise to two different connected components of $V_d(S_1, \ldots, S_r)$. Nori's theorem [42] implies, in particular, that $\pi_1(\mathbb{P}^2 \setminus C) = \mathbb{Z}/d\mathbb{Z}$ for any curve $C \in V_d^{irr}(S_1, \ldots, S_r)$ with $\sum_{i=1}^{r} 3\mu(S_i) < d^2$. Hence, any examples of Zariski pairs must have asymptotics of singularity invariants as in the necessary condition for existence (4.11), but not as in (4.18).

However, in [21], we showed, by an example of cuspidal curves, that the fundamental group of the complement can, in general, not distinguish between irreducible components.

5.4. RELATED PROBLEMS

5.4.1. *Curves on algebraic surfaces*
It is natural to consider the existence, T-smoothness and irreducibility problem more generally for equisingular families of curves on any smooth projective surface Σ.

The methods leading to the sufficient existence condition (4.10) can be extended to curves on arbitrary algebraic surfaces. As a result Keilen and Tyomkin obtain explicit sufficient conditions for the existence of curves with prescribed topological types of singularities for rational, direct product, ruled surfaces, surfaces in $I\!P^3$ and K3-surfaces [31].

A different approach is suggested by Chiantini and Ciliberto [7] who use Ran's deformation construction [43] in the case $\Sigma \subset I\!P^3$ being a smooth surface of general type. In particular, they show that there are *reducible* equisingular strata of *nodal* curves in Σ.

For T-smoothness basically only one general sufficient condition is known [57, 16, 17]:

$$-K_\Sigma C > \deg X' - \varepsilon', \qquad (5.19)$$

K_Σ the canonical divisor on Σ and $\varepsilon' \geq 0$ the "isomorphism defect".

So far, there are only some special cases where the condition (5.19) could be improved. We should like to mention the cases of nodal curves on rational surfaces, $K3$-surfaces and surfaces of general type [55, 57, 41, 18, 8].

Finally, concerning the irreducibility problem there are just a few results for special surfaces, e.g., for Hirzebruch surfaces [44] and for blown-up $I\!P^2$ [18]. The general problem is still widely open: *Let C be an ample divisor on Σ. Find sufficient conditions for the irreducibility of $V_{|dC|}(S_1, \ldots, S_r)$, $d \geq 1$, in the form of upper bounds to the sum of certain singularity invariants by a quadratic function of d.*

5.4.2. *Higher dimensions*

One can formulate the existence, T-smoothness, irreducibility problems for families of hypersurfaces with isolated singularities, belonging to (very) ample linear systems of projective algebraic varieties. To find an appropriate approach to these problems is the most important question.

While the construction of curves with prescribed singularities as presented above can, in principle, be generalized to higher dimensions [53, 65], for the smoothness and irreducibility problem we do not yet have an adequate approach. The first steps in this direction were made by Du Plessis and Wall [13], resp. Shustin and Tyomkin [54].

Acknowledgements

We should like to thank E. Shustin for the fruitful collaboration during the last years. Most of the results reported in this survey were obtained as joint work with him in the framework of Grant No. G 039-304.01/95 of the German Israeli Foundation for Research and Development. Work on this paper has also been supported by the DFG-Schwerpunkt "Globale Methoden in der komplexen Geometrie", Grant GR 640/9-1, of the Deutsche Forschungsgemeinschaft.

References

1. J. Alexander and A. Hirschowitz, Une lemme d'Horace différentielle: application aux singularités hyperquartiques de $I\!P^5$, J. Alg. Geom. 1 (1992) 411–426.

2. J. Alexander and A. Hirschowitz, An asymptotic vanishing theorem for generic unions of multiple points, Invent. Math. 140 (2000) 303–325.

3. V. I. Arnold, A. N. Varchenko, and S. M. Gusein-Zade, *Singularities of differentiable mappings vol I* Birkhäuser, Basel-Boston, 1985.

4. M. Artin, *Deformations of singularities*, Tata Inst. Fund. Res. Lect. Notes 54, 1976.

5. D. Barkats, Irréductibilité des variétés des courbes planes à noeuds et à cusps, preprint, Univ. de Nice-Sophia-Antipolis, 1993.

6. J. Bingener, Darstellbarkeitskriterien für analytische Funktoren, Ann. Sci. Ecole Norm. Sup. 13 (1980) 317–347.

7. L. Chiantini and C. Ciliberto, On the Severi varieties of surfaces in $I\!P^3$, J. Alg. Geom. 8 (1999) 67–83.

8. L. Chiantini and E. Sernesi, Nodal curves on surfaces of general type, Math. Ann. 307 (1997) 41–56.

9. C. Ciliberto, Geometric aspects of polynomial interpolation in more variables and of Waring's problem, preprint, 2000.

10. C. Ciliberto and R. Miranda, Linear systems of plane curves with base points of equal multiplicity, Trans. Amer. Math. Soc. 352 (2000) 4037–4050.

11. E. D. Davis, 0-dimensional subschemes of $I\!P^2$: new applications of Castelnuovo's function, Ann. Univ. Ferrara 32 (1986) 93–107.

12. S. Diaz and J. Harris, Ideals associated to deformations of singular plane curves, Trans. Amer. Math. Soc. 309:2 (1988) 433–467.

13. A. A. du Plessis and C. T. C. Wall, Versal deformations in spaces of polynomials of fixed weight, Compositio Math. 114 (1998) 113–124.

14. A. M. Gabrièlov, Bifurcations, Dynkin diagrams and modality of isolated singularities, Funct. Anal. Appl. 8 (1974) 94–98.

15. C. Giacinti-Diebolt, Variétés des courbes projectives planes de degré et lieu singulier donnés, Math. Ann. 266 (1984) 321–350.

16. G.-M. Greuel and U. Karras, Families of varieties with prescribed singularities, Compositio Math. 69 (1989) 83–110.

17. G.-M. Greuel and C. Lossen, Equianalytic and equisingular families of curves on surfaces, Manuscr. Math. 91 (1996) 323–342.

18. G.-M. Greuel, C. Lossen and E. Shustin, Geometry of families of nodal curves on the blown-up projective plane, Trans. Amer. Math. Soc. 350 (1998) 251–274.

19. G.-M. Greuel, C. Lossen and E. Shustin, Plane curves of minimal degree with prescribed singularities, Invent. Math. 133 (1998) 539–580.

20. G.-M. Greuel, C. Lossen and E. Shustin, New asymptotics in the geometry of equisingular families of curves, Int. Math. Res. Not. 13 (1997) 595–611.

21. G.-M. Greuel, C. Lossen and E. Shustin, Castelnuovo function, zero-dimensional schemes and singular plane curves, J. Alg. Geom. **9** (2000) 663-710.

22. G.-M. Greuel and E. Shustin, Geometry of equisingular families of curves, *Singularity Theory* (eds Bill Bruce and David Mond), London Math. Soc. Lecture Notes **263** (Cambridge Univ. Press, 1999) pp 79-108.

23. B. Harbourne, Complete linear systems on rational surfaces, Trans. Amer. Math. Soc. **289** (1985) 213-226.

24. J. Harris, On the Severi problem, Invent. Math. **84** (1985) 445-461.

25. R. Hartshorne, *Algebraic Geometry*, Springer, New York, 1977.

26. A. Hirano, Constructions of plane curves with cusps, Saitama Math. J. **10** (1992) 21-24.

27. A. Hirschowitz, La méthode d'Horace pour l'interpolation à plusieurs variables, Manuscripta Math. **50** (1985) 337-388.

28. A. Hirschowitz, Une conjecture pour la cohomologie des diviseurs sur les surfaces rationelles génériques. J. reine angew. Math. **397** (1989) 208-213.

29. A. A. Iarrobino, *Punctual Hilbert schemes*, Memoir Amer. Math. Soc. **188**, no. 10, 1977.

30. K. Ivinskis, *Normale Flächen und die Miyaoka-Kobayashi-Ungleichung*, Diplomarbeit, Univ. Bonn, 1985.

31. T. Keilen and I. Tyomkin, Existence of curves with prescribed topological singularities, preprint, Univ. Kaiserslautern, 2000.

32. A. Laudal, *Formal moduli of algebraic structures*, Springer Lecture Notes in Math **754**, 1979.

33. R. Lazarsfeld, Lectures on linear series, *Complex Algebraic Geometry* (ed J. Kollár), Amer. Math. Soc., 1997.

34. C. Lossen, The geometry of equisingular and equianalytic families of curves on a surface, thesis, Univ. Kaiserslautern, 1999.

35. C. Lossen, New asymptotics for the existence of plane curves with prescribed singularities, Comm. in Alg. **27** (1999) 3263-3282.

36. C. Lossen, Asymptotically proper results on smoothness of equisingular families, preprint, Univ. Kaiserslautern, 2000.

37. T. Mignon, Systèmes de courbes planes à singularités imposées: le cas des multiplicités inférieures ou égales à quatre, prépublication de l'ENS Lyon *230*, 1998.

38. T. Mignon, Courbes lisses sur les surfaces rationnelles génériques: un lemme d'Horace géométrique, prépublication de l'ENS Lyon *245*, 1999.

39. D. Mumford, *Lectures on curves on an algebraic surface*, Princeton Univ. Press, 1966.

40. N. Nagata, On the fourteenth problem of Hilbert, Amer. J. Math. **81** (1959) 766-772.

41. A. Nobile, Families of curves on surfaces, Math. Zeits. **187**:4 (1984) 453-470.

42. M. Nori, Zariski conjecture and related problems, Ann. Sci. Ec. Norm. Sup. (4) **16** (1983) 305-344.

43. Z. Ran, Enumerative geometry of singular plane curves, Invent. Math. **97** (1989) 447-465.

44. Z. Ran, Families of plane curves and their limits: Enriques' conjecture and beyond, Ann. of Math. **130** (1989) 121-157.

45. F. Sakai, Singularities of plane curves, *Geometry of complex projective varieties*, Seminars and Conferences **9** (Mediterranean Press, Rende, 1993) pp 257-273.

46. M. Schlessinger, Functors of Artin rings, Trans. Amer. Math. Soc. **130** (1968) 208–222.

47. B. Segre, Esistenza e dimensione di sistemi continui di curve piane algebriche con dati caraterri, Atti Acad. naz. Lincei Rendiconti ser. 6, **10** (1929) 31–38.

48. E. Sernesi, On the existence of certain families of curves, Invent. Math. **75** (1984) 25–57.

49. F. Severi, *Vorlesungen über algebraische Geometrie*, Teubner, 1921. Reprinted Johnson, 1968.

50. E. Shustin, Real plane algebraic curves with prescribed singularities, Topology **32** (1993) 845–856.

51. E. Shustin, Smoothness of equisingular families of plane algebraic curves, Int. Math. Res. Not. **2** (1997) 67–82.

52. E. Shustin, Gluing of singular and critical points, Topology **37** (1998) 195–217.

53. E. Shustin, Lower deformations of isolated hypersurface singularities, Algebra i Analiz **10**:5 (1999) 221–249.

54. E. Shustin and I. Tyomkin, Versal deformations of algebraic hypersurfaces with isolated singularities, Math. Ann. **313**:2 (1999) 297–314.

55. A. Tannenbaum, Families of algebraic curves with nodes, Compositio Math. **41** (1980) 107–126.

56. A. Tannenbaum, Families of curves with nodes on $K3$-surfaces, Math. Ann. **260** (1982) 239–253.

57. A. Tannenbaum, On the classical characteristic linear series of plane curves with nodes and cuspidal points: two examples of Beniamino Segre, Compositio Math. **51** (1984) 169–183.

58. B. Teissier, The hunting of invariants in the geometry of the discriminant, *Real and complex singularities, Oslo 1976* (ed Per Holm), (Sijthoff and Noordhoff, Alphen aan den Rijn, 1977) pp 567–677.

59. A. N. Varchenko, On semicontinuity of the spectrum and an upper estimate for the number of singular points of a projective hypersurface, Sov. Math. Doklady **27** (1983) 735–739.

60. O. Ya. Viro, Gluing of algebraic hypersurfaces, smoothing of singularities and construction of curves, *Proc. Leningrad Int. Topological Conf.* (1983) pp 149–197 (Russian).

61. O. Ya. Viro, Gluing of plane real algebraic curves and construction of curves of degrees 6 and 7, Springer Lecture Notes in Math. **1060** (1984) 187–200.

62. O. Ya. Viro, Real algebraic plane curves: constructions with controlled topology, Leningrad Math. J. **1** (1990) 1059–1134.

63. J. Wahl, Deformations of plane curves with nodes and cusps, Amer. J. Math. **96** (1974) 529–577.

64. J. Wahl, Equisingular deformations of plane algebroid curves, Trans. Amer. Math. Soc. **193** (1974) 143–170.

65. E. Westenberger, Applications of the Viro glueing method in algebraic geometry and singularity theory, Diplomarbeit, Univ. Kaiserslautern, 2000.

66. G. Xu, Ample line bundles on smooth surfaces, J. reine angew. Math. **469** (1995) 199–209.

67. O. Zariski, *Algebraic Surfaces*, 2nd ed. Springer Verlag, 1971.

On the preparation theorem for subanalytic functions

Adam Parusiński
Département de Mathématiques, Université d'Angers
(parus@tonton.univ-angers.fr)

1. Introduction

We present an introduction to the preparation theorem for subanalytic functions as a technique for studying the metric properties of subanalytic sets. The preparation theorem was introduced in [17] as one of the main tools in proving the existence of a Lipschitz stratification for subanalytic sets. Later it was used by J.-M. Lion and J.-P. Rolin to study various properties of singular sets such as for instance: integration on subanalytic sets, o-minimality, order of contact between solutions of differential equations, see [11], [12].

The preparation theorem in its original statement, see [17], [11], says that each subanalytic function $f(x_1, \ldots, x_n)$ of n real variables can be prepared with respect to the variable x_n in the following sense. The domain of f can be decomposed into a locally finite union of subanalytic sets such that on each of them

$$f(x) = (x_n - \varphi(x'))^r a(x') u(x), \tag{1.1}$$

where $x = (x', x_n) = (x_1, \ldots, x_n)$, φ and a are subanalytic functions of x', $u(x)$ is a unit in a sense which will be clear later, and $r \in \mathbb{Q}$. We stress two important features of this result. Firstly, in contrast to the Weierstrass preparation theorem, no assumption on the genericity of the system of coordinates is made. Secondly, the exponent r may be negative even if f is a solution function of an algebraic equation.

Our purpose is to introduce the preparation theorem for subanalytic functions as a multivariable Puiseux Theorem, or as a primitive version of a rectilinearization theorem. This will be done through Sections 2-4. The preparation theorem in a strengthened form is stated and proven in Section 5. Our proof is close in spirit to the one proposed in [17]. In Section 3 we propose also a complex counterpart of the theorem.

In the second part of this paper we present some applications of the preparation theorem; in particular the following beautiful result due to Lion and Rolin [12]. Suppose given a subanalytic family of k-dimensional compact subanalytic sets X_t, where $t \in \mathbb{R}$ is a parameter. Then the integral $\varphi(t) = \int_{X_t} f(t, x) \mathrm{d}\,\mathrm{vol}_k$ of a subanalytic function

D. Siersma et al. (eds.), *New Developments in Singularity Theory*, 193–215.
© 2001 K̲   r Academic Publishers. Printed in the Netherlands.

$f(x,t)$ admits an asymptotic expansion at $t = 0$ of the form $\varphi(t) = \sum_{i=0}^{k} g_i(t)(\ln t)^i$, where the $g_i(t)$ are fractional power series. We shall also discuss the density and the Lipschitz structure of subanalytic sets.

ASSUMPTION

In order to keep our arguments elementary we shall restrict to the case of semi-algebraic functions, that is, the solution functions $f(x)$ $x = (x_1, \ldots, x_n)$, of equations of the type

$$G(x, f(x)) = 0, \qquad (1.2)$$

where $G(x, z)$ is a real polynomial. Moreover we shall always assume that

$$G(x, z) = z^d + \sum_{i=1}^{d} a_i(x) z^{d-i}, \qquad (1.3)$$

although this assumption is not essential, see Exercise 2.4 below.

For subanalytic sets the ideas of the proof of the preparation theorem are similar but we need more technology. In [17, section 7] we used the local flattening theorem (see also [3] for the use of the local flattening theorem in a similar context). Alternatively, as in [11], one may use the finiteness theorem, see [11, p. 870], and the technique of [6].

1.1. TERMINOLOGY.

Consider a power series in $x = (x_1, \ldots, x_n)$ with complex (or real) coefficients a_α

$$f(x) = \sum_{\alpha} a_\alpha x^\alpha, \qquad (1.4)$$

where $\alpha = (\alpha_1, \ldots, \alpha_n)$. We call $f(x)$ a *fractional power series in x* if there is a positive integer q such that all $\alpha_i \in \frac{1}{q}\mathbb{Z}$ and all $\alpha_i \geq 0$. We call f *meromorphic* if the exponents α_i are allowed to be negative, but are bounded from below. A fractional power series (1.4) is *a unit* if the constant term a_0 of (1.4) is nonzero.

We say that $f(x)$ is *a meromorphic fractional monomial*, resp. *a fractional monomial*, if

$$f(x) = a \prod x_i^{\alpha_i}, \qquad a \neq 0, \qquad (1.5)$$

and all $\alpha_i \in \mathbb{Q}$, resp. all $\alpha_i \in \mathbb{Q}$ and $\alpha_i \geq 0$.

Finally we say that a fractional power series $f(x)$ is *a meromorphic fractional normal crossings*, resp. *a fractional normal crossings*, if f is the product of a meromorphic fractional monomial, resp. a fractional monomial, by a unit. Note, in particular, that a meromorphic fractional normal crossings is not identically zero.

In this paper we consider only series which are convergent in a neighborhood of the origin so usually we skip the word 'convergent' in the statements.

By a function with bounded partial derivatives we mean a function with all partial derivatives of the first order bounded.

1.2. NOTATION.

For $x = (x_1, \ldots, x_n)$ we shall often denote $x' = (x_1, \ldots, x_{n-1})$ and write $x = (x', x_n)$.

2. Puiseux Theorem

Consider a solution function $f(x)$, $x \in \mathbb{C}^n$, of equation (1.2), where $G(x, z)$ is of the form (1.3) with analytic coefficients $a_i(x)$. We may suppose that G is reduced, and hence that the discriminant $\Delta(x)$ of G is not identically zero. We call *the first non-zero coefficient of G* the coefficient a_d if $a_d \not\equiv 0$ and a_{d-1} otherwise. Note that in the latter case $a_{d-1} \not\equiv 0$ since G is reduced.

If $n = 1$ then, by the Newton-Puiseux Theorem, f can be expanded at the origin as a fractional power series in x. If $n > 1$ this is no longer true without additional assumptions on G. We present some such possible statements below. But in general in order to express $f(x)$ as a fractional power series we have to blow up the variables x, as follows from the work of Hironaka, see [8] for details. The idea of the preparation theorem follows in fact from this approach.

Our first statement, Lemma 2.1, is often called the Jung-Abhyankar Theorem, see [1] or [13]. We set $U_\delta = \{x \in \mathbb{C}^n : |x_i| < \delta_i\}$ for $\delta = (\delta_1, \ldots, \delta_n) \in \mathbb{R}_+^n$.

LEMMA 2.1. *Let $G(x, z) = z^d + \sum_{i=1}^d a_i(x)z^{d-i}$ be a complex analytic function defined on $U_\delta \times \mathbb{C}$, and assume that the discriminant $\Delta(x)$ of G is normal crossings. Then there exist positive integers r_i such that*

$$G(y_1^{r_1}, y_2^{r_2}, \ldots, y_n^{r_n}, z) = \prod_{i=1}^d (z - b_i(y)) \qquad (2.1)$$

on $U_{\delta'} \times \mathbb{C}$, where $\delta' = (\delta_1^{1/r_1}, \ldots, \delta_n^{1/r_n})$ and the b_i are complex analytic functions. Moreover, all the differences of the b_i are normal crossings.

If, furthermore, the first non-zero coefficient of G is normal cross-ings, then so are all the b_i's that are not identically zero.

Proof. By assumption, the projection of the zero set Z of G on U_δ is finite. Fix $i = 1, \ldots, n$ and take a point $x^0 \in H_i = \{x_i = 0, x_j \neq 0 \text{ for } j \neq i\}$. Then, by the Puiseux Lemma and the assumption on the discriminant, we may find r_i such that the substitution

$$x_i = y_i^{r_i},$$
$$x_j = y_j \text{ if } j \neq i$$

gives (2.1) near x^0. Such r_i does not depend on the choice of $x^0 \in H_i$ (in the sense that if it is good at one point it is so at the others). Fix such r_i's. Then the functions b_i satisfying (2.1) are well-defined as an (unordered) set of analytic functions on the complement C of the union of coordinate subspaces of codimension 2. Since C is simply connected, the b_i are in fact well-defined complex analytic functions on C. Moreover the b_i's are bounded, and therefore defined everywhere. They are analytic by Cauchy's integral formula. Since the discriminant of $G(y_1^{r_1}, y_2^{r_2}, \ldots, y_n^{r_n}, z)$ is normal crossings, so are all the differences of the b_i.

The product of all the b_i equals $(-1)^d a_d(y_1^{r_1}, y_2^{r_2}, \ldots, y_n^{r_n})$, and consequently if a_d is normal crossings so are all of the b_i. If a_d is identically equal to 0, then so is exactly one b_i, and the product of the others equals $(-1)^{d-1} a_{d-1}(y_1^{r_1}, y_2^{r_2}, \ldots, y_n^{r_n})$, which is by assumption normal crossings. The proof of the lemma is complete.

LEMMA 2.2. *Let $f(x)$ be as above and suppose that the discriminant $\Delta(x)$ of G is non-zero on the set*

$$V = \{(x', x_n) : 0 < |x_n - \varphi(x')| < |\psi(x')|\}, \tag{2.2}$$

where $\varphi(x')$ is a fractional power series and $\psi(x')$ a fractional normal crossings in $x' = (x_1, \ldots, x_{n-1})$. Then f is a fractional power series in $x_1, \ldots, x_{n-1}, (x_n - \varphi)/\psi$ on $\hat{V} = \{(x', x_n) : |x_n - \varphi(x')| < |\psi(x')|\}$.

If, furthermore, the first non-zero coefficient of G does not vanish on V, then f is a fractional normal crossings on $\hat{V}$ in the same system of coordinates.

Proof. After the change of coordinates $x_n = \varphi(x') + w\psi(x')$, $|w| < 1$, the zero set of the pull-back $\Delta(x', w)$ of the discriminant of G is contained in the union $\{\psi = 0\} \cup \{w = 0\}$ and hence is normal crossings. Thus the statement follows from Lemma 2.1. More precisely, f becomes a power series in $x_1^{r_1}, \ldots, x_{n-1}^{r_{n-1}}, w^{r_n}$, convergent for x_i, $i = 1, \ldots, n-1$, small, and $|w| < 1$.

LEMMA 2.3. *Let $f(x)$ be as above and suppose that the discriminant $\Delta(x)$ of G is non-zero on the set*

$$V = \{(x', x_n) : C|\hat{\psi}(x')| < |x_n - \varphi(x')| < c|\psi(x')|\}, \qquad (2.3)$$

$0 < c, C < \infty$, *where $\varphi(x'), \psi(x'), \hat{\psi}(x')$ are fractional power series in $x_1, \ldots, x_{n-1}$. Suppose, moreover, that $\psi, \hat{\psi}$ are fractional normal crossings in x' and that there exist $i_0 \in \{1, \ldots, n-1\}$ and $\gamma > 0$ such that*

$$\hat{\psi}(x') = x_{i_0}^{\gamma} \psi(x') \cdot unit. \qquad (2.4)$$

Then, on the set V, f is equal to a fractional power series in

$$x_1, \ldots, \widehat{x_{i_0}}, \ldots, x_{n-1}, \hat{\psi}/(x_n - \varphi), (x_n - \varphi)/\psi.$$

If, furthermore, the first non-zero coefficient of G does not vanish on V, then f is a fractional normal crossings in the same system of coordinates.

Proof. Consider the following coordinates on V

$$x_1, \ldots, \widehat{x_{i_0}}, \ldots, x_{n-1}, w^- = \frac{\hat{\psi}}{x_n - \varphi}, w^+ = \frac{x_n - \varphi}{\psi},$$

where $|w^-| < 1/C$ and $|w^+| < c$.

The zero set of the discriminant $\Delta(x_1, \ldots, \widehat{x_{i_0}}, \ldots, x_{n-1}, w^-, w^+)$ is contained in the zero set of $\hat{\psi}$, which is normal crossings in the new coordinates (use $(x_{i_0})^{\gamma} = w^- \cdot w^+ \cdot unit$). Hence the lemma again follows from Lemma 2.1.

EXERCISE 2.4. Suppose $f(x)$, $x \in \mathbb{C}^n$, is a solution function of (1.2) with $G(x, z)$ given by

$$G(x, z) = a_0(x)z^d + a_1(x)z^{d-1} + \cdots + a_d(x), \qquad (2.5)$$

with analytic coefficients a_i. Then $h(x) = a_0(x)f(x)$ satisfies the equation

$$h^d + \sum_{i=1}^{d} a_i a_0^{i-1} h^{d-i} = 0. \qquad (2.6)$$

Show that if $a_0(x)$, $a_d(x)$, and the discriminant of G are simultaneously normal crossings then f is a meromorphic fractional normal crossings. Generalize lemmas 2.2 and 2.3 to this case.

REMARK 2.5. In each of the cases considered in Lemmas 2.2 and 2.3 above, f becomes a power series

$$f(y_1, \ldots, y_n) = \sum_\gamma c_\gamma y_1^{\gamma_1} \cdots y_n^{\gamma_n}. \tag{2.7}$$

in $y = (y_1, \ldots, y_n)$, where $y_i = x_i$, $i = 1, \ldots, n-1$, and $y_n = x_n - \varphi$. The exponents γ_i of this series are not necessarily bounded from below. On the other hand these exponents cannot be arbitrary. Indeed, under the assumptions of Lemma 2.2, let $\psi(x') = unit \cdot \prod_{i=1}^{n-1} x_i^{\alpha_i}$. Then $\gamma_n \geq 0$ and the exponets γ_i, $i = 1, \ldots, n-1$, satisfy

$$\gamma_i + \gamma_n \alpha_i \geq 0. \tag{2.8}$$

Similarly, under assumptions of Lemma 2.3, let $\psi = unit \cdot \prod_{i=1}^{n-1} x_i^{\alpha_i}$, $\hat{\psi} = unit \cdot \psi \cdot x_{i_0}^{\gamma}$. Then (2.8) holds for $i = 1, \ldots, \hat{i_0}, \ldots, n-1$ and

$$\gamma_{i_0} + \gamma_n \alpha_{i_0} \geq 0, \quad \gamma_{i_0} + \gamma_n(\alpha_{i_0} + \gamma) \geq 0. \tag{2.9}$$

3. Rectilinearization in horns

This section contains an elementary study of the solution functions of (1.2) for $n = 2$. Given a solution function $f(x_1, x_2)$ of an equation of type (1.2), with G complex analytic, we will cover a neighborhood of the origin by a finite number of horns in $\mathbb{C}^2$ at the origin (see the definition below) such that f becomes a fractional normal crossings on each horn. We ignore the fact that f is only a multivalued function.

By a *solid horn* in $\mathbb{C}^2$ at the origin we mean the germ of a subset X such that

$$X = \{(x_1, x_2) | |x_2 - \varphi(x_1)| < |\psi(x_1)|, x_1 \in D\}, \tag{3.1}$$

where $\varphi(x_1)$, $\psi(x_1)$ are fractional power series defined on an open disc D centered at $0 \in \mathbb{C}$.

By a *hollow horn* in $\mathbb{C}^2$ at the origin we mean the germ of a subset X such that

$$X = \{(x_1, x_2) | |\hat{\psi}(x_1)| < |x_2 - \varphi(x_1)| < |\psi(x_1)|, x_1 \in D\}, \tag{3.2}$$

where $\varphi(x_1)$, $\psi(x_1)$, $\hat{\psi}(x_1)$ are fractional power series defined on an open disc D centered at $0 \in \mathbb{C}$ and we suppose

$$\hat{\psi}(x_1)/\psi(x_1) \to 0 \quad \text{as } x_1 \to 0. \tag{3.3}$$

Let X be a horn at the origin. By *the associated system of coordinates* $v = (v_1, v_2)$ we mean

(i) $v_1 = x_1$ and $v_2 = (x_2 - \varphi)/\psi$ for a solid horn.

(ii) $v_1 = \hat{\psi}/(x_2 - \varphi)$, $v_2 = (x_2 - \varphi)/\psi$ for a hollow horn.

THEOREM 3.1. *Let $f(x_1, x_2)$ be a solution function of an algebraic equation (1.2) with $G(x, z)$ complex analytic defined in a neighborhood of the origin in $\mathbb{C}^2 \times \mathbb{C}$.*

Then there exists a neighborhood of the origin which can be covered by a finite number of horns such that on each of them f is a meromorphic fractional normal crossings in the associated system of coordinates.

Proof. The proof is elementary. We sketch the main points.

Let $\Delta(x)$ denote the discriminant of $G(x, z)$ which we assume not identically zero (G is reduced). Let $a(x)$ denote the first non-zero coefficient of G. Denote $g(x) = \Delta(x) \cdot a(x)$. We suppose $G(x, z)$ of the form (1.3) and similarly that $g(x) = x_2^{d'} + \sum b_i(x_1) x_2^{d'-i}$, leaving the general case to the reader. These extra assumptions will imply that f is actually a fractional normal crossings on each horn.

Let $H = \{h_1(x_1), \ldots, h_s(x_1)\}$ denote the set of solution functions of

$$g(x_1, h(x_1)) = 0. \qquad (3.4)$$

After a substitution x_1 by x_1^m we may suppose each h_i analytic. Define the order of contact between the branches h_i and h_j to be the $r \in \mathbb{Q}$ such that

$$h_i(x_1) - h_j(x_1) = x_1^r \cdot unit, \qquad (3.5)$$

Let r_i be the maximal order of contact between h_i and the other functions in H. We may embed the graph of h_i in a solid horn X_i by setting $\varphi = h_i$, $\psi = \varepsilon x_1^{r_i}$, $\varepsilon > 0$ small. Then f is a fractional normal crossings on X_i by Lemma 2.2.

Now we show how to cover the complement of $\bigcup_i X_i$.

First consider the case when all orders of contact between different h_i are equal to r. Let $V = \{(x_1, x_2) : |x_2 - h(x_1)| < C|x_1^r|\}$ with $C > 0$ fixed and large. We may cover $V \setminus \bigcup_i X_i$ by finitely many solid horns defined by $\varphi = h + cx_1^r$ and $\psi = \varepsilon x_1^r$, $c \in \mathbb{C}$ being a constant and ε being positive and small. Again by Lemma 2.2, f is a fractional normal crossings on every such horn. Finally, the complement of V (in a neighborhood of the origin) is a hollow horn, $\varphi = h$, $\hat{\psi} = Cx_1^r$ and $\psi = c$, c being a small positive constant. Then f is a fractional normal crossings on this horn by Lemma 2.3.

In general let r denote the maximal order of contact between different h_i, and let $H_0 = \{h_{i_1}, \ldots, h_{i_l}\}$ be a maximal set of solutions such that (3.5) holds for $i, j \in \{i_1, \ldots, i_l\}$, $i \neq j$. Clearly $l \geq 2$. Fix an $h \in H_0$. Now we replace the whole H_0 in H by a single solution function h. The set of solutions H' obtained has cardinality strictly smaller than that of H. We apply to this set the same procedure as to H, which leads to the construction of a finite set of horns. There is precisely one solid horn X_h such that $\varphi = h$ and $\psi = \varepsilon x_1^s$, $s < r$. The other horns do not touch the solution functions from H_0 and hence we may suppose that they satisfy the statement of theorem. Now we decompose X_h into solid horns and one hollow horn as in the special case considered before.

4. Rectilinearization of subanalytic functions

Let $f : U \to \mathbb{R}$ be a bounded subanalytic function defined on a subanalytic subset $U \subset \mathbb{R}^n$. Our aim is to simplify (rectilinearize) f by composing it with some standard mappings which are defined below. The results presented below are based on [2], [3], [19], and [18].

A mapping $\sigma : U \to \mathbb{R}^n$, $U \subset \mathbb{R}^n$ open, is called *a standard affine modification* if, maybe after a permutation of variables,

$$\sigma(x_1, \ldots, x_n) = (x_1, \ldots, x_k, x_{k+1}x_n, \ldots, x_{n-1}x_n, x_n). \qquad (4.1)$$

or σ is an open embedding.

A mapping $\sigma : U \to \mathbb{R}^n$, $U \subset \mathbb{R}^n$ open, is called *a substitution of powers* if there are positive integers $r_1, \ldots, r_n$ and signs $\varepsilon_i \in \{-1, 1\}$, $i = 1, \ldots, n$, such that

$$\tau(x_1, \ldots, x_n) = (\varepsilon_1 x_1^{r_1}, \ldots, \varepsilon_n x_n^{r_n}). \qquad (4.2)$$

A mapping $\sigma : U \to \mathbb{R}^n$, $U \subset \mathbb{R}^n$ open, is called *a shift* if, maybe after a permutation of variables,

$$\sigma(x_1, \ldots, x_n) = (x_1, \ldots, x_{n-1}, x_n - \varphi(x')) \qquad (4.3)$$

where $\varphi(x')$ is a continuous subanalytic function.

THEOREM 4.1. *Let* $f : U \to \mathbb{R}$ *be a bounded subanalytic function defined on a relatively compact subanalytic subset* $U \subset \mathbb{R}^n$. *Then there exist a finite number of mappings* $\tau_i : U_i \to \mathbb{R}^n$ *such that*

(i) U_i is an open subset of $\mathbb{R}^n$, and τ_i is a finite composition of standard affine modifications, substitutions of powers, and shifts.

(ii) $\overline{U} = \cup \overline{\tau_i(U_i)}$ and $f \circ \tau_i$ is an analytic function and a normal crossings.

Proof. As always we suppose additionally that f satisfies an equation of type (1.2) with $G(x, z)$ reduced and in the form (1.3).

The proof is by induction on n and we establish the statement for all (complex-valued) solution functions of (1.2) and their differences simultaneously. Let $\Delta(x)$ denote the discriminant of $G(x, z)$ and $a(x)$ the first non-zero coefficient of G. Set $g(x) = \Delta(x) \cdot a(x)$, and assume that g is of the form (1.3). We may apply the inductive assumption to all complex valued solution functions $h(x')$ of the equation

$$g(x', h(x')) = 0, \tag{4.4}$$

and their differences. Thus by the inductive hypothesis g is of the form

$$g(x) = a(x)\Delta(x) = \prod_{j=1}^{s} (x_n - a_j(x')),$$

where the a_j are analytic (maybe complex valued) and all not identically zero differences $a_i - a_j$ are normal crossings. Now our goal is to make g normal crossings and then use Lemma 2.1.

LEMMA 4.2. *Let $V \subseteq \mathbb{R}^n$ be an open subset and let $g\colon V \to \mathbb{R}$ be a real analytic function on V such that*

$$g(x) = (x')^{\delta} \prod_{j=1}^{s} (x_n - a_j(x')). \tag{4.5}$$

where the a_j are analytic (maybe complex valued) and all not identically zero differences $a_i - a_j$ are normal crossings.

Then there exist a composition of global blowings-up with smooth nowhere dense centers $\tau\colon W \to V$ and a finite covering of W by open sets W_λ such that:

(i) for each λ there is an open embedding $W_\lambda \subset \mathbb{R}^n$ such that the induced coordinate functions on W_λ are the pull-backs by τ of functions of one of the following forms:

- $x_1, \ldots, x_{n-1}, \dfrac{x_n - \varphi(x')}{\psi(x')}$;

- $x_1, \ldots, \widehat{x_{i_0}}, \ldots, x_{n-1}, \dfrac{\hat{\psi}(x')}{x_n - \varphi(x')}, \dfrac{x_n - \varphi(x')}{\psi(x')}$

where $\varphi(x')$ is the real part of one of a_j or is identically zero,

$$\psi(x') = \prod_{i=1}^{n-1} x_i^{\gamma_i}, \quad \hat{\psi}(x') = x_{i_0} \psi(x'), \tag{4.6}$$

and $\gamma_i \in \mathbb{N}$, $i_0 \in \{1, \ldots, n-1\}$;

(ii) For every λ, the restriction of $g \circ \tau$ to W_λ is normal crossings in these coordinates.

For the proof we need the following elementary lemma which uses the following notation. Let $\alpha = (\alpha_1, \ldots, \alpha_n) \in \mathbb{N}^n$, $\beta = (\beta_1, \ldots, \beta_n) \in \mathbb{N}^n$. We write $\alpha \leq \beta$ if $\alpha_i \leq \beta_i$ for all i.

LEMMA 4.3. ([2, Lemma 4.7], [19, 6.VI])
Let $\alpha, \beta, \gamma \in \mathbb{N}^n$ and let $a(x), b(x), c(x)$ be invertible elements of $\mathbb{C}\{x\}$. If

$$a(x)x^\alpha - b(x)x^\beta = c(x)x^\gamma,$$

then either $\alpha \leq \beta$ or $\beta \leq \alpha$.

Proof. We show that up to permutation of α, β, and γ, we have $\beta = \gamma \leq \alpha$.

Let $\delta_k = \min\{\alpha_k, \beta_k, \gamma_k\}$. Suppose, for instance, that $\alpha_j > \delta_j$. Then, $x^{\alpha - \delta}$ vanishes identically on $\{x_j = 0\}$, and so on $\{x_j = 0\}$ we have

$$-b(x)x^{\beta - \delta} = c(x)x^{\gamma - \delta}. \tag{4.7}$$

Neither side of (4.7) is identically zero on $\{x_j = 0\}$ otherwise $\beta_j > \delta_j$ and $\gamma_j > \delta_j$. Hence $\beta_j = \gamma_j = \delta_j$. On the other hand (4.7) gives $\beta_k - \delta_k = \gamma_k - \delta_k$ for $k \neq j$. Consequently $\beta = \gamma \leq \alpha$ as claimed.

Proof of Lemma 4.2. We assume for simplicity that $0 \in V$ and we will work in a neighborhood of the origin. First we note that if $a_i(0) \neq a_j(0)$ then $a_i - a_j$, as normal crossings, vanishes nowhere on V. In particular if $a_j(0)$ is not real then $a_j - \bar{a}_j$ is nowhere zero and consequently a_j is not real anywhere. In this case the factor $(x_n - a_j(x'))(x_n - \bar{a}_j(x'))$ of g is a unit. Therefore it suffices to consider only those a_i whose values at 0 are real.

Let A denote the set of all $a_i(0)$ (assumed real). For every $a \in A$ denote $I_a = \{i : a_i(0) = a\}$ and define

$$Y_a = \bigcup_{i \in I_a} \{x : x_n = a_i(x')\} \quad V_a = V \setminus \bigcup_{a' \neq a} Y_{a'}. \tag{4.8}$$

Note that different Y_a's are disjoint and it follow from the construction below that each center of the blowings-up we consider lies over one of the Y_a's. Thus for a fixed a it suffices to blow-up V_a, and the blowings-up of the various V_a glue to one global blowing-up. Therefore we may assume that all values of $a_i(0)$ are equal.

The proof is by induction on the number of distinct a_i. If this number is 1 there is nothing to prove, g is already normal crossings. Suppose that this number, say m, is bigger than 1.

First we consider the case when at least one of the a_i's, say a_1 is real. Then after a shift $y_k = x_k$, $k = 1, \ldots, n-1$ and $y_n = x_n - a_1(x')$ we can assume that $a_1 \equiv 0$.

By Lemma 4.3 the exponents α^i, of the nonzero $a_i(x') \sim (x')^{\alpha^i}$ are totally ordered. Let ξ be the smallest such exponent. We proceed by induction on $(m, |\xi|)$. Choose (arbitrarily) $k = 1, \ldots, n-1$ such that $\xi_k \neq 0$. Let $\sigma : V' \to V$ be the blowing-up of $\{x_k = x_n = 0\}$. Then V' can be covered by two coordinate charts V'_n and V'_k such that the restrictions $\sigma_n = \sigma_{|V'_n}$ and $\sigma_k = \sigma_{|V'_k}$ are given in coordinates $y_1, \ldots, y_n$ on V'_n (V'_k resp.) by:

$$\sigma_n = \begin{cases} x_k = y_k y_n \\ x_i = y_i & \text{if } i \neq k \end{cases} \tag{4.9}$$

$$\sigma_k = \begin{cases} x_n = y_k y_n \\ x_i = y_i & \text{if } i \neq n. \end{cases} \tag{4.10}$$

In particular

$$g \circ \sigma_n(y) = y_n^p (y')^{\tilde{\delta}} \prod_{i=1}^{s} (1 - \tilde{a}_i(y)), \tag{4.11}$$

where $p = \xi_k + s$. Since each $\tilde{a}_i$ vanishes identically on $V'_n \setminus V'_k$, $g \circ \sigma_k$ is normal crossings in a neighborhood of $V'_n \setminus V'_k$ in coordinates $y_i = x_i$, $1 \leq i \leq n-1, i \neq k$, $y_k = x_k / (x_n - a_1(x'))$, $y_n = (x_n - a_1(x'))$.

On V'_k

$$g \circ \sigma_k(y) = (y')^{\delta} (y'_k)^s \prod_{i=1}^{s} (y_n - \tilde{a}_i(y')), \tag{4.12}$$

where, as it is easy to check, all nonzero $\tilde{a}_i$ and their differences are normal crossings and $\tilde{a}_1 \equiv 0$. Hence either m decreases or all $\tilde{a}_i$ vanish at 0. In the latter case, the exponents $\tilde{\alpha}^i$ of $\tilde{a}_i$ satisfy $\tilde{\alpha}^i_m = \alpha^i_m$ for $m \neq k$ and $\tilde{\alpha}^i_k = \alpha^i_k - 1$. Therefore, the smallest such exponent $\tilde{\xi}$ satisfies $|\tilde{\xi}| < |\xi|$ which concludes the inductive step. Note that the new coordinates of this step are $y' = x'$, $y_n = (x_n - a_1(x'))/x_k$. In particular the original x' coordinates are preserved up to the last step when they are only modified on the chart V'_n.

The case where all the a_i's are non-real is similar. Suppose as before that all the a_i vanish at 0. Suppose their number is $m > 0$, so m must be even. Fix one of the a_i's, say a_1 and let b_1 denote the real part of a_1. After a coordinate transformation $y_k = x_k$, $k = 1, \ldots, n-1$ and $y_n = x_n - b_1(x')$ we can assume that $b_1 \equiv 0$. Again by lemma 4.3 the exponents of $a_i - a_1 \sim (x')^{\alpha^i}$ are totally ordered; again we consider the smallest of them, ξ. We proceed by induction on $|\xi|$ to decrease m in the same way as above, that is to say we blow-up $\{x_k = x_n = 0\}$, $\xi_k \neq 0$. In order to obtain formulae analogous to (4.11) and (4.12), we need to show that all a_j are divisible by x_k. This indeed follows from our assumptions. Since all the differences of the a_j's are divisible by x_k so is the imaginary part $c_1 = (a_1 - \bar{a}_1)/2i$ of a_1. But $a_1 = ic_1$ and hence for any j, $a_j = (a_j - a_1) + ic_1$ is divisible by x_k as claimed. The rest of the construction is exactly the same as in the previous case. Therefore after this blowing-up either m decreases or if it remains the same then $|\xi|$ decreases. The process ends either on V'_n as in the previous case or on V'_k if all $\bar{a}_j(0)$ become non-real, which must happen eventually.

Thus $g(x)$ becomes normal crossings after the changes of coordinates given by Lemma 4.2. Consequently Theorem 4.1 follows from Lemma 2.1.

5.   The Preparation Theorem

The preparation theorem for subanalytic functions, Theorem 5.1 below, is no other than Theorem 4.1 written 'down-stairs' in terms of the original coordinates x on $\mathbb{R}^n$ without referring to blowings-up. Each map $\tau : U_i \to \mathbb{R}^n$ is replaced by its image, a subset of $\mathbb{R}^n$ we shall call a cusp. Then we recover the coordinates, say $v = (v_1, \ldots, v_n)$ on U_i, from the original coordinates x on $\mathbb{R}^n$ by an inductive construction. As a result we obtain a real analytic version of Theorem 3.1. Moreover, the construction gives us a more precise statement, since we are able to decompose the domain of f into cusps on each of which f has the required properties.

This translation is not difficult but requires quite a complex terminology in order to encode correctly the inductive procedure of construction of new coordinates. Therefore we first state the theorem and then explain the meaning of the terms used in the statement.

THEOREM 5.1. *Let $f(x)$, $x = (x_1, \ldots, x_n) \in \mathbb{R}^n$, be a bounded sub-analytic function defined on a relatively compact domain. Then the domain of f can be decomposed into a finite number of open pre-pared cusps on each of which f is a fractional normal crossings in the associated system of coordinates.*

(In the statement by 'decompose' we mean that the domain of f is expressed as the union of cusps, disjoint up to a set of dimension smaller than n.)

Let $y = (y_1, \ldots, y_n)$. In the real analytic category we consider fractional power series in $|y_i|$, rather than in y_i as in the complex case, in order for the radicals to be well-defined. In particular these series are not multi-valued.

In general by *an open cusp in $\mathbb{R}^n$* we mean a bounded subanalytic subset $X \in \mathbb{R}^n$ such that

$$X = \{(x', x_n) | \hat{\psi}(x') < \varepsilon(x_n - \varphi(x')) < \psi(x'), x' \in X'\}, \quad \varepsilon = \pm 1,$$
$$(5.1)$$

where X' is an open cusp in $\mathbb{R}^{n-1}$, $\varphi(x')$, $\psi(x')$, $\hat{\psi}(x')$ are continuous bounded subanalytic functions defined on X' and $0 \le \hat{\psi} < \psi$ on X'. One dimensional cusps are open segments.

The following definition is motivated by Lemmas 2.2 and 2.3, the associated system of coordinates of Section 3, and Lemma 4.2.

DEFINITION 5.2. *Suppose given $\varphi(w')$, $\hat{\psi}(w')$, $\psi(w')$, where $w' = w'(x')$ is a system of coordinates on $\mathbb{R}^{n-1}$ and $\hat{\psi}(w') = w_{i_0}\psi$, where $i_0 \in \{1, \ldots n-1\}$, or $\hat{\psi} \equiv 0$. Then the standard extension of coordinates w' is the system of coordinates on $\mathbb{R}^n$ given by*

Case 1: $w_1, \ldots, w_{n-1}, w_n = (x_n - \varphi)/\psi$ *if* $\hat{\psi} \equiv 0$;

Case 2: $w_1, \ldots, \widehat{w_{i_0}}, \ldots, w_{n-1}, w_n^- = \hat{\psi}/(x_n - \varphi), w_n^+ = (x_n - \varphi)/\psi$ *otherwise.*

DEFINITION 5.3. *An open prepared cusp in $\mathbb{R}^n$ is a subset $X \subset \mathbb{R}^n$, a system of functions $y_i(x_1, \ldots, x_i)$ on X, $i = 1, \ldots, n$, and bounded subanalytic functions $\varphi(y'), \psi(y'), \hat{\psi}(y')$ of $y' = (y_1, \ldots, y_{n-1})$ such that*

$$y_n = x_n - \varphi(y') \tag{5.2}$$

is of constant sign $\varepsilon = \pm 1$ on X. The variables y' depend only on x' and so we may consider $\varphi, \psi, \hat{\psi}$ as functions of x'. Then we require that

$$X = \{(x', x_n) | \hat{\psi}(x') < \varepsilon(x_n - \varphi(x')) < \psi(x'), x' \in X'\}, \tag{5.3}$$

where ε is a sign, X' is a prepared cusp with system of functions $y_1, \ldots, y_{n-1}$, and one of the following cases holds:

Case 1: $\hat{\psi} \equiv 0$ *and ψ is a meromorphic fractional monomial in y', positive on X';*

Case 2: ψ *and $\hat{\psi}$ are meromorphic fractional monomials in y', positive on X'.*

We also require that there is a system of coordinates $v = v(x) = (v_1(x), \ldots, v_n(x))$ on X, called the associated system of coordinates, constructed in the following way. Let $w' = (w'_1, \ldots, w'_{n-1})$ be the associated system of coordinates for X'. Then we suppose that φ is a fractional power series in w', $\hat{\psi}$ and ψ are fractional monomials in w', and v is a standard extension of coordinates in the sense of Definition 5.2.

REMARK 5.4.   The coordinates v are introduced to handle the problem of convergence. There is another equivalent approach to this problem. Note that each v_i is a meromorphic fractional monomial in y and vice versa. Suppose f is a fractional power series

$$f(y_1, \ldots, y_n) = \sum_{\gamma} c_\gamma y_1^{\gamma_1} \cdots y_n^{\gamma_n}. \tag{5.4}$$

with exponents γ_i not necessarily bounded from below. Then f is a fractional power series in v if and only if the exponents γ_i satisfy a sequence of inequalities defined recursively as in (2.8), (2.9) of Remark 2.5. One may easily check that there are exactly n such inequalities. Consequently the exponents $\gamma = (\gamma_1, \ldots, \gamma_n)$ are contained in a strongly convex rational cone $P \subset \mathbb{Q}^n$ spanned by the vectors of exponents of $v_j(y)$, $j = 1, \ldots, n$.

Proof of Theorem 5.1. The proof follows the idea of proof of Theorem 3.1 and Theorem 4.1. We sketch the main points.

We proceed by induction on n. We suppose that $f(x)$ is a semialgebraic function satisfying an equation of type (1.2) with $G(x, z)$ a real polynomial of the form (1.3). By the argument of the first part of the proof of Theorem 4.1 we may suppose that $g(x)$, the product of the discriminant of G and the first non-zero coefficient of G, is of the form (4.5). Thus we may place ourselves in the assumptions of Lemma 4.2. Let the functions $a_j(x')$ be given by (4.5) and let as before A denote the set of all real values $a_j(0)$. We divide the cylinder over a small neighborhood of the origin in $\mathbb{R}^{n-1}$ into finitely many prepared cusps.

Clearly in the complement of the union of the graphs of the a_j, g is normal crossings in the variables x'.

Choose one solution function, say a_1. Suppose for instance $a_1(0) = 0$. If a_1 is real we may again suppose $a_1 \equiv 0$. If a_1 is the only solution function a_i which vanishes at 0 then we divide a neighborhood of the origin into the two cusps:

$$X_+ = \{(x', x_n) : 0 < x_n < c\}$$
$$X_- = \{(x', x_n) : 0 < -x_n < c\},$$

c being a small positive constant. Clearly the statement of the theorem holds on each of those cusps.

Suppose there is more than one solution function a_i which vanishes at 0. Let ξ be the smallest of the exponents of such a_i, and let $\xi_k \neq 0$. We divide a neighborhood of the origin into three cusps:

$$X_+ = \{(x', x_n) : Cx_k < x_n < c\}$$
$$X_- = \{(x', x_n) : Cx_k < -x_n < c\}$$
$$X_0 = \{(x', x_n) : -Cx_k < x_n < Cx_k\},$$

where c, resp C, is a small, resp. large, positive constant. Clearly the statement of the theorem holds on the first two cusps. After the blowing up σ_k, X_0 becomes of the form $\{-C < y_n < C\}$. So we continue to subdivide using an inductive procedure.

The case where all a_i's vanishing at 0 are non-real is similar, but we make a shift $y_n = x_n - b_i(x)$, where b_i denotes the real part of a_i. The details are left to the reader.

EXERCISE 5.5.   Suppose that f satisfies (1.2) with G of the form (2.5), but no longer suppose that f is bounded. Use Exercise 2.4 to show that Theorem 4.1 still holds if we only require $f \circ \tau_i$ to be a meromorphic fractional normal crossings. Similarly Theorem 5.1 holds if we only require f to be a meromorphic fractional normal crossings on each cusp.

EXERCISE 5.6.   State and prove a multivariable version $(n > 2)$ of Theorem 3.1.

6. Integration with parameter

In this section we discuss the following theorem due to J.-M. Lion et J.-P. Rolin [12], see also [5].

THEOREM 6.1. *Let* $f(t, x)$, $t = (t_1, \ldots, t_m)$, $x = (x_1, \ldots, x_n)$, *be a bounded subanalytic function defined on the subanalytic set* $X \subset \mathbb{R}^m \times \mathbb{R}^n$. *Suppose that the fibres* $X_t = X \cap (\{t\} \times \mathbb{R}^n)$ *are bounded and of dimension at most* k. *Then the integral with parameter*

$$\varphi(t) = \int_{X_t} f(t, x) \, d \operatorname{vol}_k \qquad (6.1)$$

with respect to the k-*dimensional volume is of the form*

$$P(\tilde{t}_1, \ldots, \tilde{t}_d, \ln \tilde{t}_1, \ldots, \ln \tilde{t}_d), \qquad (6.2)$$

where $\tilde{t}_1, \ldots, \tilde{t}_d$ *are subanalytic functions in* t *and* P *is a real polynomial, of degree at most* k *with respect to the logarithms.*

The following special case is crucial for the proof.

LEMMA 6.2. *The statement of Theorem 6.1 holds if* $k = n$.

Proof. By the definition of subanalytic functions, the problem is local with respect to t. For the purpose of exposition, we consider only the case $m = 1$, $t \in \mathbb{R}$: the general case follows by a similar argument. We apply theorem 5.1 to $f(t, x)$ as a function of the variables $t, x_1, \ldots, x_n$. Thus we may suppose that the domain of f is a prepared cusp X, and that $f(t, x)$ is a fractional power series in the associated coordinates which we denote $t, v_1, \ldots, v_n$. We denote by t, y the system of functions (y in Definition 5.3) on X and by $\psi(y)$, $\hat{\psi}(y)$ the associated fractional power series.

The idea of the proof is very simple if we use t, y as a system of coordinates on X. For instance, for each t the change of variables $x \to y$ is 'triangular', in particular, of Jacobian identically equal to 1. Therefore

$$\varphi(t) = \int_{X_t} f(t, y) \, dy, \qquad (6.3)$$

The only inconvenience of the coordinates y is that $f(t, y)$ is not necessarily a meromorphic fractional power series in t, y and we have to worry about convergence. That is why we prefer to work in the associated system of coordinates v.

The proof is by induction on n. For the inductive step we need to consider the following larger family of functions on prepared cusps. Fix $p \in \mathbb{N}$ and consider finite sums of functions on X of the form

$$h(t, y) = f(t, y) \cdot \prod (\ln y_k)^{p_k}, \qquad (6.4)$$

where

(i) f is a convergent fractional power series in the variables t, v.

(ii) p_k are non-negative integers and $\sum p_k \le p$.

For the inductive step it suffices to show that for each such h the integral

$$\int_{\hat{\psi}}^{\psi} h(t, y)\, dy_n \tag{6.5}$$

is a function of similar type on X', with p replaced by $p + 1$.

In Case 1, $v_n = w_n = y_n/\psi$ and $dy_n = \psi\, dw_n$. Integration gives

$$\int_0^{\psi} h(t, y)\, dy_n = \int_0^1 h\psi\, dw_n \tag{6.6}$$

$$= \psi \prod_{k<n} (\ln y_k)^{p_k} \int_0^1 f(t, w', w_n) \cdot (\ln w_n - \ln \psi)^{p_n}\, dw_n.$$

Expand $(\ln w_n - \ln \psi)^{p_n}$ and then integrate by parts to show that the result of integration belongs to the desired class.

In Case 2 the change of coordinates is more complicated, see Definition 5.2. Denote the new associated coordinates by $w_n^- = \hat{\psi}/y_n$, $w_n^+ = y_n/\psi$. Recall that $w_n^- \cdot w_n^+ = \hat{\psi}/\psi = w_{i_0}$. We need the following elementary lemma.

LEMMA 6.3. *(see [12], 1.6) Let $g(u, w^-, w^+)$ be a convergent power series. Then*

$$g(u, w^-, w^+) = g_1(u, w^-w^+, w^+) + w^- g_2(u, w^-w^+, w^-), \tag{6.7}$$

where g_1 and g_2 are convergent power series.

Let f be a fractional power series in the variables t, v. Since $w_n^- \cdot w_n^+ = w_{i_0}$, by Lemma 6.3 we may split

$$f = f_- + y_n^{-1} f_0 + f_+ \tag{6.8}$$

so that:

(i) f_0 is a fractional power series in t, w',

(ii) $f_- = (w_n^-)^{\beta+1} \tilde{f}_-(t, w', w_n^-)$, where $\beta > 0$ and $\tilde{f}_-(t, w', w_n^-)$ is a fractional power series,

(iii) $f_+ = (w_n^+)^{\beta-1} \tilde{f}_+(t, w', w_n^+)$, where $\beta > 0$ and $\tilde{f}_+(t, w', w_n^+)$ is a fractional power series.

The splitting of f gives a splitting of h

$$h(t, y) = h_-(t, y) + y_n^{-1} h_0(t, y) + h_+(t, y), \qquad (6.9)$$

where $h_\pm(t, y) = f_\pm(t, y) \cdot \prod (\ln y_k)^{p_k}$ and $h_0(t, y) = f_0(t, y') \cdot \prod (\ln y_k)^{p_k}$.

Then we integrate each summand of (6.9) separately. We shall compute the integrals of h_0 and h_-. The case of h_+, being similar to (6.6), is left to the reader.

Recall that $w_n^- = \hat\psi/y_n$ and $dy_n = -\hat\psi \cdot (w_n^-)^{-2} \, dw_n^-$. Then

$$\int_{\hat\psi}^{\check\psi} h_-(t, y) \, dy_n = \int_{w_{i_0}}^1 h_-(t, (y(w))) \cdot \hat\psi \cdot (w_n^-)^{-2} \, dw_n^-$$

$$= \hat\psi \prod_{k<n} (\ln y_k)^{p_k} \int_{w_{i_0}}^1 \tilde f_-(t, w', w_n^-) \cdot (\ln \hat\psi - \ln w_n^-)^{p_n} \cdot (w_n^-)^{\beta-1} \, dw_n^-,$$

and we argue as in the case of the integral (6.6).

Finally

$$\int_{\hat\psi}^{\check\psi} y_n^{-1} h_0(t, y) \, dy_n = \prod_{k<n} (\ln y_k)^{p_k} \cdot f_0(t, y') \cdot \int_{w_{i_0}}^1 y_n^{-1} (\ln y_n)^{p_n} \, dy_n$$

$$\hspace{10cm} (6.10)$$

is again of the required form, as is easy to see. This ends the proof of the lemma.

Proof of Theorem 6.1. Suppose first that X is the graph of $\varphi(t, x)$: $U \to \mathbb{R}^{n-k}$, where $U \subset \mathbb{R} \times \mathbb{R}^k$ is subanalytic and φ is subanalytic with bounded partial derivatives in x. Then

$$\int_{X_t} f(t, x) \, d \operatorname{vol}_k = \int_{U_t} g(t, x) \, dx_1 \ldots dx_k, \qquad (6.11)$$

where $g(t, x)$ is bounded and subanalytic, and the theorem follows from Lemma 6.2.

In general, X can be decomposed into finitely many subanalytic pieces $X = \bigcup X_i$; each X_i being, after an appropriate orthogonal change of coordinates in $\mathbb{R}^n$, the graph of $\varphi_i(t, x_1, \ldots, x_k) : U_i \to \mathbb{R}^{n-k}$, where $U_i \subset \mathbb{R} \times \mathbb{R}^k$ is subanalytic and bounded and φ_i is a bounded subanalytic mapping having bounded derivatives with respect to the variables x. This completes the proof.

Suppose $m = 1$ and we study the asymptotic behavior of the integral (6.1) as $t \to 0^+$. Then each $\tilde t_i(t)$ is a fractional power series in t. Consequently in this case we obtain a particularly simple statement.

COROLLARY 6.4. *The integral* (6.1) *with one parameter* $t \in \mathbb{R}$ *of a bounded subanalytic function* $f(t, x)$ *admits an asymptotic expansion as* $t \to 0^+$

$$\int_{X_t} f(t, x) \, d \operatorname{vol}_k = \sum_{j=0}^{k} \sum_{p \in \frac{1}{q}\mathbb{N}} a_{p,j} t^p (\ln t)^j, \qquad (6.12)$$

where q *is a positive integer, and the coefficient of each* $(\ln t)^j$ *is a (convergent) fractional power series.*

REMARK 6.5. *Stability by limits* (see [5], [12]). Let $g(t)$ be a function of the form

$$g(t) = P(\tilde{t}_1, \dots, \tilde{t}_d, \ln \tilde{t}_1, \dots, \ln \tilde{t}_d), \qquad (6.13)$$

where $\tilde{t}_1, \dots, \tilde{t}_d$ are subanalytic functions of $t \in \mathbb{R}^m$ and P is a real polynomial. Set $t = (t', t_m)$. Then the set of those $t' \in \mathbb{R}^{m-1}$ such that the limit

$$h(t') = \lim_{\varepsilon \to 0^+} g(t', \varepsilon) \qquad (6.14)$$

exists is subanalytic. Moreover, $h(t')$ is a function of the same type, that is, a polynomial in subanalytic functions and logarithms of subanalytic functions.

7. Density

Let $X \in \mathbb{R}^n$ be subanalytic of dimension k, and $x_0 \in X$. Denote by $B(x_0, r)$ the ball centered at x_0 and of radius r, and by σ_k the k-volume of the k-dimensional unit ball. The theorem of Kurdyka and Raby [10] states that the limit

$$\lim_{r \to 0} \frac{\operatorname{vol}_k(X \cap B(x_0, r))}{\sigma_k r^k} \qquad (7.1)$$

always exists and is finite (it could be zero). This limit is called *the density of* X *at* x_0. We next obtain a more precise statement.

PROPOSITION 7.1. *The asymptotic expansion of* $\operatorname{vol}_k(X \cap B(x_0, r))$ *is of the form*

$$\operatorname{vol}_k(X \cap B(x_0, r)) = \sum_{j=0}^{l_0} a_{p_0, j} r^{p_0} (\ln r)^j + \dots, \qquad a_{p_0, l_0} \neq 0, \quad (7.2)$$

where either $p_0 = k$ *and* $l_0 = 0$ *or* $p_0 > k$ *and then* $l_0 \leq k - 2$.

Proof. For an asymptotic expansion as in (6.12) we call the term $a_{p_0,j_0} t^{p_0} (\ln t)^{j_0}$ the *leading term* and p_0, l_0 the *leading exponents*.

Decompose X into finitely many subanalytic pieces $X = \bigcup X_i$, each X_i being, after an appropriate orthogonal change of coordinates, the graph of $\varphi_i(x_1, \ldots, x_k) : U_i \to \mathbb{R}^{n-k}$, where $U_i \subset \mathbb{R}^k$ is subanalytic and φ_i is a bounded subanalytic mapping with bounded partial derivatives. Then the leading exponents of the asymptotic expansions of the volumes of X_i and U_i coincide. Thus it suffices to consider the case $n = k$. In this case, $\mathrm{vol}_k(X \cap B(x_0, r)) \leq \sigma_k r^k$ and the statement is clear if $p_0 = k$.

By the proof of proposition 6.2, each one variable integration may produce one logarithm, but only in Case 2. The first integration is that of the constant function equal to 1 so does not produce a logarithm. The last integration, which is from 0 to r, is in Case 1 and does not give an extra logarithm either.

Note that the leading exponents of the asymptotic expansion of volume are bi-lipschitz invariants. They are related to the leading exponents of the asymptotic expansion

$$\mathrm{vol}_{k-1}(X \cap S(x_0, r)), \qquad (7.3)$$

where $S(x_0, r)$ denote the sphere centered at x_0 and of radius r. Indeed, write

$$s(r) = \mathrm{vol}_{k-1}(X \cap S(x_0, r))$$
$$b(r) = \mathrm{vol}_k(X \cap B(x_0, r))$$

and let p_0', l_0' denote the leading exponents of $s(r)$. Then

$$b(r) = \int_0^r s(\rho)\, d\rho = \int_0^r a_{p_0',l_0'} \rho^{p_0'} (\ln \rho)^{l_0'}\, d\rho + \ldots$$
$$= a_{p_0',l_0'} (p_0' + 1)^{-1} r^{p_0'+1} (\ln r)^{l_0'} + \ldots$$

and hence $p_0 = p_0' + 1, l_0 = l_0'$. The asymptotic expansion of $s(r)$ for real analytic surfaces was studied in [7].

Finally we present an example when the leading term of the asymptotic expansion of volume contains a logarithm (this example is a result of discussion with D. Grieser).

EXAMPLE 7.2. Let d be an integer ≥ 4. Consider a semi-algebraic subset of $\mathbb{R}^3$

$$X = \{(x, y, z) \in \mathbb{R}^3 : 0 \leq x \leq \tfrac{1}{2},\ x^2 \leq y \leq x,\ 0 \leq z \leq x^d/y\}. \quad (7.4)$$

Then the leading term of the asymptotic expansion of $b(r) = \text{vol}_k(X \cap B(x_0, r))$ coincides with the leading term of

$$c(r) = \int_0^r \left(\int_{x^2}^x \left(\int_0^{x^d/y} 1 \, dz \right) dy \right) dx = -(d+1)^{-1} r^{d+1} \ln r + \cdots . (7.5)$$

Indeed, it is easy to see that $b(r) \leq c(r) \leq b(2r)$.

8. Towards understanding the Lipschitz structure of subanalytic sets

The study of the Lipschitz structure of analytic sets was initiated in a fundamental paper of Mostowski [14] on Lipschitz stratification of complex analytic sets. In our work we considered mostly the real analytic and subanalytic cases, see [16] for a review of this work. One of our main tools in this study was the preparation theorem. It seems that a better understanding of this theorem should lead to a complete classification of local Lipschitz structure of subanalytic sets. For such a classification in the 2-dimensional case see [4].

DEFINITION 8.1. *We call a prepared cusp $Y_n \subset \mathbb{R}^n$ L-regular if the functions $\varphi(y')$, $\psi(y')$, $\hat{\psi}(y')$ of its Definition 5.3 have bounded derivatives and Y_{n-1} also is L-regular.*

An (open) L-cusp in $\mathbb{R}^n$ is a subset $Y \subset \mathbb{R}^n$ such that there is an orthogonal system of coordinates in $\mathbb{R}^n$ in which Y is an L-regular prepared cusp as defined above.

A k-dimensional subanalytic subset $X \in \mathbb{R}^n$ will be called a k-dimensional L-cusp if, maybe after an orthogonal change of coordinates in $\mathbb{R}^n$, X is the graph of an analytic mapping $\Phi : Y \to \mathbb{R}^{n-k}$, whose graph is subanalytic in $\mathbb{R}^n$, such that Y is an open L-cusp in $\mathbb{R}^k$ and Φ has bounded partial derivatives.

Suppose Y is an L-regular prepared cusp in $\mathbb{R}^n$. Then the change of coordinates $x \to y$ defined in Definition 5.3 is bilipschitz. In particular each open L-cusp is bilipschitz homeomorphic to a set defined by inequalities in meromorphic fractional monomials in y. The same holds for an arbitrary L-cusp. Indeed any analytic mapping defined on an open L-cusp which has bounded derivatives is Lipschitz. This follows for instance from the following lemma (the proof is left to the reader).

LEMMA 8.2. *Let Y be an open L-cusp in $\mathbb{R}^n$. Then*

(1) Y is homeomorphic to an open ball;

(2) for every $p, q \in Y$ there exists an C^1 curve γ in Y joining p and q and such that $\text{length}(\gamma) \leq C|p - q|$, where C does not depend on the choice of p, q;

(3) if $\varphi : Y \to \mathbb{R}$ is a C^1 map with bounded partial derivatives, then φ is Lipschitz.

The closures of L-cusps are in particular L-regular sets in the sense of [15], [17], [9]. As proven in Proposition 2.13 of [17], each compact subanalytic set can be decomposed as a union of L-regular sets. The proof is quite technical and is based on the regular projection theorem. In particular the way in which L-regular sets coming from different system of coordinates fit together is not clear. The same technique should allow us to show that each compact subanalytic set can be decomposed as a disjoint union of L-cusps.

REMARK 8.3.   It is necessary to allow negative exponents in the monomial inequalities defining L-cusps. Indeed, one cannot expect to be able to decompose any subanalytic set as a union of L-cusps defined by monomial inequalities with non-negative exponents, since such L-cusps have no logarithm in the leading term of the asymptotic expansion (7.2). On the other hand Example 7.2 shows that logarithms may appear in the leading term already for sets of dimension ≥ 3.

References

1. R. Abhyankar, On the ramification of algebraic functions, Amer. J. Math **77** (1955) 575–592.
2. E. Bierstone and P. D. Milman, Semianalytic and subanalytic sets, Publ. Math. IHES. **67** (1988) 5–42.
3. E. Bierstone and P. D. Milman, Arc-analytic functions, Invent. math. **101** (1990) 411–424.
4. L. Birbrair, Local bi-Lipschitz classification of 2-dimensional semi-algebraic sets, Houston J. Math. **25** (1999) 453–471.
5. G. Comte, J.-M. Lion and J.-P. Rolin, Nature log-analytique du volume des sous-analytiques, Illinois J. Math, to appear.
6. J. Denef and L. van den Dries, p-adic and real subanalytic sets, Ann. of Math. **128** (1988) 79–138.
7. D. Grieser, Local geometry of singular real analytic surfaces, preprint.
8. H. Hironaka, *Introduction to real-analytic sets and real-analytic maps*, Inst. Mat. L. Tonelli, Pisa, 1973.
9. K. Kurdyka, On a subanalytic stratification satisfying a Whitney property with exponent 1, *Real Algebraic Geometry, Proceedings, Rennes 1991* (eds M. Coste et al.), Springer Lecture Notes in Math. **1524** (1992) 316–323.

10. K. Kurdyka and G. Raby, Densité des ensembles sous-analytiques, Ann. Inst. Fourier **39**:3 (1989) 753–771.

11. J.-M. Lion and J.-P. Rolin, Théorème de préparation pour les fonctions logarithmico-exponentielles, Ann. Inst. Fourier **47**:3 (1997) 859–884.

12. J.-M. Lion, J.-P. Rolin, Intégration des fonctions sous-analytiques et volumes des sous-ensembles sous-analytiques, Ann. Inst. Fourier **48**:3 (1999) 755–767.

13. I. Luengo, A new proof of the Jung-Abhyankar theorem, J. Algebra **85** (1983) 399–409.

14. T. Mostowski, Lipschitz Equisingularity, Dissertationes Math. **243** (1985).

15. A. Parusiński, Lipschitz properties of semianalytic sets, Ann. Inst. Fourier **38**:4 (1988) 189–213.

16. A. Parusiński, Lipschitz stratification, *Global analysis in modern mathematics* (proceedings of a symposium in honor of Richard Palais' sixtieth birthday), (ed K. Uhlenbeck), (Publish or Perish, Houston, 1993) pp 73–91.

17. A. Parusiński, Lipschitz stratifications of subanalytic sets, Ann. Sci. Ec. Norm. Sup. **27** (1994) 661–696.

18. A. Parusiński, Subanalytic functions, Trans. Amer. Math. Soc. **344** (1994) 583–595.

19. H. J. Sussmann, Real-analytic desingularization and subanalytic sets: an elementary approach, Trans. Amer. Math. Soc. **317** (1990) 417–461.

Computing Hodge-theoretic invariants of singularities

Mathias Schulze
Fachbereich Mathematik, Universität Kaiserslautern,
(mschulze@mathematik.uni-kl.de)

Joseph Steenbrink
Mathematical institute, University of Nijmegen,
(steenbri@sci.kun.nl)

1. Introduction

Let $Y = V(F) \subset \mathbb{P}^{n+1}$ be a smooth hypersurface of degree d. By the exact sequence

$$0 \to H^{n+1}(\mathbb{P}^{n+1} \setminus Y, \mathbb{C}) \to H^n(Y, \mathbb{C}) \to H^{n+2}(\mathbb{P}^{n+1}, \mathbb{C}) \to 0,$$

the primitive cohomology of Y in degree n is identified with the cohomology of its complement in projective space. On the other hand this group can be described, by Grothendieck's algebraic de Rham theorem [5], as the space of rational differential $n + 1$-forms on $\mathbb{P}^{n+1}$ with poles only along Y modulo exact forms. According to Griffiths [4], this space is filtered by the order of pole of representatives along X and the resulting filtration on $H^n(Y, \mathbb{C})$ is its *Hodge filtration*.

Varying the equation of Y we obtain a holomorphic vector bundle $\mathcal{H}^n$ on the space U of nonsingular hypersurfaces, together with a holomorphically varying filtration $\{\mathcal{F}^p\}$ by the Hodge bundles. The bundle $\mathcal{H}^n$ is equipped with the *Gauss-Manin connection*

$$\nabla : \mathcal{H}^n \to \Omega_U^1 \otimes \mathcal{H}^n$$

which is integrable, and satisfies the *Griffiths transversality condition*

$$\nabla(\mathcal{F}^p) \subset \Omega_U^1 \otimes \mathcal{F}^{p-1}.$$

Now consider a hypersurface Y_0 which may be singular. We put it in a family $f : \mathcal{Y} \to S$, where S is the unit disc in the complex plane, and $Y_0 = f^{-1}(0)$. We assume that 0 is the only critical value of f. Then the cohomology groups of the fibers of f form a local system over the punctured disc, and we have a monodromy transformation which is quasi-unipotent by the monodromy theorem. By passing to a ramified covering of the disc, we may assume that the monodromy of the family is in fact unipotent. Let $u_1, \dots, u_m$ be a basis for the group

D. Siersma et al. (eds.), *New Developments in Singularity Theory*, 217–233.
© 2001 Kluwer Academic Publishers. Printed in the Netherlands.

of multivalued horizontal sections of $R^n f_* \mathbb{Z}_Y$ (modulo torsion). Write $N = \log T$, and let $\tilde{u}_i = \exp(-2\pi i N)u_i$. Then $\tilde{u}_1, \ldots, \tilde{u}_m$ is a basis of sections for $\mathcal{H}^n$ over $S \setminus \{0\}$ and we can extend $\mathcal{H}^n$ to a vector bundle over the whole of S by taking these sections as a frame.

For a complex vector space V of finite dimension equipped with a nilpotent endomorphism N, and a given integer n, one defines the *weight filtration of N centered at n* as the unique increasing filtration $\{W_k V\}$ of V with the properties that $N(W_k) \subset W_{k-2}$ for each k and the induced maps $N^k : \mathrm{Gr}^W_{n+k} V \to \mathrm{Gr}^W_{n-k} V$ are isomorphisms for all $k > 0$.

The results of W. Schmid [14] imply that

1. The Hodge bundles extend to holomorphic sub-bundles $\tilde{\mathcal{F}}^p$ of $\tilde{\mathcal{H}}^n$;

2. the connection extends to a logarithmic connection
$$\nabla : \tilde{\mathcal{H}}^n \to \Omega^1_S(\log 0) \otimes \tilde{\mathcal{H}}^n$$
whose residue at 0 is nilpotent (and can be identified with N);

3. the vector space $\tilde{\mathcal{H}}^n(0)$ with its Hodge filtration $\{\tilde{\mathcal{F}}^p(0)\}$, its integral lattice given by the values at 0 of the sections $\tilde{u}_1, \ldots, \tilde{u}_m$, and the filtration $W(N, n)$ is a mixed Hodge structure, polarized by N in a suitable sense;

4. the semisimple part T_s of the monodromy (before passing to a finite covering) acts as an automorphism of this mixed Hodge structure.

The latter mixed Hodge structure is called the *limit mixed Hodge structure* of the family.

Suppose that we have an isolated hypersurface singularity
$$f : (\mathbb{C}^{n+1}, 0) \to (\mathbb{C}, 0).$$

The coordinates on $\mathbb{C}^{n+1}$ are denoted as $(z_0, \ldots, z_n)$ and t is the coordinate on the target $\mathbb{C}$. As f is finitely determined, we may represent it by a polynomial of arbitrarily high degree d. By results of Brieskorn [2] and Scherk [13], we may assume that f has moreover the following properties:

1. Let $F(Z_0, \ldots, Z_{n+1}) = Z_{n+1}^d f(Z_0/Z_{n+1}, \ldots, Z_n/Z_{n+1})$ be the usual homogenization of f. Then the hypersurface $Y_0 = V(F)$ in $\mathbb{P}^{n+1}$ has a unique singular point at $x = [0, \ldots, 0, 1]$.

2. Let $Y_t = V(F - tZ_{n+1}^d)$, and let X_t be the intersection of Y_t with a sufficiently small ball around x, i.e. a Milnor fibre of f at x. Then the restriction map $H^n(Y_t, \mathbb{C}) \to H^n(X_t, \mathbb{C})$ is surjective.

One may consider this restriction map "in the limit", i.e. with the limit mixed Hodge structure of the family as its source. Then under the given conditions, this identifies the cohomology of the Milnor fibre, now denoted by $H^n(X_\infty, \mathbb{C})$, with the quotient of the limit mixed Hodge structure, which we denote by $H^n(Y_\infty, \mathbb{C})_0$, by the part which is invariant under the monodromy. This equips $H^n(X_\infty, \mathbb{C})$ with a mixed Hodge structure, such that its weight filtration is $W(N, n)$ on the part on which T acts with eigenvalues different from 1 and is $W(N, n+1)$ on the unipotent part.

Let us explain this in some more detail. One has the exact *specialization sequence*

$$0 \to H^n(Y_0) \to H^n(Y_t) \to H^n(X_t) \to 0,$$

where the first map is the composition of the isomorphism $H^n(Y_0) \simeq H^n(\mathcal{Y})$ (valid because the inclusion of Y_0 in $\mathcal{Y}$ is a homotopy equivalence) with the restriction map $H^n(\mathcal{Y}) \to H^n(Y_t)$. By the *local invariant cycle theorem* the image of $H^n(Y_0) \to H^n(Y_t)$ is equal to the part fixed by the monodromy.

It is this mixed Hodge structure on $H^n(X_t)$ with its action of T_s which contains a wealth of invariants of the singularity. It was first introduced by the second author in [17].

The discrete invariants of the mixed Hodge structures are encoded in the spectrum (more strongly: the spectral pairs). The computation of this spectrum was possible in many cases where one disposed of an embedded resolution of the singularity. A first description of a mixed Hodge structure without reference to a resolution was given by Varchenko [19]. This description was translated into the language of $\mathcal{D}$-modules by Scherk and the second author [13], and was the starting point of M. Saito's theory of mixed Hodge modules [10, 11]. In this paper we describe a genuine algorithm for the computation of the spectrum and the spectral pairs, based on the approach via $\mathcal{D}$-modules. The algorithm can also be used to calculate invariants which are closely related to the spectrum, such as the geometric genus [9] and the irregularity [18]. It has been implemented by the first author in SINGULAR.

Hertling has recently formulated an intriguing conjecture concerning the *variance* of the spectrum, see his paper [6] in this volume. It has been verified by Dimca [3] for weighted homogeneous polynomials, and by M. Saito (unpublished) for irreducible plane curve singularities. There also exists a spectrum for polynomials, as contrasted to germs. The contribution of Dimca in this volume [3] gives interesting analogues between the local and global cases.

2. The Brieskorn lattice and its V-filtration

We consider an isolated hypersurface singularity $f : (\mathbb{C}^{n+1}, 0) \to (\mathbb{C}, 0)$. We abbreviate

$$\mathcal{O} := \mathcal{O}_{\mathbb{C},0}, \quad \mathcal{D} := \mathcal{D}_{\mathbb{C},0} = \mathcal{O}_{\mathbb{C},0}[\partial_t], \quad \Omega^p := \Omega^p_{\mathbb{C}^{n+1},0}.$$

Since f is an isolated singularity, the *Milnor module*

$$\Omega_f := \Omega^{n+1}/df \wedge \Omega^n$$

has finite $\mathbb{C}$-dimension equal to the *Milnor number*

$$\mu := \mu_f := \dim_{\mathbb{C}} \Omega_f.$$

2.1. THE GAUSS-MANIN SYSTEM

By [12], the *Brieskorn lattices*

$$\begin{aligned}
\mathcal{H}' &:= \mathcal{H}'_f := df \wedge \Omega^n/df \wedge d\Omega^{n-1}, \\
\mathcal{H}'' &:= \mathcal{H}''_f := \Omega^{n+1}/df \wedge d\Omega^{n-1}
\end{aligned}$$

are free $\mathcal{O}$-modules of rank μ and $\mathcal{H}' \subset \mathcal{H}''$. Note that $\Omega_f = \mathcal{H}''/\mathcal{H}'$.

We have an operator $\partial_t : \mathcal{H}' \to \mathcal{H}''$ which is defined as follows. For $x \in \mathcal{H}'$ choose $\eta \in \Omega^n$ such that $df \wedge \eta$ represents x, and define $\partial_t(x)$ to be the class of $d\eta$ in $\mathcal{H}''$. Then $\partial_t(gx) = g'x + g\partial_t(x)$ for $g \in \mathcal{O}$ so ∂_t is a differential operator of degree one.

The modules $\mathcal{H}'$ and $\mathcal{H}''$ are lattices in the so-called *Gauss-Manin system* of f. This is a $\mathcal{D}$-module which is defined in the following way.

By [13, p. 645] the complex $\Omega[D]$, with differential $\mathbf{d}$ defined by

$$\mathbf{d}(\omega D^i) := d\omega D^i - df \wedge \omega D^{i+1},$$

is a complex of $\mathcal{D}$-modules with $\mathcal{D}$-action

$$\begin{aligned}
\partial_t \omega D^i &:= \omega D^{i+1}, \\
t \omega D^i &:= f\omega D^i - i\omega D^{i-1}.
\end{aligned}$$

The $\mathcal{D}$-module

$$\mathcal{H} := \mathcal{H}_f := H^{n+1}(\Omega[D], \mathbf{d}) = \Omega^{n+1}[D]/d\Omega^n[D]$$

is called the *Gauss-Manin system of f*.

2.2. MONODROMY

Let $k : \mathbb{C} \to \mathbb{C}$ be given by $k(s) = \exp(s)$. We have the inclusion $\mathcal{O} \hookrightarrow (k_*\mathcal{O}_\mathbb{C})_0 =: \mathcal{A}$, and the $\mathcal{D}$-action on $\mathcal{O}$ extends to a $\mathcal{D}$-action on $\mathcal{A}$ by

$$\partial_t g := \exp(-s)\partial_s g, \quad tg := \exp(s)g$$

and induces an action of $\mathcal{D}$ on the space of *microfunctions*

$$\mathcal{M} := \mathcal{A}/\mathcal{O}.$$

We have an automorphism M of $\mathcal{A}$ given by $Mg := g(s - 2\pi i)$, which is called the *monodromy operator*. It is the identity on $\mathcal{O}$ and hence induces an automorphism of $\mathcal{M}$.

For any $\mathcal{D}$-module $\mathcal{E}$ we have its *solution space* $E(\mathcal{E}) := \operatorname{Hom}_{\mathcal{D}}(\mathcal{E}, \mathcal{A})$ and its *microsolution space* $F(\mathcal{E}) := \operatorname{Hom}_{\mathcal{D}}(\mathcal{E}, \mathcal{M})$. These also carry monodromy operators.

The analytic monodromy operator M_f on $E(\mathcal{H}_f)$ is called the *monodromy operator of f*. One has a natural identification of $E(\mathcal{H}_f)$ with $H^n(X_\infty, \mathbb{C})$ such that M_f corresponds to the monodromy T from the introduction so, by the monodromy theorem, the eigenvalues of M_f are roots of unity and the Jordan blocks of M_f have size at most $n + 1$.

2.3. REGULARITY

A finitely generated $\mathcal{D}$-module $\mathcal{E}$ is called *regular* if there exists a free $\mathcal{O}$-submodule $\mathcal{F}$ of $\mathcal{E}$ of finite rank which generates $\mathcal{E}$ as a $\mathcal{D}$-module (i.e. an $\mathcal{O}$-*lattice in $\mathcal{E}$*) and which is stable under the operator $t\partial_t$. This is the case if there exists a free finite rank $\mathcal{O}$-submodule $\mathcal{F}_0$ of $\mathcal{E}$ which generates $\mathcal{E}$ as a $\mathcal{D}$-module and such that the sequence $\mathcal{F}_0 \subset \mathcal{F}_1 \subset \cdots$ defined by $\mathcal{F}_k := \mathcal{F}_{k-1} + t\partial_t\mathcal{F}_{k-1}$ becomes stationary after a finite number of steps. The $\mathcal{O}$-module $\mathcal{F}_k$ for k sufficiently large is then called the *saturation* of $\mathcal{F}_0$.

By [2], the module $\mathcal{H}$ is regular. Moreover the operator ∂_t is invertible on $\mathcal{H}$ by [13, 3.5]. By the classification of regular $\mathcal{D}$-modules (cf. [1]) this implies

PROPOSITION 2.1. *There is an isomorphism of $\mathcal{D}$-modules*

$$\mathcal{H} \cong \bigoplus_{j=1}^{s} \mathcal{D}/\mathcal{D}(t\partial_t - a_j)^{r_j},$$

where $-1 < a_j \leq 0$ and $(\exp(-2\pi i a_j))_{1 \leq j \leq s}$ are the eigenvalues and $(r_j)_{1 \leq j \leq s}$ the corresponding Jordan block sizes of the monodromy operator M_f.

Note that $a_j \in \mathbb{Q}$ and $r_j \leq n + 1$ for $1 \leq j \leq s$ by the monodromy theorem.

2.4. V-FILTRATION

For a regular $\mathcal{D}$-module $\mathcal{E}$ with quasi-unipotent monodromy there exists a unique collection $(V^a\mathcal{E})_{a \in \mathbb{Q}}$ of lattices in $\mathcal{E}$ such that

1. $V^a\mathcal{E} \subset V^b\mathcal{E}$ if $a > b$;

2. each $V^a\mathcal{E}$ is stable by $t\partial_t$;

3. the induced action of $t\partial_t - a$ on $\mathrm{Gr}_a^V \mathcal{E} := V^a\mathcal{E}/\bigcup_{b>a} V^b\mathcal{E}$ is nilpotent.

We use the notation $V^{>a}\mathcal{E} = \bigcup_{b>a} V^b\mathcal{E}$. If $C^b := \bigcup_{r>0} \ker(t\partial_t - b)^r$, then

$$V^a\mathcal{E} = \sum_{b \geq a} \mathcal{O}C^b \text{ and } V^a\mathcal{E} = C^a \oplus V^{>a}\mathcal{E}.$$

In this way we get a V-filtration on $\mathcal{H}$, which induces ones on $\mathcal{H}'$ and $\mathcal{H}''$ by intersection. Note that $\partial_t^k V^a\mathcal{H} = V^{a-k}\mathcal{H}$ for all $k \in \mathbb{Z}$ and $tV^a\mathcal{H} \subset V^{a+1}\mathcal{H}$. The induced V-filtration on the subquotient $\mathcal{H}''/\partial_t^{-1}\mathcal{H}'' = \Omega_f$ of $\mathcal{H}$ is given by

$$V^a\Omega_f = (V^a\mathcal{H} \cap \mathcal{H}'' + \mathcal{H}')/\mathcal{H}',$$
$$V^{>a}\Omega_f = (V^{>a}\mathcal{H} \cap \mathcal{H}'' + \mathcal{H}')/\mathcal{H}'.$$

2.5. THE HODGE FILTRATION

Following [8, p. 160], we give the following definition. For $k \geq 0$, let

$$F_k\Omega^{n+1}[D] := \bigoplus_{i=0}^{k} \Omega^{n+1}D^i$$

and let $F_k\mathcal{H}$ be the image of $F_k\Omega^{n+1}[D]$ under the canonical map

$$\Omega^{n+1}[D] \to \Omega^{n+1}[D]/d\Omega^n[D] = \mathcal{H}.$$

Moreover we put $F_k\mathcal{H} = 0$ for $k < 0$. This defines a filtration F on $\mathcal{H}$ called the *Hodge filtration*. It is a *good filtration* in the sense that it turns $\mathcal{H}$ into a graded $\mathcal{D}$-module. Here $\mathcal{D}$ carries the filtration, also denoted by F, by the order of the differential operators:

$$F_k\mathcal{D} = \{\sum_{i=0}^{k} g_i(t)\partial_t^i\}.$$

The associated graded ring $\mathrm{Gr}^F \mathcal{D}$ is isomorphic to $\mathcal{O}[\xi]$ where ξ is the symbol of ∂_t.

We summarize properties of the Hodge filtration in the

PROPOSITION 2.2.

1. $F_k \mathcal{H} = \partial_t{}^k F_0 \mathcal{H}$ for $k \geq 0$. [13, 3.4]

2. $\mathcal{H}' = \partial_t{}^{-1} F_0 \mathcal{H} \subset F_0 \mathcal{H} = \mathcal{H}''$ [8, p. 160-161]

The induced Hodge filtration on the subquotient $\mathrm{Gr}_V^a \mathcal{H} = C^a$ of $\mathcal{H}$ is given by

$$F_k C^a = (F_k \mathcal{H} \cap V^a \mathcal{H} + V^{>a}\mathcal{H})/V^{>a}\mathcal{H}.$$

2.6. SINGULARITY SPECTRUM

The singularity spectrum is defined in terms of the V-filtration on Ω_f as follows: for $b \in \mathbf{Q}$ let $d(b) := \dim_{\mathbf{C}} \mathrm{Gr}_V^b \Omega_f$. We put

$$\mathrm{Sp}(f) := \sum_{b \in \mathbf{Q}} d(b)(b) \in \mathbf{Z}[\mathbf{Q}]$$

where the latter is the integral group ring of the additive group of the rational numbers. Then $\mathrm{Sp}(f)$ is called the *singularity spectrum of f*.

By [13, 7.3 (i)], we have

PROPOSITION 2.3.

1. $d(b) \neq 0$ implies that $-1 < b < n$.

2. $d(n - 1 - b) = d(b)$

Proposition 2.3 leads to a description of the singularity spectrum in terms of the Hodge filtration on C^a.

COROLLARY 2.4. *For $-1 < a \leq 0$ and $b = a + k$, $\partial_t{}^k$ induces an isomorphism*

$$\mathrm{Gr}_V^b \Omega_f \simeq \mathrm{Gr}_k^F C^a.$$

Proof. For $k \geq 0$, we have

$$\begin{aligned}
\mathrm{Gr}_k^F C^a &= (V^a\mathcal{H} \cap F_k\mathcal{H} + V^{>a}\mathcal{H})/(V^a\mathcal{H} \cap F_{k-1}\mathcal{H} + V^{>a}\mathcal{H}) \\
&= (V^a\mathcal{H} \cap F_k\mathcal{H})/(V^a\mathcal{H} \cap F_{k-1}\mathcal{H} + V^{>a}\mathcal{H} \cap F_k\mathcal{H}) \\
&= \partial_t{}^k(V^b\mathcal{H} \cap F_0\mathcal{H})/\partial_t{}^k(V^b\mathcal{H} \cap \partial_t{}^{-1}F_0\mathcal{H} + V^{>b}\mathcal{H} \cap F_0\mathcal{H})
\end{aligned}$$

by Proposition 2.2 and

$$\operatorname{Gr}^b_V \Omega_f = (F_0 \mathcal{H} \cap V^b \mathcal{H} + \partial_t^{-1} F_0 \mathcal{H})/(F_0 \mathcal{H} \cap V^{>b} \mathcal{H} + \partial_t^{-1} F_0 \mathcal{H})$$
$$= (V^b \mathcal{H} \cap F_0 \mathcal{H})/(V^b \mathcal{H} \cap \partial_t^{-1} F_0 \mathcal{H} + V^{>b} \mathcal{H} \cap F_0 \mathcal{H}).$$

For $k < 0$, we have $\operatorname{Gr}^F_k C^a = 0$ by definition and $\operatorname{Gr}^b_V \Omega_f = 0$ by proposition 2.3.

Remark. In [13], the space $\operatorname{Gr}^F_k C^a$ is identified with the eigenspace of the semisimple monodromy belonging to the eigenvalue $\exp(-2\pi i a)$ in the p-th graded part for the Hodge filtration on the cohomology of the Milnor fibre of f.

The following corollary will be the key to the computation of the singularity spectrum.

COROLLARY 2.5. $V^{>-1}\mathcal{H} \supset \mathcal{H}'' \supset V^{n-1}\mathcal{H}$
    Proof. Since $d(b) = 0$ for $b \geq n$, we have $V^n \mathcal{H} \subset \partial_t^{-1} \mathcal{H}''$ and hence

$$V^{n-1}\mathcal{H} = \partial_t V^n \mathcal{H} \subset \mathcal{H}''.$$

Since $d(b) = 0$ for $b \leq -1$, we also have $\mathcal{H}'' \subset V^{>-1}\mathcal{H}$.

2.7. MICROLOCALIZATION

We define

$$\mathbb{C}\{\{s\}\} := \left\{ \sum_{i \geq 0} a_i s^i \in \mathbb{C}[s] \,\Big|\, \sum_{i \geq 0} \frac{a_i}{i!} t^i \in \mathbb{C}\{t\} \right\}.$$

Then $\mathbb{C}\{\{\partial_t^{-1}\}\}$ is the ring of *microdifferential* operators with constant coefficients. By [13, Lem. 8.3], we have

LEMMA 2.6. For $a \neq -1, -2, \ldots,$ $([(t\partial_t - a)^j])_{0 \leq j < r}$ *is a basis of* $\mathcal{D}/\mathcal{D}(t\partial_t - a)^r$ *as a* $\mathbb{C}\{\{\partial_t^{-1}\}\}[\partial_t]$*-module.*

According to [7, Prop. 2.5] we have

PROPOSITION 2.7. $\mathcal{H}''$ *is a free* $\mathbb{C}\{\{\partial_t^{-1}\}\}$*-module of rank* μ *and*

$$\mathcal{H} = \mathcal{H}'' \otimes_{\mathbb{C}\{\{\partial_t^{-1}\}\}} \mathbb{C}\{\{\partial_t^{-1}\}\}[\partial_t].$$

¿From now on we abbreviate $s := \partial_t^{-1}$. Note that $\mathcal{H}' = \partial_t^{-1}\mathcal{H}'' = s\mathcal{H}''$, so $\Omega_f = \mathcal{H}''/s\mathcal{H}''$. Moreover, the operator $s^{-2}t$ satisfies

$$[s^{-2}t, s] = \partial_t^2 t \partial_t^{-1} - \partial_t t = 1,$$

so $\mathcal{H}$ becomes a $\mathbb{C}\{\{s\}\}[\partial_s]$-module if ∂_s acts as $s^{-2}t = \partial_t^2 t$. Note that

$$s\partial_s = s^{-1}t = \partial_t t$$

hence we see that the saturation sequence of $\mathcal{H}''$ as an $\mathcal{O}$-module coincides with the saturation sequence of $\mathcal{H}''$ as a $\mathbb{C}\{\{s\}\}$-module! However, the latter is much easier to compute, for the following reason.

By a result of Briançon, $f^{n+1}\Omega^{n+1} \subset df \wedge \Omega^n$, and this exponent is sometimes really the minimal one. This implies that in general

$$\partial_t \mathcal{H}'' \subset t^{-n-1}\mathcal{H}''$$

i.e. ∂_t has a pole of order at most $n+1$ on $\mathcal{H}''$. However, if we consider the microlocal structure on $\mathcal{H}''$, then we observe that $\partial_s \mathcal{H}'' = s^{-2}t\mathcal{H}'' \subset s^{-2}\mathcal{H}''$, so ∂_s has at most a pole of order two on $\mathcal{H}'''$!

3. Algorithms

For $g = \sum_{i\in\mathbb{Z}} g_i s^i \in \mathbb{C}\{\{s\}\}[s^{-1}]$ and integers n_1, n_2 we put

$$\mathrm{jet}_{n_1}^{n_2}(g) = \sum_{i=n_1}^{n_2} g_i s^i$$

where $s = \partial_t^{-1}$. We abbreviate

$$\mathrm{jet}_{n_1} := \mathrm{jet}_{n_1}^\infty, \quad \mathrm{jet}^{n_2} := \mathrm{jet}_{-\infty}^{n_2}.$$

3.1. The operator t

In this subsection, we explain how to compute the operator t in a $\mathbb{C}\{\{\partial_t^{-1}\}\}$-basis of $\mathcal{H}''$.

Let $m = (m_1, \ldots, m_\mu)^t$ represent a section $v \in \mathrm{Hom}_\mathbb{C}(\Omega_f, \mathcal{H}'')$ of $\pi : \mathcal{H}'' \to \Omega_f$. Then m induces an isomorphism $\mathbb{C}\{\{s\}\}^\mu \cong \mathcal{H}''$ by Nakayama's lemma. We consider m as a $\mathbb{C}\{\{s\}\}$-basis of $\mathcal{H}''$ and a $\mathbb{C}\{\{s\}\}[s^{-1}]$-basis of $\mathcal{H}$. We define the matrix

$$A = \sum_{k\geq 0} A_k s^k \in \mathrm{Mat}(\mu, \mathbb{C}\{\{s\}\})$$

of the operator t with respect to m by

$$tm =: Am.$$

Since $[t, s] = s^2$, we have

$$tg = gt + s^2 \partial_s(g)$$

for $g = \sum_{k=0}^{\infty} g^k s^k \in \mathbb{C}\{\{s\}\}[s^{-1}]$. Hence, for $g \in \mathbb{C}\{\{s\}\}[s^{-1}]^\mu$, we have

$$\begin{aligned} tgm &= gtm + s^2 \partial_s(g)m \\ &= (gA + s^2 \partial_s(g))m \end{aligned}$$

and

$$A + s^2 \partial_s : \mathbb{C}\{\{s\}\}[s^{-1}]^\mu \to \mathbb{C}\{\{s\}\}[s^{-1}]^\mu$$

is the basis representation of t with respect to m.

If $U \in \mathrm{GL}(\mu, \mathbb{C}\{\{s\}\}[s^{-1}])$ is a change of basis for the module $\mathcal{H}''$ and $A' \in \mathrm{Mat}(\mu, \mathbb{C}\{\{s\}\})$ is the matrix of the operator t with respect to $m' := Um$, then

$$\begin{aligned} Am' &= tm' \\ &= tUm \\ &= Utm + s^2 \partial_s Um \\ &= UAm + s^2 \partial_s Um \\ &= (UA + s^2 \partial_s(U))U^{-1}m' \end{aligned}$$

and hence

$$A' = (UA + s^2 \partial_s(U))U^{-1}.$$

Since f is an isolated singularity, we may assume $\deg f < \infty$ by the finite determinacy theorem, and replace $\mathbb{C}\{z\}$ by $\mathbb{C}[z]_{(z)}$ for the computation. By definition, we have

$$\Omega_f = \Omega^{n+1}/df \wedge \Omega^n$$

and $df \wedge \Omega^n = J(f)\Omega^{n+1}$, where $J(f)$ is the Jacobian ideal of f. Hence one can compute a monomial $\mathbb{C}\{\{s\}\}$-basis m of $\mathcal{H}''$ using standard basis methods.

In the following Lemma 3.1 and Proposition 3.2, we show how to compute basis representations with respect to m.

LEMMA 3.1. *For $\omega \in \Omega^{n+1}$ with $[\omega] \in s\mathcal{H}''$, one can compute $\omega' \in \Omega^{n+1}$ such that $[\omega] = s[\omega']$.*

Proof. We have

$$\mathcal{H}''/s\mathcal{H}'' = \Omega^{n+1}/df \wedge \Omega^n = \Omega^{n+1}/J(f)\Omega^{n+1}.$$

Since $[\omega] \in s\mathcal{H}''$, we have $\omega \in J(f)\Omega^{n+1}$ and one can compute $\omega'' \in \Omega^n$ such that

$$\omega = df \wedge \omega''$$

using standard basis methods. Then we have

$$[\omega] = [df \wedge \omega'] = s[d\omega''] = s[\omega'],$$

where $\omega' := d\omega'' \in \Omega^{n+1}$.

PROPOSITION 3.2. *For a $\mathbb{C}\{\{s\}\}$-basis m of $\mathcal{H}''$, one can compute basis representations with respect to m up to arbitrarily high order.*

Proof. Let $h \in \mathcal{H}''$ and $K \geq 0$. By induction, one can compute $v_k \in \mathbb{C}^\mu$, $0 \leq k < K$, and $h_K \in \mathcal{H}''$ such that

$$h - \sum_{k=0}^{K-1} v_k s^k m = s^K h_K.$$

Since m represents a monomial $\mathbb{C}$-basis of

$$\mathcal{H}''/s\mathcal{H}'' = \Omega^{n+1}/df \wedge \Omega^n = \Omega^{n+1}/(\partial_z f)\Omega^{n+1},$$

one can compute $v_K \in \mathbb{C}^\mu$ such that

$$h_K - v_K m \in s\mathcal{H}''$$

using standard basis methods. By Lemma 3.1, one can compute $h_{K+1} \in \mathcal{H}''$ such that

$$h_K - v_K m = s h_{K+1}.$$

Then

$$h - \sum_{k=0}^{K} v_k s^k m = s^{K+1} h_{K+1}$$

and $v = \sum_{k=0}^{K} v_k s^k$ is the basis representation of h with respect to m up to order K.

By the proof of Lemma 3.1 and Proposition 3.2, one can compute the matrix A of the operator t with respect m up to order K using the

ALGORITHM 1.

```
\PROC |tmat|(f,m,K) \BODY
w:=fm;
A:=0;
k:=-1;
\WHILE k<K \DO
C(m\mod\rd f\wedge\Omega^n):=w\mod\rd f\wedge\Omega^n;
k:=k+1;
A:=A+Cs^k;
\IF k<K \THEN
w:=\rd\bigl((w-Cm)/\rd f\bigr);
\FI
\OD;
A\ENDPROC
```

3.2. MONODROMY OPERATOR

In this subsection, we give an algorithm to compute the eigenvalues of the monodromy operator M_f.

Let $\mathcal{L} \subset \mathcal{H}$ be a $\mathbb{C}\{t\}$-lattice. Then

$$\mathcal{L}_0 := \mathcal{L},$$
$$\mathcal{L}_{k+1} := \mathcal{L}_k + \partial_t t \mathcal{L}_k = \mathcal{L}_k + s\partial_s \mathcal{L}_k$$

defines an increasing sequence of $\mathbb{C}\{t\}$-lattices

$$\mathcal{L}_0 \subseteq \mathcal{L}_1 \subseteq \mathcal{L}_2 \subseteq \cdots.$$

Note that

$$\mathcal{L}_k = \sum_{j=0}^{k} (\partial_t t)^j \mathcal{L}.$$

If $\mathcal{L}$ is a $\mathbb{C}\{\{s\}\}$-lattice then also $\mathcal{L}_k$ is a $\mathbb{C}\{\{s\}\}$-lattice. Since $\mathcal{H}$ is regular, the *saturation*

$$\mathcal{L}_\infty := \sum_{j \geq 0} (\partial_t t)^j \mathcal{L}$$

of $\mathcal{L}$ is a $\mathbb{C}\{t\}$-lattice. Hence

$$\mathcal{L}_\infty = \mathcal{L}_k$$

for some $k \geq 0$. If $\mathcal{L} = \mathcal{L}_\infty$, or equivalently

$$\partial_t t \mathcal{L} \subset \mathcal{L},$$

then $\mathcal{L}$ is called *saturated*. Note that $\mathcal{L}_\infty$ is saturated. If $\mathcal{L}$ is saturated, then $\partial_t t$ induces an endomorphism

$$\overline{\partial_t t} \in \operatorname{End}_{\mathbb{C}}(\mathcal{L}/t\mathcal{L}).$$

If $\mathcal{L}$ is saturated and a $\mathbb{C}\{\{s\}\}$-lattice, then

$$t\mathcal{L} \subset \partial_t^{-1} \mathcal{L} = s\mathcal{L}$$

and hence $t\mathcal{L} = s\mathcal{L}$. Note that this holds for $\mathcal{H}_\infty$.

The following well known fact [15] will allow us to compute the eigenvalues of the monodromy.

PROPOSITION 3.3. *If $\mathcal{L} \subset \mathcal{H}$ is a saturated $\mathbb{C}\{t\}$-lattice and*

$$\overline{\partial_t t} \in \operatorname{End}_{\mathbb{C}}(\mathcal{L}/t\mathcal{L})$$

is the endomorphism induced by $\partial_t t$ on $\mathcal{L}/t\mathcal{L}$, then the spectrum of the map $\exp(2\pi i \overline{\partial_t t})$ is the spectrum of the monodromy operator M_f.

For the computation, we replace $\mathbb{C}\{\{s\}\}$ by $\mathbb{C}[s]_{(s)}$ and use standard basis methods. First, we compute a monomial $\mathbb{C}$-basis m of Ω_f. The matrix A of t with respect to m given by

$$tm = Am.$$

We replace $\mathcal{H}''$, ∂_t^{-1}, and t by the basis representations $H := \mathbb{C}[s]_{(s)}^\mu$, s, and $A + s^2\partial_s$ with respect to m. Since

$$H \subset H_k \subset s^{-k}H,$$

the lattices H_k are given by

$$\begin{aligned}
\delta_0 H &:= H_0 = H, \\
\delta_{k+1}H &:= ((s^{-1}A^{\mathrm{jet}^k} + s\partial_s)\delta_k H)^{\mathrm{jet}^{-1}}, \\
H_{k+1} &= H_k + \delta_{k+1}H.
\end{aligned}$$

We compute sets of generators of the lattices H_k and check if $H_k = H_{k+1}$ until $H_k = H_\infty$. Then we compute a basis m' of H_∞. The matrix of t with respect to m' is given by

$$(A + s^2\partial_s)m' = A'm'.$$

We define

$$\delta(m') := \max\{\mathrm{ord}((m'_{i_1})_{j_1})) - \mathrm{ord}((m'_{i_2})_{j_2})\,|\,(m'_{i_1})_{j_1} \neq 0 \neq (m'_{i_2})_{j_2}\}$$

and the matrices $\mathrm{jet}^k A'$ by

$$(A^{\mathrm{jet}^{k+\delta(m')}} + s^2\partial_s)m' =: (\mathrm{jet}^k A')m'.$$

Note that $\delta(m') \leq k$ if $H_\infty = H_k$. The matrix of $A^{\mathrm{jet}_{k+\delta(m')}}m'$ with respect to m' has order at least k, and hence

$$(A')^{\mathrm{jet}^k} = (\mathrm{jet}^k A')^{\mathrm{jet}^k}.$$

Finally, we compute

$$\overline{s^{-1}A' + s\partial_s} = (\mathrm{jet}^1 A')_1 \in \mathrm{End}_\mathbb{C}(\mathbb{C}[s]_{(s)}^\mu / s\mathbb{C}[s]_{(s)}^\mu).$$

One can compute the eigenvalues of monodromy using the

ALGORITHM 2.

```
\PROC |monospec|(f) \BODY
m:=|basis|(\Omega_f);
w:=fm;
```

```
A:=0;
H':=0;
H:=\BC[s]_{(s)}^\mu;
\delta H:=H;
k:=-1;
K:=0;
\WHILE k<K \OR H'\ne H \DO
C(m\mod\rd f\wedge\Omega^n):=w\mod\rd f\wedge\Omega^n;
k:=k+1;
A:=A+Cs^k;
\IF H'\ne H \THEN
H':=H;
\delta H:=\bigl(\bigl(s^{-1}A+s\ds\bigr)\delta H\bigr)^{\jet^{-1}};
H:=H+\delta H;
\IF H=H' \THEN
m':=|basis|(H');
K:=|delta|(m')+1;
\FI
\FI
\IF k<K \OR H'\ne H \THEN w:=\rd\bigl((w-Cm)/\rd f\bigr) \FI
\OD;
A'm':=\bigl(A+s^2\ds\bigr)m;
|spectrum|(A'_1)\ENDPROC
%\end{program}
```

3.3. THE V-FILTRATION

In this subsection, we give an algorithm to compute the V-filtration on Ω_f. Since

$$V^{>-1}\mathcal{H} \supset \mathcal{H}'' \supset V^{n-1}\mathcal{H},$$

we have

$$V^{>0}\mathcal{H} \supset \partial_t^{-1}\mathcal{H}'' \supset V^n\mathcal{H},$$

$$V^{>-1}\mathcal{H} \supset \mathcal{H}''_\infty \supset V^{n-1}\mathcal{H}.$$

We choose $N \geq 0$ such that

$$\partial_t^{-N}\mathcal{H}''_\infty \subset V^n\mathcal{H}.$$

It suffices to choose $N \geq n + 1$. Then

$$\mathcal{H}''_\infty/\partial_t^{-N}\mathcal{H}''_\infty = \bigoplus_{a<n}(C^a \cap \mathcal{H}''_\infty) \oplus (V^n\mathcal{H}/\partial_t^{-N}\mathcal{H}''_\infty),$$

and we obtain induced endomorphisms $\overline{\partial_t t}, \overline{t\partial_t} \in \text{End}_{\mathbb{C}}(\mathcal{H}''_\infty/\partial_t^{-N}\mathcal{H}''_\infty)$ with

$$\ker(\overline{\partial_t t} - (a+1))^{n+1} = \ker(\overline{t\partial_t} - a)^{n+1} = C^a \cap \mathcal{H}''_\infty$$

for $a < n$ and

$$\ker(\overline{\partial_t t} - (a+1))^{n+1} = \ker(\overline{t\partial_t} - a)^{n+1} \subset V^n\mathcal{H}/\partial_t^{-N}\mathcal{H}''_\infty$$

for $a \geq n$. We choose $N \geq k_\infty + 1$ where

$$k_\infty := \min\{k \geq 0 \mid \mathcal{H}''_k = \mathcal{H}''_\infty\}.$$

Then $\partial_t t$ preserves the outer terms of the flag

$$\partial_t^{-N}\mathcal{H}''_\infty \subset \partial_t^{-1}\mathcal{H}'' \subset \mathcal{H}'' \subset \mathcal{H}''_\infty$$

and, since

$$\partial_t^{-1}\mathcal{H}''/\partial_t^{-N}\mathcal{H}''_\infty \supset V^n\mathcal{H}/\partial_t^{-N}\mathcal{H}''_\infty,$$

the V-filtration $V_{\overline{t\partial_t}} = V^{\bullet+1}_{\overline{\partial_t t}}$ defined by $\overline{t\partial_t}$ on $\mathcal{H}''_\infty/\partial_t^{-N}\mathcal{H}''_\infty$ induces the V-filtration on the subquotient $\mathcal{H}''/\partial_t^{-1}\mathcal{H}'' = \Omega_f$.

For the computation, we replace $\mathbb{C}\{\{s\}\}$ by $\mathbb{C}[s]_{(s)}$ and use standard basis methods. As in subsection 3.2, we compute the matrix $\text{jet}^N A'$ such that

$$\overline{s^{-1}A' + s\partial_s} = \overline{s^{-1}\text{jet}^N A' + s\partial_s} \in \text{End}_{\mathbb{C}}(\mathbb{C}[s]^\mu_{(s)}/s^N\mathbb{C}[s]^\mu_{(s)}).$$

Then we compute the V-filtration $V^{\bullet+1}_{\overline{s^{-1}A'+s\partial_s}}$, a basis of the lattice $H' \subset \mathbb{C}[s]^\mu_{(s)}$ defined by

$$H'm' := H,$$

and the induced V-filtration on the subquotient H'/sH'.

One can compute the V-filtration on Ω_f using the

ALGORITHM 3.

```
\PROC |vfilt|(f) \BODY
m:=|basis|(\Omega_f);
w:=fm;
A:=0;
H':=0;
H:=\BC[s]_{(s)}^\mu;
\delta H:=H;
```

```
k:=-1;
K:=0;
\WHILE k<K \OR H'\ne H \DO
C(m\mod\rd f\wedge\Omega^n):=w\mod\rd f\wedge\Omega^n;
k:=k+1;
A:=A+Cs^k;
\IF H'\ne H \THEN
H':=H;
\delta H:=\bigl(\bigl(s^{-1}A+s\ds\bigr)\delta H\bigr)^{\jet^{-1}};
H:=H+\delta H;
\IF H=H' \THEN
m':=|basis|(H');
K:=|delta|(m')+\max\{k,n\}+1;
\FI
\FI
\IF k<K \OR H'\ne H \THEN w:=\rd\bigl((w-Cm)/\rd f\bigr) \FI
\OD;
A'm':=\bigl(A+s^2\ds\bigr)m;
H'm':=\BC[s]_{(s)}^\mu;
V^{\bullet+1}_{\overline{\inv sA'+s\ds}}(H'/sH')\ENDPROC
```

References

1. J. Briançon and P. Maisonobe, Idéaux de germes d'opérateurs différentiels à une variable, l'Enseignement Math. **30** (1984) 7–38.

2. E. Brieskorn, Die Monodromie der isolierten Singularitäten von Hyperflächen, Manuscr. math. **2** (1970) 103–161.

3. A. Dimca, Monodromy and Hodge theory of regular functions, this volume

4. P. A. Griffiths, On the periods of certain rational integrals I,II, Ann. of Math. **90** (1969) 460–541.

5. A. Grothendieck, On the de Rham cohomology of algebraic varieties, Publ. Math. IHES. **29** (1966) 95–103.

6. C. Hertling, Frobenius manifolds and variance of the spectral numbers, this volume

7. F. Pham, Caustiques, phase stationnaire et microfonctions, Acta Math. Vietn. **2** (1977) 35–101.

8. F. Pham, *Singularités des systèmes de Gauss-Manin*, Progress in Math. **2**, Birkhäuser, 1979.

9. M. Saito, On the exponents and the geometric genus of an isolated hypersurface singularity, *Singularities* (ed Peter Orlik), Proc. Symp. Pure Math. **40**:2, (Amer. Math. Soc., 1983) pp 465–472.

10. M. Saito, Modules de Hodge polarisables, Publ. Res. Inst. Math. Sci. **24** (1989) 849–995.

11. M. Saito, Mixed Hodge modules, Publ. Res. Inst. Math. Sci. **26** (1990) 221–333.

12. M. Sebastiani, Preuve d'une conjecture de Brieskorn, Manuscr. math. **2** (1970) 301–308.

13. J. Scherk and J. H. M. Steenbrink, On the mixed Hodge structure on the cohomology of the Milnor fibre, Math. Ann. **271** (1985) 641–665.

14. W. Schmid, Variation of Hodge structure: the singularities of the period mapping, Invent. Math. **22** (1973) 211-320.

15. M. Schulze, Computation of the monodromy of an isolated hypersurface singularity, Diplomarbeit, Universität Kaiserslautern, 1999.
    http://www.mathematik.uni-kl.de/ mschulze

16. M. Schulze, Algorithms to compute the singularity spectrum, Master Class thesis, University of Utrecht, 2000.
    http://www.mathematik.uni-kl.de/ mschulze

17. J. H. M. Steenbrink, Mixed Hodge structure on the vanishing cohomology, *Real and complex singularities, Oslo 1976* (ed. Per Holm), (Sijthoff and Noordhoff, Alphen aan den Rijn, 1977) pp 525–563.

18. D. van Straten and J. H. M. Steenbrink, Extendability of holomorphic differential forms near isolated hypersurface singularities, Abh. Math. Sem. Univ. Hamburg **55** (1985) 97–110.

19. A. Varchenko, The asymptotics of holomorphic forms determine a mixed Hodge structure, Sov. Math. Dokl. **22** (1980) 248–252.

Frobenius manifolds and variance of the spectral numbers

Claus Hertling
Mathematisches Institut der Universität Bonn,
(hertling@math.uni-bonn.de)

1. Introduction

A Frobenius manifold is a complex manifold with a multiplication and a metric on the holomorphic tangent bundle and two distinguished vector fields, satisfying a series of natural conditions.

This notion was introduced 1991 by Dubrovin [5], motivated by topological field theory. It has been studied since then by him, Manin, Kontsevich and many others. It plays a role in quantum cohomology and mirror symmetry.

But the first big class had been constructed already 1983 by K. Saito [25][27] and M. Saito [28]: The base space M of a semiuniversal unfolding of an isolated hypersurface singularity can be equipped with the structure of a Frobenius manifold.

The multiplication alone is already a rich and interesting structure. Each tangent space $T_t M$, $t \in M$, is as an algebra isomorphic to the sum of the Jacobi algebras of the singularities above the parameter t. Manifolds with such a multiplication are called F-manifolds [13][23] and are studied in [14].

The construction of the metric is not unique and much more difficult than that of the multiplication. It uses the Gauß-Manin connection, polarized mixed Hodge structures, K. Saito's primitive forms and his higher residue pairings. A simplified version is presented in [15].

Two hopes are related to this construction of Frobenius manifolds in singularity theory: first, to find new results on singularities with it, and second, that it is a gate through which other even richer structures from quantum cohomology or topological field theory enter into singularity theory.

The first purpose of this paper is to provide in a concise form the basic properties of the Frobenius manifolds in singularity theory and to discuss one additional structure, their G-functions.

The G-function of a semisimple Frobenius manifold was defined by Dubrovin and Zhang [7] and independently by Givental [11]. It is a fascinating function with several origins. One is the τ-*function* of the isomonodromic deformations, which are associated to such a Frobenius

D. Siersma et al. (eds.), *New Developments in Singularity Theory*, 235–255.
© 2001 Kluwer Academic Publishers. Printed in the Netherlands.

manifold. Another is, that in the case of a semisimple Frobenius manifold coming from quantum cohomology the G-function is the genus one Gromov-Witten potential.

The second purpose is to present an application: the formula (1.3) for the variance of the spectral numbers of a quasihomogeneous singularity.

The spectrum of an isolated hypersurface singularity $f : (C^{n+1}, 0) \to (C, 0)$ is an important discrete invariant of the singularity. Its main properties have been established by Steenbrink and Varchenko. It consists of μ rational numbers $\alpha_1, ..., \alpha_\mu$ with $-1 < \alpha_1 \leq ... \leq \alpha_\mu < n$ and $\alpha_i + \alpha_{\mu+1-i} = n - 1$. The numbers $e^{-2\pi i \alpha_1}, ..., e^{-2\pi i \alpha_\mu}$ are the eigenvalues of the monodromy.

The spectral numbers come from a Hodge filtration on the cohomology of a Milnor fiber [33] or, more instructively, from the Gauß-Manin connection of f ([1], cf. also [32]).

In the case of a quasihomogeneous singularity f of weighted degree 1 with weights $w_0, ..., w_n \in (0, \frac{1}{2}] \cap \mathbb{Q}$ they can be calculated easily [34][1]: Then the Jacobi algebra $\mathcal{O}_{C^{n+1}, 0}/(\frac{\partial f}{\partial x_0}, ..., \frac{\partial f}{\partial x_n}) =: \mathcal{O}/J_f$ has a natural grading $\mathcal{O}/J_f = \bigoplus_\alpha (\mathcal{O}/J_f)_\alpha$, and

$$\sharp(i \mid \alpha_i = \alpha) = \dim(\mathcal{O}/J_f)_{\alpha - \alpha_1}, \tag{1.1}$$

$$\alpha_1 = -1 + \sum_{i=0}^{n} w_i. \tag{1.2}$$

Because of the symmetry $\alpha_i + \alpha_{\mu+1-i} = n - 1$ one can consider $\frac{n-1}{2}$ as their *expectation value*. Then $\frac{1}{\mu} \sum_{i=1}^{\mu} \left(\alpha_i - \frac{n-1}{2} \right)^2$ is their *variance*.

THEOREM 1.1. *In the case of a quasihomogeneous singularity the variance is*

$$\frac{1}{\mu} \sum_{i=1}^{\mu} \left(\alpha_i - \frac{n-1}{2} \right)^2 = \frac{\alpha_\mu - \alpha_1}{12}. \tag{1.3}$$

CONJECTURE 1.2. *For any isolated hypersurface singularity*

$$\frac{1}{\mu} \sum_{i=1}^{\mu} \left(\alpha_i - \frac{n-1}{2} \right)^2 \leq \frac{\alpha_\mu - \alpha_1}{12}. \tag{1.4}$$

The proof of (1.3) in chapter 7 is a very short application of the properties of the G-function of a Frobenius manifold. When I presented this result at the summer school on singularity theory in Cambridge in August 2000, A. Dimca found an elementary proof, see Remark 7.3 a) and [4].

Conjecture 1.2 was originally based only on some series of examples. Recently M. Saito proved it for irreducible plane curve singularities [30] and A. Dimca formulated a dual conjecture for tame polynomials [4]. I would appreciate a general proof as well as counterexamples, and also applications, for example on deformations of singularities or on their topology.

2. F-manifolds

First we fix some notations:

1) In the whole paper M is a complex manifold of dimension $m \geq 1$ (with $m = \mu$ in the singularity case) with holomorphic tangent bundle TM, sheaf $\mathcal{T}_M$ of holomorphic vector fields, and sheaf $\mathcal{O}_M$ of holomorphic functions.

2) A (k,l)-tensor T is an $\mathcal{O}_M$-linear map $\mathcal{T}_M^{\otimes k} \to \mathcal{T}_M^{\otimes l}$. The Lie derivative $\mathrm{Lie}_X T$ of it by a vector field $X \in \mathcal{T}_M$ is again a (k,l)-tensor. For example, a vector field $Y \in \mathcal{T}_M$ yields a $(0,1)$-tensor $\mathcal{O}_M \to \mathcal{T}_M$, $1 \mapsto Y$, with $\mathrm{Lie}_X Y = [X,Y]$.

3) If ∇ is a connection on M then the covariant derivative $\nabla_X T$ of a (k,l)-tensor by a vector field X is again a (k,l)-tensor. As $\nabla_X T$ is $\mathcal{O}_M$-linear in X (in contrast to $\mathrm{Lie}_X T$) ∇T is a $(k+1,l)$-tensor.

4) A *multiplication* $\circ$ on the tangent bundle TM of a manifold M is a symmetric and associative $(2,1)$-tensor $\circ : \mathcal{T}_M \otimes \mathcal{T}_M \to \mathcal{T}_M$. It equips each tangent space $T_t M$, $t \in M$, with the structure of a commutative and associative $\mathbb{C}$-algebra. We will be interested only in a multiplication with a global unit field e.

5) A *metric* g on a manifold M is a symmetric and nondegenerate $(2,0)$-tensor. It equips each tangent space with a symmetric and nondegenerate bilinear form. Its Levi-Civita connection ∇ is the unique connection on TM which is torsion free, i.e. $\nabla_X Y - \nabla_Y X = [X,Y]$, and which satisfies $\nabla g = 0$, i.e. $X\, g(Y,Z) = g(\nabla_X Y, Z) + g(Y, \nabla_X Z)$.

The notion of an F-manifold was first defined in [13] (cf. [23] I §5). It is studied extensively in [14].

DEFINITION 2.1. *a) An F-manifold $(M, \circ, e)$ is a manifold M together with a multiplication $\circ$ on the tangent bundle and a global unit field e such that the multiplication satisfies the following integrability condition:*

$$\forall\, X, Y \in \mathcal{T}_M \qquad \mathrm{Lie}_{X \circ Y}(\circ) = X \circ \mathrm{Lie}_Y(\circ) + Y \circ \mathrm{Lie}_X(\circ)\,. \qquad (2.1)$$

b) *Let* $(M, \circ, e)$ *be an F-manifold. An Euler field of weight* $c \in \mathbb{C}$ *is a vector field* $E \in \mathcal{T}_M$ *with*

$$\text{Lie}_E(\circ) = c \cdot \circ . \tag{2.2}$$

An Euler field of weight 1 is simply called an Euler field.

One reason why this is a natural and good notion is that any Frobenius manifold is an F-manifold ([13], [14, chapter 5]). Another is given in Proposition 2.2 and Theorem 2.3.

PROPOSITION 2.2. *([14, Prop. 4.1]) The product* $(M_1 \times M_2, \circ_1 \oplus \circ_2, e_1 + e_2)$ *of two F-manifolds is an F-manifold. The sum (of the lifts to* $M_1 \times M_2$*) of two Euler fields* E_i, $i = 1, 2$, *on* $(M_i, \circ_i, e_i)$ *of the same weight* $c \in \mathbb{C}$ *is an Euler field of weight* $c \in \mathbb{C}$ *on* $M_1 \times M_2$.

Theorem 2.3 describes the decomposition of a germ of an F-manifold. In order to state it properly we need the following classical and elementary fact: each tangent space of an F-manifold decomposes as an algebra $(T_t M, \circ, e)$ uniquely into a direct sum

$$(T_t M, \circ, e) = \bigoplus_{k=1}^{l(t)} ((T_t M)_k, \circ, e_k) \tag{2.3}$$

of local subalgebras $(T_t M)_k$ with units e_k and with $(T_t M)_j \circ (T_t M)_k = 0$ for $j \neq k$ (cf. e.g. [14, Lemma 1.1]). One can obtain this decomposition as the simultaneous eigenspace decomposition of the commuting endomorphisms $X \circ : T_t M \to T_t M$ for $X \in T_t M$.

THEOREM 2.3. *([14, Theorem 4.2]) Let* $(M, \circ, e)$ *be an F-manifold and* $t \in M$. *The decomposition (2.3) extends to a unique decomposition*

$$(M, t) = \prod_{k=1}^{l(t)} (M_k, t) \tag{2.4}$$

of the germ (M, t) *into a product of germs of F-manifolds. These germs are irreducible germs of F-manifolds as already the algebras* $T_t(M_k) \cong (T_t M)_k$ *are irreducible (as they are local algebras).*

   An Euler field of weight c *decomposes accordingly.*

The proof in [14] uses (2.1) in a way which justifies calling it an *integrability condition*.

   Consider an F-manifold $(M, \circ, e)$. The function $l : M \to \mathbb{N}$ defined in (2.3) is lower semicontinuous ([14, Proposition 2.3]): write l_{max} for its maximum, hence generic value. The *caustic* $\mathcal{K} := \{t \mid l(t) < l_{max}\}$ is empty or a hypersurface (the proof in [14, Proposition 2.4] for the case $l_{max} = m$ works for any value of l_{max}).

The multiplication on $T_t M$ is *semisimple* if $l(t) = m$. The F-manifold is *massive* if the multiplication is generically semisimple.

Up to isomorphism there is only one germ of a 1-dimensional F-manifold, $(\mathbb{C}, \circ, e)$ with $e = \frac{\partial}{\partial u}$ for u a coordinate on $(\mathbb{C}, 0)$. The space of Euler fields of weight 0 is $\mathbb{C} \cdot e$, an Euler field of weight 1 is $u\, e$. This germ of an F-manifold is called A_1.

By Theorem 2.3, any germ of a semisimple F-manifold is a product A_1^m, that means, there are local coordinates $u_1, ..., u_m$ with $e_i = \frac{\partial}{\partial u_i}$ and $e_i \circ e_j = \delta_{ij} e_i$. They are unique up to renumbering and shift and are called canonical coordinates, following Dubrovin. The vector fields e_i are called *idempotent*. Also by Theorem 2.3, each Euler field of weight 1 takes the form $\sum_{i=1}^{m}(u_i + r_i) e_i$ for some $r_i \in \mathbb{C}$.

EXAMPLE 2.4.   Fix $m \geq 1$ and $n \geq 2$. The manifold $M = \mathbb{C}^m$ with coordinate fields $\delta_i = \frac{\partial}{\partial t_i}$ and multiplication defined by

$$\delta_1 \circ \delta_2 = \delta_2 , \tag{2.5}$$

$$\delta_2 \circ \delta_2 = t_2^{n-2}\delta_1 , \tag{2.6}$$

$$\delta_i \circ \delta_j = \delta_{ij}\delta_i \qquad \text{if } (i,j) \notin \{(1,2),(2,1),(2,2)\} \tag{2.7}$$

is a massive F-manifold. The submanifold $\mathbb{C}^2 \times \{0\}$ is an F-manifold with the name $I_2(n)$, with $I_2(2) = A_1^2$, $I_2(3) = A_2$, $I_2(4) = B_2$, $I_2(5) = H_2$, and $I_2(6) = G_2$.

$(M, \circ, e)$ decomposes globally into a product $\mathbb{C}^2 \times \mathbb{C} \times ... \times \mathbb{C}$ of F-manifolds of the type $I_2(n)A_1^{m-2}$. The unit fields for the components are $\delta_1, \delta_3, ..., \delta_m$, the global unit field is $e = \delta_1 + \delta_3 + ... + \delta_m$, the caustic is $\mathcal{K} = \{t \mid t_2 = 0\}$. The idempotent vector fields in a simply connected subset of $M - \mathcal{K}$ are

$$e_{1/2} = \frac{1}{2}\delta_1 \pm \frac{1}{2}t_2^{-\frac{n-2}{2}}\delta_2 , \tag{2.8}$$

$$e_i = \delta_i \qquad \text{for } i \geq 3 , \tag{2.9}$$

canonical coordinates there are

$$u_{1/2} = t_1 \pm \frac{2}{n}t_2^{\frac{n}{2}} , \tag{2.10}$$

$$u_i = t_i \qquad \text{for } i \geq 3 . \tag{2.11}$$

An Euler field of weight 1 is

$$E = t_1\delta_1 + \frac{2}{m}t_2\delta_2 + \sum_{i \geq 3} t_i\delta_i . \tag{2.12}$$

The space of global Euler fields of weight 0 is $\sum_{i\neq 2} C \cdot \delta_i$.

The classification of 3-dimensional irreducible germs of massive F-manifolds is already vast ([14, chapter 20]). But the classification of 2-dimensional irreducible germs of massive F-manifolds is nice ([14, Theorem 12.1]): they are precisely the germs at 0 for $m = 2$ and $n \geq 3$ in Example 2.4 with the names $I_2(n)$.

3. Frobenius manifolds

Frobenius manifolds were first defined by Dubrovin [5]. They turn up now in many places, see [6] and [23] (also for more references), especially in quantum cohomology and mirror symmetry. In this paper we will only be concerned with the Frobenius manifolds in singularity theory (chapter 4).

DEFINITION 3.1. *A Frobenius manifold $(M, \circ, e, E, g)$ is a manifold M with a multiplication $\circ$ on the tangent bundle, a global unit field e, another global vector field E, which is called the Euler field, and a metric g, subject to the following conditions:*

1) the metric is multiplication invariant, $g(X \circ Y, Z) = g(X, Y \circ Z)$,

2) (potentiality) the $(3,1)$-tensor $\nabla\circ$ is symmetric (here ∇ is the Levi-Civita connection of the metric),

3) the metric g is flat,

4) the unit field e is flat, $\nabla e = 0$,

5) the Euler field satisfies $\mathrm{Lie}_E(\circ) = 1 \cdot \circ$ and $\mathrm{Lie}_E(g) = D \cdot g$ for some $D \in C$.

REMARK 3.2.   a) Condition 2) implies (2.1) ([14, Theorem 5.2]). Therefore a Frobenius manifold is an F-manifold.

  b) The $(3,0)$-tensor A with $A(X,Y,Z) := g(X \circ Y, Z)$ is symmetric by 1). Then 2) is equivalent to the symmetry of the $(4,0)$-tensor ∇A. If X,Y,Z,W are local flat fields then 2) is equivalent to the symmetry in X,Y,Z,W of $X A(Y,Z,W)$. This is equivalent to the existence of a local *potential* $\Phi \in \mathcal{O}_M$ with $A(X,Y,Z) = XYZ(\Phi)$ for flat fields X,Y,Z.

  c) $\mathrm{Lie}_E(g) = D \cdot g$ means that E is a sum of an infinitesimal dilation, rotation and shift. Therefore ∇E maps a flat field X to a flat field $\nabla_X E = [X,E] = -\mathrm{Lie}_E X$, i.e. it is a flat $(1,1)$-tensor, $\nabla(\nabla E) = 0$.

Its eigenvalues are denoted $d_1, ..., d_m$. Now $\text{Lie}_E(\circ) = 1 \cdot \circ$ implies $\nabla_e E = [e, E] = e$, and $\nabla E - \frac{D}{2}\text{id}$ is an infinitesimal isometry because of $\text{Lie}_E(g) = D \cdot g$. One can order the eigenvalues so that $d_1 = 1$ and $d_i + d_{m+1-i} = D$.

4. Hypersurface singularities

Let $f : (\mathbb{C}^{n+1}, 0) \to (\mathbb{C}, 0)$ be a holomorphic function germ with an isolated singularity at 0 and with Milnor number μ.

An unfolding of f is a holomorphic function germ $F : (\mathbb{C}^{n+1} \times \mathbb{C}^m, 0) \to (\mathbb{C}, 0)$ with $F|(\mathbb{C}^{n+1} \times \{0\}, 0) = f$. The coordinates on $(\mathbb{C}^{n+1} \times \mathbb{C}^m, 0)$ are denoted $(x_0, ..., x_n, t_1, ..., t_m)$.

The germ $(C, 0) \subset (\mathbb{C}^{n+1} \times \mathbb{C}^m, 0)$ of the critical space is defined by the ideal $J_F = (\frac{\partial F}{\partial x_0}, ..., \frac{\partial F}{\partial x_n})$. The projection $pr_C : (C, 0) \to (\mathbb{C}^m, 0)$ is finite and flat of degree μ. The map

$$\mathbf{a} : \mathcal{T}_{\mathbb{C}^m, 0} \longrightarrow \mathcal{O}_{C,0} \tag{4.1}$$

$$\frac{\partial}{\partial t_i} \mapsto \frac{\partial F}{\partial t_i}|_{(C,0)} \tag{4.2}$$

is the Kodaira-Spencer map.

The unfolding is semiuniversal iff $\mathbf{a}$ is an isomorphism. Consider a semiuniversal unfolding. Then $m = \mu$, and we set $(M, 0) := (\mathbb{C}^\mu, 0)$. The map $\mathbf{a}$ induces a multiplication $\circ$ on $\mathcal{T}_{M,0}$ by $\mathbf{a}(X \circ Y) = \mathbf{a}(X) \cdot \mathbf{a}(Y)$ with unit field $e = \mathbf{a}^{-1}([1])$ and a vector field $E := \mathbf{a}^{-1}([F])$.

This multiplication and the field E were first considered by K. Saito [25][27]. Because the Kodaira-Spencer map $\mathbf{a}$ behaves well under morphisms of unfoldings, the tuple $((M, 0), \circ, e, E)$ is essentially independent of the choice of the semiuniversal unfolding. This and the following fact are discussed in [14, chapter 16].

THEOREM 4.1. *The base space $(M, 0)$ of a semiuniversal unfolding F of an isolated hypersurface singularity f is a germ $((M, 0), \circ, e, E)$ of a massive F-manifold with Euler field $E = \mathbf{a}^{-1}([F])$. For each $t \in M$ there is a canonical isomorphism*

$$(T_t M, \circ, E|_t) \cong (\bigoplus_{x \in \text{Sing}(F_t)} \text{Jacobi algebra of } (F_t, x), \text{mult.}, [F_t]) . \tag{4.3}$$

At generic points of the caustic the germ of the F-manifold is of the type $A_2 A_1^{\mu-2}$.

The base spaces of two semiuniversal unfoldings are canonically isomorphic as germs of F-manifolds with Euler fields.

By work of K. Saito [25][27] and M. Saito [28] one can even construct a metric g on M such that $((M,0), \circ, e, E, g)$ is the germ of a Frobenius manifold. The construction uses the Gauß-Manin connections for f and F, K. Saito's higher residue pairings, a polarized mixed Hodge structure, and results of Malgrange on deformations of microdifferential systems.

A more elementary and much broader version of the construction, which does not use Malgrange's results, is given in [15]. Here we restrict ourselves to a formulation of the result. This uses the *spectral numbers* of f [33][1].

Let $f : X \to \Delta$ be a representative of the germ f as usual, with $\Delta = B_\delta \subset C$ and $X = f^{-1}(\Delta) \cap B_\varepsilon^{n+1} \subset C^{n+1}$ (with $1 \gg \varepsilon \gg \delta > 0$). The cohomology bundle $H^n = \bigcup_{t \in \Delta^*} H^n(f^{-1}(t), C)$ is flat. Denote by $\Delta^\infty \to \Delta^*$ a universal covering. A *global flat multivalued section* in H^n is a map $\Delta^\infty \to H^n$ which factors locally into the map $\Delta^\infty \to \Delta^*$ and a flat section of H^n. The μ-dimensional space of global flat multivalued sections in H^n is denoted by H^∞. Steenbrink's Hodge filtration $F^\bullet$ on H^∞ [33] together with topological data yields a polarized mixed Hodge structure on it (see [12][15] for definitions and a discussion of this).

The *spectral numbers* $\alpha_1, ..., \alpha_\mu$ of f are μ rational numbers with

$$\sharp(i \mid \alpha_i = \alpha) = \dim \mathrm{Gr}_F^{[n-\alpha]} H^\infty_{e^{-2\pi i \alpha}} . \tag{4.4}$$

Here $H^\infty_{e^{-2\pi i \alpha}}$ is the generalized eigenspace of the monodromy on H^∞ with eigenvalue $e^{-2\pi i \alpha}$. So $e^{-2\pi i \alpha_1}, ..., e^{-2\pi i \alpha_\mu}$ are the eigenvalues of the monodromy. The spectral numbers satisfy $-1 < \alpha_1 \leq ... \leq \alpha_\mu < n$ and $\alpha_i + \alpha_{\mu+1-i} = n - 1$.

Essential for understanding them and for the whole construction of the metric g is Varchenko's way to construct a mixed Hodge structure on H^∞ with the Gauß-Manin connection of f ([35][1], cf. also [31][28][12][15] [32]).

It turns out that a metric g such that $((M,0), \circ, e, E, g)$ is a Frobenius manifold is in general not unique. By work of K. Saito and M. Saito each choice of a filtration on H^∞ which is *opposite* to $F^\bullet$ (see [28][15] for the definition) yields a metric which gives a Frobenius manifold structure as in Theorem 4.2. A more precise statement and a detailed proof of it are given in [15, chapter 6].

THEOREM 4.2. *One can choose a metric g on the base space $(M, 0)$ of a semiuniversal unfolding of an isolated hypersurface singularity f such that $((M,0), \circ, e, E, g)$ is a germ of a Frobenius manifold and ∇E is semisimple with eigenvalues $d_i = 1 + \alpha_1 - \alpha_i$ and with $D = 2 - (\alpha_\mu - \alpha_1)$.*

In fact, often one can also find metrics giving Frobenius manifold structures with $\{d_1, ..., d_\mu\} \neq \{1 + \alpha_1 - \alpha_i \mid i = 1, ..., \mu\}$ ([28, 4.4], [15, Remarks 6.7]).

5. Socle field

A Frobenius manifold has another distinguished vector field besides the unit field and the Euler field. It will be discussed in this section. We call it the *socle field*. It is used implicitly in Dubrovin's papers and in [11].

Let $(M, \circ, e, g)$ be a manifold with a multiplication $\circ$ on the tangent bundle, with a unit field, and with a multiplication invariant metric g. We do not need flatness and potentiality and an Euler field for the moment.

Each tangent space $T_t M$ is a Frobenius algebra. This means (more or less by definition) that the splitting (2.3) is now a splitting into a direct sum of Gorenstein rings (cf. e.g. [14, Lemma 1.2])

$$T_t M = \bigoplus_{k=1}^{l(t)} (T_t M)_k \tag{5.1}$$

They have maximal ideals $\mathbf{m}_{t,k} \subset (T_t M)_k$ and units e_k such that $e = \sum e_k$. They satisfy

$$(T_t M)_j \circ (T_t M)_k = \{0\} \qquad \text{for } j \neq k , \tag{5.2}$$

and thus

$$g((T_t M)_j, (T_t M)_k) = \{0\} \qquad \text{for } j \neq k . \tag{5.3}$$

The socle $\mathrm{Ann}_{(T_t M)_k}(\mathbf{m}_{t,k})$ is 1-dimensional and has a unique generator $H_{t,k}$ which is normalized such that

$$g(e_k, H_{t,k}) = \dim(T_t M)_k . \tag{5.4}$$

The following lemma shows that the vectors $\sum_k H_{t,k}$ glue to a holomorphic vector field, the *socle field* of $(M, \circ, e, g)$.

LEMMA 5.1. *For any dual bases* $X_1, ..., X_m$ *and* $\tilde{X}_1, ..., \tilde{X}_m$ *of* $T_t M$, *that means,* $g(X_i, \tilde{X}_j) = \delta_{ij}$, *one has*

$$\sum_{k=1}^{l(t)} H_{t,k} = \sum_{i=1}^{m} X_i \circ \tilde{X}_i . \tag{5.5}$$

Proof. One sees easily that the sum $\sum X_i \circ \tilde{X}_i$ is independent of the choice of the basis $X_1, ..., X_m$. One can suppose that $l(t) = 1$ and that $X_1, ..., X_m$ are chosen such that they yield a splitting of the filtration $T_t M \supset \mathbf{m}_{t,1} \supset \mathbf{m}_{t,1}^2 \supset ....$ Then $g(e, X_i \circ \tilde{X}_i) = 1$ and

$$g(\mathbf{m}_{t,1}, X_i \circ \tilde{X}_i) = g(X_i \circ \mathbf{m}_{t,1}, \tilde{X}_i) = 0 .$$

Thus $X_i \circ \tilde{X}_i = \frac{1}{m} H_{t,1}$.

It will be useful to fix the multiplication and vary the metric.

LEMMA 5.2. *Let* $(M, \circ, e, g)$ *be a manifold with multiplication* $\circ$ *on the tangent bundle, unit field* e *and multiplication invariant metric* g. *For each multiplication invariant metric* $\tilde{g}$ *there exists a unique vector field* Z *such that multiplication it is everywhere invertible and for all vector fields* X, Y,

$$\tilde{g}(X, Y) = g(Z \circ X, Y) . \tag{5.6}$$

The socle fields H *and* $\tilde{H}$ *of* g *and* $\tilde{g}$ *satisfy*

$$H = Z \circ \tilde{H} . \tag{5.7}$$

Proof. The situation for a Frobenius algebra is described for example in [14, Lemma 1.2]. It yields (5.6) immediately. (5.7) follows by comparison of (5.4) and (5.6).

Denote by

$$H_{op} : \mathcal{T}_M \to \mathcal{T}_M, \quad X \mapsto H \circ X \tag{5.8}$$

the multiplication by the socle field H of $(M, \circ, e, g)$ as above. The socle field is especially interesting if the multiplication is generically semisimple, that means, generically $l(t) = m$. Then the caustic $\mathcal{K} = \{ t \in M \mid l(t) < m \}$ is the set where the multiplication is not semisimple. It is the hypersurface

$$\mathcal{K} = \det(H_{op})^{-1}(0) . \tag{5.9}$$

In an open subset of $M - \mathcal{K}$ with basis $e_1, ..., e_m$ of idempotent vector fields the socle field is

$$H = \sum_{i=1}^{m} \frac{1}{g(e_i, e_i)} e_i . \tag{5.10}$$

It determines the metric g everywhere because (5.10) determines the metric at semisimple points.

THEOREM 5.3. *Let $(M, \circ, e, g)$ be a massive F-manifold with multipli-cation invariant metric g. Suppose that at generic points of the caustic the germ of the F-manifold is of the type $I_2(n)A_1^{m-2}$.*

Then the function $\det(H_{op})$ vanishes with multiplicity $n - 2$ along the caustic.

Proof. It is sufficient to consider Example 2.4. A multiplication invariant metric g is uniquely determined by the 1-form $\varepsilon = g(e, .)$. Because of (5.7) it is sufficient to prove the claim for one metric. We choose the metric with 1-form

$$\varepsilon(\delta_i) = 1 - \delta_{i1} . \tag{5.11}$$

The bases $\delta_1, \delta_2, \delta_3, ..., \delta_m$ and $\delta_2, \delta_1, \delta_3, ..., \delta_m$ are dual with respect to this metric. Its socle field is by Lemma 5.1

$$H = 2\delta_2 + \delta_3 + ... + \delta_m \tag{5.12}$$

and satisfies $\det(H_{op}) = -4t_2^{n-2}$.

6. G-function of a massive Frobenius manifold

Associated to any simply connected semisimple Frobenius manifold is a fascinating and quite mysterious function. Dubrovin and Zhang [7][8] called it the G-function and proved the deepest results on it. But Givental [11] studied it, too, and it originates in much older work. It takes the form

$$G(t) = \log \tau_I - \frac{1}{24} \log J \tag{6.1}$$

and is determined only up to addition of a constant. First we explain the simpler part, $\log J$. Let $(M, \circ, e, E, g)$ be a semisimple Frobenius mani-fold with canonical coordinates $u_1, ..., u_m$ and flat coordinates $\tilde{t}_1, ..., \tilde{t}_m$. Then

$$J = \det\left(\frac{\partial \tilde{t}_i}{\partial u_j}\right) \cdot \text{constant} \tag{6.2}$$

is the determinant of the base change matrix between flat and idem-potent vector fields.

One can rewrite it using the socle field. Denote $\eta_i := g(e_i, e_i)$ and consider the basis $v_1, ..., v_m$ of vector fields with

$$v_i = \frac{1}{\sqrt{\eta_i}} e_i \tag{6.3}$$

(for some choice of the square roots). The matrix $\det(g(v_i, v_j)) = 1$ is constant as is the corresponding matrix for the flat vector fields. Therefore

$$\text{constant} \cdot J = \prod_{i=1}^{m} \sqrt{\eta_i} = \det(H_{op})^{-\frac{1}{2}} . \tag{6.4}$$

Here $H = \sum v_i \circ v_i$ is the socle field.

One of the origins of the first part $\log \tau_I$ are isomonodromic deformations. The second structure connections and the first structure connections of the semisimple Frobenius manifold are isomonodromic deformations over $\mathbb{P}^1 \times M$ of restrictions to a slice $\mathbb{P}^1 \times \{t\}$. The function τ_I is their τ-*function* in the sense of [18][19][17][22]. See [24] for other general references on this.

The situation for Frobenius manifolds is discussed and put into a Hamiltonian framework in [6, Lecture 3], [23] (II §2), and in [16]. The coefficients H_i of the 1-form $d \log \tau_I = \sum H_i du_i$ are certain Hamiltonians and motivate the definition of this 1-form. Hitchin [16] compares the realizations of this for the first and the second structure connections.

Another origin of the whole G-function comes from quantum cohomology. Getzler [10] studied the relations between cycles in the moduli space $\overline{\mathcal{M}}_{1,4}$ and derived from it recursion relations for genus one Gromov-Witten invariants of projective manifolds and differential equations for the genus one Gromov-Witten potential.

Dubrovin and Zhang [7, chapter 6] investigated these differential equations for any semisimple Frobenius manifold and found that they have always one unique solution (up to addition of a constant), the G-function. Therefore in the case of a semisimple Frobenius manifold coming from quantum cohomology, the G-function is the genus one Gromov-Witten potential.

They also proved part of the conjectures in [11] concerning $G(t)$. Finally, they found that the potential of the Frobenius manifold (for genus zero) and the G-function (for genus one) are the basements of full free energies in genus zero and one and give rise to Virasoro constraints [8]. Exploiting this for singularities will be a big task for the future.

In chapter 7 we need only the definition of $\log \tau_I$ and the behaviour of $G(t)$ with respect to the Euler field and the caustic in a massive Frobenius manifold. We recall some known formulas related to the canonical coordinates of a semisimple Frobenius manifold ([6], [23], also [11]).

The 1-form $\varepsilon = g(e,.)$ is closed and can be written as $\varepsilon = d\eta$. One defines

$$\eta_i := e_i\eta = g(e_i, e) = g(e_i, e_i) , \tag{6.5}$$

$$\eta_{ij} := e_ie_j\eta = e_i\eta_j = e_j\eta_i , \tag{6.6}$$

$$\gamma_{ij} := \frac{1}{2}\frac{\eta_{ij}}{\sqrt{\eta_i}\sqrt{\eta_j}} , \tag{6.7}$$

$$V_{ij} := -(u_i - u_j)\gamma_{ij} , \tag{6.8}$$

$$d\log\tau_I := \frac{1}{2}\sum_{i\neq j}\frac{V_{ij}^2}{u_i - u_j}du_i = \frac{1}{2}\sum_{i\neq j}(u_i - u_j)\gamma_{ij}^2 du_i \tag{6.9}$$

$$= \frac{1}{8}\sum_{i\neq j}(u_i - u_j)\frac{\eta_{ij}^2}{\eta_i\eta_j}du_i . \tag{6.10}$$

THEOREM 6.1. *Let* $(M, \circ, e, E, g)$ *be a semisimple Frobenius manifold with global canonical coordinates* $u_1, ..., u_m$.

a) The rotation coefficients γ_{ij} *(for* $i \neq j$*) satisfy the Darboux-Egoroff equations*

$$e_k\gamma_{ij} = \gamma_{ik}\gamma_{kj} \qquad for \ k \neq i \neq j \neq k , \tag{6.11}$$

$$e\gamma_{ij} = 0 \qquad for \ i \neq j . \tag{6.12}$$

b) The connection matrix of the flat connection for the basis $v_1, ..., v_m$ *from (6.3) is the matrix* $\Gamma := (\gamma_{ij}d(u_i - u_j))$. *The Darboux-Egoroff equations are equivalent to the flatness condition* $d\Gamma + \Gamma \wedge \Gamma = 0$.

c) The 1-form $d\log\tau_I$ *is closed and comes from a function* $\log\tau_I$.

d) $E(\eta_i) = (D-2)\eta_i$ *and* $E(\gamma_{ij}) = -\gamma_{ij}$.

e) If the canonical coordinates are chosen such that $E = \sum u_ie_i$ *then the matrix* $-(V_{ij})$ *is the matrix of the endomorphism* $\mathcal{V} = \nabla E - \frac{D}{2}\text{id}$ *on* T_M *with respect to the basis* $v_1, ..., v_m$.

Proof. a)+b) See [6, pp. 200–201] or [23, I §3].

c) This can be checked easily with the Darboux-Egoroff equations.

d) It follows from $\text{Lie}_E(g) = D \cdot g$ and from $[e_i, E] = e_i$.

e) This is implicit in [6, pp. 200–201]. One can check it with a)+b)+d).

The endomorphism $\mathcal{V}$ is skew-symmetric with respect to g, and flat with eigenvalues $d_i - \frac{D}{2}$; the numbers d_i can be ordered such that $d_1 = 1$ and $d_i + d_{m+1-i} = D$ (cf. Remark 3.2 c)).

COROLLARY 6.2. *([7, Theorem 3]) Suppose that* $E = \sum u_ie_i$. *Then*

$$E\log\tau_I = -\frac{1}{4}\sum_{i=1}^{m}\left(d_i - \frac{D}{2}\right)^2 , \tag{6.13}$$

$$E\, G(t) \;=\; -\frac{1}{4}\sum_{i=1}^{m}\left(d_i-\frac{D}{2}\right)^2 + \frac{m(2-D)}{48} =: \gamma\,. \qquad (6.14)$$

Proof.

$$E\log\tau_I \;=\; \frac{1}{2}\sum_{i\neq j}\frac{u_i V_{ij}^2}{u_i-u_j} = \frac{1}{2}\sum_{i<j}V_{ij}^2 \qquad (6.15)$$

$$= -\frac{1}{4}\sum_{ij}V_{ij}V_{ji} = -\frac{1}{4}\mathrm{trace}(V^2)$$

$$= -\frac{1}{4}\sum_{i=1}^{m}\left(d_i-\frac{D}{2}\right)^2\,.$$

(6.4) shows $E(J) = m\frac{D-2}{2}J$. Now (6.14) follows from the definition of the G-function.

If M is a massive Frobenius manifold with caustic $\mathcal{K}$, one may ask which kind of poles the 1-form $d\log\tau_I$ has along $\mathcal{K}$ and when the G-function extends over $\mathcal{K}$.

All the F-manifolds $I_2(n)$ (cf. Example 2.4) are in a natural way (up to the choice of a scalar) equipped with a metric g, such that they get Frobenius manifolds ([6, Lecture 4], cf. [14, chapter 19]). These Frobenius manifolds are also denoted $I_2(n)$.

In [7, chapter 6] the G-function is calculated for them with coordinates (t_1, t_2) on $M = \mathbb{C}^2$ and $e = \frac{\partial}{\partial t_1}$. It turns out to be

$$G(t) \;=\; -\frac{1}{24}\frac{(2-n)(3-n)}{n}\log t_2\,. \qquad (6.16)$$

In particular, for the case $I_2(3) = A_2$, the G-function is $G(t) = 0$. This was checked independently in [11]. Givental concluded that in the case of singularities the G-function of the base space of a semiuniversal unfolding with some Frobenius manifold structure extends holomorphically over the caustic. This is a good guess, but it does not follow from the case A_2, because a Frobenius manifold structure on a germ of an F-manifold of type $A_2 A_1^{m-2}$ for $m \geq 3$ is never the product of the Frobenius manifolds A_2 and A_1^{m-2} (the numbers $d_1, ..., d_m$ would not be symmetric). Anyway, it is true, as the following result shows.

THEOREM 6.3. *Let $(M, \circ, e, E, g)$ be a simply connected massive Frobenius manifold. Suppose that at generic points of the caustic $\mathcal{K}$ the germ of the underlying F-manifold is of type $I_2(n) A_1^{m-2}$ for some fixed number $n \geq 3$.*

a) *The form* $d\log\tau_I$ *has a logarithmic pole along* $\mathcal{K}$ *with residue* $-\frac{(n-2)^2}{16n}$ *along* $\mathcal{K}_{reg}$.

b) *The G-function extends holomorphically over* $\mathcal{K}$ *iff* $n = 3$.

Proof. Theorem 5.3 and (6.4) say that the form $-\frac{1}{24}d\log J$ has a logarithmic pole along $\mathcal{K}$ with residue $\frac{n-2}{48}$ along $\mathcal{K}_{reg}$. This equals $\frac{(n-2)^2}{16n}$ iff $n = 3$. So b) follows from a).

It is sufficient to prove a) for the F-manifold in Example 2.4, equipped with some metric which makes a Frobenius manifold out of it (we do not need an Euler field here). Unfortunately we do not have an identity for $d\log\tau_I$ like (5.7) for the socle field which would allow to consider only a convenient metric.

We use (2.5) - (2.11) and (6.5) - (6.10) and consider a neighborhood of $0 \in \mathbb{C}^m = M$. Denote for $j \geq 3$

$$T_{1j} := (u_1 - u_j)\frac{\eta_{j1}^2}{\eta_j\eta_1} + (u_2 - u_j)\frac{\eta_{j2}^2}{\eta_j\eta_2}, \tag{6.17}$$

$$T_{2j} := (u_1 - u_j)\frac{\eta_{j1}^2}{\eta_j\eta_1} - (u_2 - u_j)\frac{\eta_{j2}^2}{\eta_j\eta_2},$$

$$T_{12} := (u_1 - u_2)\frac{\eta_{12}^2}{\eta_1\eta_2}d(u_1 - u_2).$$

With $\eta_j(0) \neq 0$ for $j \geq 3$, (6.10) and (2.10) one calculates

$$8d\log\tau_I = \text{holomorphic 1-form} + T_{12}$$
$$+ \sum_{j\geq3}T_{1j}dt_1 + \sum_{j\geq3}T_{2j}t_2^{\frac{n-2}{2}}dt_2 \tag{6.18}$$
$$- \sum_{j\geq3}T_{1j}du_j.$$

From (2.8) one obtains

$$\eta_{1/2} = \frac{1}{2}\delta_1(\eta) \pm \frac{1}{2}\delta_2(\eta)t_2^{-\frac{n-2}{2}}, \tag{6.19}$$

$$\eta_1 \cdot \eta_2 = \frac{1}{4}t_2^{-n+2}(-\delta_2(\eta)^2 + t_2^{n-2}\delta_1(\eta)^2), \tag{6.20}$$

$$\eta_{12} = \frac{1}{4}\delta_1\delta_1(\eta) + \frac{1}{4}\frac{n-2}{2}t_2^{-n+1}\delta_2(\eta) - \frac{1}{4}t_2^{-n+2}\delta_2\delta_2(\eta). \tag{6.21}$$

The vector $\delta_2|_0$ is a generator of the socle of the subalgebra in T_0M which corresponds to $I_2(n)$. Therefore $\delta_2(\eta)(0) \neq 0$. It is not hard to see with (6.19) and (2.10) that the terms T_{1j} and $T_{2j}t_2^{\frac{n-2}{2}}$ for $j \geq 3$ are holomorphic at 0. The term T_{12} is

$$T_{12} = \frac{4}{n} \cdot t_2^{\frac{n}{2}} \cdot \frac{\eta_{12}^2}{\eta_1\eta_2} \cdot d(\frac{4}{n} \cdot t_2^{\frac{n}{2}})$$

$$= \frac{8}{n} \cdot t_2^{n-1} \cdot \frac{\eta_{12}^2}{\eta_1 \eta_2} \cdot dt_2 \qquad (6.22)$$

$$= -\frac{(n-2)^2}{2n} \cdot \frac{dt_2}{t_2} + \text{holomorphic 1-form} .$$

This proves part a).

REMARK 6.4. It might be interesting to look for massive Frobenius manifolds satisfying the hypothesis of Theorem 6.3 with $n = 3$, but where the underlying F-manifolds are not locally products of those arising from hypersurface singularities. In view of [14, Theorem 16.6] the *analytic spectra* $\text{Specan}(\mathcal{T}_M, \circ) \subset T^*M$ of such F-manifolds would have singularities, but only in codimension ≥ 2, as the analytic spectrum of A_2 is smooth.

The analytic spectrum is Cohen-Macaulay and even Gorenstein and a Lagrange variety ([14, chapter 6]). P. Seidel (Ecole Polytechnique) showed me a normal and Cohen-Macaulay Lagrange surface. But it seems to be unclear whether there exist normal and Gorenstein Lagrange varieties which are not smooth.

7. Variance of the spectrum

By Theorem 6.3 the germ $(M, 0)$ of a Frobenius manifold as in Theorem 4.2 for an isolated hypersurface singularity f has a holomorphic G-function $G(t)$, unique up to addition of a constant. By Corollary 6.2 and Theorem 4.2 this function satisfies

$$E \, G(t) = -\frac{1}{4} \sum_{i=1}^{\mu} \left(\alpha_i - \frac{n-1}{2} \right)^2 + \frac{\mu(\alpha_\mu - \alpha_1)}{48} =: \gamma . \qquad (7.1)$$

So it has a very peculiar strength: it gives a hold on the squares of the spectral numbers $\alpha_1, ..., \alpha_\mu$ of the singularity. Because of the symmetry $\alpha_i + \alpha_{\mu+1-i} = n - 1$, the spectral numbers are scattered around their *expectation value* $\frac{n-1}{2}$. One may ask about their *variance* $\frac{1}{\mu} \sum_{i=1}^{\mu} (\alpha_i - \frac{n-1}{2})^2$.

CONJECTURE 7.1. *The variance of the spectral numbers of an isolated hypersurface singularity is*

$$\frac{1}{\mu} \sum_{i=1}^{\mu} \left(\alpha_i - \frac{n-1}{2} \right)^2 \leq \frac{\alpha_\mu - \alpha_1}{12} , \qquad (7.2)$$

or, equivalently,

$$\gamma \geq 0 . \tag{7.3}$$

THEOREM 7.2.   *In the case of a quasihomogeneous singularity f*

$$\frac{1}{\mu} \sum_{i=1}^{\mu} (\alpha_i - \frac{n-1}{2})^2 = \frac{\alpha_\mu - \alpha_1}{12} , \tag{7.4}$$

and

$$\gamma = 0 . \tag{7.5}$$

Proof. $(\mathcal{O}/J_f, mult., [f]) \cong (T_0 M, \circ, E|_0)$. Here one has $f \in J_f$ and $E|_0 = 0$ and therefore $E\, G(t) = 0$.

REMARK 7.3.   a) When I presented Theorem 7.2 at the summer school on singularity theory in Cambridge in August 2000, I asked for an elementary proof of it. This was found by A. Dimca. It uses the characteristic function

$$\chi_f := \frac{1}{\mu} \sum_{i=1}^{\mu} T^{\alpha_i + 1} \tag{7.6}$$

of the spectral numbers. The variance is

$$\frac{d}{dT}(T \cdot \frac{d}{dT}(\chi_f \cdot T^{-\frac{n+1}{2}}))|_{T=1} . \tag{7.7}$$

In the case of a quasihomogeneous singularity with weights $w_0, ..., w_n \in (0, \frac{1}{2}]$ and degree 1 the characteristic function is

$$\chi_f = \frac{1}{\mu} \prod_{i=0}^{n} \frac{T - T^{w_i}}{T^{w_i} - 1} , \tag{7.8}$$

as is well known. Using this product formula A. Dimca [4] showed that in the case of a quasihomogeneous singularity the variance is $\sum_{i=0}^{n} \frac{1-2w_i}{12} = \frac{\alpha_\mu - \alpha_1}{12}$. He also made a conjecture dual to Conjecture 7.1 for the case of tame polynomials: there the inverse inequality to (7.2) should hold. The conjectures intersect in the case of quasihomogeneous singularities and there give the equality (7.4).

b) Recently M. Saito proved Conjecture 7.1 in the case of irreducible plane curve singularities [30].

c) If the singularity is Newton nondegenerate then the spectrum is closely related to the Newton filtration [29][20]. This generalizes (7.8) in a weak sense. Once could try to prove Conjecture 7.1 in this case with a refinement of Dimca's calculations. But in the general case one needs probably richer structures.

d) The only unimodal or bimodal families of non-semiquasihomogeneous singularities are the cusp singularities T_{pqr} and the 8 bimodal series. The spectral numbers are given in [1]. One finds

$$\gamma(T_{pqr}) = \frac{1}{24}(1 - \frac{1}{p} - \frac{1}{q} - \frac{1}{r}) \geq 0 \qquad (7.9)$$

with equality only for the simple elliptic singularities. In the case of the 8 bimodal families one obtains

$$\gamma = \frac{p}{48 \cdot \kappa} \cdot \left(1 - \frac{1}{p+\kappa}\right) \geq 0 \qquad (7.10)$$

with $\kappa := 9, 7, 6, 6, 5$ for $E_{3,p}, Z_{1,p}, Q_{2,p}, W_{1,p}, S_{1,p}$, respectively, and

$$\gamma = \frac{p}{48 \cdot \kappa} \cdot \left(1 + \frac{1}{p+2\kappa}\right) \geq 0 \qquad (7.11)$$

with $\kappa := 6, 5, \frac{9}{2}$ for $W_{1,p}^{\sharp}, S_{1,p}^{\sharp}, U_{1,p}$, respectively.

e) At the summer school in Cambridge in August 2000 Conjecture 7.1 was confirmed for many other singularities using the computer algebra system Singular and especially the program of M. Schulze for computing spectral numbers, which is presented in [32].

f) In [26] K. Saito studied the distribution of the spectral numbers and their characteristic function χ_f heuristically and formulated several questions about them. The G-function might help with these problems.

g) One can speculate that the Conjecture 7.1, if true, comes from a deeper hidden interrelation between the Gauß-Manin connection and polarized mixed Hodge structures. The existence of Frobenius manifolds and G-functions alone is not sufficient, as the following shows.

In [15, Remark 6.7 b] an example of M. Saito [28, 4.4] is sketched which leads for the semiquasihomogeneous singularity $f = x^6 + y^6 + x^4 y^4$ to Frobenius manifold structures with $\{d_1, .., d_\mu\} \neq \{1 + \alpha_1 - \alpha_i \mid i = 1, ..., \mu\}$. The number γ in that case is $\gamma = -\frac{1}{144} < 0$.

h) In the case of the simple singularities A_k, D_k, E_6, E_7, E_8, the parameters $t_1, ..., t_\mu$ of a suitably chosen unfolding are weighted homogeneous with positive degrees with respect to the Euler field. Therefore $G = 0$ in these cases (cf. [11]).

LEMMA 7.4. *The number γ of the sum $f(x_0, ..., x_n) + g(y_0, ..., y_m)$ of two singularities f and g satisfies*

$$\gamma(f+g) = \mu(f) \cdot \gamma(g) + \mu(g) \cdot \gamma(f) . \qquad (7.12)$$

Proof. Let $\alpha_1, ..., \alpha_{\mu(f)}$ and $\beta_1, ..., \beta_{\mu(g)}$ denote the spectral numbers of f and g. Then the spectrum of $f + g$ as an unordered tuple is [1][31]

$$(\alpha_i + \beta_j + 1 \mid i = 1, ..., \mu(f), \ j = 1, ..., \mu(g)) \, . \qquad (7.13)$$

This and the symmetry of the spectra yields (7.6).

REMARK 7.5. For any Frobenius manifold the variance $\frac{1}{m} \sum_{i=1}^{m} (d_i - \frac{D}{2})^2$ of the eigenvalues $d_1, ..., d_m$ of ∇E is interesting. It turns up not only as in Corollary 6.2 related to the G-function in the semisimple case, but also in the operator L_0 of the Virasoro constraints in [8](2.30) for any Frobenius manifold.

Prior to [8] the Virasoro constraints were postulated in the case of quantum cohomology of projective manifolds with $h^{p,q} = 0$ for $p \neq q$ in [9]. There a formula for the variance was considered which turned out to be a special case of the following formula from [21] (cf. also [3]), which is valid for any projective manifold:

$$\sum_{p,q} (-1)^{p+q} h^{p,q} \left(p - \frac{n}{2} \right)^2 = \frac{n}{12} c_n + \frac{1}{6} c_1 c_{n-1} \, . \qquad (7.14)$$

Here c_l is the lth Chern class of the manifold, n is its dimension. The proof uses the Hirzebruch-Riemann-Roch theorem. The formula is generalized to projective varieties with at most Gorenstein canonical singularities in [2]. (The formula (7.14) and the references [2][3][9][21] were pointed out to me by A. Takahashi.) Comparing the right hand side with the singularity case, one can speculate $n \sim \alpha_\mu - \alpha_1$, $c_n \sim \mu$ and ask about $\frac{1}{6} c_1 c_{n-1} \sim ?$.

References

1. V. I. Arnold, S. M. Gusein-Zade and A. N. Varchenko, *Singularities of differentiable maps, volume II*, Birkhäuser, Boston, 1988.
2. V. V. Batyrev, Stringy Hodge numbers and Virasoro algebra, preprint alg-geom/9711019.
3. L. A. Borisov, On the Betti numbers and Chern classes of varieties with trivial odd cohomology groups, preprint alg-geom/9703023.
4. A. Dimca, Monodromy and Hodge theory of regular functions, this volume.
5. B. Dubrovin, Integrable systems in topological field theory, Nucl. Phys. **B 375** (1992) 627–685.
6. B. Dubrovin, Geometry of 2D topological field theories, *Integrable sytems and quantum groups, Montecatini Terme 1993*, (eds. M Francoviglia and S. Greco), Springer Lecture Notes in Math. **1620** 120–348.
7. B. Dubrovin and Y. Zhang, Bihamiltonian hierarchies in 2D topological field theory at one-loop approximation, Commun. Math. Phys. **198** (1998) 311–361.

8. B. Dubrovin and Y. Zhang, Frobenius manifolds and Virasoro constraints, Sel. math., New ser. **5** (1999) 423–466.
9. T. Eguchi, K. Hori and Ch.-Sh. Xiong, Quantum cohomology and Virasoro algebra, Phys. Lett. **B402** (1997) 71–80.
10. E. Getzler, Intersection theory on $\bar{\mathcal{M}}_{1,4}$ and elliptic Gromov-Witten invariants, J. Amer. Math. Soc. **10** (1997) 973–998.
11. A. B. Givental, Elliptic Gromov-Witten invariants and the generalized mirror conjecture, *Integrable systems and algebraic geometry. Proceedings of the Taniguchi Symposium 1997*, (eds. M.-H. Saito, Y. Shimizu and K. Ueno), (World Scientific, River Edge NJ, 1998) pp. 107–155.
12. C. Hertling, Classifying spaces and moduli spaces for polarized mixed Hodge structures and for Brieskorn lattices, Compositio Math. **116** (1999) 1–37.
13. C. Hertling and Yu. Manin, Weak Frobenius manifolds, Int. Math. Res. Notices **6** (1999) 277–286.
14. C. Hertling, Multiplication on the tangent bundle, first part of the habilitation (also math.AG/9910116).
15. C. Hertling, Frobenius manifolds and moduli spaces for hypersurface singularities, second part of the habilitation (2000).
16. N. J. Hitchin, Frobenius manifolds (notes by D. Calderbank), *Gauge Theory and symplectic geometry, Montreal 1995*, (eds J. Hurtubise and F. Lalonde), (Kluwer Academic Publishers, Netherlands, 1997) pp 69–112.
17. M. Jimbo and T. Miwa, Monodromy preserving deformations of linear ordinary differential equations with rational coefficients II, Physica **2D** (1981) 407–448.
18. M. Jimbo, T. Miwa, Y. Mori and M. Sato, Density matrix of impenetrable Bose gas and the fifth Painlevé transcendent, Physica **1D** (1980) 80–158.
19. M. Jimbo, T. Miwa and K. Ueno, Monodromy preserving deformations of linear ordinary differential equations with rational coefficients I, Physica **2D** (1981) 306–352.
20. A. G. Khovanskii and A. N. Varchenko, Asymptotics of integrals over vanishing cycles and the Newton polyhedron, Sov. Math. Dokl. **32** (1985) 122–127.
21. A. S. Libgober and J.W. Wood, Uniqueness of the complex structure on Kähler manifolds of certain homotopy types, J. Diff. Geom. **32** (1990) 139–154.
22. B. Malgrange, Sur les déformations isomonodromiques, I, II, *Séminaire de l'ENS, Mathématique et Physique, 1979–1982*, Progress in Mathematics **37**, (Birkhäuser, Boston, 1983) pp. 401–438.
23. Yu. Manin, *Frobenius manifolds, quantum cohomology, and moduli spaces*, American Math. Soc. Colloquium Publ. **47**, 1999.
24. C. Sabbah, *Déformations isomonodromiques et variétés de Frobenius, une introduction*, Centre Math. de l'Ecole Polytechnique, U.M.R. 7640 du C.N.R.S., no. 2000-05, 251 pages.
25. K. Saito, Primitive forms for a universal unfolding of a function with an isolated critical point, J. Fac. Sci. Univ. Tokyo, Sect. IA **28** (1982) 775–792.
26. K. Saito, The zeroes of characteristic function χ_f for the exponents of a hypersurface isolated singular point, *Algebraic varieties and analytic varieties*, Advanced Studies in Pure Math. **1**, (North-Holland 1983), pp 195–217.
27. K. Saito, Period mapping associated to a primitive form, Publ. RIMS, Kyoto Univ. **19** (1983) 1231–1264.
28. M. Saito, On the structure of Brieskorn lattices, Ann. Inst. Fourier **39** (1989) 27–72.
29. M. Saito, Exponents and Newton polyhedra of isolated hypersurface singularities, Math. Ann. **281** (1988) 411–417.

30. M. Saito, Exponents of an irreducible plane curve singularity, preprint, 14 pages, math.AG/0009133.

31. J. Scherk and J.H.M. Steenbrink, On the mixed Hodge structure on the cohomology of the Milnor fibre, Math. Ann. **271** (1985) 641-665.

32. M. Schulze and J.H.M Steenbrink, Computing Hodge-theoretic invariants of singularities, this volume.

33. J. H. M. Steenbrink, Mixed Hodge structure on the vanishing cohomology, *Real and complex singularities, Oslo 1976* (ed P. Holm), (Sijthoff and Noordhoff, Alphen aan den Rijn, 1977) pp 525-562.

34. J. H. M. Steenbrink, Intersection form for quasi-homogeneous singularities, Compositio Math. **34** (1977) 211-223.

35. A. N. Varchenko, The asymptotics of holomorphic forms determine a mixed Hodge structure, Sov. Math. Dokl. **22** (1980) 772-775.

Monodromy and Hodge Theory of Regular Functions

Alexandru Dimca
Laboratoire de Mathématiques Pures de Bordeaux,
(dimca@math.u-bordeaux.fr)

1. Introduction: local versus global questions on regular functions

In the last 35 years or so a lot of effort was devoted to and great success achieved in the study of the singularities of an analytic function germ $f : (\mathbb{C}^{n+1}, 0) \to (\mathbb{C}, 0)$. Indeed, when f defines an isolated hypersurface singularity (IHS for short in the sequel) the topology of the situation was studied by Milnor [36] and Brieskorn [9] who have obtained fascinating relations to the exotic differentiable structures on spheres. Arnold and his school have classified the simplest IHS and brought into light unexpected relations to the classification of simple Lie algebras, du Val rational double points of surfaces and other classical objects in algebraic geometry, see for a complete presentation the monograph [2].

The work on analytic singularities evolved along two natural paths (the selection below reflects personal interests related to this paper and is by no means exhaustive).

(A) the detailed study of specific classes of IHS as for example

(a1) plane curve singularities, see for instance A'Campo [1] and Lê [30];

(a2) singularities which are non degenerate with respect to their Newton polyhedra at the origin, see Koushnirenko [28] and various sections in [2] ;

(a3) Yomdin singularities, i.e. the IHS defined by a function f whose Taylor series at the origin looks like $f = f_d + f_{d+k} +$ higher terms and such that $V(f_d)_{sing} \cap V(f_{d+k}) = \emptyset$, where $V(g)$ is the hypersurface in $\mathbb{P}^n$ defined by the homogeneous polynomial g, see Artal-Bartolo [3] , Luengo [32], Melle-Hernandez [34], and

(B) the study of various classes of non-isolated or arbitrary singularities as for example

(b1) the results of Kato and Matsumoto [26] on the connectivity of the Milnor fiber F of a singularity in terms of the dimension of the singular locus;

D. Siersma et al. (eds.), New Developments in Singularity Theory, 257–278.
© 2001 Kluwer Academic Publishers. Printed in the Netherlands.

(b2) the results by Barlet [7], Malgrange [33], Navarro-Aznar [37] and Siersma [49] on the monodromy of non-isolated hypersurface singularities.

A high point in this success story was the construction of a natural mixed Hodge structure (MHS for short in the sequel) on the cohomology of the Milnor fiber of an IHS by Steenbrink [52] and Varchenko [56]. A key numerical invariant obtained out of this MHS is the spectrum of an IHS, see for related results the papers by Hertling and by Schulze and Steenbrink in this volume.

In the last 20 years there was an ever increasing interest in studying the global setting of a polynomial function $f : \mathbb{C}^{n+1} \to \mathbb{C}$ in analogy to the local setting described above. To get a first idea why such a study is important, one can have a look at the Bourbaki report by Kraft [27] where a lot of important open problems in affine algebraic geometry are listed.

For general facts about this global setting we refer to the papers by Broughton [10] and by Gusein-Zade, Luengo and Melle Hernández [24]. It is known that for any polynomial f there is a finite bifurcation set $B \subset \mathbb{C}$ such that f induces a locally trivial fibration over the complement of B. For $s \notin B$ fixed, we call the fiber $F = f^{-1}(s)$ the generic fiber of the polynomial f. It was realized from the beginning that for a large class $\mathcal{C}$ of polynomials with good behavior at infinity there is a strong analogy between the properties of IHS and properties of these polynomials, e.g. the generic fiber F is homotopy equivalent to a bouquet of n-dimensional spheres exactly as the Milnor fiber of an IHS. The class $\mathcal{C}$ above contains the class of tame polynomials introduced by Broughton [10], latter generalized to M-tame polynomials by Némethi and Zaharia [41] and to polynomials with no singularities at infinity by Siersma and Tibăr [50]. When one looks only at cohomological questions over $\mathbb{Q}$ one can work as well with the class of cohomologically tame polynomials introduced by Sabbah [44]. To shorten the discussion (and also because it is not known whether one of the above classes contains all the others!), we call a polynomial in any of these classes a weakly tame polynomial as in [39].

As in the local case, the work on polynomial functions can be divided into two main streams:

(A′) the detailed study of specific classes of polynomials as for example

(a1′) polynomials of two variables, see for instance Artal-Bartolo, Cassou-Noguès and Dimca [4, 14], Michel and Weber [35], Neumann [42];

(a2′) polynomials which are non-degenerate with respect to their Newton polyhedra at infinity, see Cassou-Noguès [11], Artal-Bartolo, Luengo and Melle Hernández [5];

(a3′) Yomdin-at-infinity polynomials, with a definition dual to the definition of Yomdin singularities given in (a3) above, see the paper by Gusein-Zade, Luengo and Melle Hernández in this volume and the series of papers by Garcia-Lopez and Némethi [21, 22, 23], and

(B′) general results on arbitrary polynomials as for example

(b1′) connectivity results on the generic and special fibers of f in terms of the dimensions of some algebraic varieties describing the behavior of f at infinity, see Dimca and Păunescu [16] and Tibăr [55];

(b2′) results on the monodromy representation as in Dimca and Némethi [15], Neumann and Norbury [43] or Siersma and Tibăr [51].

The aim of this paper is twofold:

(i) to draw the attention of people working in nearby areas to this beautiful subject of research rich in challenging open questions;

(ii) to exhibit a certain duality (some readers may prefer the word "symmetry" instead of the word "duality" and they are of course free to use the word they like best) that exists between the local setting of IHS and the global setting of weakly tame polynomials.

This duality is sometimes obvious, as in the definitions of Yomdin singularities and Yomdin-at-infinity polynomials, but some other times it is hidden and difficult to grasp. As an instance of this latter case we discuss the limit MHS at infinity on the cohomology of the generic fiber of a weakly tame polynomial following the results by Sabbah [44], Némethi and Sabbah [39] and Garcia-Lopez and Némethi [21, 22, 23]. The papers [21, 22, 23] discuss the special case of *-polynomials, which are nothing else but Yomdin-at-infinity polynomial with $k = 1$. The duality in this case can be expressed by saying that the size of the Jordan blocks for the eigenvalue 1 in the monodromy acting on the cohomology of the Milnor fiber (resp. generic fiber) is smaller (resp. greater) by one than the corresponding blocks in the monodromy of the associated proper fibrations.

In the first section we recall basic facts on limit MHS, spectra and spectral pairs.

In Section 2 we describe in detail the spectrum and the spectral pairs for polynomials in two variables in order to develop basic geometric intuition. We compute several key examples: the (local) elliptic fibrations and the Briançon polynomial. We obtain also a necessary and sufficient condition for such a polynomial to have a symmetric spectrum at infinity.

The third section is the main one and contains the discussion announced above on the spectra of weakly tame polynomials.

The final section is related to a recent conjecture by C. Hertling on the distribution of spectral numbers: we prove a special case and state a dual global analog.

2. Limits of MHS, spectra and spectral pairs

Let $f : X \to S$ be a morphism of algebraic varieties such that $\dim X = n + 1$ for some $n \geq 0$ and $\dim S = 1$. We assume that S is a smooth curve and that S' is a smooth compactification of S.

Let $B \subset S$ be a finite set such that if we set $S^* = S \setminus B$ and $X^* = X \setminus f^{-1}(B)$, then $f : X^* \to S^*$ is a topologically locally trivial fibration with generic fiber Y. We set $B' = B \cup (S' \setminus S)$ and $X_s = f^{-1}(s)$ for any $s \in S$. Hence Y is homeomorphic to X_s for $s \in S^*$.

For any $b \in B'$ there is a limit MHS on the cohomology $H^*(Y, \mathbb{Q})$, see Deligne [13], Elzein [19] and Steenbrink and Zucker [54]. When $H^*(Y, \mathbb{Q})$ is endowed with this limit MHS, the corresponding MHS is denoted by $H^*_{lim,b}(Y, \mathbb{Q})$. The cohomology groups $H^*(X_s, \mathbb{Q})$ are endowed with the Deligne MHS, see Deligne [12]. It follows that for any $j \geq 0$ the Hodge filtration F satisfies the following.

$$F^0 H^j(X_s, \mathbb{Q}) = H^j(X_s, \mathbb{Q}), \quad F^{j+1} H^j(X_s, \mathbb{Q}) = 0 \; if \; j \leq n \quad (1a)$$

and

$$F^{j-n} H^j(X_s, \mathbb{Q}) = H^j(X_s, \mathbb{Q}), \quad F^{n+1} H^j(X_s, \mathbb{Q}) = 0 \; for \; j \geq n. \;(1b)$$

These equalities hold for the limit MHS $H^*_{lim,b}(Y, \mathbb{Q})$ and this implies the following equalities for the weight filtration M on $H^j_{lim,b}(Y, \mathbb{Q})$

$$M_{-1} H^j_{lim,b}(Y, \mathbb{Q}) = 0, \quad M_{2j} H^j_{lim,b}(Y, \mathbb{Q}) = H^j_{lim,b}(Y, \mathbb{Q}) \; if \; j \leq n \;(2a)$$

and

$$M_{2(j-n)-1} H^j_{lim,b}(Y, \mathbb{Q}) = 0, \quad M_{2n} H^j_{lim,b}(Y, \mathbb{Q}) = H^j_{lim,b}(Y, \mathbb{Q}) \; for \; j \geq n. \tag{2b}$$

Let N_b be the logarithm of the unipotent part of the monodromy operator m_b acting on $H^*(Y, \mathbb{Q})$ associated with going once anti-clockwise around the bifurcation point b. The filtration M, being the weight filtration of N_b relative to the weight filtration W, the limit of the weight filtrations of the MHS on $H^*(X_s, \mathbb{Q})$, satisfies $N_b(M_k) \subset M_{k-2}$ for any k, see [54], (2.5). In particular, the existence of the limit MHS and the remark (2a), (2b) imply the following Monodromy Theorem (for other proofs see Landman [29] and Fried [20]):

COROLLARY 2.1. *Any Jordan block of the monodromy operator m_b acting on $H^j(Y, Q)$ has size at most $min(j+1, 2n-j+1)$.*

Let (H, T_s, m) be a triple consisting of a MHS H, an automorphism of finite order T_s of H and an integer m.

DEFINITION 2.2. *The spectrum of the triple (H, T_s, m) is the sum*

$$Sp(H, T_s, m) = \sum_a d(H, T_s, m)_a \cdot a \in \mathbb{N}^{(Q)}$$

given by $d(H, T_s, m)_a = \dim Gr_F^{[m+1-a]} H_\lambda$ where $\lambda = exp(-2\pi i a)$ and $H_\lambda = Ker(T_s - \lambda Id)$ has the induced Hodge filtration $F^p H_\lambda = F^p H \cap H_\lambda$.

We use the shift $Sp(H, T_s, m)[n] = Sp(H, T_s, m+n)$ for an integer n and we set $Sp(H, m) = Sp(H, T_s, m)$ when $T_s = Id$. Note that $Sp(H(-n), T_s, m) = Sp(H, T_s, m)[-n]$ where $H(-n)$ is the Tate twist.

In this paper we are interested mainly in the following two cases.

The Local Case: Let $f : (\mathbb{C}^{n+1}, x) \to \mathbb{C}$ be an *isolated hypersurface singularity*. If we take $H = \tilde{H}^n(F_x, Q)$, the reduced cohomology of the Milnor fiber endowed with the Steenbrink MHS (see [52, 47]), and T_s the semisimple part of the local monodromy, then the spectrum $Sp(H, T_s, n)$ is denoted in the sequel by $Sp(f, x)$ and the corresponding coefficients $d(H, T_s, m)_a$ are denoted by $d(f, x)_a$. When $d(f, x)_a \neq 0$, a is called a spectral number of the singularity f. This spectrum is usually shifted to the left by 1, see for instance [47], so that it is contained in the interval $(-1, n)$ and is symmetric with respect to the middle point $(n-1)/2$, but we will not use this shift (essentially due to the conventions in [44] and in order to simplify our computations in section 4).

The Global Case: Let $f : X \to S$ be a (global) *regular function* as above and $b \in B'$. Then we take

$$(H, T_s, m) = (H_{lim,b}^j(Y, Q), S_b, j)$$

where S_b is the semisimple part of the monodromy operator m_b and we denote the corresponding spectrum by $Sp(f, b, j)$ and let $d(f, b, j)_a$ stand for $d(H, T_s, m)_a$.

Using (1a), (1b) it follows that for $j \leq n$ one has

$$d(f, b, j)_a = 0 \quad if \quad a \leq 0 \quad or \quad a > j + 1. \tag{3}$$

We express this property in the following intuitive way

$$Supp(Sp(f, b, j)) \subset (0, j + 1]. \tag{4}$$

In some cases one has the stronger claim

$$Supp(Sp(f, b, j)) \subset (0, j + 1) \tag{5}$$

e.g. when $b = \infty$ for a cohomologically tame polynomial $f : \mathbb{C}^{n+1} \to \mathbb{C}$ by C. Sabbah [44] and for any polynomial $f : \mathbb{C}^2 \to \mathbb{C}$, see (3.6,i) below.
    Similarly for $j \geq n$ one gets

$$Supp(Sp(f, b, j)) \subset (j - n, j + 1] \tag{6}$$

We want to emphasize the fact that the MHS occurring in the local case and in the global case for a cohomologically tame polynomial enjoy some special properties which we single out in the following.

DEFINITION 2.3. *We say that a triple (H, T_s, m) (or, for short, that the MHS H) is symmetric if there is a nilpotent MHS morphism N : $H \to H$ of type $(-1, -1)$ and commuting with T_s such that either:*
    (i) T_s has no eigenvalue equal to 1 and the weight filtration of H is the weight filtration associated to the nilpotent operator N with center m;
    (ii) T_s has all eigenvalues equal to 1 and the weight filtration of H is the weight filtration associated to the nilpotent operator N with center $m + 1$; or
    (iii) the triple (H, T_s, m) is a direct sum of two triples, the first of type (i) above, the second of type (ii) above.

We have the following linear algebra result (whose proof is left to the reader) which explains this definition.

LEMMA 2.4. *Let (H, T_s, m) be a symmetric triple. Then the corresponding spectrum $Sp(H, T_s, m)$ is symmetric with respect to $(m + 1)/2$.*

On the other hand it is known that the MHS occuring in the local case and in the global case for a cohomologically tame polynomial are symmetric, the nilpotent operator N being in these cases the unipotent logarithm N_b of the monodromy, see [47] and [44]. We will see below that in general the spectrum $Sp(f, b, j)$ associated to a regular function is not symmetric, hence in particular the corresponding MHS are not symmetric. This means that the relation between the relative weight filtration M and the operator N_b is in general more complicated.

REMARK 2.5.

Claude Sabbah considers in the paper [44] the spectrum of the triple $(H^{n+1}(X, Y; \mathbb{Q}), T_s, n)$ the relative cohomology group $H^{n+1}(X, Y; \mathbb{Q})$ being given a natural limit MHS and T_s denoting the semisimple part of the monodromy at infinity. In case $H^n(X, \mathbb{Q}) = H^{n+1}(X, \mathbb{Q}) = 0$ this spectrum coincides with the spectrum $Sp(f, \infty, n)$.

To conclude this discussion on spectra, we establish the relation between $Sp(H, T_s, m)$ and $Sp(H, T_s^{-1}, m)$. It is clear that $H_\lambda(T_s) = Ker(T_s - \lambda Id)$ and $H_{\overline{\lambda}}(T_s^{-1})$, which is defined in the same way, are identical. Hence, if $\lambda = \overline{\lambda}$ (i.e. if $\lambda = exp(2\pi i a)$ and $2a \in \mathbb{Z}$), then the corresponding contributions to the two spectra are the same. In the opposite case, one has

$$v \in Gr_F^{[m+1-a]} H_\lambda(T_s) \quad \text{iff} \quad v \in Gr_F^{[m+1-a']} H_{\overline{\lambda}}(T_s^{-1})$$

where a and a' are in the same interval $(k, k+1)$, $k \in \mathbb{Z}$ and symmetric with respect to the middle point $k + \frac{1}{2}$. Hence we get the following.

LEMMA 2.6. *To get the spectrum $Sp(H, T_s^{-1}, m)$ from the spectrum $Sp(H, T_s, m)$ we have to*
   (i) leave unchanged all the terms $d_a \cdot a$ for $2a \in \mathbb{Z}$;
   (ii) replace all the terms $d_a \cdot a$ for $2a \notin \mathbb{Z}$ by the terms $d_a \cdot a'$, where a' is determined as above.

In particular, if the spectrum $Sp(H, T_s, m)$ is symmetric with respect to a point p such that $2p \in \mathbb{Z}$, then the spectrum $Sp(H, T_s^{-1}, m)$ is also symmetric with respect to the point p.

In the theory of isolated hypersurface singularities there is a very useful invariant, called the *spectral pairs*, which is finer than the spectrum, see Steenbrink [52], Némethi [38], Némethi and Steenbrink [40]. Our definition below is very similar but not identical to their definitions (which also differ slightly from paper to paper).

DEFINITION 2.7. *The spectral pairs of the triple* (H, T_s, m) *is the sum*

$$Spp(H, T_s, m) = \sum_{a,w} d(H, T_s, m)_{a,w} \cdot (a, w) \in \mathbb{N}^{(\mathbb{Q} \times \mathbb{Z})}$$

given by $d(H, T_s, m)_{a,w} = \dim Gr_F^{[m+1-a]} Gr_w^M H_\lambda$ *where* $\lambda = exp(-2\pi i a)$ *and*
$Gr_w^M H_\lambda = Ker(Gr_w^M T_s - \lambda Id)$ *has the induced Hodge filtration.*

We extend in an obvious way all the notations used for the spectrum of a triple (resp. singularity, resp. regular function) to the case of the corresponding spectral pairs.

It is clear that the spectral pairs determine the spectrum since

$$d(H, T_s, m)_a = \sum_w d(H, T_s, m)_{a,w}.$$

We express this fact in the following shortened way.

$$Sp(H, T_s, m) = pr_{1,*}(Spp(H, T_s, m))$$

On the other hand, the information contained in the spectral pairs of a triple (H, T_s, m) is equivalent to the set of the equivariant Hodge numbers

$$h_\lambda^{p,q}(H, T_s, m) = \dim Gr_F^p Gr_{p+q}^W H_\lambda \qquad (7)$$

plus the integer m. Indeed we have

$$h_\lambda^{p,q}(H, T_s, m) = d(H, T_s, m)_{a,w} \qquad (8)$$

where $p = [m + 1 - a]$, $q = w - p$, $\lambda = exp(-2\pi i a)$ and, conversely, $w = p + q$ and $a = m + 1 - p - \alpha$ with $0 \le \alpha < 1$ and $\lambda = exp(-2\pi i a)$.

REMARK 2.8.
   (i) By conjugation we get $h_\lambda^{p,q} = h_{\bar\lambda}^{q,p}$. However, note that in general $h_\lambda^{p,q} \neq h_\lambda^{q,p}$, see Example (2.9) below.
   (ii) The spectral pairs in the case of a symmetric MHS give exactly the spectrum and the size of the Jordan blocks for N.

In the following example the spectral pairs are computed for two polynomials illustrating two classes of cohomologically tame polynomials relatively easy to handle, namely the *-polynomials studied by García-López and Némethi [21, 22, 23] and the convenient Newton nondegenerate polynomials studied by Sabbah [44] and Douai [18].

EXAMPLE 2.9.

(i) Let $f : \mathbb{C}^2 \to \mathbb{C}$ be the $*$-polynomial given by $f(x,y) = (x+y)^3 + x^2 y^2$, see [23], (7.1). Then the only nonzero equivariant Hodge numbers for the monodromy at infinity are the following: $h_1^{1,1} = h_{-1}^{1,1} = h_{-1}^{0,0} = 1$ and $h_\lambda^{0,1} = h_{\bar\lambda}^{1,0} = 2$ for $\lambda = exp(-\frac{\pi}{3})$. Hence

$$Spp(f, \infty, 1) = (\frac{3}{2}, 0) + 2 \cdot (\frac{1}{6}, 1) + 2 \cdot (\frac{11}{6}, 1) + (\frac{1}{2}, 2) + (1, 2)$$

and

$$Sp(f, \infty, 1) = 2 \cdot \frac{1}{6} + \frac{1}{2} + 1 + \frac{3}{2} + 2 \cdot \frac{11}{6}$$

In particular, there is a Jordan block of size two for $\lambda = -1$.

(ii) Let $f : \mathbb{C}^3 \to \mathbb{C}$ be given by $f(x,y,z) = x + y + z + x^2 y^2 z^2$, see [18]. Then f is convenient and Newton nondegenerate with respect to the Newton polyhedron at infinity and the results of Sabbah [44] give the following result.

$$Spp(f, \infty, 2) = (\frac{5}{2}, 0) + (\frac{3}{2}, 2) + (2, 2) + (\frac{1}{2}, 4) + (1, 4)$$

and there is a Jordan block of size two for $\lambda = 1$ and a Jordan block of size three for $\lambda = -1$.

3. The case of smooth curve families

In this section we assume that $\dim X = 2$ and that the morphism $f : X^* \to S^*$ is smooth and has a connected general fiber. By replacing S^* by a smaller dense Zariski open subset, we may assume that f has a compactification $f' : X' \to S'$ such that

(i) X' is a smooth proper surface and if we set $X'^* = X' \cap f'^{-1}(S^*)$, then $f' : X'^* \to S^*$ is smooth;

(ii) $D = X' \setminus X$ is a normal crossing divisor and $D^* = D \cap f'^{-1}(S^*)$ is proper and smooth over S^*.

Then we have an exact sequence of variations of MHS on S^*, namely

$$0 \to V' \to V \to V''(-1) \to 0 \tag{9}$$

where

(a) $V' = R^1 f'_*(Q_{X'*})$ is a variation of HS of weight 1 such that $V'_s = H^1(X'_s, Q)$ for any $s \in S^*$ with $X'_s = f'^{-1}(s)$;

(b) $V = R^1 f_*(Q_{X*})$ is the variation of MHS of interest to us, and

(c) $V'' = \tilde{R}^0 f'_*(Q_{D*})$ is a variation of HS of weight 0 such that $V''_s = \tilde{H}^0(A_s, Q)$ where $A_s = X'_s \setminus X_s$.

For $b \in B'$ let V'_b, V_b, V''_b be the corresponding limit MHS at the point b. Note that both relative weight filtrations on V'_b and V''_b are in fact weight filtrations of nilpotent operators.

Lemma (5.26) in [54] and the compatibility of the MHS with the semisimple operators S_b, S'_b and S''_b (where the latter two are defined in an obvious way), give the following exact sequence

$$0 \to Gr_F^p Gr_{p+q}^M V'_{b,\lambda} \to Gr_F^p Gr_{p+q}^M V_{b,\lambda} \to Gr_F^p Gr_{p+q}^M V''(-1)_{b,\lambda} \to 0 \tag{10}$$

for any p, q and any eigenvalue λ. This implies

$$Spp(V_b, S_b, 1) = Spp(V'_b, S'_b, 1) + Spp(V''(-1)_b, S''_b, 1) \tag{11}$$

In other words, to understand the spectral pairs $Spp(f, b, 1) = Spp(V_b, S_b, 1)$ it is enough to study each of the two parts, say $A(f, b) = Spp(V'_b, S'_b, 1)$ and $B(f, b) = Spp(V''_b(-1), S''_b, 1)$.

PROPOSITION 3.1.   *There is an element* $A'(f, b) \in N^{(Q \times Z)}$ *such that* $pr_{1,*}A'(f, b)$ *is contained in the interval* $(0, 2)$, *is symmetric with respect to (and does not contain) the middle point 1 and* $A(f, b) = A'(f, b) + d \cdot (2, 0) + c \cdot (1, 1) + c \cdot (2, 1) + d \cdot (1, 2)$. *Here* $2c + 2d = \dim V'_{b,1}$ *and* d *equals the number of Jordan blocks of size two in* m'_b *for* $\lambda = 1$.

Proof.

We can write $V' = V'_1 \oplus V'_{\neq 1}$ to separate the generalized eigenspaces corresponding to the eigenvalues $\lambda = 1$ and $\lambda \neq 1$ in the monodromy operator m'_b. This decomposition gives a decomposition at the level of limit MHS $V'_b = V'_{b,1} \oplus V'_{b,\neq 1}$ and the part $A'(f, b)$ corresponds to the spectral pairs of the MHS $V'_{b,\neq 1}$ and the corresponding spectrum $pr_{1*}A'(f, b)$ is seen to be symmetric by Lemma (2.4).

The claim concerning the MHS $V'_{b,=1}$, i.e. the coefficients c and d, is also obvious.

EXAMPLE 3.2.

(i) For a polynomial $f : C^2 \to C$ and for $b = \infty$ we know that $m'_{b,1}$ is semisimple without the eigenvalue 1, see [14]. Therefore in this case the partial spectrum $pr_{1,*}A(f, \infty)$ is contained in the interval $(0, 2)$ and is symmetric with respect to 1, as in this case $c = d = 0$.

(ii) Let now $f : C^2 \to C$ be the Briançon polynomial described in [4], [14], and let $b = 0$. In this case one has an opposite situation: $A'(f, 0) = 0$ and $A(f, 0) = (2, 0) + (1, 2)$ (to see this use the fact that in this case $V_b' = V_{b,1}'$ as follows from the description of the monodromy given in loc. cit.).

To compute the part $A'(f, b)$ of the spectral pairs seems to be a difficult question in general. We give the answer below for some local elliptic fibration, using the Kodaira notation for the special fibers, see [8], p.150.

PROPOSITION 3.3. *Assume that* f' *induces an elliptic fibration over a small disc centered at* b. *Then we have the following table where under the type of the central special fiber we give the corresponding spectral pairs in* $A(f, b)$.

I_k	I_k^*	II	II^*
$(2, 0) + (1, 2)$	$(\frac{3}{2}, 0) + (\frac{1}{2}, 2)$	$(\frac{1}{6}, 1) + (\frac{11}{6}, 1)$	$(\frac{5}{6}, 1) + (\frac{7}{6}, 1)$

III	III^*	IV	IV^*
$(\frac{1}{4}, 1) + (\frac{7}{4}, 1)$	$(\frac{3}{4}, 1) + (\frac{5}{4}, 1)$	$(\frac{1}{3}, 1) + (\frac{5}{3}, 1)$	$(\frac{2}{3}, 1) + (\frac{4}{3}, 1)$

Proof. In the case of types I_k and I_k^* $(k > 0)$ the result follows from (3.1) and the description of the monodromy given in [8], p. 159. Note that here occur the only non-semisimple monodromy operators.

In half of the remaining cases, using the fact that there is unicity for a local elliptic fibration of a given type, see [8] p. 155, we can use the following (affine) models for f'.

II: $f(x, y) = x^2 + y^3$ III: $f(x, y) = x^2 + y^4$ IV: $f(x, y) = x^3 + y^3$

The Hodge filtration on the cohomology of the Milnor fiber Y of such a weighted homogeneous polynomial is described by Steenbrink in [53]. In fact we have

(i) $V_0' = H^1_{lim,0}(Y, \mathbb{Q})_{\neq 1}$ as in [53];

(ii) $H^1_{lim,0}(Y, \mathbb{Q}) = H^1(Y, \mathbb{Q})$ as MHS, since in this case $N_0 = 0$, see [54], (2.14).

The result follows now by using Lemma (1.12) since passing to the dual fibration replaces the monodromy by its inverse, see [8], pp. 157-159.

EXAMPLE 3.4. For the Briançon polynomial again, but this time for the special fiber corresponding to $b = -16/9$, we have a local elliptic fibration of type II and hence for this fiber we have $c = d = 0$ and

$$A'(f, b) = (1/6, 1) + (11/6, 1)$$

Now we turn to the study of the remaining part of the spectral pairs, namely $B(f, b)$. To do this, let D_i for $i = 1, ..., \delta(f)$ denote the horizontal (or dicritical) components of the exceptional divisor D. Set $D^h = \cup_{i=1,...,\delta(f)} D_i$. Each fiber X'_b (which may have several irreducible components) intersects any component D_i in j_i points and let $\sum k_{i,j} a_{i,j}$ be the divisor on D_i describing this intersection, i.e. $a_{i,j} \in D_i$ for $j = 1, ..., j_i$ and k_{ij} are strictly positive integers.

This implies that the monodromy m''_b (which is semisimple as a special case of (2.1) or from a direct geometric picture) has the following characteristic polynomial

$$\Delta(m''_b)(t) = (\prod (t^{k_{ij}} - 1))(t - 1)^{-1}$$

As a result the limit MHS V''_b is pure of weight 0 and we get the following.

PROPOSITION 3.5.

$$B(f, b) = \sum B(k_{ij}) + (|Dic_b| - 1) \cdot (1, 2)$$

where the sum is over all integers $k_{ij} > 1$, $B(k) = (\frac{1}{k}, 2) + ... + (\frac{k-1}{k}, 2)$ and $Dic_b = X'_b \cap D^h$.

In particular $pr_{1*} B(f, b)$ is always contained in the interval $(0, 1]$.

EXAMPLE 3.6.
(i) For a polynomial $f : \mathbb{C}^2 \to \mathbb{C}$ and for $b = \infty$ we know that $j_i = 1$ for all i, and hence one has $k_{i1} = deg(D_i)$, the degree of the corresponding dicritical component. We get the following result:

For a polynomial $f : \mathbb{C}^2 \to \mathbb{C}$ and for $b = \infty$ the spectrum $Sp(f, b)$ is always contained in the interval $(0, 2)$ and it is symmetric with respect to 1 if and only if all the dicritical components of f have degree 1.

It is known that for a (cohomologically) tame polynomial $f : \mathbb{C}^2 \to \mathbb{C}$ all the dicritical components have degree 1, but the converse is not true, e.g. consider the polynomial $f(x, y) = x + x^2 y^2 + x^2 y^3$, see [6].

(ii) Let now $f : \mathbb{C}^2 \to \mathbb{C}$ be the Briançon polynomial described in [4], [14], and let $b = 0$. Using [4] it follows that $B(f, 0) = 2 \cdot (1, 2)$, hence in this case we get

$$Spp(f, 0) = (2, 0) + 3 \cdot (1, 2)$$

For the second special fiber corresponding to $b = -\frac{16}{9}$ we get $B(f, b) = (\frac{1}{2}, 2) + (1, 2)$ and hence

$$Spp(f, -\frac{16}{9}) = (\frac{1}{6}, 1) + (\frac{11}{6}, 1) + (\frac{1}{2}, 2) + (1, 2)$$

since the central fiber is of type II. Finally, for the monodromy at infinity of this polynomial we get

$$Spp(f, \infty) = (\frac{2}{3}, 1) + (\frac{4}{3}, 1) + (\frac{1}{2}, 2) + (1, 2).$$

Indeed, the special fiber X'_∞ is of type IV^* and $B(f, \infty) = (\frac{1}{2}, 2) + (1, 2)$ as before.

Note that all the three corresponding spectra of f are not symmetric, and hence the corresponding limit MHS are not symmetric as well.

4. The case of weakly tame polynomials

In this section we discuss the limit MHS arising from a weakly tame polynomial $f : \mathbb{C}^{n+1} \to \mathbb{C}$, $n > 0$ (see [44]S, [39] for the relevant definitions and results) under the assumption that f has a compactification $f' : X' \to \mathbb{C}$ which is equisingular in the following sense.

Set $X = \mathbb{C}^{n+1}$ and $D = X' \setminus X$ the divisor at infinity. This divisor has a natural decomposition $D = D_h \cup D_v$ where D_h (resp. D_v) is the union of horizontal (resp. vertical) irreducible components in D. Recall that a component E of D is called horizontal (resp. vertical) if $f'(E) = \mathbb{C}$ (resp. $f'(E)$ is a point). Then we require the following properties to hold.

(P1) for any $s \in S^* = C \setminus B$, the corresponding fiber $X'_s = f'^{-1}(s)$ is an orientable Q-manifold;

(P2) the restriction $f'|D_h : D_h \to C$ is a topological trivial fibration.

These conditions are clearly satisfied for the *-polynomials of R. García López and A. Némethi, see [21, 22, 23] as well as for a *-polynomial with respect to some positive integer weights, e.g. $f = z^3 + x^2yz + x^4y + y^5$ for the weights $(2,2,3)$. (In such a case the required compactifications can be constructed using weighted projective spaces).

For each $s \in S^*$ we have an exact sequence of MHS (the coefficients are in Q and are omitted)

$$0 \to H^n(X'_s, X_s) \to H^n(X'_s) \to$$
$$\to H^n(X_s) \to H^{n+1}(X'_s, X_s) \to H^{n+1}(X'_s) \to 0 \qquad (12)$$

giving rise to an exact sequence of geometric variations of MHS on S^*

$$0 \to V^n(X', X) \to V^n(X') \to V^n(X) \to V^{n+1}(X', X) \to V^{n+1}(X') \to 0. \qquad (13)$$

LEMMA 4.1. The local systems $V^k(X', X)$ for $k = n, n+1$ and $V^{n+1}(X')$ are trivial.

Proof. The claim for the local system $V^k(X', X)$ follows from the isomorphism $H^m(X'_s, X_s) = H_{2n-m}(D_s)$ given by Alexander duality for the oriented Q-manifold X'_s combined with the assumption $P2$. Here $D_s = D \cap X'_s = X'_s \setminus X_s$.

The claim for $V^{n+1}(X')$ in the case $n > 2$ follows from the exact sequence

$$0 = H_c^{n+1}(X_s) \to H^{n+1}(X'_s) \to H^{n+1}(D_s) \to H_c^{n+2}(X_s) = 0.$$

The remaining cases $n = 1, 2$ can be treated in a similar way. See also [21], (4.4).

COROLLARY 4.2. For any bifurcation point $b \in B'$ we have

(i) an isomorphism of MHS $H_{lim,b}^n(Y)_{\neq 1} = H_{lim,b}^n(Y')_{\neq 1}$, where Y' is the generic fiber of f'. In particular the weight filtration on $H_{lim,b}^n(Y)_{\neq 1}$ is the weight filtration of N_b centered at n and $N_b : H_{lim,b}^n(Y)_{\neq 1} \to H_{lim,b}^n(Y)_{\neq 1}$ is a MHS morphism of type $(-1, -1)$.

(ii) under the isomorphism in (i), an equality of monodromy operators $m_{b,\neq 1} = m'_{b,\neq 1}$.

In the exact sequence (12), $H^n(X_s', X_s)$ and $H^n(X_s')$ are pure HS of weight n, $H^{n+1}(X_s')$ is a pure HS of weight $n+1$, while $H^n(X_s)$ is a MHS of weight $\geq n$ with $W_n H^n(X_s) = Im(H^n(X_s') \to H^n(X_s))$. It follows that $H^{n+1}(X_s', X_s)$ has weights $> n$ and we have exact sequences

$$0 \to H^n(X_s', X_s) \to H^n(X_s')_1 \to W_n H^n(X_s)_1 \to 0 \qquad (14)$$

$$0 \to Gr_{n+1}^W H^n(X_s) \to Gr_{n+1}^W H^{n+1}(X_s', X_s) \to H^{n+1}(X_s') \to 0 \qquad (15)$$

and isomorphisms $Gr_m^W H^n(X_s) = Gr_m^W H^{n+1}(X_s', X_s)$ for any $m > n+1$.

COROLLARY 4.3. *The local systems $Gr_m^W H^n(X_s)$ are trivial for any $m > n$.*

It follows from our results below that the extensions of local systems

$$0 \to W_{m-1} H^n(X_s) \to W_m H^n(X_s) \to Gr_m^W H^n(X_s) \to 0$$

are not trivial in general.

For a variation V of MHS on S^* and for a point $b \in B'$ we denote by V_b the corresponding limit MHS at b if it exists. Applying this to the exact sequence (13) we get the following exact sequence of MHS.

$$0 \to V^n(X', X)_b \to V^n(X')_{b,1} \to V^n(X)_{b,1} \to$$
$$\to V^{n+1}(X', X)_b \to V^{n+1}(X')_b \to 0. \qquad (16)$$

This can be written as a short exact sequence of MHS

$$0 \to C \to V^n(X)_{b,1} \to K \to 0 \qquad (17)$$

where C is the cokernel of the first (non-trivial) morphism in (16) and K is the kernel of the last morphism in (16). Now we have:

(i) $V^n(X', X)_b$ is pure of weight n and contained in the kernel of N_b';

(ii) the weight filtration on $V^n(X')_{b,1}$ is the weight filtration of N_b' centered at n.

These two facts imply that the weight filtration on C is just the weight filtration of N_b' centered at n. On the other hand, K has weights $> n$ such that

$$\dim Gr_m^M(K) = \dim Gr_m^W(K) = \dim Gr_m^W H^n(X_s) \qquad (18)$$

for any $m > n$.

In general it is not yet clear how the weight filtrations on C and on K glue together to give the weight filtration on $V^n(X)_{b,1}$. However in the most interesting case, i. e. when $b = \infty$ this follows from the work of C. Sabbah [44]. Indeed, there is exactly one way in which a weight filtration associated to a nilpotent operator N' centered at n can become a weight filtration associated to a nilpotent operator N centered at $n+1$ by adding some terms as in the exact sequence (17): each Jordan block of size k in N' becomes a Jordan block of size $k+1$ in N and there are possibly some 1-blocks in N as well corresponding exactly to the elements in $Gr^W_{n+1}K$. In fact, the situation at hand here is *dual* to the situation taking place for an isolated hypersurface singularity, where the shift in the center is realized by passing to a quotient rather than to a larger object.

Formally, we have the following situation. For $r \geq 0$ let

$$P_{n+r} = Ker N' : Gr^M_{n+r}C \to Gr^M_{n-r-2}C.$$

We denote by P'_m a vector subspace in C projecting isomorphically onto P_m. The diagram (17) implies the existence of a direct sum decomposition $H = C \oplus \oplus_{r>0}Q_{n+r}$ where $H = V^n(X)_{b,1}$ and Q_m are vector subspaces in H such that the projection $H \to K$ induces isomorphisms $Q_m = Gr^M_m K$ and $N(Q_{n+1}) = 0$, $N(Q_m) \subset P'_{m-2}$ for all $m > n+1$.

The claim above that for $b = \infty$ the filtration M on H is the weight filtration associated to the nilpotent operator N centered at $n+1$ is equivalent to the fact that the induced linear maps $N : Q_m \to P'_{m-2}$ are isomorphisms for $m > n+1$.

In the case of *-polynomials, the surjectivity of these maps was shown essentially in [21], section 4. using the variation map. The injectivity in this case was proved in [23], section 3., using the *-condition to relate the global situation to the local case of isolated hypersurface singularities.

Using the above decomposition of H, we can describe the W-filtration on H as follows for any b.

$$W_m H = 0 \ \ for \ \ m < n, \ \ W_n H = C \ \ and \ \ W_{n+r}H = C \oplus \oplus_{1 \leq k \leq r}Q_{n+k}.$$
$$(19)$$

In particular, note that $M = W$ if and only if $NW_m \subset W_{m-1}$ according to [54], (2.14) and in our case this happens exactly when $N(C) = 0$, i.e. $m'_{b,1} = 1$. Moreover it follows that in general N is not strictly compatible with the filtration W, i.e. $NW_m H = NH \cap W_m H$ does not hold for all m, though this is the case for $n = 1$ or more generally when W has length 2, according to [54], (2.16).

Note also that the claim $Gr^W_m H = 0$ for $m \neq n+1$ is definitely false, but that a statement of this type for the Fourier transform of the sheaf $R^n f_* Q_X$ is one of the key facts in [44].

These remarks give the following results, most of which where proved for *-polynomials by R. García López and A. Némethi, see [21, 22, 23].

The first result expresses the relation between the number $\sharp_k m_{\infty,1}$ of Jordan blocks of size k for the eigenvalue 1 in the monodromy operator at infinity and the Deligne MHS on $H^n(X_s)$ for $s \in S^*$. Note that there is no analog of this result in the case of an IHS.

PROPOSITION 4.4. *For any $k > 0$ we have*

$$\sharp_k m_{\infty,1} = \dim Gr^W_{n+k} H^n(X_s).$$

In particular $\sharp_{n+1} m_{\infty,1} = 0$ as a special case of [17] and $\dim Ker(m_\infty - 1) = b_n(X_s) - b_n(X'_s) + b_n(D_s)$.

Note that for a cohomologically tame or M-tame polynomial, we have an equality $\dim KerS = \dim Ker(m_\infty - 1)$, where S denotes the intersection form on $H^n(X_s)$, see for details [15], (4.2).

The second result compares the numbers of Jordan blocks of size k for the eigenvalue 1 in the monodromy operator at infinity m_∞ with the same numbers for the compactified monodromy operator at infinity m'_∞, see [21], Thm. (4.6) in the case of *-polynomials.

PROPOSITION 4.5. *(i) $\sharp_2 m_{\infty,1} = \sharp_1 m'_{\infty,1} - b_n(D_s)$*
    (ii) $\sharp_k m_{\infty,1} = \sharp_{k-1} m'_{\infty,1}$ for any $k > 2$.
    Moreover, 1 is not an eigenvalue for m'_∞ iff $m_{\infty,1}$ is the identity.

5. Hertling's local conjecture and a global analog

Recently C. Hertling has proposed the following conjecture. Let $f : (\mathbb{C}^{n+1}, 0) \to (\mathbb{C}, 0)$ be an IHS with Milnor number μ. Let $0 < a_1 \le a_2 \le \dots \le a_\mu < n + 1$ be the spectrum of f with an obvious notation. Then one has the following.

CONJECTURE 5.1.

$$\sum_{i=1,\mu} \left(a_i - \frac{n+1}{2} \right)^2 \le \frac{(a_\mu - a_1)\mu}{12}.$$

Hertling has checked this inequality for large classes of singularities and proved that in fact this is an equality for a weighted homogeneous singularity. His proof uses the recent theory of Frobenius manifols, see Hertling's paper [25] in this volume. Morihiko Saito has a proof for (5.1) for irreducible plane curve singularities, see [46].

We first give an elementary proof of the equality for a weighted homogeneous singularity.

PROPOSITION 5.2. *With the above notations, if in addition f is a weighted homogeneous polynomial, then*

$$\sum_{i=1,\mu} \left(a_i - \frac{n+1}{2} \right)^2 = \frac{(a_\mu - a_1)\mu}{12}.$$

Proof. Let $(w_0, w_1, ..., w_n)$ be a system of positive rational weights with respect to which the polynomial f is weighted homogeneous of degree 1. Let

$$p(t) = \sum_{i=1,\mu} t^{a_i}$$

be the generating function of the spectrum. It is known that

$$p(t) = \prod_{k=0}^{n} \frac{t^{w_k} - t}{1 - t^{w_k}}, \qquad a_1 = w \qquad and \qquad a_\mu = n + 1 - w$$

where $w = \sum_{k=0}^{n} w_k$, see [52] or [45]. Consider now the function $q(t) = p(t)t^{-\frac{n+1}{2}}$. It follows that

$$\sum_{i=1}^{\mu} \left(a_i - \frac{n+1}{2} \right)^2 = lim_{t\to 1}(tq'(t))'.$$

On the other hand we can write $q(t) = \prod_{k=0,n} q_k(t)$ where

$$q_k(t) = \frac{t^{w_k - \frac{1}{2}} - t^{\frac{1}{2}}}{1 - t^{w_k}}$$

Note that

$$q'(t) = q(t) \sum_{k=0}^{n} \frac{q_k'(t)}{q_k(t)} \qquad and \qquad lim_{t\to 1}q(t) = \mu.$$

To conclude the proof, it is enough to show that

(*) $lim_{t\to 1}\frac{q_k'(t)}{q_k(t)} = 0 \qquad and \qquad lim_{t\to 1}\frac{q_k'(t)}{(t-1)q_k(t)} = \frac{1-2w_k}{12}.$

Indeed, the first equality in (∗) implies that

$$\lim_{t\to1}(tq'(t))' = \lim_{t\to1}\frac{q'(t)}{t-1}$$

and this and the second equality give the result directly.

Finally, to prove (∗), we consider the function

$$r(t) = \frac{t^{w-\frac{1}{2}} - t^{\frac{1}{2}}}{1 - t^w}$$

To prove the first claim in (∗), it is enough to show that

$$\lim_{t\to1}r'(t) = 0$$

If we write $r'(t) = \frac{A(t)}{B(t)}$ with $B(t) = (1 - t^w)^2$, a direct computation shows that: $A(1) = A'(1) = A''(1) = 0$ and $B(1) = B'(1) = 0$ but $B''(1) \neq 0$.

To prove the second claim, one obtains after some tedious computations

$$A'''(1) = \frac{(2w - 1)(w - 1)w}{2}$$

and $((t - 1)B(t))'''|_{t=1} = 6w^2$. These last equalities imply that

$$\lim_{t\to1}\frac{r'(t)}{(t - 1)r(t)} = \frac{1 - 2w}{12}$$

and Proposition (5.2) follows by summation.

Morihiko Saito has observed that the above computations of higher order derivatives can be replaced by the use of Taylor series expansions up to the second order at $s = 0$ for functions of the form $(1 + s)^a$, via the coordinate change $t = 1 + s$.

To conclude we offer the following conjecture showing clearly the duality alluded to in the introduction.

CONJECTURE 5.3. *Let f be a weakly tame polynomial, let $0 < a_1 \leq a_2 \leq ... \leq a_\mu < n + 1$ be the spectrum of f at infinity where $\mu = b_n(F)$. Then*

$$\sum_{i=1,\mu}\left(a_i - \frac{n+1}{2}\right)^2 \geq \frac{(a_\mu - a_1)\mu}{12}.$$

We have verified this dual inequality for several examples. Thomas Brélivet from Bordeaux University has recently proved both (5.1) and (5.3) for polynomials in 2 variables which are Newton non-degenerate with respect to a polyhedron having up to 5 edges. Note also that a weighted homogeneous polynomial having an IHS at the origin can be regarded at the same time as being a tame polynomial and the two associated spectra coincide. Hence the equality in (5.2) can in fact be regarded as a natural consequence of the inequalities (5.1) and (5.3).

References

1. N. A'Campo, Sur la monodromie des singularités isolées d'hypersurfaces complexes, Invent. Math. **20** (1973) 147–169.
2. V. I. Arnold, S. M. Gusein-Zade and A. N. Varchenko, *Singularities of differentiable maps, vol 1,2*, Birkhäuser, Boston, 1985, 1988.
3. E. Artal Bartolo, Forme de Jordan de la monodromie des singularités superisolées, Memoirs Amer. Math. Soc. **109** (1994).
4. E. Artal Bartolo, Pi. Cassou-Noguès and A. Dimca, Sur la topologie des polynômes complexes, *Singularities: the Brieskorn anniversary volume* (ed V. I. Arnold, G.-M. Greuel and J. H. M. Steenbrink), Progress in Math. **162**, (Birkhäuser, Basel, 1998), 317–343.
5. E. Artal Bartolo, I. Luengo and A. Melle Hernández, Milnor number at infinity, topology and Newton boundary of a polynomial function, Math. Zeits. **233** (2000) 679–696.
6. G. Bailly-Maitre, Sur le système local de Gauss-Manin d'un polynôme de deux variables, Bull. Soc. Math. France **128** (2000) 87–101.
7. D. Barlet, Construction du cup-produit de la fibre de Milnor aux pôles de $|f|^{2\lambda}$, Ann. Inst. Fourier, **34** (1984) 75–107.
8. W. Barth, C. Peters and A. Van de Ven, *Compact Complex Surfaces*, Ergebnisse der Math. (3) **4**, Springer-Verlag, 1984.
9. E. Brieskorn, Beispiele zur Differentialtopologie von Singularitäten, Invent. Math. **2** (1966) 1–14.
10. S.A. Broughton, Milnor numbers and the topology of polynomial hypersurfaces, Invent. Math. **92** (1988) 217–241.
11. Pi. Cassou-Noguès, Generalization of a theorem of Kushnirenko, Compositio. Math. **103** (1996) 95–121.
12. P. Deligne, Théorie de Hodge I, Actes Congrès Intern. Math., 1970, vol. 1, 425–430; II, Publ. Math. IHES **40** (1971) 5–57; III ibid., **44** (1974), 5–77.
13. P. Deligne, La conjecture de Weil, Publ. Math. IHES **52** (1980) 137–252.
14. A. Dimca, Monodromy at infinity for polynomials in two variables, Jour. Alg. Geom. **7** (1998) 771–779.

15. A. Dimca and A. Némethi, On the monodromy of complex polynomials, Duke Math. J. (to appear).
    math.AG 9912072.

16. A. Dimca and L. Paunescu, On the connectivity of complex affine hypersurfaces, II, Topology **39** (2000), 1035–1043.

17. A. Dimca and M. Saito, Monodromy at infinity and the weights of cohomology, preprint.
    math.AG/0002214.

18. A. Douai, Très bonnes bases du réseau de Brieskorn d'un polynôme modéré, Bull. Soc. Math. France **127** (1999) 255–287.

19. F. El Zein, Théorie de Hodge des cycles évanescents, Ann. Sci. Ecole Norm. Sup. (4) **19** (1986) 107–184.

20. D. Fried, Monodromy and dynamical systems, Topology **25** (1986) 443–453.

21. R. García López and A. Némethi, On the monodromy at infinity of a polynomial map, Compositio Math. **100** (1996) 205–231.

22. R. García López and A. Némethi, On the monodromy at infinity of a polynomial map II, Compositio Math. **115** (1999) 1–20.

23. R. García López and A. Némethi, Hodge numbers attached to a polynomial map, Ann. Inst. Fourier, Grenoble **49** (1999) 1547–1579.

24. S. M. Gusein-Zade, I. Luengo and A. Melle Hernández, Bifurcations and topology of meromorphic functions, this volume.

25. C. Hertling, Frobenius manifolds and variance of the spectral numbers, this volume.

26. M. Kato and Y. Matsumoto, On the connectivity of the Milnor fiber of a holomorphic function at a critical point, *Manifolds Tokyo 1973*, (Univ. of Tokyo Press, 1975) 131–136.

27. H. Kraft, Challenging problems on affine n-space, Séminaire Bourbaki 802 (1994/95), Astérisque **237** (1996) 295–318.

28. A.G. Kushnirenko, Polyèdres de Newton et nombres de Milnor, Invent. Math. **32** (1976) 1–31.

29. A. Landman, On the Picard-Lefschetz transformation for algebraic manifolds acquiring general singularities, Trans. Amer. Math. Soc. **181** (1973) 89–126.

30. D. T. Lê, Sur les noeuds algébriques, Compositio Math. **25** (1972) 281–321.

31. D. T. Lê and C. Weber, Polynômes à fibres rationnelles et conjecture jacobienne à 2 variables, Comptes Rendus Acad. Sci. Paris, **320** (1995) 581–584.

32. I. Luengo, The μ-constant stratum is not smooth, Invent. Math. **90** (1987) 139–152.

33. B. Malgrange, Polynôme de Bernstein-Sato et cohomologie évanescente, Astérisque **101–102** (1983) 243–267.

34. A. Melle Hernández, Milnor numbers for surface singularities, Israel J. Math. **115** (2000) 29–50.

35. F. Michel and C. Weber, On the monodromies of a polynomial map from C^2 to C, preprint, 1998.

36. J. Milnor, *Singular points of complex hypersurfaces*, Annals of Math. Studies **61**, Princeton Univ. Press, 1968.

37. V. Navarro Aznar, Sur la théorie de Hodge-Deligne, Invent. Math. **90** (1987) 11–76.

38. A. Némethi, The real Seifert form and the spectral pairs of isolated hypersurface singularities, Compositio Math. **98** (1995) 23–41.

39. A. Némethi and C. Sabbah, Semicontinuity of the spectrum at infinity, Abh. Math. Sem. Univ. Hamburg **69** (1999) 25–35.

40. A. Némethi and J. Steenbrink, Spectral pairs, mixed Hodge modules and series of plane curve singularities, New York J. Math. 1995. (http://nyjm.albany.edu:8000/j/v1/Nemethi-Steenbrink.html)

41. A. Nemethi and A. Zaharia, Milnor fibration at infinity, Indag. Math. **3** (1992) 323–335.

42. W. Neumann, Complex algebraic curves via their links at infinity, Invent. Math. **98** (1989) 445–489.

43. W. Neumann and P. Norbury, Unfolding polynomial maps at infinity, Math. Ann. **318** (2000) 149–180.

44. C. Sabbah, Hypergeometric periods for a tame polynomial, Comptes Rendus Acad. Sci. Paris, Sér. I, **328** (1999) 603–608.

45. M. Saito, On the structure of Brieskorn lattice, Ann. Inst. Fourier **39** (1989) 27–72.

46. M. Saito, Exponents of an irreducible plane curve singularity, preprint.

47. J. Scherk and J. Steenbrink, On the mixed Hodge structure on the cohomology of the Milnor fiber, Math. Ann. **271** (1985) 641–665.

48. M. Schulze and J. Steenbrink, Computing Hodge-theoretic invariants of singularities, this volume.

49. D. Siersma, Non-isolated singularities, this volume.

50. D. Siersma and M. Tibăr, Singularities at infinity and their vanishing cycles, Duke Math. J. **80** (1995) 771–783.

51. D. Siersma and M. Tibăr, Vanishing cycles and singularities of meromorphic functions, preprint no. 1105, Utrecht University, 1999. math.AG/9905108.

52. J. Steenbrink, Mixed Hodge structures on the vanishing cohomology, *Real and complex singularities, Oslo 1976* (ed Per Holm), (Sijthoff and Noordhoff, Alphen aan den Rijn, 1977) 525–563.

53. J. Steenbrink, Intersection form for quasi-homogeneous singularities, Compositio Math. **34** (1977) 211–223.

54. J. Steenbrink and S. Zucker, Variation of mixed Hodge structure I, Invent. Math. **80** (1985) 485–542.

55. M. Tibăr, Topology at infinity of polynomial mappings and Thom regularity condition, Compositio Math. **111** (1998) 89–109.

56. A. N. Varchenko, The asymptotics of holomorphic forms determine a mixed Hodge structure, Sov. Math. Dokl. **22** (1980) 248–252.

Bifurcations and topology of meromorphic germs

Sabir Gusein-Zade [*]
Moscow State University, Faculty of Mathematics and Mechanics,
(sabir@mccme.ru)

Ignacio Luengo [†]
Departamento de Álgebra, Universidad Complutense de Madrid,
(iluengo@eucmos.sim.ucm.es)

Alejandro Melle Hernández [‡]
Departamento de Álgebra, Universidad Complutense de Madrid,
(amelle@eucmos.sim.ucm.es)

Introduction: polynomial and meromorphic functions

Maps defined by polynomial functions are traditional objects of interest in algebraic geometry and singularity theory. A polynomial P in n complex variables defines a map $P : \mathbb{C}^n \to \mathbb{C}$. The map P is not a locally trivial fibration over critical values of P. However, since the source $\mathbb{C}^n$ is not compact, the map P fails to be a locally trivial fibration over some other values as well. It is well known that a polynomial map defines a locally trivial fibration over the complement to a finite set in $\mathbb{C}$ (the *bifurcation set of P*): [41, 45, 47].

To describe the atypical values which detect an irregular behaviour of a polynomial at infinity is an important and unsolved problem. There are several known regularity conditions which guarantee that a value is not atypical at infinity or that there are no atypical values at infinity (see, e.g., [8, 22, 28, 30, 37, 42, 43, 44, 34]).

A number of papers are devoted to the study of the topology of generic fibres (e.g. theorems of bouquet type) and their difference from non-generic ones (see, e.g., [11, 37, 4, 9, 5, 6]).

Important invariants of a polynomial map are monodromy operators corresponding to small loops around atypical values and (usually the most important) the monodromy operator at infinity corresponding to a big loop around all atypical values. They are related to a number of properties of the polynomial, including arithmetic ones (see, e.g., [27, 26, 15, 14, 38, 13, 12]).

[*] Partially supported by RFBR 98–01–00612, NWO-RFBR 047.008.005.
[†] Partially supported by DGCYT PB97-0284-C02-01.
[‡] Partially supported by DGCYT PB97-0284-C02-01.

D. Siersma et al. (eds.), New Developments in Singularity Theory, 279–304.
© 2001 *Kluwer Academic Publishers. Printed in the Netherlands.*

A natural way to understand the behaviour of a polynomial at infinity is to consider it as a rational function on the projective space $\mathbb{CP}^n$. In this case one has to study its behaviour at points of the infinite hyperplane $\mathbb{CP}^{n-1}_\infty$. The problem is that at such points a polynomial map defines not a holomorphic germ, but a meromorphic one (of a particular type). Thus in order to use local considerations for a description of global properties of a polynomial map one needs to describe related properties of meromorphic germs.

It is thus natural to reconsider the entire local theory in the more general context of arbitrary meromorphic germs P/Q. A first step in this direction was taken by V.I.Arnold [2], who gave classifications of simple meromorphic germs with respect to certain equivalence relations. One may now seek to establish and study analogues of notions already well-understood for analytic germs and by now also developed in some detail for the behaviour of polynomial maps at infinity, such as Milnor fibres and monodromy, for meromorphic germs in general. We will thus not attempt to survey the by now considerable literature on polynomial maps, and we refer to the paper [12] by Dimca in this volume for an account of the Hodge-theoretic aspects of this study.

Despite a number of close parallels with the earlier results, certain new features require attention. Their study was begun by the authors in [18] and further developed in [20] and by Siersma and Tibăr in [39]. Although the problem is very closely related to the study of pencils $sP + tQ = 0$, for which also a number of interesting results are known — for example, the characterisation by Lê and Weber [25] of atypical fibres — this aspect also is slightly different.

In the first chapter of this article we present the basic definitions, and then study the monodromy by calculating its zeta-function. For results on homology splitting, and for bouquet type theorems, we refer to [38]. In the second chapter we give some corresponding results for the global case of meromorphic functions on compact complex manifolds. The traditional case is that of rational functions on $\mathbb{CP}^n$ (see e.g. [19, 21, 39]), including in particular polynomial functions on $\mathbb{C}^n$, and we give applications to this in the final section.

1. Local theory

A *meromorphic germ* at the origin in the complex space $\mathbb{C}^n$ is a ratio $f = \frac{P}{Q}$ of two holomorphic germs P and Q on $(\mathbb{C}^n, 0)$.

The following equivalence relation (first introduced by V.I. Arnold) is essential for further considerations. Two meromorphic germs $f = \frac{P}{Q}$ and $f' = \frac{P'}{Q'}$ are *equal* if and only if $P' = P \cdot U$ and $Q' = Q \cdot U$

for a holomorphic germ U not equal to zero at the origin: $U(0) \neq 0$. According to this definition $\frac{x}{y} \neq \frac{x^2}{xy}$, but $\frac{x}{y} = \frac{x \exp(x)}{y \exp(x)}$.

A meromorphic germ $f = \frac{P}{Q}$ defines a map from the complement of the indeterminacy locus $I(f) := \{P = Q = 0\}$ to the complex projective line $\mathbb{CP}^1$. Unfortunately, this map is not a locally trivial fibration over the complement of a finite set in $\mathbb{CP}^1$. Roughly speaking, f fails to be a locally trivial fibration over values c for which the level set $f^{-1}(c)$ is not transversal to the sphere $S_\varepsilon^{2n-1} = \partial B_\varepsilon^{2n}$ (among others). This is a "real condition" and thus it can occur at points of the projective line $\mathbb{CP}^1$ forming a set of real codimension 1. Thus one cannot define a generic fibre of a meromorphic germ in this way.

EXAMPLE 1.1.   Let $f = \frac{x^2 - y^3}{y^2}$. One can see that $f : B_\varepsilon^4 \setminus \{0\} \to \mathbb{CP}^1$ fails to be a locally trivial fibration over neighbourhoods of 0, ∞, and points c ($=(c : 1)$) such that $\|c\| = \frac{3}{2}\varepsilon$.

However if one fixes a value c in $\mathbb{CP}^1$, this does not happen in a neighbourhood of c provided the radius ε is small enough.

THEOREM 1.2.   ([18]) *For any value $c \in \mathbb{CP}^1$, there exists $\varepsilon_0 > 0$ ($\varepsilon_0 = \varepsilon_0(c)$) such that for any positive $\varepsilon \leq \varepsilon_0$ the sphere S_ε^{2n-1} is transversal to all strata of the level set $f^{-1}(c)$, and the map $f : B_\varepsilon^{2n} \setminus I(f) \to \mathbb{CP}^1$ is a locally trivial fibration over a punctured neighbourhood of c.*

DEFINITION 1.3.   The fibration described is called *the c-Milnor fibration* of the meromorphic germ f. A fibre of the c-Milnor fibration, i.e.,

$$\mathcal{M}_f^c = \{z \in B_\varepsilon^{2n} \setminus I(f) : f(z) = \frac{P(z)}{Q(z)} = c'\}$$

for ε small enough and for c' close enough to c (in $\mathbb{CP}^1$), is a (non-compact) $(n-1)$-dimensional complex manifold with boundary, and will be called *the c-Milnor fibre* of the meromorphic germ f.

EXAMPLE 1.4.   For the f of Example 1.1, $\mathcal{M}_f^0$ is a (2-dimensional) disk minus two points; for $c \neq 0$, $\mathcal{M}_f^c$ is the disjoint union of two punctured disks.

REMARK 1.5.   1) In particular, the definition means that $\mathcal{M}_f^0$ is equal to

$$\{z \in B_\varepsilon : P(z) = c' \cdot Q(z), \quad P(z) \neq 0\}$$

(c' close to 0, $c' \neq 0$) and thus, if $R(0) = 0$, the Milnor fibres of the functions $\frac{P}{Q}$ and $\frac{RP}{RQ}$ are, generally speaking, different.

2) For $f = \frac{P}{Q}$, let $f^{-1} = \frac{Q}{P}$. It is not difficult to see that $\mathcal{M}^c_{f^{-1}} = \mathcal{M}^{c^{-1}}_f$, in particular $\mathcal{M}^0_{f^{-1}} = \mathcal{M}^\infty_f$, $\mathcal{M}^\infty_{f^{-1}} = \mathcal{M}^0_f$. Let $f - c = \frac{P - cQ}{Q}$. Then $\mathcal{M}^c_f = \mathcal{M}^0_{f-c}$. The same properties hold for the monodromy transformations and for the zeta-functions discussed below.

A fibration over a punctured neighbourhood of a point in the projective line $\mathbb{CP}^1$ defines a monodromy transformation, which is a diffeomorphism of the fibre (well defined up to isotopy).

DEFINITION 1.6. The monodromy transformation $h^c_f : \mathcal{M}^c_f \to \mathcal{M}^c_f$ of the c-Milnor fibration is called *the c-monodromy transformation* of the meromorphic germ f.

EXAMPLE 1.7. For the f of Example 1.1, h^c_f is trivial (i.e. isotopic to the identity) for all $c \neq 0, \infty$. The 0-monodromy transformation is a self-map of a disk without two points which interchanges these points. The ∞-monodromy transformation interchanges two punctured disks.

One can show that for almost all values c (i.e. for all but a finite number) the c-monodromy transformation h^c_f is trivial, i.e., isotopic to the identity.

DEFINITION 1.8. A value $c \in \mathbb{CP}^1$ is called a *typical* value of the meromorphic germ f if for ε small enough the map $f : B^{2n}_\varepsilon \setminus I(f) \to \mathbb{CP}^1$ is a locally trivial (and thus trivial) fibration over a neighbourhood (not punctured) of the point c. Otherwise the value c is called *atypical*. The set $B(f)$ of atypical values is called the *bifurcation set* of the germ f.

Note that if a value c is typical, then the corresponding monodromy transformation h^c_f is isotopic to identity. Moreover we have the following.

THEOREM 1.9. ([20]) *There exists a finite set $\Sigma \subset \mathbb{CP}^1$ such that for all $c \in \mathbb{CP}^1 \setminus \Sigma$ the c-Milnor fibres of f are diffeomorphic to each other and the c-monodromy transformations are trivial (i.e., isotopic to identity). In particular, the set of atypical values is finite.*

EXAMPLE 1.10. The meromorphic germ f of Example 1.1 has two atypical values: 0 and ∞.

To prove Theorems 1.2 and 1.9 we use resolution of singularities. A *resolution of the germ f* is a modification of the space $(\mathbb{C}^n, 0)$ (i.e., a proper analytic map $\pi : \mathcal{X} \to \mathcal{U}$ of a smooth analytic manifold $\mathcal{X}$ onto a neighbourhood $\mathcal{U}$ of the origin in $\mathbb{C}^n$, which is an isomorphism outside of a proper analytic subspace in $\mathcal{U}$) such that the total transform $\pi^{-1}(H)$ of the hypersurface $H = \{P = 0\} \cup \{Q = 0\}$ is a normal crossing divisor at each point of the manifold $\mathcal{X}$. We assume that the map π is an isomorphism outside of the hypersurface H.

The fact that the preimage $\pi^{-1}(H)$ is a divisor with normal crossings implies that in a neighbourhood of any of its points there exists a local system of coordinates $y_1, \ldots, y_n$ such that the liftings $\tilde{P} = P \circ \pi$ and $\tilde{Q} = Q \circ \pi$ of the functions P and Q to the total space $\mathcal{X}$ of the modification have the forms $u \cdot y_1^{k_1} \cdots y_n^{k_n}$ and $v \cdot y_1^{\ell_1} \cdots y_n^{\ell_n}$ respectively, where $u(0) \neq 0$, $v(0) \neq 0$, and the k_i and ℓ_i are nonnegative integers.

Note that the values 0 and ∞ in the projective line $\mathbb{CP}^1$ are used as distinguished points for convenience, so as to use the usual notion of a resolution of a function.

Proof (of Theorem 1.9). One can make additional blow-ups along intersections of pairs of irreducible components of the divisor $\pi^{-1}(H)$ so that the lifting $\tilde{f} = f \circ \pi = \frac{\tilde{P}}{\tilde{Q}}$ of the function f defines a holomorphic map from the manifold $\mathcal{X}$ to the complex projective line $\mathbb{CP}^1$. This means that $\tilde{P} = V \cdot P'$, $\tilde{Q} = V \cdot Q'$ where V is a section of a line bundle, say $\mathcal{L}$, over $\mathcal{X}$, P' and Q' are sections of the line bundle $\mathcal{L}^{-1}$, and P' and Q' have no common zeroes on $\mathcal{X}$. Let $\tilde{f}' = \frac{P'}{Q'}$.

On each component of the divisor $\pi^{-1}(H)$ and on each intersection of several of the components, $\tilde{f}'$ defines a map to the projective line $\mathbb{CP}^1$. These maps have a finite number of critical values, say $a_1, a_2, \ldots, a_s$.

If the function $\tilde{f}'$ is constant on a component of the divisor $\pi^{-1}(H)$, or on an intersection of components, then this constant value is critical. The value of the function $\tilde{f}'$ on an intersection of n components (this intersection is zero-dimensional) should also be considered as a critical value.

Let $c \in \mathbb{CP}^1$ be different from $a_1, a_2, \ldots, a_s$. We shall show that for all c' in a neighbourhood of the point c (including c itself) the c'-Milnor fibres of the meromorphic function f are diffeomorphic to each other and the c'-monodromy transformations are trivial.

Let $r^2(z)$ be the square of the distance from the origin in the space $\mathbb{C}^n$ and let $\tilde{r}^2(x) = r^2(\pi(x))$ be the lifting of this function to the total space $\mathcal{X}$ of the modification. In order to obtain a c'-Milnor fibre one has to choose $\varepsilon_0 > 0$ (a Milnor radius) small enough so that the level manifold $\{\tilde{r}^2(x) = \varepsilon^2\}$ is transversal to $\{\tilde{f}'(x) = c'\}$ for all ε with

$0 < \varepsilon \leq \varepsilon_0$. Let $\varepsilon_0 = \varepsilon_0(c)$ be a Milnor radius for the value c. Since $\{\tilde{f}'(x) = c\}$ is transversal to all components of the divisor $\pi^{-1}(H)$ and to all their intersections, ε_0 is also a Milnor radius for all c' in a neighbourhood of the point $c \in \mathbb{CP}^1$ (and the level manifold $\{\tilde{f}'(x) = c'\}$ is transversal to all components of the divisor $\pi^{-1}(H)$ and to their intersections). This implies that for all such c' the c'-Milnor fibres of the meromorphic germ f are diffeomorphic to each other and the c'-monodromy transformations are trivial.

The c-Milnor fibre for a generic value $c \in \mathbb{CP}^1$ can be called *the generic Milnor fibre* of the meromorphic germ f. One can see that the generic Milnor fibre of a meromorphic germ can be considered as embedded in the c-Milnor fibre for any value $c \in \mathbb{CP}^1$.

THEOREM 1.11. *Let $f = \frac{P}{Q}$ be a meromorphic germ on the space $(\mathbb{C}^n, 0)$ such that the numerator P has an isolated critical point at the origin and, if $n = 2$, assume also that the germs of the curves $\{P = 0\}$ and $\{Q = 0\}$ have no common irreducible components. Then, for a generic $t \in \mathbb{C}$,*

$$\chi(\mathcal{M}_f^0) = (-1)^{(n-1)}(\mu(P, 0) - \mu(P + tQ, 0)).$$

Here $\mu(g, 0)$ stands for the usual Milnor number of the holomorphic germ g at the origin.

Remark. Similar results for polynomials (i.e., for meromorphic germs of the form P/z^d) can be found in [30] and [37].

Proof. The Milnor fibre $\mathcal{M}_f^0$ of the meromorphic germ f has the following description. Let ε be small enough (and thus be a Milnor radius for the holomorphic germ P). Then

$$\mathcal{M}_f^0 = B_\varepsilon(0) \cap (\{P + tQ = 0\} \setminus I(f))$$

for $t \neq 0$ with $|t|$ small enough (and thus t generic). Note that the zero-level set $\{P + tQ = 0\}$ is non-singular outside the origin for $|t|$ small enough. The space $B_\varepsilon(0) \cap \{P = Q = 0\}$ is homeomorphic to a cone and therefore its Euler characteristic is equal to 1. Therefore

$$\chi(\mathcal{M}_f^0) = \chi(B_\varepsilon(0) \cap \{P + tQ = 0\}) - 1.$$

Now Theorem 1.11 is a consequence of the following well known statement (see, e.g., [16]).

STATEMENT 1.12. *Let $P : (\mathbb{C}^n, 0) \to (\mathbb{C}, 0)$ be a germ of a holomorphic function with an isolated critical point at the origin and let P_t be* a deformation of P (i.e., $P_0 = P$). Let ε be small enough. Then for $|t|$ small enough,

$$(-1)^{n-1}(\chi(B_\varepsilon(0) \cap \{P_t = 0\}) - 1)$$

is equal to the number of critical points of P (counted with multiplicities) which split from the zero level set, i.e., to

$$\mu(P, 0) - \sum_{Q \in \{P_t = 0\} \cap B_\varepsilon} \mu(P_t, Q).$$

EXAMPLE 1.13. The example $f = \frac{xy}{x}$ shows the necessity of the condition that, for $n = 2$, the curves $\{P = 0\}$ and $\{Q = 0\}$ have no common components.

EXAMPLE 1.14. In Theorem 1.11, for the difference of Milnor numbers we can substitute the difference of Euler characteristics of the corresponding Milnor fibres (of the germs P and $P + tQ$) (up to sign). However the formula obtained this way is not correct if the germ P has a nonisolated critical point at the origin. This is shown by the example $f = \frac{x^2 + z^2 y}{z^2}$.

THEOREM 1.15. Let $f = \frac{P}{Q}$ be the germ of a meromorphic function on $(\mathbb{C}^n, 0)$ such that the numerator P has an isolated critical point at the origin. The value 0 is typical for the meromorphic germ f if and only if $\chi(\mathcal{M}_f^0) = 0$.

Proof. " Only if" follows from the definition and 1.11.

" If" is a consequence of a result of A. Parusiński [31] (or rather of its proof). He proved that, if $\mu(P) = \mu(P + tQ)$ for $|t|$ small enough, then the family of maps $P_t = P + tQ$ is topologically trivial. In particular the family of germs of hypersurfaces $\{P_t = 0\}$ is topologically trivial. For $n \neq 3$ this was proved by Lê D.T. and C.P. Ramanujam [24]. However in order to apply the result to the present situation, it is necessary to have a topological trivialization of the family $\{P_t = 0\}$ which preserves the subset $\{P = Q = 0\}$ and is smooth outside the origin. For the family $P_t = P + tQ$, such a trivialization was explicitly constructed in [31] without any restriction on the dimension.

EXAMPLE 1.16. If the germ of the function P has a non-isolated critical point at the origin, then this characterization is no longer true. Take, for example, $P(x, y) = x^2 y^2$ and $Q(x, y) = x^4 + y^4$.

1.1. THE MONODROMY AND ITS ZETA-FUNCTION

DEFINITION 1.17. *For a transformation* $h : X \to X$ *of a topological space* X *its zeta-function* $\zeta_h(t)$ *is the rational function defined by*

$$\zeta_h(t) = \prod_{q \geq 0} \{ \det [id - t\, h_* |_{H_q(X;\mathbb{C})}] \}^{(-1)^q}.$$

This definition coincides with that in [3] and differs by a sign in the exponent from that in [1].

Let $\zeta_f^c(t)$ be the zeta-function of the c-monodromy transformation h_f^c of the meromorphic germ f. The degree of the rational function $\zeta_f^c(t)$ (i.e., the degree of the numerator minus the degree of the denominator) is equal to the Euler characteristic of the c-Milnor fibre $\mathcal{M}_f^c$. In the general case one has the following statement.

THEOREM 1.18. ([20]) *If the value* c *is typical then the Euler characteristic of the* c-*Milnor fibre is equal to* 0 *and its zeta-function* $\zeta_f^c(t)$ *is equal to* 1.

EXAMPLE 1.19. For the f of Example 1.1, $\zeta_f^0(t) = \frac{1}{1+t}$ and $\zeta_f^c(t) = 1$ for $c \neq 0$.

In the holomorphic case, resolution is an important tool for understanding the Milnor fibration. An excellent example of this fact is the formula of A'Campo, [1]. We also have an A'Campo formula in the meromorphic case.

Let $f = \frac{P}{Q}$ be a germ of a meromorphic function on $(\mathbb{C}^n, 0)$ and let $\pi : \mathcal{X} \to \mathcal{U}$ be a resolution of the germ f. The preimage $\mathcal{D} = \pi^{-1}(0)$ of the origin of $\mathbb{C}^n$ is a normal crossing divisor. Let $S_{k,\ell}$ be the set of points of $\mathcal{D}$ in a neighbourhood of which the functions $P \circ \pi$ and $Q \circ \pi$ in some local coordinates have the form $u \cdot y_1^k$ and $v \cdot y_1^\ell$ respectively ($u(0) \neq 0$, $v(0) \neq 0$). A slight modification of the arguments of A'Campo ([1]) permits us to obtain the following version of his formula for the zeta-function of the monodromy of a meromorphic function.

THEOREM 1.20. ([18]) *Let the resolution* $\pi : \mathcal{X} \to \mathcal{U}$ *be an isomorphism outside the hypersurface* $H = \{P = 0\} \cup \{Q = 0\}$. *Then*

$$\zeta_f^0(t) = \prod_{k > \ell} (1 - t^{k-\ell})^{\chi(S_{k,\ell})},$$
$$\zeta_f^\infty(t) = \prod_{k < \ell} (1 - t^{\ell-k})^{\chi(S_{k,\ell})}.$$

Remark. A resolution π of the germ $f' = \frac{RP}{RQ}$ is at the same time a resolution of the germ $f = \frac{P}{Q}$. Moreover the multiplicities of any component C of the exceptional divisor in the zero divisors of the liftings $(RP) \circ \pi$ and $(RQ) \circ \pi$ of the germs RP and RQ are obtained from those of the germs P and Q by adding one and the same integer, the multiplicity $m = m(C)$ of R. Nevertheless the meromorphic functions f and f' can have different zeta-functions. The reason why the formulae in the previous theorem give different results for f and f' consists in the fact that if an open part of the component C lies in $S_{k,\ell}(f)$ then, generally speaking, its part which lies in $S_{k+m,\ell+m}(f')$ is smaller.

The A'Campo theorem for germs of holomorphic functions has been generalized to the case when the modification $\pi : (\mathcal{X}, \mathcal{D}) \to (\mathbb{C}^n, 0)$ is not a resolution of the singularity, see [17]. This can also be done in the present set-up.

In order to have somewhat more attractive and unified formulae here and below it will be convenient to use the notion of the integral with respect to the Euler characteristic ([48]). The main property of a traditional (say, Lebesgue) measure, which, together with positivity, permits one to define a notion of integral, is the property $\sigma(X \cup Y) = \sigma(X) + \sigma(Y) - \sigma(X \cap Y)$. For spaces that can be represented as finite unions of cells, say semialgebraic spaces, the Euler characteristic defined as the alternating sum of numbers of cells of different dimensions also possesses this property. In this sense it can be considered as a measure, though nonpositive. Nonpositivity of the Euler characteristic imposes restrictions on the class of functions for which the integral with respect to the Euler characteristic can be defined.

Let A be an Abelian group with group operation $*$, and let X be a semianalytic subset of a complex manifold. Let $\Psi : X \to A$ be a function on X with values in A for which there exists a finite partition $\mathcal{S}$ of X into semianalytic sets (strata) Ξ such that Ψ is constant on each stratum Ξ (and equal to ψ_Ξ). Then by definition the integral with respect to the Euler characteristic of Ψ over X is equal to

$$\int_X \Psi(x)\, d\chi = \sum_{\Xi \in \mathcal{S}} \chi(\Xi)\psi_\Xi,$$

where $\chi(\Xi)$ is the Euler characteristic of the stratum Ξ. In the above formula we use additive notation for the operation $*$. In what follows, this definition will be used for integer valued functions and also for local zeta-functions $\zeta_x(t)$, which are elements of the Abelian group of non-zero rational functions of the variable t with respect to multiplication.

In the latter case, in multiplicative notation, the above formula becomes

$$\int_X \zeta_x(t)\, d\chi = \prod_{\Xi \in \mathcal{S}} (\zeta_\Xi(t))^{\chi(\Xi)}.$$

Let $f = \frac{P}{Q}$ be the germ of a meromorphic function on $(\mathbb{C}^n, 0)$, and let $\pi : (\mathcal{X}, \mathcal{D}) \to (\mathbb{C}^n, 0)$ be an arbitrary modification of $(\mathbb{C}^n, 0)$, which is an isomorphism outside the hypersurface $H = \{P = 0\} \cup \{Q = 0\}$ (i.e. π is not necessarily a resolution). Let $\tilde{f} = f \circ \pi$ be the lifting of f to the space of the modification, i.e., the meromorphic function $\frac{P \circ \pi}{Q \circ \pi}$. For a point $x \in \pi^{-1}(H)$, let $\zeta^0_{\tilde{f},x}(t)$ be the zeta-function of the 0-monodromy of the germ of the function $\tilde{f}$ at x. Let $\mathcal{S} = \{\Xi\}$ be a pre-stratification of $\mathcal{D} = \pi^{-1}(0)$ (that is, a partitioning into semianalytic subspaces without any regularity conditions) such that, for each stratum Ξ of $\mathcal{S}$, the zeta-functions $\zeta^0_{\tilde{f},x}(t)$ do not depend on x, for $x \in \Xi$. We denote these zeta-functions by ζ^0_{Ξ}.

THEOREM 1.21.

$$\zeta^0_f(t) = \int_{\mathcal{D}} \zeta^0_{\tilde{f},x}(t)\, d\chi,$$

In the holomorphic case, the Newton diagram of a function gives a lot of information about the singularity (for a singularity non-degenerate with respect to its Newton diagram). It determines an embedded resolution of the singularity and one can read the zeta-function of the singularity from this resolution (by a formula of Varchenko [46]). There is a version of this formula also for meromorphic germs.

For a germ $R = \sum a_k x^k : (\mathbb{C}^n, 0) \to (\mathbb{C}, 0)$ of a holomorphic function $(k = (k_1, \ldots, k_n), x^k = x_1^{k_1} \cdot \ldots \cdot x_n^{k_n})$, its Newton diagram $\Gamma = \Gamma(R)$ is the union of the compact faces of the polytope $\Gamma_+ = \Gamma_+(R)$ which is the convex hull of $\bigcup_{k : a_k \neq 0}(k + \mathbb{R}^n_+) \subset \mathbb{R}^n_+$.

Let $f = \frac{P}{Q}$ be a germ of a meromorphic function on $(\mathbb{C}^n, 0)$ and let $\Gamma_1 = \Gamma(P)$ and $\Gamma_2 = \Gamma(Q)$ be the Newton diagrams of P and Q. We call the pair $\Lambda = (\Gamma_1, \Gamma_2)$ of Newton diagrams Γ_1 and Γ_2 *the Newton pair of f*. We say that the meromorphic germ f is *non-degenerate with respect to its Newton pair* $\Lambda = (\Gamma_1, \Gamma_2)$ if the pair of germs (P, Q) is non-degenerate with respect to the pair $\Lambda = (\Gamma_1, \Gamma_2)$ in the sense of [29] (which is an adaptation for *germs* of complete intersections of the definition of A.G. Khovanskii [23]). Almost all meromorphic germs with Newton pair Λ are non-degenerate with respect to it.

Let us define zeta-functions $\zeta^0_\Lambda(t)$ and $\zeta^\infty_\Lambda(t)$ for a Newton pair $\Lambda = (\Gamma_1, \Gamma_2)$. Let $1 \le \ell \le n$ and let $\mathcal{I}$ be a subset of $\{1, \ldots, n\}$ with cardinality $\#\mathcal{I} = \ell$. Let $L_\mathcal{I}$ be the coordinate subspace $L_\mathcal{I} =$

$\{k \in \mathbb{R}^n : k_i = 0 \text{ for } i \notin \mathcal{I}\}$ and $\Gamma_{i,\mathcal{I}} = \Gamma_i \cap L_{\mathcal{I}} \subset L_{\mathcal{I}}$. Let $L_{\mathcal{I}}^*$ be the dual of $L_{\mathcal{I}}$ and $L_{\mathcal{I}+}^*$ its positive orthant (the set of covectors with positive values on $L_{\mathcal{I} \geq 0} = \{k \in L_{\mathcal{I}} : k_i \geq 0 \text{ for } i \in \mathcal{I}\}$). For a primitive integer covector $a \in (\mathbb{R}^*)_+^n$, let $m(a, \Gamma) = \min_{x \in \Gamma}(a, x)$ and let $\Delta(a, \Gamma) = \{x \in \Gamma : (a, x) = m(a, \Gamma)\}$. We denote by $m_{\mathcal{I}}$ and $\Delta_{\mathcal{I}}$ the corresponding objects for the diagram $\Gamma_{\mathcal{I}}$ and a primitive integer covector $a \in L_{\mathcal{I}+}^*$. Let $E_{\mathcal{I}}$ be the set of primitive integer covectors $a \in L_{\mathcal{I}+}^*$ such that $\dim(\Delta(a, \Gamma_1) + \Delta(a, \Gamma_2)) = \ell - 1$ (the Minkowski sum $\Delta_1 + \Delta_2$ of two polytopes Δ_1 and Δ_2 is the polytope $\{x = x_1 + x_2 : x_1 \in \Delta_1, \quad x_2 \in \Delta_2\}$). There exist only a finite number of such covectors. For $a \in E_{\mathcal{I}}$, let $\Delta_1 = \Delta(a, \Gamma_1)$, $\Delta_2 = \Delta(a, \Gamma_2)$ and

$$V_a = \sum_{s=0}^{\ell-1} V_{\ell-1}(\underbrace{\Delta_1, \ldots, \Delta_1,}_{s \text{ terms}} \underbrace{\Delta_2, \ldots, \Delta_2}_{\ell-1-s \text{ terms}}),$$

where the definition of the (Minkowski) mixed volume $V(\Delta_1, \ldots, \Delta_m)$ can be found, e.g., in [7] or [29]. The $(\ell - 1)$-dimensional volume in a rational $(\ell - 1)$-dimensional affine subspace of $L_{\mathcal{I}}$ has to be normalized in such a way that the volume of the unit cube spanned by any integer basis of the corresponding linear subspace is equal to 1. Recall that $V_m(\underbrace{\Delta, \ldots, \Delta}_{m \text{ terms}})$ is simply the m-dimensional volume of Δ. We have to

define $V_0(\text{nothing}) = 1$ (this is necessary to define V_a for $\ell = 1$). Let:

$$\zeta_{\mathcal{I}}^0(t) = \prod_{a \in E_{\mathcal{I}}: m(a, \Gamma_1) > m(a, \Gamma_2)} (1 - t^{m(a, \Gamma_1) - m(a, \Gamma_2)})^{(\ell-1)! V_a},$$

$$\zeta_{\mathcal{I}}^\infty(t) = \prod_{a \in E_{\mathcal{I}}: m(a, \Gamma_1) < m(a, \Gamma_2)} (1 - t^{m(a, \Gamma_2) - m(a, \Gamma_1)})^{(\ell-1)! V_a},$$

$$\zeta_{\ell}^\bullet(t) = \prod_{\mathcal{I}: \#(\mathcal{I}) = \ell} \zeta_{\mathcal{I}}^\bullet(t),$$

$$\zeta_\Lambda^\bullet(t) = \prod_{\ell=1}^{n} (\zeta_{\ell}^\bullet(t))^{(-1)^{\ell-1}},$$

where $\bullet = 0$ or ∞.

THEOREM 1.22. ([18]) *Let $f = \frac{P}{Q}$ be a germ of a meromorphic function on $(\mathbb{C}^n, 0)$ non-degenerated with respect to its Newton pair $\Lambda = (\Gamma_1, \Gamma_2)$. Then*

$$\zeta_f^0(t) = \zeta_\Lambda^0(t) \quad and \quad \zeta_f^\infty(t) = \zeta_\Lambda^\infty(t).$$

1.2. EXAMPLES

For the germ of a meromorphic function of two variables, a resolution can be obtained by a sequence of blow-ups at points.

EXAMPLE 1.23. The minimal resolution of the germ f of Example 1.1 can be described by Fig. 1.

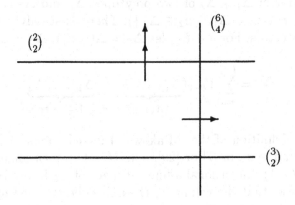

Figure 1

Here lines correspond to components of the exceptional divisor $\mathcal{D}$. Each component is isomorphic to the complex projective line $\mathbb{CP}^1$. The pairs of numbers near them are the multiplicities of the liftings of the numerator P and of the denominator Q along these components. The arrow (respectively the double arrow) corresponds to the strict transform of the curve $\{P = 0\}$ (respectively, of the curve $\{Q = 0\}$). Then $S_{2,2}$ (respectively $S_{3,2}$ and $S_{6,4}$ is the complex projective line minus two points (minus one and three points respectively). Thus

$$\zeta_f^0(t) = (1-t)(1-t^2)^{-1} = \frac{1}{1+t},$$
$$\zeta_f^\infty(t) = 1.$$

EXAMPLE 1.24. Let $f = \frac{x^3 - xy}{y}$. The Milnor fibre $\mathcal{M}_f^0$ (respectively $\mathcal{M}_f^\infty$) is $\{(x,y) : \|(x,y)\| < \varepsilon,\ x^3 - xy = cy\} \setminus \{(0,0)\}$, where $\|c\|$ is small (respectively large). From the equation $x^3 - xy = cy$ one has $y = \frac{x^3}{x+c}$ and thus $\mathcal{M}_f^0$ is diffeomorphic to the disk $\mathcal{D}$ in the x-plane with two points removed: $-c$ and the origin. In the same way $\mathcal{M}_f^\infty$ is diffeomorphic to the punctured disk $\mathcal{D}^*$. It is not difficult to see that the action of the monodromy transformation on the homology groups is trivial in both cases. Thus

$$\zeta_f^0(t) = (1-t)^{-1} \quad \text{and} \quad \zeta_f^\infty(t) = 1.$$

Now let us calculate these zeta-functions from their Newton diagrams.

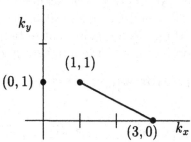

Figure 2

We have $\zeta_1^\bullet(t) = 1$ since each coordinate axis intersects only one Newton diagram. There is only one linear function (namely $a = k_x + 2k_y$) such that $\dim \Delta(a, \Gamma_1) = 1$. The one-dimensional volume $V_1(\Delta(a, \Gamma_1))$ of $\Delta(a, \Gamma_1)$ is equal to 1 and $V_1(\Delta(a, \Gamma_2)) = 0$. We have $m(a, \Gamma_1) = 3$ and $m(a, \Gamma_2) = 2$. Thus $\zeta_2^0(t) = (1-t)$, $\zeta_2^\infty(t) = 1$, $\zeta_{(\Gamma_1, \Gamma_2)}^0(t) = (1-t)^{-1}$ and $\zeta_{(\Gamma_1, \Gamma_2)}^\infty(t) = 1$, which coincides with the formulae for f written above.

EXAMPLE 1.25. Let $P = xyz + x^p + y^q + z^r$ be a $T_{p,q,r}$ singularity, $\frac{1}{p} + \frac{1}{q} + \frac{1}{r} < 1$, and let $Q = x^d + y^d + z^d$ be a homogeneous polynomial of degree d. Suppose that $p > q > r > d > 3$ and that p, q, and r are pairwise coprime. Let us compute the zeta-functions of $f = \frac{P}{Q}$ using the methods described above.

(a) It is clear that f is non-degenerate with respect to its Newton pair $\Lambda = (\Gamma_1, \Gamma_2)$. Thus

$$\zeta_f^\bullet(t) = \zeta_\Lambda^\bullet(t) = \zeta_1^\bullet(\zeta_2^\bullet)^{-1}\zeta_3^\bullet \quad (\bullet = 0 \text{ or } \infty).$$

One has $\zeta_1^\infty = \zeta_2^\infty = 1$ and the only covector which is necessary for computing ζ_3^∞ is $a = (1,1,1)$. In this case $m(a, \Gamma_1) = 3$, $m(a, \Gamma_2) = d$, $\Delta(a, \Gamma_1) = \{(1,1,1)\}$ and $\Delta(a, \Gamma_2)$ is the simplex $\{k_x + k_y + k_z = d, k_x \geq 0, k_y \geq 0, k_z \geq 0\}$, its two-dimensional volume is equal to $\frac{d^2}{2}$. Thus $\zeta_f^\infty = (1 - t^{d-3})^{d^2}$.

We have

$$\zeta_1^0 = (1 - t^{p-d})(1 - t^{q-d})(1 - t^{r-d}),$$

$$\zeta_2^0 = (1 - t^{r(q-d)})(1 - t^{r(p-d)})(1 - t^{q(p-d)})(1 - t^{r-d})^{2d}(1 - t^{q-d})^d.$$

To compute ζ_3^0 one has to take into account both the covectors $(rq - q - r, r, q)$, $(r, pr - p - r, p)$, and $(q, p, qp - p - q)$ corresponding to two-dimensional faces of Γ_1, and the covectors $(1, r - 2, 1)$, $(r - 2, 1, 1)$, and $(q - 2, 1, 1)$ corresponding to pairs of the form (one-dimensional face of Γ_1, one-dimensional face of Γ_2). E.g., for $a = (1, r - 2, 1)$, $\Delta(a, \Gamma_1)$ (respectively $\Delta(a, \Gamma_2)$) is the segment between $(0, 0, r)$ to $(1, 1, 1)$ (respectively between $(d, 0, 0)$ and $(0, 0, d)$). Note the absence of symmetry: the latter three covectors are not obtained from each other by permuting the coordinates and the numbers p, q, and r. Then

$$\zeta_3^0 = (1 - t^{r(q-d)})(1 - t^{r(p-d)})(1 - t^{q(p-d)})(1 - t^{r-d})^{2d}(1 - t^{q-d})^d$$

and

$$\zeta_f^0 = (1 - t^{p-d})(1 - t^{q-d})(1 - t^{r-d}).$$

(b) For computing the zeta-functions of f with the help of partial resolutions, let $\pi : (\mathcal{X}, \mathcal{D}) \to (\mathbb{C}^3, 0)$ be the blowing-up of the origin in $\mathbb{C}^3$ and let $\tilde{f}$ be the lifting $f \circ \pi$ of f to the space $\mathcal{X}$. The exceptional divisor $\mathcal{D}$ is the complex projective plane $\mathbb{CP}^2$. Let H_1 and H_2 be the strict transforms of the hypersurfaces $\{P = 0\}$ and $\{Q = 0\}$, $D_i = \mathcal{D} \cap H_i$. The curve D_1 consists of three transversal lines ℓ_1, ℓ_2, ℓ_3 and has three singular points $S_1 = \ell_2 \cap \ell_3 = (0, 0, 1)$, $S_2 = \ell_1 \cap \ell_3 = (0, 1, 0)$, and $S_3 = \ell_1 \cap \ell_3 = (1, 0, 0)$. The curve D_2 is a smooth curve of degree d, it intersects D_1 at $3d$ distinct points $\{P_1, \ldots, P_{3d}\}$.

One has the following natural stratification of the exceptional divisor $\mathcal{D}$:

(i) 0-dimensional strata Λ_i^0 ($i = 1, 2, 3$), each consisting of one point S_i;

(ii) 0-dimensional strata Ξ_i^0 consisting of one point P_i each ($i = 1, \ldots, 3d$);

(iii) 1-dimensional strata $\Xi_i^1 = \ell_i \setminus \{D_2 \cup \ell_j \cup \ell_k\}$ ($i = 1, 2, 3$) and $\Xi_4^1 = D_2 \setminus D_1$;

(iv) the 2-dimensional stratum $\Xi^2 = \mathcal{D} \setminus (D_1 \cup D_2)$.

It is not difficult to see that $\zeta_{\Xi_2}^0(t) = 1$, $\zeta_{\Xi_2}^\infty(t) = 1 - t^{d-3}$, and for each stratum Ξ from Ξ_i^0 ($1 \le i \le 3d$), Ξ_i^1 ($1 \le i \le 4$) one has $\zeta_\Xi^\bullet(t) = 1$ ($\bullet = 0$ or ∞).

In what follows, the exceptional divisor $\mathcal{D}$ has the local equation $u = 0$. At the point S_1 the lifting $\tilde{f}$ of the function f is of the form $\frac{u^3 x_1 y_1 + u^r + x_i^p u^p + y_i^q u^q}{u^d x_1^d + u^d y_1^d + u^d}$. This germ has the same Newton pair as the germ

$\frac{u^3 x_1 y_1 + u^r}{u^d}$. Using the Newton diagram formula one has $\zeta_{\Lambda_1^\infty}^\infty = 1$, $\zeta_{\Lambda_1^0}^\infty = 1 - t^{r-d}$. At the point S_2 the function $\tilde{f}$ has the form $\frac{u^3 x_1 z_1 + z_1^r u^r + x_1^p u^p + u^q}{u^d x_1^d + u^d + z_1^d u^d}$, and the same Newton pair as $\frac{u^3 x_1 z_1 + z_1^r u^r + u^q}{u^d}$. Again from the Newton diagrams one has $\zeta_{\Lambda_2^\infty}^\infty(t) = 1$, $\zeta_{\Lambda_2^0}^0(t) = 1 - t^{q-d}$. In the same way, $\zeta_{\Lambda_3^\infty}^\infty(t) = 1$, $\zeta_{\Lambda_3^0}^0(t) = 1 - t^{p-d}$. Combining these computations, one has the same results as we obtained above without using a partial resolution.

2. Global theory

We want to consider fibrations defined by meromorphic functions on manifolds. In order to have more general statements we prefer to use a notion of a meromorphic function slightly different from the standard one. Let M be an n-dimensional compact complex analytic manifold.

DEFINITION 2.1. *A meromorphic function f on the manifold M is a ratio $\frac{P}{Q}$ of two non-zero sections of a line bundle $\mathcal{L}$ over M. Two meromorphic functions $f = \frac{P}{Q}$ and $f' = \frac{P'}{Q'}$ (where P' and Q' are sections of a line bundle $\mathcal{L}'$) are equal if $P = U \cdot P'$ and $Q = U \cdot Q'$ where U is a section of the bundle $\mathrm{hom}(\mathcal{L}', \mathcal{L}) = \mathcal{L} \otimes \mathcal{L}'^*$ without zeroes (in particular, this implies that the bundles $\mathcal{L}$ and $\mathcal{L}'$ are isomorphic).*

A particular important case of meromorphic functions is given by rational functions $\frac{P(x_1,...,x_n)}{Q(x_1,...,x_n)}$ on the projective space $\mathbb{CP}^n$, where P and Q are homogeneous polynomials of the same degree.

A meromorphic function $f = \frac{P}{Q}$ defines a map f from the complement $M \setminus \{P = Q = 0\}$ of the set of common zeros of P and Q to the complex line $\mathbb{CP}^1$. The indeterminacy set $I(f) := \{P = Q = 0\}$ may have components of codimension one. For $c \in \mathbb{CP}^1$, let $F_c = f^{-1}(c)$.

Standard arguments (using a resolution of singularities; see, e.g., [45]) give the following statement.

THEOREM 2.2. *The map $f : M \setminus \{P = Q = 0\} \to \mathbb{CP}^1$ is a C^∞ locally trivial fibration outside a finite subset of the projective line $\mathbb{CP}^1$.*

Any fibre $F_{gen} = f^{-1}(c_{gen})$ of this fibration is called *a generic fibre* of the meromorphic function f. The smallest subset $B(f) \subset \mathbb{CP}^1$ for which f is a C^∞ locally trivial fibration over $\mathbb{CP}^1 \setminus B(f)$ is called *the bifurcation set of the meromorphic function f. Its elements are called atypical values of the meromorphic function f.*

A loop in the complement $\mathbb{CP}^1 \setminus B(f)$ to the bifurcation set $B(f)$ gives rise to a monodromy transformation of the fibre bundle. The monodromy transformation is defined only up to homotopy (or rather, up to isotopy), but the monodromy operator (the action of the monodromy transformation on a homology group of the generic fibre of the meromorphic function f) is well defined. Therefore the fundamental group $\pi_1(\mathbb{CP}^1 \setminus B(f))$ of the complement to the bifurcation set acts on the homology groups $H_*(F_{gen}; \mathbb{C})$ of the generic fibre of the meromorphic function f. The image of the group $\pi_1(\mathbb{CP}^1 \setminus B(f))$ in the group of automorphisms of $H_*(F_{gen}; \mathbb{C})$ is called *the monodromy group* of the meromorphic function f. It is generated by local monodromy operators corresponding to simple loops around the atypical values of f (see [3]).

Let $\zeta_f^c(t)$ be the zeta-function of the local monodromy corresponding to the value $c \in \mathbb{CP}^1$ (i.e., defined by a simple loop around the value c).

Note that the local monodromy and the corresponding zeta-function are defined for any value $c \in \mathbb{CP}^1$, not only for atypical ones. For a generic value of the meromorphic function f, the local monodromy is the identity and its zeta-function is equal to $(1 - t)^{\chi(F_{gen})}$.

Let $f = \frac{P}{Q}$ be a meromorphic function on the complex manifold M. The following statement is a direct consequence of the definitions.

STATEMENT 2.3. Let $\pi : \widetilde{M} \to M$ be a proper analytic map of an n-dimensional compact complex manifold $\widetilde{M}$ which is an isomorphism outside the union of the indeterminacy set $I(f)$ of the meromorphic function f and a finite number of level sets $f^{-1}(c_i)$. Let $\widetilde{f} = \frac{P \circ \pi}{Q \circ \pi}$ be the lifting of the meromorphic function $f = \frac{P}{Q}$ to $\widetilde{M}$. Then the generic fibre of $\widetilde{f}$ coincides with that of f, and for each $c \in \mathbb{CP}^1$ one has

$$\zeta_{\widetilde{f}}^c(t) = \zeta_f^c(t).$$

Note that, in order to ensure that the generic fibres of $\widetilde{f}$ and f coincide, one really needs to use the equivalence relation for meromorphic functions formulated above.

Let c be a point of the projective line $\mathbb{CP}^1$. For a point $x \in M$, let $\zeta_{f,x}^c(t)$ be the corresponding zeta-function of the germ of the meromorphic function f at the point x, and let $\chi_{f,x}^c$ be its degree $\deg \zeta_{f,x}^c(t)$.

THEOREM 2.4.

$$\zeta_f^c(t) = \int_{I(f) \cup F_c} \zeta_{f,x}^c(t) \, d\chi, \tag{1}$$

$$\chi(F_{gen}) - \chi(F_c) = \int_{F_c} (\chi_{f,x}^c - 1) \, d\chi + \int_{I(f)} \chi_{f,x}^c \, d\chi. \tag{2}$$

Proof. The proof follows the lines of the proof of 1.2 in [17]. Without any loss of generality one can suppose that $c = 0$. There exists a modification $\pi : \mathcal{X} \to \mathcal{M}$ of the manifold M which is an isomorphism outside the set
$$\{P = Q = 0\} \cup \{f = 0\} \cup \{f = \infty\} = \{P = 0\} \cup \{Q = 0\}$$
such that $\mathcal{D} = \pi^{-1}(\{P = 0\} \cup \{Q = 0\})$ is a normal crossing divisor in the manifold $\mathcal{X}$. Then at each point of the exceptional divisor $\mathcal{D}$ in a local system of coordinates one has $P \circ \pi = u \cdot y_1^{k_1} \cdot \ldots \cdot y_n^{k_n}$, $Q \circ \pi = v \cdot y_1^{\ell_1} \cdot \ldots \cdot y_n^{\ell_n}$ with $u(0) \neq 0$, $v(0) \neq 0$, $k_i \geq 0$ and $\ell_i \geq 0$. There exist Whitney stratifications $\mathcal{S}$ and $\mathcal{S}^*$ of M and $\mathcal{X}$ respectively such that:

1. the map π is a stratified morphism with respect to these stratifications;

2. the set $\{P = 0\} \cup \{Q = 0\}$ is a stratified subspace of the stratified space $(M, \mathcal{S})$;

3. for each stratum $\Xi^* \in \mathcal{S}^*$, the germs of the liftings $P \circ \pi$ and $Q \circ \pi$ of the sections P and Q at points of Ξ^* have normal forms $u \cdot y_1^{k_1} \cdot \ldots \cdot y_n^{k_n}$ and $v \cdot y_1^{\ell_1} \cdot \ldots \cdot y_n^{\ell_n}$ where $(k_1, \ldots, k_n)$ and $(\ell_1, \ldots, \ell_n)$ do not depend on the point of Ξ^*;

4. for each stratum $\Xi \in \mathcal{S}$, the zeta-function $\zeta_{f,x}^c(t)$ does not depend on the point x for $x \in \Xi$.

Actually, point (4) follows from the first three. However it is convenient to include it in the list of conditions.

One applies the following version of the formula of A'Campo ([1]) and also its local variant for meromorphic germs (1.20). Let $S_{k,\ell}$ be the set of points of the manifold $\mathcal{X}$ in a neighbourhood of which the liftings $P \circ \pi$ and $Q \circ \pi$ of P and Q in some local coordinates have the forms $u \cdot y_1^k$ and $v \cdot y_1^\ell$ respectively $(u(0) \neq 0, v(0) \neq 0)$.

STATEMENT 2.5.

$$\zeta_f^0(t) = \prod_{k > \ell \geq 0} (1 - t^{k-\ell})^{\chi(S_{k,\ell})}.$$

Property (1) of the stratifications $\mathcal{S}$ and $\mathcal{S}^*$ implies that the morphism π is locally trivial over each stratum of $\mathcal{S}$: if the stratum Ξ of $\mathcal{S}$ is the image of the stratum Ξ^* of $\mathcal{S}^*$, $\Xi = \pi(\Xi^*)$, then $\pi : \Xi^* \to \Xi$ is a smooth locally trivial fibre bundle. In particular,

$$\chi(\Xi^*) = \chi(\Xi) \cdot \chi(\pi^{-1}(x) \cap \Xi^*), \qquad (x \in \Xi).$$

Let $\Xi_{k,\ell}$ be the set of strata from S^* such that the germs of the liftings $P \circ \pi$ and $Q \circ \pi$ of P and Q at their points are equivalent to y_1^k and y_1^ℓ respectively; $S_{k,\ell} = \bigcup\limits_{\Xi^* \in \Xi_{k,\ell}} \Xi^*$. We have

$$\zeta_f^0(t) = \prod_{k > \ell \geq 0} (1 - t^{k-\ell})^{\sum\limits_{\Xi^* \in \Xi_{k,\ell}} \chi(\Xi^*)} = \prod_{k > \ell \geq 0} \prod_{\Xi^* \in \Xi_{k,\ell}} (1 - t^{k-\ell})^{\chi(\Xi^*)} =$$

$$= \prod_{k > \ell \geq 0} \prod_{\Xi \in S} \prod_{\Xi^* \in \Xi_{k,\ell} \cap \pi^{-1}(\Xi)} (1 - t^{k-\ell})^{\chi(\Xi^*)} =$$

$$= \prod_{\Xi \in S} \prod_{k > \ell \geq 0} \prod_{\Xi^* \in \Xi_{k,\ell} \cap \pi^{-1}(\Xi)} (1 - t^{k-\ell})^{\chi(\Xi) \cdot \chi(\pi^{-1}(x) \cap \Xi^*)} =$$

$$= \prod_{\Xi \in S} \left(\prod_{k > \ell \geq 0} \prod_{\Xi^* \in \Xi_{k,\ell} \cap \pi^{-1}(\Xi)} (1 - t^{k-\ell})^{\chi(\pi^{-1}(x) \cap \Xi^*)} \right)^{\chi(\Xi)} =$$

$$= \prod_{\Xi \in S} [\zeta_{f,x}^0(t)]^{\chi(\Xi)} = \int_{\{P = Q = 0\} \cup F_0} \zeta_{f,x}^0(t) \, d\chi.$$

As usual, the formula for the Euler characteristic of the generic fibre follows from the formula for the zeta-function, since the Euler characteristic is the degree of the zeta-function.

The difference between $(\chi_{f,x}^c - 1)$ and $\chi_{f,x}^c$ in the two integrals in (2) reflects the fact that the Euler characteristic of the local level set $F_c \cap B_\varepsilon(x)$ (where $B_\varepsilon(x)$ is the ball of small radius ε centred at the point x) of the germ of the function f is equal to 1 for a point x of the level set F_c and is equal to 0 for a point x of the indeterminacy set $I(f)$. In the first case this local level set is contractible, and in the second it is the difference between two contractible sets. $\square$

Let us denote $(-1)^{n-1}$ times the first and the second integrals in (2) by $\mu_f(c)$ and $\lambda_f(c)$ respectively. Let $\mu_f = \sum\limits_{c \in \mathbb{CP}^1} \mu_f(c)$, $\lambda_f = \sum\limits_{c \in \mathbb{CP}^1} \lambda_f(c)$ (in each sum only a finite number of summands are different from zero).

THEOREM 2.6.

$$\mu_f + \lambda_f = (-1)^{n-1} \left(2 \cdot \chi(F_{gen}) - \chi(M) + \chi(I(f)) \right).$$

Proof. One has

$$\int_{\mathbb{CP}^1} \chi(F_c) \, d\chi = \chi(M \setminus \{P = Q = 0\}) = \chi(M) - \chi(I(f)).$$

Therefore

$$\chi(M)-\chi(\{P=Q=0\}) = \int_{\mathbb{CP}^1} \chi(F_{gen})\, d\chi + \int_{\mathbb{CP}^1} (\chi(F_c)-\chi(F_{gen}))\, d\chi =$$

$$= 2\cdot\chi(F_{gen})-(-1)^{n-1} \sum_{c\in\mathbb{CP}^1} (\mu_f(c)+\lambda_f(c)) = 2\cdot\chi(F_{gen})+(-1)^n(\mu_f+\lambda_f).$$

Let $\tilde{f}$ be the restriction of f to $M \setminus \{Q = 0\}$,
$$\tilde{f} : M \setminus \{Q = 0\} \to \mathbb{C} = \mathbb{CP}^1 \setminus \{\infty\}.$$
Note that the fibres of f and $\tilde{f}$ over values $c \in \mathbb{C}$ coincide.

COROLLARY 2.7.

$$\chi(F_{gen}) = \chi(M) - \chi(\{Q = 0\}) + (-1)^{n-1}(\lambda_f - \lambda_f(\infty) + \mu_f - \mu_f(\infty)).$$

For the meromorphic function on the complex projective space $\mathbb{CP}^n$ defined by a polynomial P in n variables with isolated critical points, $\mu_f(c)$ is the sum of the Milnor numbers of the critical points of the polynomial P with critical value c, and $\lambda_f(c)$ is equal to the invariant $\lambda_P(c)$ studied in [5]. Therefore $\mu_f(c)$ and $\lambda_f(c)$ can be considered as generalizations of those invariants (they have sense also in the case when critical points of the polynomial P are not isolated). One has $\mu_f = \mu_P + \mu_f(\infty)$, $\lambda_f = \lambda_P + \lambda_f(\infty)$, where $\mu_P = \sum_{c\in\mathbb{C}} \mu_P(c)$, $\lambda_P = \sum_{c\in\mathbb{C}} \lambda_P(c)$. Note that in this case Corollary 1 turns into the well known formula $\chi(F_{gen}) = 1 + (-1)^{n-1}(\lambda_P + \mu_P)$.

3. Applications

3.1. POLYNOMIAL FUNCTIONS

A polynomial $P : \mathbb{C}^n \to \mathbb{C}$ defines a meromorphic function $f = \dfrac{\tilde{P}}{x_0^d}$ on the projective space $\mathbb{CP}^n$ ($d = \deg P$, $\tilde{P}$ is the homogenization of P). For any $c \in \mathbb{CP}^1$, the local monodromy of the polynomial P and its zeta-function $\zeta_P^c(t)$ are defined (in fact they coincide with those of the meromorphic function f). The technique described can be applied to this case. For instance,

1. For $c \in \mathbb{C} \subset \mathbb{CP}^1$,

$$\zeta_{\tilde{P}}^c(t) = \left(\int_{\{\tilde{P}=0\}\cap\mathbb{CP}_\infty^{n-1}} \zeta_{f,x}^c(t)\, d\chi \right) \cdot \left(\int_{\{P=c\}} \zeta_{P,x}^c(t)\, d\chi \right). \qquad (3)$$

2. For the infinite value,

$$\zeta_P^\infty(t) = \int_{\mathbb{CP}_\infty^{n-1}} \zeta_{f,x}^\infty(t)\, d\chi.$$

Note that the zeta-function of the monodromy at infinity of the polynomial P is nothing but $\zeta_P^\infty(t)$.

The bifurcation set consists of critical values of the polynomial P (in the affine part) and of atypical ("critical") values at infinity.

In order to study the level set $\{P = c\}$, one can consider the zero level set of the polynomial $(P - c)$. Thus let us consider the zero level set $V_0 = \{P = 0\} \subset \mathbb{C}^n$ of the polynomial P. Let us suppose that V_0 has only isolated singular points (in the affine part $\mathbb{C}^n$). For $\rho > 0$, let B_ρ be the open ball of radius ρ centred at the origin in $\mathbb{C}^n$ and $S_\rho = \partial B_\rho$ be the $(2n-1)$-dimensional sphere of radius ρ with the centre at the origin. There exists $R > 0$ such that, for all $\rho \geq R$, the sphere S_ρ is transversal to the level set $V_0 = \{P = 0\}$. The restriction $P|_{\mathbb{C}^n \setminus B_R} : \mathbb{C}^n \setminus B_R \to \mathbb{C}$ of the function P to the complement of the ball B_R defines a C^∞ locally trivial fibration over a punctured neighbourhood of the origin in $\mathbb{C}$. The loop $\varepsilon_0 \cdot \exp(2\pi i \tau)$ $(0 \leq \tau \leq 1, \|\varepsilon_0\|$ small enough) defines the monodromy transformation $h : V_{\varepsilon_0} \setminus B_R \to V_{\varepsilon_0} \setminus B_R$. Let us denote its zeta-function $\zeta_h(t)$ by $\widetilde{\zeta}_P^0(t)$. We use the following definition.

DEFINITION 3.1. *The value 0 is atypical at infinity for the polynomial P if the restriction $P|_{\mathbb{C}^n \setminus B_R}$ of the function P to the complement of the ball B_R is not a C^∞ locally trivial fibration over a neighbourhood of the origin in $\mathbb{C}$.*

This definition does not depend on choice of coordinates, i.e., it is invariant with respect to polynomial diffeomorphisms of the space $\mathbb{C}^n$. One can see that an atypical value at infinity is atypical, i.e. it belongs to $B(P)$. Moreover the bifurcation set $B(P)$ is the union of the set of critical values of the polynomial P (in $\mathbb{C}^n$) and of the set of values atypical at infinity in the sense described. If the singular locus of the level set $V_0 = \{P = 0\}$ is not finite, the value 0 can hardly be considered as typical at infinity.

THEOREM 3.2. *The zeta-function near infinity $\widetilde{\zeta}_P^0(t)$ of the local monodromy (corresponding to the value 0) of the polynomial P is the first factor in formula (3), with $c = 0$. If this zeta-function is different from 1, then the value 0 is atypical at infinity.*

EXAMPLE 3.3. Let $P(x, y, z) = x^a y^b(x^c y^d - z^{c+d}) + z$, $(ad - bc) \neq 0$, and let $\delta = \deg(P) = a + b + c + d$. The curve $\{P_\delta = 0\} \subset \mathbb{CP}_\infty^2$

consists of three components: the line $C_1 = \{x = 0\}$ with multiplicity a, the line $C_2 = \{y = 0\}$ with multiplicity b, and the reduced curve $C_3 = \{x^c y^d - z^{c+d} = 0\}$. Let $Q_1 = C_2 \cap C_3 = (1 : 0 : 0)$, $Q_2 = C_1 \cap C_3 = (0 : 1 : 0)$, $Q_3 = C_1 \cap C_2 = (0 : 0 : 1)$. At each point x of the infinite hyperplane $\mathbb{CP}^2_\infty$ except Q_1 and Q_2, one has $\zeta^0_{P,x}(t) = 1$. At the point Q_1, the germ of the meromorphic function P has the form $\dfrac{y^b(y^d - z^{c+d}) + zu^{\delta-1}}{u^\delta}$. Its zero zeta-function can be obtained by the Varchenko type formula (1.22). If $(ad - bc) < 0$, then $\zeta^0_{P,Q_1}(t) = 1$. If $(ad - bc) > 0$, then

$$\zeta^0_{P,Q_1}(t) = (1 - t^{\frac{ad-bc}{h}})^h,$$

where $h = g.c.d(c, d) \cdot g.c.d.(\frac{ad-bc}{g.c.d.(c,d)}, \delta - 1)$. The situation at the point Q_2 is given by symmetry. Finally,

$$\tilde{\zeta}^0_P(t) = (1 - t^{\frac{|ad-bc|}{h}})^h.$$

Thus the value 0 is atypical at infinity. In the same way $\tilde{\zeta}^0_{P-c}(t) = 1$, for $c \neq 0$.

EXAMPLE 3.4.   The polynomial function $P(x,, y, z) = x + x^2yz$ has 0 as an atypical value at infinity and $\tilde{\zeta}^0_P(t) = 1$. Hence the converse of the above theorem does not hold.

3.2. YOMDIN-AT-INFINITY POLYNOMIALS

For a polynomial $P \in \mathbb{C}[z_1, \ldots, z_n]$, let P_i be its homogeneous part of degree i. Let P be of the form $P = P_d + P_{d-k} + terms\ of\ lower\ degree$, $k \geq 1$. Consider the hypersurfaces in $\mathbb{CP}^{n-1}_\infty$ defined by $\{P_d = 0\}$ and $\{P_{d-k} = 0\}$. Let $Sing(P_d)$ be the singular locus of the hypersurface $\{P_d = 0\}$ (including all points where $\{P_d = 0\}$ is not reduced). One says that P is a *Yomdin-at-infinity polynomial* if $Sing(P_d) \cap \{P_{d-k} = 0\} = \emptyset$ (in particular this implies that $Sing(P_d)$ is finite).

   Y. Yomdin ([49]) considered critical points of holomorphic functions which are local versions of such polynomials. He gave a formula for their Milnor numbers. In [35] the zeta-function of the classical monodromy transformation of such a germ was described; see also the contribution of D. Siersma in this volume, [36]. The generic fibre (level set) of a Yomdin-at-infinity polynomial is homotopy equivalent to the bouquet of n-dimensional spheres ([10]). Its Euler characteristic χ_P (or rather the (global) Milnor number) was determined in [6]. For $k = 1$ the zeta-function of such a polynomial was obtained in [15].

Let $P(z_1, \ldots, z_n) = P_d + P_{d-k} + \ldots$ be a Yomdin-at-infinity polynomial. Let $Sing(P_d)$ consist of s points $Q_1, \ldots, Q_s$. One has the following natural stratification of the infinite hyperplane $\mathbb{CP}^{n-1}_\infty$:

1. the $(n-1)$-dimensional stratum $\Xi^{n-1} = \mathbb{CP}^{n-1}_\infty \setminus \{P_d = 0\}$;

2. the $(n-2)$-dimensional stratum $\Xi^{n-2} = \{P_d = 0\} \setminus \{Q_1, \ldots, Q_s\}$;

3. the 0-dimensional strata Ξ^0_i $(i = 1, \ldots, s)$, each consisting of one point Q_i.

The Euler characteristic of the stratum Ξ^{n-1} is equal to

$$\chi(\mathbb{CP}^{n-1}_\infty) - \chi(\{P_d = 0\}) = n - \chi(n-1, d) + (-1)^{n-2} \sum_{i=1}^{s} \mu_i,$$

where $\chi(n-1, d) = n + \frac{(1-d)^n - 1}{d}$ is the Euler characteristic of a non-singular hypersurface of degree d in the complex projective space $\mathbb{CP}^{n-1}_\infty$, μ_i is the Milnor number of the germ of the hypersurface $\{P_d = 0\} \subset \mathbb{CP}^{n-1}_\infty$ at the point Q_i. At each point of the stratum Ξ^n, the germ of the meromorphic function P has (in some local coordinates $u, y_1, \ldots, y_n$ where $u = 0$ defines $\mathbb{CP}^{n-1}$) the form $\frac{1}{u^d}$, and its infinite zeta-function $\zeta^\infty_{\Xi^n}(t)$ is equal to $(1 - t^d)$.

At each point of the stratum Ξ^{n-2}, the germ of the polynomial P has (in some local coordinates $u, y_1, \ldots, y_{n-1}$) the form $\frac{y_1}{u^d}$. Its infinite zeta-function $\zeta^\infty_{\Xi^{n-2}}(t)$ is equal to 1 and thus it does not contribute a factor to the zeta-function of the polynomial P.

At a point Q_i $(i = 1, \ldots, s)$, the germ of the meromorphic function P has the form $\varphi(u, y_1, \ldots, y_{n-1}) = \dfrac{g_i(y_1, \ldots, y_{n-1}) + u^k}{u^d}$, where g_i is a local equation of the hypersurface $\{P_d = 0\} \subset \mathbb{CP}^{n-1}_\infty$ at the point Q_i. Thus μ_i is its Milnor number.

In order to compute the infinite zeta-function $\zeta^\infty_\varphi(t)$ of the meromorphic germ φ, let us consider a resolution $\pi : (\mathcal{X}, \mathcal{D}) \to (\mathbb{C}^{n-1}, 0)$ of the singularity g_i, i.e., at each point of the exceptional divisor $\mathcal{D}$, the lifting $g_i \circ \pi$ of the function g_i to the space $\mathcal{X}$ of the modification has (in some local coordinates) the form $y_1^{m_1} \cdot \ldots \cdot y_{n-1}^{m_{(n-1)}}$ $(m_i \geq 0)$. Let us consider the modification

$$\tilde{\pi} = id \times \pi : (\mathbb{C}_u \times \mathcal{X}, 0 \times \mathcal{D}) \to (\mathbb{C}^n, 0) = (\mathbb{C}_u \times \mathbb{C}^{n-1}, 0)$$

of the space $(\mathbb{C}^n, 0)$. Let $\tilde{\varphi} = \varphi \circ \tilde{\pi}$ be the lifting of the meromorphic function φ to the space $\mathbb{C}_u \times \mathcal{X}$ of the modification $\tilde{\pi}$. At points of $\{0\} \times \mathcal{D}$ the function $\tilde{\phi}$ has the form $\dfrac{y_1^{m_1} \cdot \ldots \cdot y_{n-1}^{m_{n-1}}) + u^k}{u^d}$. Let $\mathcal{M}^\infty_{\tilde{\varphi}} = \tilde{\pi}^{-1}(\mathcal{M}^\infty_\varphi)$ ($\mathcal{M}^\infty_\varphi$ is the infinite Milnor fibre of the germ φ) be the local level set of the meromorphic function $\tilde{\varphi}$ (close to the infinite one). In the natural way one has the (infinite) monodromy h_{φ}^{∞} acting on $\mathcal{M}_{\varphi}^{\infty}$ and its zeta-function $\zeta_{\varphi}^{\infty}(t)$.

The problem is that the modification $\tilde{\pi}$ is not an isomorphism outside the union of the zero sets of the numerator and denominator of φ. Therefore $\zeta_{\varphi}^{\infty}(t)$ does not coincide with $\zeta_{\tilde{\varphi}}^{\infty}(t)$. However one can show that

$$\zeta_{\tilde{\varphi}}^{\infty}(t) = (1 - t^{d-k})^{\chi(\mathcal{D})-1} \zeta_{\varphi}^{\infty}(t).$$

The infinite zeta-function of the germ φ can be computed using the Varchenko type formula (1.22), by taking into account the local normal form of the germ of $\tilde{\phi}$ described above and the fact that the zeta-function $\zeta_h(t)$ of a self-transformation $h : X \to X$ of a space X determines the zeta-function $\zeta_{h^k}(t)$ of the k-th power of h. In particular, if $\zeta_h(t) = \prod_{m \geq 1} (1 - t^m)^{a_m}$, then

$$\zeta_h^k(t) = \prod_{m \geq 1} \left(1 - t^{\frac{m}{g.c.d.(k,m)}}\right)^{g.c.d.(k,m) \cdot a_m}.$$

THEOREM 3.5. ([21]) *For a Yomdin-at-infinity polynomial $P = P_d + P_{d-k} + \ldots$, its zeta-function at infinity is equal to*

$$\zeta_P(t) = (1 - t^d)^{\chi(\Xi^{n-1})}(1 - t^{d-k})^s \left(\prod_{i=1}^{s} \zeta_{g_i}^k(t^{d-k})\right)^{-1},$$

where $\chi(\Xi^{n-1}) = \frac{1-(1-d)^n}{d} + (-1)^{n-2} \sum_{i=1}^{s} \mu(g_i)$.

3.3. THE EULER CHARACTERISTIC OF A SINGULAR HYPERSURFACE

Let X be a compact complex manifold and let $\mathcal{L}$ be a holomorphic line bundle on X. Let s be a section of the bundle $\mathcal{L}$ not identically equal to zero, $Z := \{s = 0\}$ its zero locus (a hypersurface in the manifold X). Let s' be another section of the bundle $\mathcal{L}$, whose zero locus Z' is nonsingular and transversal to a Whitney stratification of the hypersurface Z. A. Parusiński and P. Pragacz have proved (see [32], Proposition 7) a statement which can be written as follows:

$$\chi(Z') - \chi(Z) = \int_{Z \setminus Z'} (\chi_x(Z) - 1) \, d\chi, \tag{4}$$

where $\chi_x(Z)$ is the Euler characteristic of the Milnor fibre of the germ of the section s at the point x. We shall indicate a more general formula which includes this one as a particular case.

THEOREM 3.6. *Let s be as above and let s' be a section of the bundle $\mathcal{L}$ whose zero locus Z' is non-singular. Let f be the meromorphic function s/s' on the manifold X. Then*

$$\chi(Z') - \chi(Z) = \int_{Z\backslash Z'} (\chi_x(Z) - 1)\, d\chi + \int_{Z\cap Z'} \chi^0_{f,x} d\chi, \qquad (5)$$

where $\chi^0_{f,x}$ is the Euler characteristic of the 0-Milnor fibre of the meromorphic germ f at the point x.

Proof. Let F_t be the level set $\{f = t\}$ of the (global) meromorphic function f on the manifold X (with indeterminacy set $\{s = s' = 0\}$), i.e., $F_t = \{s - ts' = 0\} \backslash \{s = s' = 0\}$. We know that, for a generic value t, one has

$$\chi(F_{gen}) - \chi(F_0) = \int_{F_0} (\chi^0_{f,x}(Z) - 1)\, d\chi + \int_{\{s=s'=0\}} \chi^0_{f,x} d\chi,$$

where $\chi^0_{f,x}$ is the Euler characteristic of the 0-Milnor fibre of the meromorphic germ f at the point x. One has $F_0 = Z \backslash (Z \cap Z')$, $F_\infty = Z' \backslash (Z \cap Z')$, and in this case F_∞ is a generic level set of the meromorphic function f (since its closure is non-singular). Therefore $\chi(F_0) = \chi(Z) - \chi(Z \cap Z')$, $\chi(F_{gen}) = \chi(Z') - \chi(Z \cap Z')$. Finally, for $x \in F_0$, the germ of the function f at the point x is holomorphic and thus $\chi^0_{f,x} = \chi_x(Z)$.

If the hypersurface Z' is transversal to all strata of a Whitney stratification of the hypersurface Z, then, for $x \in Z \cap Z'$, the Euler characteristic $\chi^0_{f,x} = 0$ (Proposition 5.1 from [33]) and therefore formula (5) reduces to (4).

References

1. N. A'Campo, La fonction zeta d'une monodromie, Comment. Math. Helv. **50** (1975) 233–248.

2. V. I. Arnold, Singularities of fractions and behaviour of polynomials at infinity, Proc. Steklov Math. Inst. **221** (1998) 48–68.

3. V. I. Arnold, S. M. Gusein-Zade, and A. N. Varchenko, *Singularities of Differentiable Maps, Vol. II*, Birkhäuser, Boston–Basel–Berlin, 1988.

4. E. Artal Bartolo, P. Cassou-Noguès and A. Dimca, Sur la topologie des polynômes complexes, *Singularities: the Brieskorn anniversary volume* (eds V. I. Arnold, G.-M. Greuel and J. H. M. Steenbrink), Progress in Math. **162** (Birkhäuser, 1998) pp 317–343.

5. E. Artal Bartolo, I. Luengo and A. Melle Hernández, Milnor number at infinity, topology and Newton boundary of a polynomial function, Math. Zeits. **233**:4 (2000) 679–696.

6. E. Artal Bartolo, I. Luengo and A. Melle Hernández, On the topology of a generic fibre of a polynomial function, Comm. Algebra **28**:4 (2000) 1767–1787.
7. D. N. Bernshtein, The number of roots of a system of equations, Funct. Anal. Appl. **9**:3 (1975) 1–4.
8. S. A. Broughton, Milnor numbers and the topology of polynomial hypersurfaces, Invent. Math. **92** (1988) 217–241.
9. P. Cassou-Noguès and A. Dimca, Topology of complex polynomials via polar curves, Kodai Math. J. **22** (1999) 131–139.
10. A. Dimca, On the connectivity of complex affine hypersurfaces, Topology **29** (1990) 511–514.
11. A. Dimca, On the Milnor fibrations of weighted homogeneous polynomials, Compositio Math. **76** (1990) 19–47.
12. A. Dimca, Monodromy and Hodge theory of regular functions, this volume.
13. A. Dimca and A. Némethi, On the monodromy of complex polynomials, preprint math.AG/9912072, Duke Math. Jour., to appear.
14. R. García López, Exponential sums and singular hypersurfaces, Manuscripta Math. **97** (1998) 45–58.
15. R. García López and A. Némethi, On the monodromy at infinity of a polynomial map I, Compositio Math. **100** (1996) 205–231.
16. S. M. Gusein-Zade, On a problem of B. Teissier, *Topics in singularity theory. V.I. Arnold's 60th anniversary collection* (eds A. Khovanskiĭ, A. Varchenko and V. Vassiliev), A.M.S. Translations, Ser. 2, **180**, (Amer. Math. Soc., Providence, RI, 1997) pp 117–125.
17. S. M. Gusein-Zade, I. Luengo and A. Melle Hernández, Partial resolutions and the zeta-function of a singularity, Comment. Math. Helv. **72**:2 (1997) 244-256.
18. S. M. Gusein-Zade, I. Luengo and A. Melle Hernández, Zeta-functions for germs of meromorphic functions and Newton diagrams, Funct. Anal. Appl. **32**:2 (1998) 26–35.
19. S. M. Gusein-Zade, I. Luengo and A. Melle Hernández, On atypical values and local monodromies of meromorphic functions, Tr. Mat. Inst. Steklova (Proc. Steklov Math. Inst.) **225** (1999) (Dedicated to Academician Sergeĭ Petrovich Novikov on the occasion of his 60th birthday)168–176.
20. S. M. Gusein-Zade, I. Luengo and A. Melle Hernández, On the topology of germs of meromorphic functions, Algebra i Analiz **11**:5 (1999) 92–99; translation in : St. Petersburg Math. Journal **11**:5, 775–780.
21. S. M. Gusein-Zade, I. Luengo and A. Melle Hernández, On the zeta-function of a polynomial at infinity, Bull. Sci. Math. **124**:3 (2000) 213–224.
22. Há Huy Vui and Lê Dũng Tràng, Sur la topologie des polynômes complexes, Acta Math. Vietnam. **9**:5 (1984) 21–32 (1985).
23. A. G. Khovanskiĭ, Newton polyhedra and toroidal varieties, Funct. Anal. Appl. **11**:4 (1977) 56–64.
24. D. T. Lê and C.P. Ramanujam, The invariance of Milnor's number implies the invariance of the topological type, Amer. J. Math. **98** (1976) 67–78.
25. D. T. Lê and Claude Weber, Equisingularité dans les pinceaux de germes de courbes planes et C^0-suffisance, l'Enseignement Math. **43** (1997) 355–380.
26. A. Libgober and S. Sperber, On the zeta-function of monodromy of a polynomial map, Compositio Math. **95** (1995) 287–307.
27. A. Némethi, Global Sebastiani-Thom theorem for polynomial maps, J. Math. Soc. Japan **43** (1991) 213–218.
28. A. Némethi and A. Zaharia, Milnor fibration at infinity, Indag. Math. (new ser.) **3** (1992) 323–335.

29. M. Oka, Principal zeta-function of non-degenerate complete intersection singularity, J. Fac. Sci. Uni. Tokyo Sect. IA **37** (1990) 11–32.

30. A. Parusiński, On the bifurcation set of a complex polynomial with isolated singularities at infinity, Compositio Math. **97** (1995) 369–38.

31. A. Parusiński, Topological triviality of μ-constant deformations of type $f(x) + tg(x)$, Bull. London Math. Soc. **31** (1999) 686–692.

32. A. Parusiński and P Pragacz, A formula for the Euler characteristic of singular hypersurfaces, J. Alg. Geom. **4** (1995) 337–351.

33. A. Parusiński and P Pragacz, Characteristic classes of hypersurfaces and characteristic cycles, J. Algebraic Geom. **10** (2001), 63–79.

34. C. Sabbah, Hypergeometric periods for a tame polynomial, Comptes Rendus Acad. Sci. Paris Ser. I **328** (1999) 603–608.

35. D. Siersma, The monodromy of a series of hypersurface singularities, Comment. Math. Helvet. **65** (1990) 181–197.

36. D. Siersma, The vanishing topology of non-isolated singularities, this volume.

37. D. Siersma and M. Tibăr, Singularities at infinity and their vanishing cycles, Duke Math. J. **80** (1995) 771–783.

38. D. Siersma and M. Tibăr, Topology of polynomial functions and monodromy dynamics, Comptes Rendus Acad. Sci. Paris Ser. I **327** (1998) 655–660.

39. D. Siersma and M. Tibăr, Vanishing cycles and singularities of meromorphic functions, preprint math.AG/9905108.

40. D. Siersma and M. Tibăr, Singularities at infinity and their vanishing cycles, II. Monodromy, preprint **1088** (Jan. 1999), Utrecht University. http://www.mi.aau.dk/ esn/abstracts.html/144/144.html.

41. R. Thom, Ensembles et morphismes stratifiées, Bull. Amer. Math. Soc. **75** (1969) 240–284.

42. M. Tibăr, Topology at infinity of polynomial mappings and Thom regularity condition, Compositio Math. **111** (1998) 89–109.

43. M. Tibăr, Regularity at infinity of real and complex polynomial functions, *Singularity Theory* (eds Bill Bruce and David Mond), London Math. Soc. Lecture Notes **263** (Cambridge University Press, 1999) pp 249–264.

44. M. Tibăr, On the monodromy fibration of polynomial functions with singularities at infinity, Comptes Rendus Acad. Sci. Paris Ser. I **324** (1997) 1031–1035.

45. A. N. Varchenko, Theorems on the topological equisingularity of families of algebraic varieties and families of polynomial mappings, Math USSR Izvestija **6** (1972) 949–1008.

46. A. N. Varchenko, Zeta-function of monodromy and Newton's diagram, Invent. Math. **37** (1976) 253–262.

47. J. L. Verdier, Stratifications de Whitney et théorème de Bertini-Sard, Invent. Math. **36** (1976) 295–312.

48. O. Y. Viro, Some integral calculus based on Euler characteristic, *Topology and Geometry — Rohlin seminar* Springer Lecture Notes in Math. **1346** 127–138.

49. Y. N. Yomdin, Complex surfaces with a one-dimensional set of singularities, Siberian Math. J. **5** (1975) 748–762.

Unitary reflection groups
and automorphisms
of simple hypersurface singularities

Victor V. Goryunov
Department of Mathematical Sciences, The University of Liverpool
(goryunov@liv.ac.uk)

In paper [7], generalising Arnold's approach to boundary singularities, we studied smoothings of simple hypersurfaces invariant under a unitary reflection of finite order. The reflection splits the homology with complex coefficients of a symmetric Milnor fibre into a direct sum of the character subspaces H_χ. The monodromy in the space of hypersurfaces with the same symmetry preserves the splitting. It was observed in [7] that the monodromy on each of the H_χ is a finite group generated by unitary reflections [13]. This way unitary reflection groups, $G(m, 1, k)$ and seven exceptional groups, made their first appearance in singularity theory addressing one of the long-standing questions of Arnold on realisations of the Shephard-Todd groups [1].

Later, in [8], some other Shephard-Todd groups were shown to be the monodromy groups of simple function singularities equivariant with respect to the action of finite order elements of $SU(2)$.

In the present paper we continue this programme and study arbitrary finite order automorphisms of the zero levels of simple function-germs whose action can be extended to some of the smoothings. This generalises the approaches of [7, 8] and, as it is explained by Slodowy in his paper [14] in this volume, is directly related to the classification of Springer's regular elements in Coxeter groups [15]. We show that the monodromy in the space of the symmetric smoothings is still a Shephard-Todd group. In addition to [7, 8], we obtain the group G_{10} (new in singularity theory) and two other realisations of the series $G(m, 1, k)$. The relation between the two realisations of $G(m, 1, k)$ is analogous to the relation between the Weyl groups B_k and C_k. The realisation of $G(m, 1, k)$ obtained in [7] was of the B-type.

For the cases not considered in [7, 8], we construct distinguished sets of generators of the H_χ. We show how symmetric Dynkin diagrams of the simple functions can be folded into diagrams of the equivariant singularities. We also describe the skew-Hermitian analogues of the unitary reflection groups under consideration.

D. Siersma et al. (eds.), New Developments in Singularity Theory, 305–328.
© 2001 Kluwer Academic Publishers. Printed in the Netherlands.

We finish the paper by giving a singularity theory interpretation of the rank 2 groups G_{12}, G_{20} and G_{22}, and showing how their Dynkin diagrams can be obtained by folding those of E_6 and E_8.

1. The unitary reflection groups

A *complex reflection* in $\mathbf{C}^k$ is a unitary transformation identical on a hyperplane, which is called the *mirror* of the reflection. The complete list of finite irreducible groups generated by complex reflections was obtained by Shephard and Todd [13]. It contains the Coxeter groups as a proper subset.

The Shephard-Todd list consists of three infinite series (Weyl groups A_k, cyclic $\mathbf{Z}_m$ and three-index $G(p, q, k)$) and 34 exceptional groups. Now we shall briefly recall the description of the groups we will deal with in detail in this paper. In our considerations a mirror is identified by its normal, which we call a *root*.

1.1. Groups $G(p, q, k)$

The group $G(p, q, k)$ (all the parameters are natural numbers, q divides p, and $k \geq 2$) is a subgroup of $U(k)$. It is generated by the rotation through $2\pi q/p$ corresponding to the root u_1 (the u_i are mutually orthogonal unit coordinate vectors) and by k reflections of order 2 defined by the roots

$$u_2 - u_1, \ u_3 - u_2, \ \ldots, u_k - u_{k-1} \text{ and } u_2 - e^{2\pi i/p} u_1 .$$

For example, in two cases, when either $q = p$ or $q = 1$, just k reflections are sufficient to generate the group.

The series contains Coxeter groups:

$$G(2, 2, k) = D_k , \qquad G(2, 1, k) = B_k , \qquad G(p, p, 2) = I_2(p) .$$

Information about generating reflections of a group can be represented by a graph analogous to a Dynkin diagram of a Coxeter group (cf. [4, 11]). Our conventions are as follows:

- a vertex of a graph represents a root;

- the Hermitian square of a root is written beside the vertex (square 2 is omitted);

- the order of the reflection is written at the vertex (order 2 is omitted);

- the Hermitian product (v_1, v_2) of the roots is written on an oriented edge $v_1 \to v_2$;

- there is no edge between two orthogonal roots;

- the product -1 is not written;

- orientation of an edge equipped with a real number is omitted.

Figure 1 shows graphs of the groups $G(m, 1, k)$ and $G(2m, 2, k)$.

$G(m,1,k)$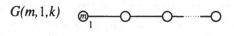

$G(2m,2,k)$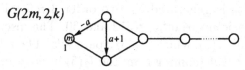

Figure 1. Graphs of the groups $G(m, 1, k)$ (k vertices) and $G(2m, 2, k)$ ($k + 1$ vertices). Notation: $a = e^{\pi i/m}$.

REMARK 1.1. It is convenient to include the groups $G(1, 1, k)$ in the series. According to the above description, the group $G(1, 1, k)$ is the permutation group of the coordinates in C^k. Hence C^k splits into the standard representation of A_{k-1} on the hyperplane $x_1 + \ldots + x_k = 0$ and the one-dimensional trivial representation.

It is also natural to set $G(m, 1, 1) = G(mq, q, 1) = \mathbb{Z}_m$.

The orbit space of any Shephard-Todd group is smooth. It contains the *discriminant* Σ of the group, that is, the space of its irregular orbits.

The basic invariants of $G(p, q, k)$ have degrees $p, 2p, \ldots, (k-1)p, kp/q$ (hence the order of the group is $p^k k!/q$). In particular, for $q = 1$ and $p > 1$, the degrees are proportional to those of the Weyl group B_k. Moreover, explicit consideration of the invariants easily implies that the discriminants of $G(p, 1, k)$ and B_k are isomorphic.

REMARK 1.2. Usually an exceptional Shephard-Todd group is denoted G_s, where s is its number in the original list in [13] (the second line in the list is occupied by a series, therefore, the Weyl group $G_2 = G(6, 6, 2)$ provides no confusion). Of course, such a notation is not very illustrative. For this reason, we used mainly the notation of [3] in [7]: $G(m, 1, k) = B_k^{(m)}$, $G_4 = A_2^{(3)}$, $G_5 = B_2^{(3,3)}$, $G_8 = A_2^{(4)}$, $G_{16} = A_2^{(5)}$, $G_{25} = A_3^{(3)}$, $G_{26} = C_3^{(3)}$ and $G_{32} = A_4^{(3)}$ (see also the table in the next section). For example, in $A_k^{(m)}$, the order 2 of the standard generators (transpositions) of the group $A_k = S_{k+1}$ is changed to m. See [7] for further details.

1.2. GROUPS G_{10} AND G_{20}

Both groups are of rank 2. The group G_{10} is generated by two reflections, one of order 3 and the other of order 4. The degrees of its basic invariants are 12 and 24, and $\Sigma(G_{10}) \simeq \Sigma(B_2)$. As an abstract group, G_{10} is a quotient group of Brieskorn's [2] braid group associated to the Weyl group B_2 (that is, generated by two elements, a and b, subject to the relation $(ab)^2 = (ba)^2$) by the relations $a^3 = id$ and $b^4 = id$. Therefore, it is natural to denote G_{10} by $B_2^{(4,3)}$.

The group G_{20} is generated by two order 3 reflections in $\mathbf{C}^2$. The degrees of the basic invariants are 12 and 30. The discriminant is isomorphic to that of $I_2(5)$, which is a 5/2 parabola. The relations defining G_{20} are $ababa = babab$ (coming from the $I_2(5)$ braids) and $a^3 = b^3 = id$. In a natural sense, $G_{20} = I_2(5)^{(3)}$.

Consider the general case of a finite rank 2 group generated by two reflections, a and b, subject to the relations $abab\ldots = baba\ldots$ (q factors on each side) and $a^r = b^s = id$. It has a graph as shown in Figure 2. One of the possible choices of the real weight w is

$$w = -\sqrt{2\frac{\cos(\frac{\pi}{r} - \frac{\pi}{s}) + \cos\frac{2\pi}{q}}{\sin\frac{\pi}{r}\sin\frac{\pi}{s}}}, \tag{1}$$

see [5].

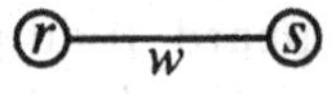

Figure 2. The standard diagram of a group generated by two reflections.

For G_{10}, another choice of w exists, with $\cos\frac{5\pi}{12}$ in (1) instead of $\cos(\frac{\pi}{3} - \frac{\pi}{4}) = \cos\frac{\pi}{12}$. The two choices correspond to two rank 2 representations of G_{10} each: the eigenvalue of a can be either $\omega = e^{2\pi i/3}$ or $\bar{\omega}$, and that of b either i or $-i$.

For G_{20}, another choice also exists: with $\cos\frac{4\pi}{5}$ instead of $\cos\frac{2\pi}{5}$ in (1). Again, the choices provide two representations each: the operators a and b must have the same eigenvalue which can be either ω or $\bar{\omega}$.

The four exact representations of each of the two groups described here will be referred to as the *standard* representations.

1.3. GROUPS G_{12} AND G_{22}

These are also rank 2 groups. Each is generated by three order 2 reflections. The degrees of the basic invariants are respectively 6 and 8,
and 12 and 20. The discriminants are 4/3 and 5/3 parabolas (that is,
E_6 and E_8 curves).

Figure 3 shows that the two groups form a mini-series of four with
A_2 and $G(4,2,2)$ (cf. Figure 1 for the latter). The sum of the three
roots in each case is zero. The diagrams of the figure can easily be
obtained from those in [11].

The change of the sign of i in Figure 3 gives one more exact representation of each of the groups except for A_2. Two further exact
representations of G_{22} correspond to the angle $2\pi/5$ used instead of
$\pi/5$.

These two representations of G_{12} and four of G_{22} will be also referred
to as their *standard* representations.

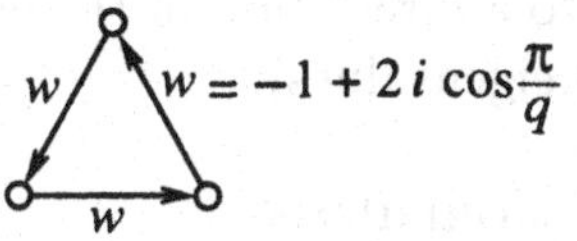

$$w = -1 + 2i\cos\frac{\pi}{q}$$

Figure 3. The diagrams of A_2, $G(4,2,2)$, G_{12} and G_{22}, provided q is 2, 3, 4 or 5
respectively.

2. Automorphisms of simple singularities

2.1. MILNOR REGULAR AUTOMORPHISMS

Let f be a holomorphic function-germ on $(\mathbb{C}^{r+1}, 0)$. Consider a diffeo-
morphism-germ g of $(\mathbb{C}^{r+1}, 0)$ sending the hypersurface $f = 0$ into
itself. It multiplies the function f by a certain function c not vanishing
at the origin. In what follows we assume g has finite order, hence c is
just a constant, a root of unity.

Consider now the space $\mathcal{O}(g, c)$ of all holomorphic function-germs
on $(\mathbb{C}^{r+1}, 0)$ multiplied by c under the action of g. The group $\mathcal{R}_g$ of
biholomorphism-germs of $(\mathbb{C}^{r+1}, 0)$ commuting with g acts on $\mathcal{O}(g, c)$.
The equivalence relation induced on $\mathcal{O}(g, c)$ is a nice geometric equivalence in the sense of Damon [6]. Therefore, the standard theorem on versal deformations holds for it. In particular, the base of an $\mathcal{R}_g$-miniversal
deformation of a function in $\mathcal{O}(g, c)$ is smooth.

DEFINITION 2.1. (cf. [14]) An automorphism g of a hypersurface
$f = 0$ is *Milnor regular* if an $\mathcal{R}_g$-versal deformation of function f
contains members with smooth zero sets.

This implies that either f is g-invariant (hence $c = 1$ and an $\mathcal{R}_g$-miniversal deformation contains the free term) or c is an eigenvalue of the operator dual to g (hence an $\mathcal{R}_g$-miniversal deformation contains a linear form). The latter condition is not sufficient for regularity.

Notice that an $\mathcal{R}_g$-miniversal deformation F_g of a function $f \in \mathcal{O}(g, c)$ can be extended to an ordinary $\mathcal{R}$-miniversal deformation F of f. To get F_g back from F one has to take only those members of F which are in $\mathcal{O}(g, c)$. Similarly, the *equivariant discriminant* $\Sigma_g(f)$ of f, that is, the set of those members of the family F_g whose zero-sets are singular, can be viewed as a section of the ordinary discriminant $\Sigma(f)$ of the family F.

Milnor regularity implies that the zero set V of a generic member of an $\mathcal{R}_g$-versal deformation is a Milnor fibre of f and allows one to consider the intersection form on the middle homology $H_r(V)$ without any complications. The automorphism g induces an automorphism of $H_r(V)$, which we also denote by g, slightly abusing the notations. The space $H_r(V; \mathbb{C})$ splits into a direct sum of the eigenspaces H_χ of g (the χ are the corresponding eigenvalues).

2.2. THE LIST OF AUTOMORPHISMS

Now assume f is a simple function-germ on $(\mathbb{C}^{r+1}, 0)$. We shall be more interested in the splitting $H_r = \oplus_\chi H_\chi$ of its vanishing homology by an automorphism g than in the element g itself. Therefore, it is sufficient for our purposes to consider only the case $r + 1 = \operatorname{corank} f$ and then extend the automorphisms to the stably equivalent functions in any consistent way. Recall that, in this minimal dimension and with f in its quasi-homogeneous normal form, the reductive automorphism group $Aut(f)$ of $f = 0$ is just the group $\mathbb{C} \setminus \{0\}$ of quasi-homogeneous dilations of $\mathbb{C}^{r+1}$ if f is of the type A_n, D_{odd} or E_n. For the other singularities, the group $Aut(f)$ is a direct product of the group of the quasi-homogeneous dilations and either S_3 (for D_4) or $\mathbb{Z}_2$ (for all the other D_{even}).

A case-by-case analysis (which is too straightforward, lengthy and boring to write about) provides the following result.

THEOREM 2.2. *The complete list of Milnor regular automorphisms of simple hypersurface singularities in $\mathbb{C}^3$ is given in the two tables below.*

The completeness here is in the following sense. One starts with $\mathbb{C}^{\operatorname{corank} f}$ and the automorphisms considered up to the groups they generate. In this setting, a function and an automorphism are brought to a simultaneous normal form (f', g') by a choice of coordinates. After this, the function is stabilised to three variables and the automorphism extended to provide a table pair (f, g). Any other, stabilisation and extension of (f', g') results in absolutely the same decomposition of the homology of a Milnor fibre into the sum of character subspaces as does (f, g). A priori, the character assignments to individual subspaces may change at this point. However, this does not affect the monodromy on the subspaces, which will be our main concern.

In the table, in each of the cases we write out monomials $e_1, \ldots, e_\tau$ spanning an equivariant miniversal deformation (here τ is the *equivariant Tjurina number*). We list all the characters for which dim $H_\chi = \tau$. A kth root of unity is denoted by ϵ_k if it is completely arbitrary, and by ε_k if it is arbitrary primitive.

For the characters listed, the table gives the monodromy representations on the H_χ (the order of χ is denoted o_χ). If $\tau = 1$, we give the eigenvalue of the only Picard-Lefschetz operator on H_χ. The equivariant monodromy is studied in more detail in the next section.

We keep the notation of [13] for the unitary reflection groups. The notation used for $\tau > 1$ invariant functions is that from [7]. The notation for the equivariant singularities studied in [8] is slightly altered.

The singularity D_4 occurs in the table in two normal forms. One of them is more convenient for expressing automorphisms involving order 3 elements of S_3 and the other for those involving the transpositions.

In all the series, $\tau = k \geq 1$.

REMARK 2.3. One of the immediate observations the table provides, as an extension of properties of Arnold's simple boundary singularities, is as follows.

Let $D(f)$ be the set of weights of parameters of a quasi-homogeneous $\mathcal{R}$-miniversal deformation F of a simple singularity f. Assume that the weights are normalised so that the largest of them is the Coxeter number of the related Weyl group X (then, of course, $D(f)$ coincides with the set $D(X)$ of degrees of basic invariants of X).

Now, for an automorphism g of $f = 0$, denote by $D(f, g) \subset D(f)$ the set of weights of the parameters of the g-equivariant miniversal deformation F_g of f. Let X_g be the unitary reflection group which is the monodromy group of a character subspace H_χ, where χ is of the maximal order, $X_g \subset X$. In all the table cases, $D(f, g) = D(X_g)$ (for all other χ, such that rk $H_\chi = \tau$, the set $D(f, g)$ is a multiple of the set of the degrees of the monodromy group) and hence $D(X_g) \subset D(X)$. Moreover, the discriminants of the function and of the group, $\Sigma_g(f)$ and $\Sigma(X_g)$, are diffeomorphic.

See [14] for an explanation of these facts.

Table I. Automorphisms of A_n and D_n singularities

function	$g : x, y, z \mapsto$	$|g|$	$\{e_i\}$	χ	monodromy on H_χ	notation
A_n : x^{n+1} $+yz$	$e^{2\pi i/m}x,\ y,\ z$ $n+1 = km$	m	$1, x^m, x^{2m},$ $\ldots, x^{(k-1)m}$	$\epsilon_m \neq 1$	$G(o_\chi, 1, k);$ $\langle \chi \rangle$ if $k=1$	$B_k^{(m)}$
	$e^{2\pi i/m}x,$ $e^{2\pi i/m}y,\ z$ $n = km$	m	$x, x^{m+1},$ $x^{2m+1}, \ldots,$ $x^{(k-1)m+1}$	ϵ_m	$G(o_\chi, 1, k);$ $\langle \chi \rangle$ if $k=1$	$A_{km}/\mathbb{Z}_m$
D_4 : $x^3 + y^3$ $+z^2$	$\omega x,\ y,\ z$	3	$1, y$	ϵ_3	G_4	$A_2^{(3)}$
	$\omega x,\ \omega^2 y,\ z$	3	$1, xy$	1	G_2	G_2
	$\omega x,\ \omega y,\ z$	3	1	ϵ_3	$\mathbb{Z}_6 = \langle -\chi \rangle$	$D_4|\mathbb{Z}_3$
	$-x,\ -\omega y,\ iz$	12	x	ϵ_{12}	$\mathbb{Z}_4 = \langle \chi^3 \rangle$	$D_4/\mathbb{Z}_{12}$
$D_n,$ $n \geq 4$: $x^2 y$ $+y^{n-1}$ $+z^2$	$e^{-2\pi i/2m}x,$ $e^{2\pi i/m}y,\ z,$ $n-1 = km$	$2m$	$1, y^m, y^{2m},$ $\ldots, y^{(k-1)m}$	$\chi^m = -1,$ rk $H_1 = 1$	$G(o_\chi, 1, k);$ $\langle -\chi^{1-m} \rangle$ if $k=1$	$D_{km+1}|\mathbb{Z}_{2m};$ C_k if $m=1$
	$e^{2\pi i/m}x,$ $e^{-2\pi i/m}y,$ $e^{2\pi i/2m}z,$ $n = km$	$2m$	$x, y^{m-1},$ $y^{2m-1}, \ldots,$ $y^{(k-1)m-1}$	$\chi^m = -1$	$G(2o_{\chi^2}, 2, k);$ $\langle \chi^{-2} \rangle$ if $k=1$	$D_{km}/\mathbb{Z}_{2m}$

3. Monodromy on the character subspaces

Now we shall demonstrate that the monodromy on the character subspaces for the automorphisms listed is that given in the table. We proceed case-by-case for the automorphisms not considered in [7, 8].

3.1. CODIMENSION ONE SINGULARITIES

The characters for the table singularities with $\tau = 1$ can easily be calculated from the cohomological point of view, using the residue-forms. The monodromy eigenvalues are obtained from weighted-homogeneous

Table II. Automorphisms of E_n singularities

function	$g : x,y,z \mapsto$	$\|g\|$	$\{e_i\}$	χ	monodromy on H_χ	notation
$E_6:$ x^3+y^4 $+z^2$	$x, -y, z$	2	$1, x, y^2, xy^2$	-1	$F_4 = G_{28}$	F_4
	$\omega x, y, z$	3	$1, y, y^2$	ε_3	G_{25}	$A_3^{(3)}$
	x, iy, z	4	$1, x$	$\pm i, -1$	G_8 on $H_{\pm i}$; A_2 on H_{-1}	$A_2^{(4)}$
	$\omega x, -y, z$	6	$1, y^2$	ε_6	G_5	$B_2^{(3,3)}$
	$\omega x, iy, z$	12	1	$\varepsilon_{12}, -\varepsilon_3$	$\langle -\chi \rangle$	$E_6\|\mathbb{Z}_{12}$
	$-x, e^{2\pi i/8}y, iz$	8	x	$\varepsilon_8, \pm 1$	$\langle \chi^3 \rangle$	$E_6/\mathbb{Z}_8$
	$e^{2\pi i/9}x, \omega y, -\omega^2 z$	18	y	ε_{18}	$\mathbb{Z}_9 = \langle \chi^4 \rangle$	$E_6/\mathbb{Z}_{18}$
$E_7:$ x^3+xy^3 $+z^2$	$x, \omega y, z$	3	$1, x, y^3$	ε_3	G_{26}	$C_3^{(3)}$
	$\omega x, e^{4\pi i/9}y, z$	9	1	$\varepsilon_9, 1$	$\langle -\chi \rangle$	$E_7\|\mathbb{Z}_9$
	$e^{6\pi i/7}x, e^{4\pi i/7}y, e^{2\pi i/7}z$	7	y	ε_7	$\langle -\chi \rangle$	$E_7/\mathbb{Z}_7$
$E_8:$ x^3+y^5 $+z^2$	$\omega x, y, z$	3	$1, y, y^2, y^3$	ε_3	G_{32}	$A_4^{(3)}$
	$x, e^{2\pi i/5}y, z$	5	$1, x$	ε_5	G_{16}	$A_2^{(5)}$
	$\omega x, e^{2\pi i/5}y, z$	15	1	ε_{15}	$\mathbb{Z}_{30} = \langle -\chi \rangle$	$E_8\|\mathbb{Z}_{15}$
	$-x, -y, iz$	4	y, x, y^3, xy^2	$\pm i$	G_{31}	$E_8/\mathbb{Z}_4$
	$-ix, iy, e^{2\pi i/8}z$	8	y, xy^2	ε_8	G_9	$E_8/\mathbb{Z}_8$
	$-\omega x, -y, iz$	12	y, y^3	ε_{12}	G_{10}	$B_2^{(4,3)}$
	$-x, e^{3\pi i/5}y, -iz$	20	x	ε_{20}	$\mathbb{Z}_{20} = \langle \chi \rangle$	$E_8/\mathbb{Z}_{20}$

considerations. We do this in detail for one of the automorphisms, for illustration.

EXAMPLE 3.1. $E_6/\mathbb{Z}_8$. The equivariant miniversal deformation is

$$F(x, y, z; \alpha) = x^3 + y^4 + z^2 + \alpha x \,.$$

Set $f_\alpha = F|_{\alpha = \text{const}}$ and $V_\alpha = \{f_\alpha = 0\} \subset \mathbb{C}^3$. Assume $\alpha \neq 0$. Consider the monomial residue-forms generating $H_2(V_\alpha; \mathbb{C})$:

$$\varphi \, dx dy dz / df_\alpha \,, \quad \varphi = 1, y, y^2, x, xy, xy^2 \,.$$

Put $\delta = e^{2\pi i/8}$. Our automorphism $g : (x, y, z) \mapsto (\delta^4 x, \delta y, \delta^2 z)$ multiplies the above forms respectively by $\delta^3, \delta^4, \delta^5, \delta^7, 1, \delta$. These are the characters.

The monodromy is defined by the homotopy $\alpha \cdot e^{2\pi i t}$, $t \in [0, 1]$, raised in the weighted-homogeneous way to $\mathbb{C}^4_{x,y,z,\alpha}$. This ends up with the transformation

$$h : (x, y, z) \mapsto (\delta^4 x, \delta^3 y, \delta^6 z) = g^3(x, y, z)$$

of 3-space. Hence h has eigenvalue χ^3 on H_χ.

3.2. SINGULARITIES OF HIGHER CODIMENSION

For each automorphism we shall construct distinguished sets of generators of the H_χ. After this we shall show that the corresponding Picard-Lefschetz operators do generate the unitary reflection group claimed.

We start the construction in the traditional way, fixing a generic point $*$ in the base $\mathbb{C}^\tau$ of an equivariant miniversal deformation of the function f. Let $V \subset \mathbb{C}^{\tau+1}$ be the smoothing corresponding to $*$.

Denote by m the order of our Milnor regular automorphism g. Let $V' \subset V$ be the subset of all points with non-trivial stationary subgroups under the action of $\mathbb{Z}_m = \langle g \rangle$. In all our cases, V' has positive codimension in V, not necessarily 1. Consider the quotient varieties, $W = V/\mathbb{Z}_m$ and $W' = V'/\mathbb{Z}_m$, and the integer relative homology group $H_r(W, W'; \mathbb{Z})$.

Let $\Sigma_g \subset \mathbb{C}^\tau$ be the discriminant of our equivariant singularity. Consider a generic line $\ell \subset \mathbb{C}^\tau$ passing through the point $*$, and a set of paths $\{\gamma_j\}$ in ℓ from $*$ to all the points of $\ell \cap \Sigma_g$. As usual, we assume that the paths have no mutual- or self-intersections, except for the common starting point.

In all the table cases, except for $D_4|\mathbb{Z}_3$, it turns out that exactly one relative cycle is contracted by the homotopy of (W, W') corresponding to motion along the path γ_j to its end-point. Moreover, the set of cycles vanishing along all the paths γ_j generates the torsion-free $H_r(W, W'; \mathbb{Z})$. This is what we call a *distinguished set of generators* of this homology group. Notice that, for all our table singularities with $\tau > 1$, the rank of $H_r(W, W'; \mathbb{Z})$ is τ.

Let e be one of the vanishing cycles on (W, W') just obtained. Since g acts freely on $V \setminus V'$, it has m independent preimages $e_0, \ldots, e_{m-1}$ in $H_r(V, V'; \mathbb{Z})$. Assume that they are ordered so that the automorphism g permutes them cyclically:

$$g : e_0 \mapsto e_1 \mapsto e_2 \mapsto \ldots \mapsto e_{m-1} \mapsto e_0 .$$

For each mth root χ of unity, the cycle

$$e_\chi = \sum_{s=0}^{m-1} \chi^{-s} e_s \tag{2}$$

is in the χ-eigenspace $H_{r,\chi}(V, V', \mathbb{C})$ of the relative action of g.

The action of g splits the exact sequence of the pair into the exact sequences of the character subspaces:

$$0 \to H_\chi = H_{r,\chi}(V; \mathbb{C}) \to H_{r,\chi}(V, V'; \mathbb{C}) \to H_{r-1,\chi}(V'; \mathbb{C}) \to \ldots .$$

The rank of the relative homology group here does not depend on the choice of χ, this is just the rank of $H_r(W, W'; \mathbb{Z})$. On the other hand, $H_{r-1,\chi}(V', \mathbb{C}) = 0$ if χ is a primitive mth root of unity (more generally, if $\chi^d \neq 1$, where $\mathbb{Z}_{m/d}$ is a stationary subgroup of a codimension 1 subset of points in V, $d \neq m$). Therefore, if $\chi = \varepsilon_m$ then the rank of H_χ is maximal possible, τ, for all our table singularities (except for $D_4|\mathbb{Z}_3$ when it is maximal, $2 = \tau + 1$, for $\chi = 1$). In fact, the primitive character subspaces provide the most interesting cases since the monodromy on them projects to those on the other character subspaces.

DEFINITION 3.2. The element e_χ in (2) is called a *vanishing χ-cycle on V* if it belongs to the kernel of the boundary operator $H_{r,\chi}(V, V'; \mathbb{C}) \to H_{r-1,\chi}(V'; \mathbb{C})$. The set of vanishing χ-cycles corresponding to the distinguished set of generators of $H_r(W, W'; \mathbb{Z})$ is called *distinguished*.

Of course, χ-cycles in a distinguished set are naturally ordered by the counter-clockwise order in which the corresponding paths in ℓ leave the base point. However, in all our cases the equivariant Dynkin diagrams are just straight chains and therefore an arbitrary order can be achieved (see [10]).

REMARK 3.3.   The above is in fact a generalisation of Arnold's construction used for boundary function singularities. The first step towards this generalisation was taken in [7]. The distinguished generating sets of vanishing χ-cycles obtained in [8] can also be produced by the algorithm described here.

The $\tau > 1$ singularities which did not appear in [7, 8] are $A_{km}/\mathbb{Z}_m$, $D_{km+1}|\mathbb{Z}_m$ and $B_2^{(4,3)}$. We deal with them in turn. Each of them is considered in two particular dimensions, one odd and one even. The intersection forms are Hermitian and skew-Hermitian respectively. Stabilisation by adding a generic quadratic form in two new variables to a function (with the automorphism action appropriately extended) changes the sign of the intersection form [9] but does not affect the monodromy group.

Diagrams of the equivariant functions will be obtained from Dynkin diagrams of the ordinary singularities in $\mathbb{C}^{r+1}$. The conventions to draw the latter are standard:

-   the square of each root is $(-1)^{(r+1)(r+2)/2}\big((-1)^r + 1\big)$;

-   in the skew-symmetric case, an edge $a \to b$ means $(a, b) = 1$;

-   in the symmetric case, a dashed edge is drawn for the intersection number $(-1)^{r/2}$ and a solid edge for the negative of this.

3.2.1. $A_{km}/\mathbb{Z}_m$

For $m = 3$, this singularity was studied in [8]. Now we consider arbitrary m.

Hermitian case. Take the singularity in just one variable: $f(x) = x^{km+1}$, with $g(x) = e^{2\pi i/m}x$. The family

$$x^{km+1} + \lambda_k x^{(k-1)m+1} + \ldots + \lambda_{k-1}x^{m+1} + \lambda_k x = xp_\lambda(x^m) \qquad (3)$$

is its equivariant miniversal deformation. The discriminant of the family is that of the Weyl group B_k corresponding to the polynomial p_λ having either zero or multiple roots.

We take a function $f_*(x) = xp_*(x^m)$, such that p_* has all its roots real positive and simple, for a marked generic member of (3). Consider the zero-set of f_*:

$$V = \{0, e^{2\pi is/m}a_j, \ s = 0, \ldots, m - 1, \ j = 1, \ldots, k\} \subset \mathbb{C}.$$

We order the real numbers a_j so that $0 < a_1 < a_2 < \ldots < a_k$.

A distinguished basis of vanishing χ-cycles on V can be obtained by varying the free term of p_*. For example, such a basis can be taken to consist of

(i) a *short* χ-*cycle* $\sum_{s=0}^{m-1} \chi^{-s}[e^{2\pi i s/m} a_1]$ vanishing at the origin (for $\chi = 1$, the term $m[0]$ must be subtracted from this sum);

(ii) $k - 1$ *long* χ-*cycles* $\sum_{s=0}^{m-1} \chi^{-s}([e^{2\pi i s/m} a_j] - [e^{2\pi i s/m} a_{j-1}])$, $j = 2, \ldots, k$.

The self-intersection number of a long χ-cycle is $2m$, and of a short χ-cycle either $m^2 + m$ if $\chi = 1$ or m otherwise.

The Dynkin diagram of the chosen basis is drawn top left in Figure 4. The upper half of the figure also shows how this diagram can be obtained by m-folding of the symmetric Dynkin diagram of the ordinary singularity A_{km}. According to the expressions in (i) and (ii), each of the k character cycles is a linear combination of m ordinary vanishing cycles situated on one of the concentric circles of the A_{km} diagram (these are orbits of the Z_m-action). In the linear combination, the crossed cycle is taken with the coefficient 1, the cycle counter-clockwise next to it with the coefficient χ^{-1}, the next one with the coefficient χ^{-2}, and so on.

Skew-Hermitian case. Addition of t^2 to the family (3) provides a miniversal equivariant deformation of the function $x^{km+1} + t^2$ associated with the Milnor regular automorphism $g : (x, t) \mapsto (e^{2\pi i/m} x, e^{2\pi i/2m} t)$ of order $2m$. Now $\chi^m = -1$. Similar to the Hermitian version, one can obtain a Dynkin diagram of the equivariant function by m-folding of that of the ordinary singularity as shown in the lower half of Figure 4.

Picard-Lefschetz operators. One of the possible versions of the A_{km}/Z_m singularity in $\mathbb{C}^{r+1}$ is provided by the addition of the quadratic form $\sum_{j=1}^{[r/2]} y_j z_j$ to the above one- or two-variable function, with the automorphism g multiplying all the y_j by $e^{2\pi i/m}$ and acting trivially on the z_j. The character subspaces in the homology are H_χ, $\chi^m = (-1)^r$. The intersection form on H_χ is the one we had in the one- or two-dimensional case multiplied by $(-1)^{[r/2]}$.

The Picard-Lefschetz operator on H_χ corresponding to the short χ-cycle e has the eigenvalue $(-1)^r \chi$ and acts by the formula

$$h_e : \quad c \mapsto c + (-1)^{(r+1)(r+2)/2}(1 - \chi)(c, e)e/m . \qquad (4)$$

The monodromy operator on H_χ, associated with a long χ-cycle e, is induced by the product of the m commuting Picard-Lefschetz operators of the ordinary singularity corresponding to the vertices on the related circle of the symmetric A_{km} diagram. The operator is

$$h_e : \quad c \mapsto c + (-1)^{(r+1)(r+2)/2}(c, e)e/m . \qquad (5)$$

Clearly, for $r = 0$, the unitary reflections (4) and (5), associated with the vertices of the top left diagram of Figure 4, do generate the Shep-

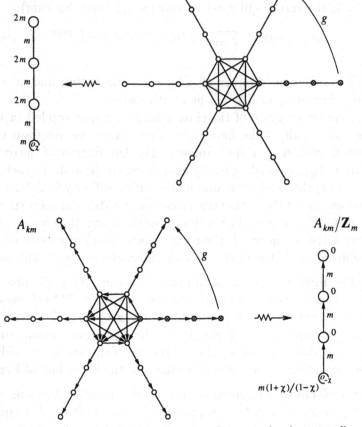

Figure 4. Folding symmetric Dynkin diagrams of A_{km} to the character diagrams of $A_{km}/\mathbf{Z}_m$, $\chi \neq 1$, $k = 4$ and $m = 6$. The Hermitian diagrams, of the one-variable singularities, are in the upper half of the figure. The skew-Hermitian diagrams, of the two-variable functions, are in the lower half.

hard-Todd group $G(o_\chi, 1, k)$ (see Remark 1.1 for the $\chi = 1$ and $k = 1$ cases).

3.2.2. $D_{km+1}|\mathbf{Z}_{2m}$

Skew-Hermitian case. We start with the two-variable singularity. It has the invariant miniversal deformation

$$x^2 y + y^{km} + \lambda_k y^{(k-1)m} + \ldots + \lambda_2 y^m + \lambda_1 = x^2 y + p_\lambda(y^m), \qquad (6)$$

with respect to the automorphism $g(x, y) = (e^{-2\pi i/2m} x, e^{2\pi i/m} y)$ of order $2m$.

The quotient W of an invariant Milnor fibre V under the action of $\mathbf{Z}_{2m}$ is diffeomorphic to a coordinate line $\mathbf{C}_v$, $v = y^m$, with a puncture at the origin. The subset $W' \subset W$ of irregular orbits consists of m

points, roots of the equation $p_\lambda(v) = 0$. Hence, the discriminant of (6) is also isomorphic to the discriminant of B_k.

Similar to what was done for the previous singularity, we mark a point $*$ in the base of deformation (6), so that the polynomial $p_*(v)$ has all its roots real positive and simple. This choice easily provides the Dynkin diagram for a character subspace $H_\chi \subset H_1(V; \mathbb{C})$, $\chi^m = -1$, which is the top left diagram of Figure 5. This diagram can also be obtained by m-folding of a symmetric Dynkin diagram of the ordinary D_{km+1} singularity as shown in the upper half of the figure. For a better symmetry, we introduced there an auxiliary cycle (drawn fine, not participating in the actual D_{km+1} diagram) which is the negative of the sum of all the cycles in the bold D_{m+1}-subdiagram. Each vanishing cycle in the $D_{km+1}|\mathbb{Z}_{2m}$ diagram is a linear combination of the m cycles in the D_{km+1} diagram situated on one of the concentric circles: they are taken with the coefficients $1, \chi^{-1}, \chi^{-2}, \ldots$, starting from the crossed cycle and going counter-clockwise.

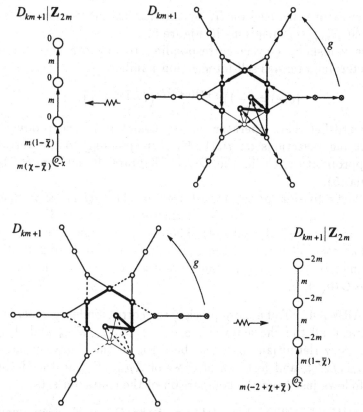

Figure 5. Folding symmetric Dynkin diagrams of the D_{km+1} singularity to $D_{km+1}|\mathbb{Z}_{2m}$ diagrams, $k = 4$ and $m = 6$. The two-variable case is in the upper half of the figure, and the three-variable is in the lower.

The lowest vertex in the character diagram corresponds to the polynomial $p_\lambda(v)$ of (6) having a zero root. Any other vertex corresponds to $p_\lambda(v)$ getting a double root.

Hermitian case. Similar diagrams for the one-variable stabilisation of the above, with the automorphism g acting trivially on the added coordinate, are drawn in the lower half of Figure 5.

Picard-Lefschetz operators. In $\mathbb{C}^{r+1}$, with g acting trivially on the 2-jet, the cycle in H_χ, $\chi^m = -1$, corresponding to the lowest vertex of the character diagram has self-intersection number

$$(e, e) = (-1)^{(r+1)(r+2)/2}\big((-1)^r\chi - 1\big)(1 - \bar{\chi})m.$$

Its Picard-Lefschetz operator has eigenvalue $(-1)^r\chi$ and, therefore, acts by the formula

$$h_e: \quad c \mapsto c + (-1)^{(r+1)(r+2)/2}\frac{(c, e)}{m(1 - \bar{\chi})}e. \qquad (7)$$

This operator is induced on H_χ by the classical monodromy defined by the bold D_{m+1}-subdiagrams in Figure 5.

The vanishing χ-cycle corresponding to any other vertex of the character diagram has self-intersection number

$$(e, e) = (-1)^{(r+1)(r+2)/2}\big((-1)^r + 1\big)m.$$

Its Picard-Lefschetz operator on H_χ comes from the product of m commuting reflections on $H_r(V; \mathbb{C})$ corresponding to the vertices on the appropriate circle in the D_{km+1} diagram. It acts on H_χ by the formula (5).

In fact, division of the lowest root in the $D_{km+1}/\mathbb{Z}_{2m}$ diagrams by $\bar{\chi} - 1$ reduces them to the character diagrams of Figure 4. The operator (7) transforms into (4). Therefore, the monodromy groups on the character subspaces of the equivariant A- and D-singularities are the same. In particular, in the Hermitian D-case, we do obtain the groups $G(o_\chi, 1, k)$.

REMARK 3.4. To be more precise, the correspondence observed here is between all the character subspaces of $D_{km+1}/\mathbb{Z}_{2m}$ and $A_{km}/\mathbb{Z}_m$ in the skew-Hermitian case, and between all the character subspaces of $D_{km+1}/\mathbb{Z}_{2m}$ and just half of those of $A_{2km}/\mathbb{Z}_{2m}$ in the Hermitian. This follows just from the comparison of the character sets.

REMARK 3.5. The above folding of the D_{km+1} diagrams could be done in two steps: via factorisation by the action of $\mathbb{Z}_2 = \langle g^m \rangle$ followed by the further $\mathbb{Z}_m$-factorisation. The intermediate output here is

Arnold's C_{km} singularity, with its $\mathbb{Z}_2$-anti-invariant homology. This, for example, immediately eliminates the generator of the rank 1 character subspace $H_{\chi=1}$ of $D_{km+1}|\mathbb{Z}_{2m}$, which is the difference of the two cycles at the whiskers of the central D_{m+1}-subdiagram in our figures.

3.2.3. $B_2^{(4,3)}$

Hermitian case. The singularity has a two-parameter equivariant miniversal deformation

$$x^3 + y^5 + \beta y^3 + \alpha y + z^2 . \tag{8}$$

This corresponds to the Milnor regular automorphism $g(x, y, z) = (-\omega x, -y, iz)$ of the E_8 surface. The bifurcation diagram of the family is isomorphic to the discriminant of B_2 and consists of two strata:

$$2A_2 = \{b^2 = 4\alpha\} \qquad \text{and} \qquad D_4|\mathbb{Z}_{12} = \{\alpha = 0\} .$$

The rectangular Dynkin diagram of the ordinary singularity E_8 folds to a diagram of the equivariant function in two steps (see Figure 6):

(i) fusion to the $A_4^{(3)}$ diagram corresponding to the 4-subspace in the vanishing homology of E_8 on which g^4 is the multiplication by ε_3 (the cycles in it have self-intersection numbers -3, and an edge $a \to b$ has weight $(a, b) = 3/(1 - \varepsilon_3)$);

(ii) further 2-folding of $A_4^{(3)}$ (this picks up the character subspace $H_{\chi=-\varepsilon_3 e_4}$ of the $B_2^{(4,3)}$ singularity; the vertices of the diagram corresponding to the reflections of orders 3 and 4 are $v_1 = e_1 + \varepsilon_4 e_4$ and $v_2 = e_2 + \chi e_3$ respectively).

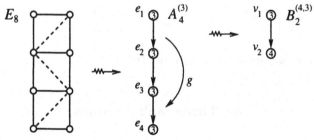

Figure 6. The two-step folding of the E_8 diagram to the diagram of the $B_2^{(4,3)}$ singularity. The weights of the vertices and edges are omitted.

The intersection form thus obtained is

$$-6 \begin{pmatrix} 1 & \frac{1}{\varepsilon_3 - 1} \\ \frac{1}{\bar{\varepsilon}_3 - 1} & 1 - \frac{\chi\varepsilon_3}{\varepsilon_3 - 1} \end{pmatrix} .$$

From this,

$$\frac{(v_1, v_2)(v_2, v_1)}{(v_1, v_1)(v_2, v_2)} = \frac{\cos \frac{\pi}{12}}{2 \sin \frac{\pi}{3} \sin \frac{\pi}{4}}, \quad \frac{\cos \frac{5\pi}{12}}{2 \sin \frac{\pi}{3} \sin \frac{\pi}{4}}$$

depending on the choice of ε_3 and ε_4. Therefore, the four options to choose ε_3 and ε_4 correspond to the four standard representations of G_{10} (see Section 1.2) acting on the subspaces $H_{\chi=-\varepsilon_3\varepsilon_4}$.

In terms of the Picard-Lefschetz operators h_{e_j} of $A_4^{(3)}$ (each with the eigenvalue ε_3), the Picard-Lefschetz operators h_{v_s} generating the representation of G_{10} on $H_{\chi=-\varepsilon_3\varepsilon_4}$ are

$$h_{v_1} = h_{e_1} h_{e_4} \quad \text{and} \quad h_{v_2} = h_{e_2} h_{e_3} h_{e_2} = h_{e_3} h_{e_2} h_{e_3}, \qquad (9)$$

with matrices

$$\begin{pmatrix} \varepsilon_3 & -\varepsilon_3 \\ 0 & 1 \end{pmatrix} \quad \text{and} \quad \begin{pmatrix} 1 & 0 \\ 1 - \chi\varepsilon_3 & \varepsilon_4 \end{pmatrix}.$$

Skew-Hermitian case. For the two-variable singularity (no z in (8)), the characters χ of g on the first homology of the curve are $\pm\varepsilon_3$. Constructing the diagram of $B_2^{(4,3)}$, we again first obtain the $A_4^{(3)}$ diagram for the 4-dimensional χ^2-eigenspace of g^2 in the first homology. This diagram looks absolutely the same as in Figure 6, but with a modified interpretation: now the self-intersection of each cycle is $3\frac{\chi^2+1}{\chi^2-1}$ and an edge $a \to b$ has weight $(a, b) = -3/(1 - \bar{\chi}^2)$. For a distinguished basis in the two-dimensional eigenspace H_χ of g one can take $v_1 = e_1 - \chi^3 e_4$ and $v_2 = e_2 - \chi e_3$. The generators (9) of the equivariant monodromy group now have the matrices

$$\begin{pmatrix} -\bar{\chi}^2 & \bar{\chi}^2 \\ 0 & 1 \end{pmatrix} \quad \text{and} \quad \begin{pmatrix} 1 & 0 \\ 1 - \bar{\chi} & \chi^3 \end{pmatrix}.$$

In particular, we get a finite group, $G(6, 1, 2)$, if $\chi = -\varepsilon_3$.

4. Three rank 2 groups

Among various relations between the sets of degrees of basic invariants of Shephard-Todd groups, similar to those observed in Remark 2.3, there are relations giving rise to singularity theory interpretations of three further rank 2 unitary reflection groups:

(i) $D(G_{20}) = D(H_4) \cap D(G_{32})$, corresponding to $G_{20} = H_4 \cap G_{32}$ in E_8;

(ii) $D(G_{22}) = D(H_4) \cap D(G_{31})$, related to $G_{22} = H_4 \cap G_{31}$ in E_8;

(iii) $D(G_{12}) \subset D(F_4) \subset D(E_6)$, corresponding to $G_{12} \subset F_4 \subset E_6$.

We write out appropriate two-parameter subfamilies of the $\mathcal{R}$-versal deformations of E_8 and E_6, and show that their monodromy groups are exactly the expected G_{20}, G_{22} and G_{12}. This time symmetry will be combined with special distribution of singular points on critical levels.

4.1. GROUP G_{20}

O.Shcherbak showed [12] that there exist two 4-parameter subfamilies in an $\mathcal{R}$-miniversal deformation of E_8 whose discriminant is that of the Coxeter group H_4. Those are either functions with all critical points of type A_2 or functions having four critical levels with two Morse points on each of them. Consider the latter subfamily:

$$x^3 + \alpha x + \beta + y^5/5 + y^3(\gamma x + \delta) + y(\gamma x + \delta)^2 + z^2. \qquad (10)$$

Under the monodromy action, the vanishing homology of E_8 breaks up into the sum of the standard representation of H_4 and its conjugate.

Intersect the family (10) with the invariant miniversal deformation of $A_4^{(3)}$ (whose monodromy group is G_{32}, see the table). This is the same as taking the invariant part of the family with respect to the $\mathbb{Z}_3$-action, $g(x, y, z) = (\omega x, y, z)$:

$$x^3 + \beta + y^5/5 + \delta y^3 + \delta^2 y + z^2. \qquad (11)$$

These are functions having two critical levels, with two A_2 points on each of them.

The family obtained has discriminant $25\beta^2 + 4\delta^5 = 0$ isomorphic to the discriminant of the group G_{20}. Therefore, the monodromy group has two generators, a and b, satisfying the relation $ababa = babab$. Moreover, $a^3 = b^3 = id$ since an elementary degeneration in (11) involves only A_2 points on the zero-level and the Coxeter number of the Weyl group A_2 is 3.

The relations obtained are those defining G_{20}. Hence the monodromy group of the family (11) is an 8-dimensional representation of G_{20}.

In fact we obtain the sum of all four standard rank 2 representations of G_{20} mentioned in Section 1.3. Indeed, consider the 4-dimensional character space $H_{\chi=\varepsilon_3}$ of g in the vanishing homology of E_8. On H_{ε_3}, the restriction of the whole $A_4^{(3)}$ monodromy (with four A_2 points situated on independent levels) to the subfamily (11) corresponds to the zig-zag folding of the diagram in the middle of Figure 6. The order 3 operators

$h_{e_1}h_{e_3}$ and $h_{e_2}h_{e_4}$ split H_{ε_3} into two invariant subspaces, generated by

$$u_1 = e_1 + \bar{\varepsilon}_3(\varepsilon_5^2 + \bar{\varepsilon}_5^2)e_3 \qquad \text{and} \qquad u_2 = e_2 - \bar{\varepsilon}_3(\varepsilon_5 + \bar{\varepsilon}_5)e_4$$

(the two subspaces differ by the choice of the values of the expressions in the brackets). For these vectors,

$$\frac{(u_1, u_2)(u_2, u_1)}{(u_1, u_1)(u_2, u_2)} = \frac{1 + \cos\frac{2\pi}{5}}{2\sin^2\frac{\pi}{3}}, \quad \frac{1 + \cos\frac{4\pi}{5}}{2\sin^2\frac{\pi}{3}}$$

depending on the value of $\varepsilon_5 + \bar{\varepsilon}_5$. Now, comparison with Section 1.3 yields the desired identification of the representations of G_{20}.

REMARK 4.1.   The sum of the spaces of two of these representations of G_{20}, with the same value of $\varepsilon_5 + \bar{\varepsilon}_5$ and different ε_3, is the space of one of the two representations of H_4 on the vanishing homology of the E_8 singularity.

4.2. Groups with equilateral triangular diagrams

4.2.1. G_{22}

The part of the H_4 family (10) contained in the equivariant versal deformation of the $E_8/\mathbb{Z}_4$ singularity (the monodromy group of the latter is G_{31}) is

$$x^3 + \alpha x + y^5/5 + \delta y^3 + \delta^2 y + z^2 . \qquad (12)$$

Its discriminant, $25\alpha^3 = 27\delta^5$, is the discriminant of the group G_{22}. A generic degeneration is 4 Morse points on the zero level $V_{\alpha,\delta} \subset \mathbb{C}^3_{x,y,z}$.

Consider the line $\ell = \{\delta = \delta_0\}$, where δ_0 is a real negative constant, in the base of deformation (12). Take V_{0,δ_0} for the marked Milnor fibre. The line ℓ intersects the discriminant at three points, with coordinates α_0, $\omega\alpha_0$ and $\omega^2\alpha_0$, where α_0 is real. We take the straight paths on ℓ from $(0, \delta_0)$ to the discriminantal points to define vanishing cycles on V_{0,δ_0}.

The real part of the plane curve $V_{\alpha_0,\delta_0} \cap \{z = 0\}$ is the sabirification of E_8 that yields the rectangular Dynkin diagram of E_8. The product of the four commuting Picard-Lefschetz operators, corresponding to the cycles on V_{0,δ_0} vanishing at the nodes of this curve, is an elementary Picard-Lefschetz operator in the family (12).

The two quadruples of cycles on V_{0,δ_0} vanishing at $(\omega\alpha_0, \delta_0)$ and $(\omega^2\alpha_0, \delta_0)$ are obtained from those vanishing at (α_0, δ_0) by multiplication of x by ω^2 and ω.

The intersection diagram of the twelve vanishing cycles on V_{0,δ_0} is given in Figure 7. The sum of three cycles on each of the four horizontal

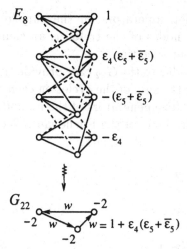

Figure 7. Folding E_8 diagram to G_{22} diagram.

levels of the prism is zero. It is easy to spot an embedding of the rectangular E_8 diagram into our diagram.

The three generators of the monodromy group of the family (12) are the products of the Euclidean reflections in the cycles in one vertical quadruple in Figure 7. The vanishing homology of E_8 splits into the sum of four rank 2 invariant subspaces. The elements spanning these subspaces are the linear combinations of the cycles in each quadruple taken with the coefficients shown in Figure 7 (the coefficients are defined by a horizontal level of the diagram). Different choices of the numbers ε_4 and $\varepsilon_5 + \bar{\varepsilon}_5$ give four different rank 2 subspaces.

Rescaling the spanning vectors to make their squares -2, we get the triangular intersection diagram shown at the bottom of Figure 7. Comparing it with the G_{22} diagram of Figure 3, we see that the four rank 2 subspaces obtained provide all four standard representations of G_{22}.

4.2.2. G_{12}
Consider now the subfamily

$$x^3 + \alpha x + y^4 + \beta y^2 + \beta^2/8 + z^2 \tag{13}$$

of the miniversal deformation of E_6. Its discriminant, $(\alpha/3)^3 + (\beta/4)^4 = 0$, is the discriminant of the group G_{12}.

This time the elementary degeneration is three Morse points on the zero-level situated symmetrically with respect to the change of sign of y (thus, one of them is on $y = 0$). For a real point of the discriminant having $\beta > 0$, the real part of the curve $V_{\alpha,\beta} \cap \{z = 0\}$ is the open trefoil yielding the rectangular Dynkin diagram of E_6. The product of the Picard-Lefschetz operators on the vanishing homology of E_6 corresponding to the nodes of the trefoil is an elementary monodromy operator in the present situation.

Considerations similar to the G_{22} case provide the prismatic diagram of Figure 8(a). Again the sum of the cycles on each of the three horizontal levels here is zero. The generators of the monodromy group Γ of the family (13) are the products of the Euclidean reflections corresponding to the vertices in the vertical triplets.

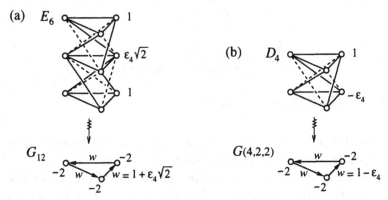

Figure 8. Obtaining G_{12} diagram from E_6, and $G(4, 2, 2)$ from D_4.

The linear combinations of the cycles in each of the vertical triplets, with the coefficients shown in the figure, span one of the three rank 2 representations of Γ. Scaling the spanning vectors to make their squares -2, we obtain the equilateral triangular intersection diagram of Figure 8(a). Comparison with Section 1.3 shows that two possible values of ε_4 provide the two standard representations of G_{12}.

The sum of the two G_{12} representations described is the anti-invariant part of the homology of E_6 with respect to the change of sign of y (that is, the F_4 vanishing homology). The sign change acts on the prismatic diagram as the reflection in the medial triangle followed by the change of orientation of all the cycles. Therefore, the third rank 2 representation of Γ in the homology of E_6 is just A_2 acting on the span of the differences of the corresponding vertices of the top and bottom faces of the prism.

4.2.3. *The mini-series*
The E_6 prism of Figure 8(a) is one store shorter than that of Figure 7. Cutting off one level more (Figure 8(b)), we get the diagram corresponding to the equivariant mini-versal deformation of $D_4/\mathbb{Z}_4$ (cf. the table where it is given in a different normal form):

$$x^3 + \alpha x + y^3 + \beta y + z^2 . \tag{14}$$

Here $g(x,y,z) = (-x,-y,iz)$. We can assume that g sends the vertices of the top face down to the underlying vertices of the bottom, and those from the bottom up to the negatives of those on the top. The monodromy group of (14) is $G(4,2,2)$, and the discriminant, $\alpha^3 = \beta^3$, of the family is that of the reflection group. The elementary degeneration is $2A_1$.

The linear combinations of the vertical pairs of the cycles with the coefficients shown in Figure 8(b) span the rank 2 subspace $H_{\chi=\varepsilon_4}$ in the homology of D_4. The result of the folding operation is the equilateral diagram of $G(4,2,2)$ (cf. Figure 3).

The degenerate prism, consisting of just one triangular level, corresponds to the A_2 $\mathcal{R}$-versal family $x^3 + \alpha x + \beta + y^2 + z^2$ and a non-generic line $\ell = \{\alpha = \text{const}\}$ in its base.

REMARK 4.2. *There is an interesting feature in all four examples of this subsection: the discriminant of the related two-dimensional unitary reflection group has the same singularity as the non-deformed singular curve $V_{0,0} \cap \{z=0\}$. Does this have any explanation avoiding explicit calculations?*

Acknowledgements. I am very grateful to Peter Slodowy for highly useful discussions. He also brought to my attention a series of papers by Yano [16, 17, 18] who studied a different, but related topic of finding Saito's flat coordinates for certain unitary reflection groups. Unfortunately, Yano abandoned the area leaving his investigations unfinished.

References

1. V. I. Arnold and his school, *Arnold's problems*, Phasis, Moscow, 2000 (in Russian).
2. E. Brieskorn, Die Fundamentalgruppe des Raumes der regülaren Orbits einer endlichen komplexen Spiegelungsgruppe, Invent. Math. **12** (1971) 57–61.
3. M. Broué and G. Malle, Zyklotomische Heckealgebren, Astérisque **212** (1993) 119–189.
4. A. M. Cohen, Finite complex reflection groups, Ann. Sci. Ecole Norm. Sup. **9** (1976) 379–436.
5. H. S. M. Coxeter, Regular complex polytopes, Cambridge Univ. Press, Cambridge, 1974.
6. J. N. Damon, *The unfolding and determinacy theorems for subgroups of $\mathcal{A}$ and $\mathcal{K}$*, Memoir Amer. Math. Soc. **50**, no.306, 1984.
7. V. V. Goryunov, Unitary reflection groups associated with singularities of functions with cyclic symmetry, Russian Math. Surveys **54**:5 (1999) 873–893.

8. V. V. Goryunov and C. E. Baines, Cyclically equivariant function singularities and unitary reflection groups $G(2m, 2, n)$, G_9, G_{31}, St. Petersburg Math. J. **11**:5 (2000) 761–774

9. S. M. Gusein-Zade, Monodromy groups of isolated singularities of hypersurfaces, Russian Math. Surveys **32**:2 (1977) 23–69.

10. O. V. Lyashko, Geometry of bifurcation diagrams. Journal of Soviet Math. **27** (1984) 2736–2759.

11. V. L. Popov, Discrete complex reflection groups, Commun. Math. Inst. Rijksuniversiteit Utrecht 15-1982 (1982), 89 pp.

12. O. P. Shcherbak, Wavefronts and reflection groups, Russian Math. Surveys **43**:3 (1988) 125–160.

13. G. C. Shephard and J. A. Todd, Finite unitary reflection groups, Canad. J. Math. **6** (1954) 274–304.

14. P. Slodowy, Simple singularities and complex reflections, this volume.

15. T. A. Springer, Regular elements of finite reflection groups, Invent. Math. **25** (1975) 159–198.

16. T. Yano, Free deformations of isolated singularities, Sci. Rep. Saitama Univ. Ser A **9** (1980) 61–70.

17. T. Yano, Invariants of finite reflection groups and flat coordinate systems of isolated singularities, RIMS Kokyuroku **444** (1981) 209–235 (in Japanese).

18. T. Yano, Deformations of rational double points associated with unitary reflection groups, Sci. Rep. Saitama Univ. Ser A **10** (1981) 7–9.

Simple Singularities and Complex Reflections

Peter Slodowy
Mathematisches Seminar der Universität Hamburg
(ms3a007@math.uni-hamburg.de)

1. Introduction

The Weyl groups of type A_k, D_k, E_k play an important role in the study of the corresponding simple singularities, e.g. they are realised as the full monodromy groups of these singularities (in dimension 2), and the base of the semiuniversal deformation of such a singularity is naturally identified with the complex orbit space of the Weyl group, cf. [1, 3], [5], [34, 35].

The remaining Weyl groups of type B_k, C_k, F_4, G_2 enter the scene, when regarding simple singularities with certain 'diagram' symmetries [34, 35] or, almost equivalently, simple boundary singularities, [2, 3], [32]. There are also interpretations for the remaining finite Coxeter groups of dihedral or icosahedral type (I_k, H_3, H_4) related to obstacles in the propagation of wavefronts generated by the singularities, cf. [23], [30]. Even earlier, T. Yano had described certain 'free' subdeformations of the semiuniversal deformation in terms of these as well as further, unitary reflection groups [46, 48].

Recently Victor Goryunov, cf. [11, 13], has exhibited a way to realise certain complex reflection groups as subgroups of the full monodromy group. He starts from very special automorphisms of the singularity and employs methods very close to those in the study of boundary singularities for the explicit construction of the complex reflection subgroups in terms of vanishing cycles.

Embeddings of complex reflection groups into finite Weyl groups had already been encountered long before the above mentioned papers in the work of T.A. Springer on regular elements in finite reflection groups [37], cf. also the subsequent papers [10], [19, 20], complementing and generalising this work.

The goal of this note is to show that these different occurences are related in a natural way. For this purpose, we shall define below the notion of a 'Milnor regular' automorphism of a simple singularity. Our main result will then show the following. The (conjugacy classes of) 'Milnor regular' automorphisms correspond exactly to the (conjugacy classes of) 'Springer regular' elements in the corresponding Weyl group

D. Siersma et al. (eds.), New Developments in Singularity Theory, 329–348.
© 2001 Kluwer Academic Publishers. Printed in the Netherlands.

(or its normalizer). Moreover, the centralizer of such an automorphism in the monodromy group will act as an irreducible complex reflection group on a subspace of the (complex) homology of the Milnor fibre and, finally, the base space of the invariant semiuniversal deformation of the singularity (with respect to that automorphism) identifies naturally with the orbit space of the complex reflection group, in particular, the respective discriminants coincide. Following the generalisations in [19, 20] one might actually relax the regularity assumption somewhat. On the singularity side, this corresponds to the contraction of certain vanishing cycles leading to 'partial' equisingular Milnor fibres and monodromy groups of corresponding fibrations. However, we shall not pursue that here.

This method, involving the realisation of a complex reflection group as (a subquotient of) a centraliser in a finite Weyl group (as in [37], [19, 20]), does not yield all finite, irreducible complex reflection groups, however. Already some of Yano's examples ([46]), e.g. H_4 and two of its subgroups, G_{20} and G_{22}, evade our construction. On the other hand, our approach involves some general constructions on the side of singularities, which one might pursue a little further. Anyhow, it would be quite interesting to hunt for (some of) the remaining groups in the realm of more general singularities and associated infinite reflection groups.

We should mention another occurence of complex reflection groups in a context of direct interest (pointed out already in [47]). In [39, 40, 41], E.B. Vinberg studied the orbital geometry and invariant theory for graded Lie algebras (i.e. for the action of the fixed group of a finite order automorphism on an eigenspace in the Lie algebra, generalizing thereby [17]). Here complex reflection groups turn up naturally as Weyl groups of these graded algebras linking in very good cases to Springer's construction. At this moment it seems that at least in some cases (e.g. type A_k) one can relate the invariant semiuniversal deformation of simple singularities to the 'graded adjoint quotient' in a manner similar to the original construction of Brieskorn and Grothendieck [5], [34] (e.g. for type A_k by means of [16]). However, that issue requires further elaboration in more cases and will hopefully be the object of some later article.

Complex reflection groups have recently attracted increased attention in the representation theory of finite groups of Lie type, cf. e.g. [6], [7]. Maybe the monodromy aspects of our correspondence are also of interest in that domain.

The ideas of this paper originated during a one-year stay at Kitami Institute of Technology (1999-2000). I would like to thank my colleagues there for stimulating discussions in a pleasant atmosphere. My thanks go also to Victor Goryunov (Liverpool) for a series of interesting exchanges (cf. also [12]). This article is an elaboration of a former very sketchy note presented at the Sapporo meeting on 'The Monster and related topics', July 1999, and distributed during the conference at the Newton Institute in August 2000. I would also like to thank the organisers and participants of these meetings for their interest.

2. Simple singularities

Simple singularities are algebraic hypersurfaces $X = \{f = 0\}$ in $\mathbb{C}^3$ with an isolated singularity at the origin.

type	equation $\{f = 0\}$	weights $(d; w_1, w_2, w_3)$
A_ℓ	$x^{\ell+1} + yz = 0$	$(\ell+1; 1, (\ell+1)/2, (\ell+1)/2)$
D_ℓ	$x^{\ell-1} + xy^2 + z^2 = 0$	$(2\ell-2; 2, \ell-2, \ell-1)$
E_6	$x^4 + y^3 + z^2 = 0$	$(12; 3, 4, 6)$
E_7	$x^3y + y^3 + z^2 = 0$	$(18; 4, 6, 9)$
E_8	$x^5 + y^3 + z^2 = 0$	$(30; 6, 10, 15)$

Fig. 1. Simple singularities

They have been studied from many points of view and are also called *Kleinian singularities, DuVal singularities*, or *rational double points*, cf. e.g. [3, 22], also for further references. Their *type* derives from the dual intersection diagram for the components of the exceptional divisor in a minimal resolution $\tilde{X} \longrightarrow X$. That diagram is a Coxeter-Dynkin diagram of the type indicated in the above list. We have also listed data on a certain $\mathbb{C}^*$-action on these spaces. Note that the defining equations f are all *weighted homogeneous* (with positive weights), i.e. if the coordinates x, y, z are given suitable positive weights w_1, w_2, w_3, then the equation is of some degree d (cf. the 3-rd column of Fig.1 for the respective values $(d; w_1, w_2, w_3)$). In geometric terms, this means that the $\mathbb{C}^*$-action on $\mathbb{C}^3$ with weights w_1, w_2, w_3 induces one on the subvariety X, thereby 'contracting' it to the origin [1].

Due to this action, all essential local properties are also global ones and conversely. For this reason, which will apply similarly for the Milnor fibre and semiuniversal deformations of these singularities, we shall

---

[1] In the case of A_ℓ, ℓ even, which will often play an exceptional role, some weights are half-integers. In that case, one has to double the weights and the degree to obtain a true and not only infinitesimal $\mathbb{C}^*$-action. We shall tacitly assume that done whenever necessary

stick to a global point of view in this paper (differing from the usual sources in singularity theory, e.g. [3, 22], which work in a local category like that of germs of complex spaces).

These $\mathbf{C}^*$-actions also extend to the semiuniversal deformation Φ : $\mathfrak{X} \longrightarrow \mathfrak{U}$ of X, i.e. there are $\mathbf{C}^*$-actions on $\mathfrak{X}$ and $\mathfrak{U}$ such that Φ becomes equivariant and the action on $\mathfrak{X}$ extends that of X. In Fig. 2, we have listed the weights of the $\mathbf{C}^*$-action on the base space $(\mathfrak{U} \cong \mathbf{C}^\ell)$. Under our convention chosen for $(d; w_1, w_2, w_3)$, they are given by the degrees $d_i, i = 1, \ldots, \ell$ of the corresponding Lie algebra $\mathfrak{g}$ of type ADE. This last fact is due to the description of $\mathfrak{U}$ as the quotient $\mathfrak{h}/W$ of a Cartan subalgebra $\mathfrak{h}$ of $\mathfrak{g}$ by the corresponding Weyl group W (cf. e.g. [34]). The coincidence of the weights and degrees can of course also be checked by direct inspection.

A_ℓ	$(2, 3, 4, \ldots, \ell + 1)$
D_ℓ	$(2, 4, 6, \ldots, 2\ell - 2, \ell)$
E_6	$(2, 5, 6, 8, 9, 12)$
E_7	$(2, 6, 8, 10, 12, 14, 18)$
E_8	$(2, 8, 12, 14, 18, 20, 24, 30)$

Fig. 2. Degrees of the simple Lie algebras

Besides the automorphisms of the simple singularities X provided by these $\mathbf{C}^*$-actions, there are further symmetries of interest to us. These symmetries commute with $\mathbf{C}^*$ and, lifted to the minimal resolution $\tilde{X}$, they realise the full symmetry of the Coxeter-Dynkin diagram (given by the dual intersection diagram of the exceptional components).

	equation	*group*	*action of generators*
$A_{2\ell-1}$	$x^{2\ell} + yz = 0$	$\mathbf{Z}/2\mathbf{Z}$	$(x, y, z) \mapsto (-x, z, y)$
$A_{2\ell}$	$x^{2\ell+1} + yz = 0$	$\mathbf{Z}/4\mathbf{Z}$	$(x, y, z) \mapsto (-x, z, -y)$
D_ℓ	$x^{\ell-1} + xy^2 + z^2 = 0$	$\mathbf{Z}/2\mathbf{Z}$	$(x, y, z) \mapsto (x, -y, -z)$
E_6	$x^4 + y^3 + z^2 = 0$	$\mathbf{Z}/2\mathbf{Z}$	$(x, y, z) \mapsto (-x, y, -z)$
D_4	$x^3 + y^3 + z^2 = 0$	S_3	$(x, y, z) \mapsto (y, x, -z)$
			$(x, y, z) \mapsto (\omega x, \omega^2 y, z), \omega^3 = 1$

Fig. 3. Diagram symmetries of simple singularities

In the next section, we shall see that these types of symmetry are all 'essential' in the sense of [45, 43, 14].

3. Milnor fibre and monodromy

Let us start with some general constructions. Let X be an arbitrary isolated singularity (of some germ of an affine algebraic variety in $\mathbb{C}^n$). By general theory ([45, 43, 14]), there is a maximal reductive subgroup $R \subset Aut(X)$ of the automorphism group of X, unique up to conjugation, and, by further general theory ([28]), there is a R-semiuniversal deformation $\Phi : \mathfrak{X} \longrightarrow \mathfrak{U}$ of X. This means in particular that R acts on $\mathfrak{X}$ and $\mathfrak{U}$ rendering Φ R-equivariant. Let $\Gamma \subset R$ be any reductive subgroup. It is then easy to see ([34]) that the restriction $\Phi^\Gamma : \mathfrak{X} \times_\mathfrak{U} \mathfrak{U}^\Gamma \longrightarrow \mathfrak{U}^\Gamma$ of Φ over $\mathfrak{U}^\Gamma$ is Γ-invariant semiuniversal, i.e. semiuniversal in the category of all Γ-equivariant deformations of X with trivial action on the base space. Of course, if $u \in \mathfrak{U}^\Gamma$, then Γ acts on the fiber $\mathfrak{X}_u = \Phi^{-1}(u)$ over u and on all its associated objects, e.g. the (co-)homology groups of $\mathfrak{X}_u$.

A singularity X is called *smoothable* if, in its semiuniversal deformation $\Phi : \mathfrak{X} \longrightarrow \mathfrak{U}$, there are smooth fibres $\mathfrak{X}_u$ for arbitrarily small $u \in \mathfrak{U}$.

Let us now restrict to the situation of our interest, that of hypersurfaces of even dimension $2m$, where smoothings and their (co-)homology are well understood (cf. e.g. [24, 3, 22], for details on the following development, one might as well take complete intersections into account). Let $\mathfrak{D} \subset \mathfrak{U}$ denote the set of points $u \in \mathfrak{U}$ such that $\mathfrak{X}_u$ is singular. It is called the *discriminant* of Φ. Then all smoothings $\mathfrak{X}_u, u \notin \mathfrak{D}$, are diffeomorphic as fibres of a differentiable fibre bundle $\dot\Phi : \dot{\mathfrak{X}} \longrightarrow \dot{\mathfrak{U}}$, $\dot{\mathfrak{U}} = \mathfrak{U} \backslash \mathfrak{D}$, $\dot{\mathfrak{X}} = \Phi^{-1}(\dot{\mathfrak{U}})$, $\dot\Phi = \Phi_{|\dot{\mathfrak{X}}}$, (sufficiently localized), the *Milnor fibration*, each of whose fibers is called 'the' *Milnor fibre* M of X. The middle homology $\Lambda = H_{2m}(M, \mathbb{Z})$ is a free lattice of finite rank μ, the *Milnor number* of X. There is a symmetric, possibly degenerate, integral bilinear form, the *intersection form* $\langle,\rangle$, on Λ. The pair $(\Lambda, \langle,\rangle)$ is called the *Milnor lattice* of X.

The geometry of the deformation also provides a distinguished class Σ of elements of Λ, that of *vanishing* cycles. In fact, there are geometric procedures to single out *geometric bases* of vanishing cycles. These are bases $\{e_1, ..., e_\mu\}$ of the lattice Λ, consisting of members of Σ, in particular, all elements have self intersection $\langle e_i, e_i \rangle = 2(-1)^m$. The remaining intersection numbers, which determine the Milnor lattice, are usually encoded in an *intersection diagram*, also called Dynkin diagram, of X. These choices are far from canonical, but different choices may be linked by means of appropriate transformations (related to the braid group on μ strings). In the case of the simple singularities, the Milnor lattice $(\Lambda, \langle,\rangle)$ may be identified with the pair $(Q, -\kappa)$, where Q is the root lattice and $-\kappa$ is the negative of the normalized Killing form on Q of the type of X. Moreover, one can find (non-canonical) geometric bases realizing the standard Coxeter-Dynkin diagram.

By means of a geometric basis $\Delta = \{e_1, ..., e_\mu\} \subset \Lambda$, one can determine an important topological invariant of the singularity, its monodromy group. Note that the lattices $H_{2m}(\mathfrak{X}_u, \mathbb{Z})$, for u varying in $\dot{\mathfrak{U}}$, form a local system with fiber Λ over $\dot{\mathfrak{U}}$, dual to the locally constant sheaf $R^{2m}\dot{\Phi}_*\mathbb{Z}_{\dot{\mathfrak{X}}}$. Correspondingly, after fixing a base point $u \in \dot{\mathfrak{U}}$, there is a *monodromy representation*

$$m : \pi_1(\dot{\mathfrak{U}}, u) \longrightarrow Aut(\Lambda, \langle, \rangle)$$

of the fundamental group of $\pi_1(\dot{\mathfrak{U}}, u)$ on the Milnor lattice $(\Lambda, \langle, \rangle)$. The image of this representation in $Aut(\Lambda, \langle, \rangle)$ shall be denoted by W. It is called the *monodromy group* of X. As a consequence of the *Picard-Lefschetz-formulae*, this group is generated by the orthogonal reflections in the elements of Δ,

$$W = \langle s_e \mid e \in \Delta \rangle , \quad s_e(\lambda) = \lambda - (-1)^m \langle \lambda, e \rangle e, \ \lambda \in \Lambda.$$

In the case of a simple singularity of type A, D or E, one thus obtains the corresponding Weyl group W acting on the root lattice $\Lambda = Q$.

DEFINITION 3.1. *An automorphism $\gamma \in R \subset Aut(X)$ is called Milnor regular if there is a point $u \in \mathfrak{U}^\gamma \cap \dot{\mathfrak{U}}$, i.e. such that $\mathfrak{X}_u$ is nonsingular.*

The action of such a γ on $M = \mathfrak{X}_u$ will induce an automorphism $\mu(\gamma)$ of the Milnor lattice $(\Lambda, \langle, \rangle)$, preserving the set of vanishing cycles Σ and thus normalizing W. The element $\mu(\gamma)$ in the normalizer $N(W) = N_{O(\Lambda, \langle, \rangle)}(W)$ depends on many non-canonical choices, only its conjugacy class is uniquely determined. Note that because of the connectivity of $\mathfrak{U}^\gamma \cap \dot{\mathfrak{U}}$ (this is a Zariski-open subset of the vector space $\mathfrak{U}^\gamma$) it does not depend on the choice of the point u in $\mathfrak{U}^\gamma \cap \dot{\mathfrak{U}}$.

DEFINITION 3.2. *Let $\gamma \in Aut(X)$ be Milnor regular. The conjugacy class of $\mu(\gamma) \in N(W)$ is called the Milnor class of γ.*

Because of the choices involved, the association of Milnor regular elements in $Aut(X)$ to $N(W)$ cannot and should in no way be compared to a homomorphism of groups! In fact, later we will exhibit a pair of commuting automorphisms, $\gamma_i, i = 1, 2$, of the simple singularity of type E_8 such that no pair of representatives of the $\mu(\gamma_i)$ in $N(W)$ will commute. The following is well known (cf. e.g. [4]).

LEMMA 3.3. *Let W be the Weyl group of a root system of type $\Delta = A, D, E$. Then the normalizer $N(W)$ of W inside the orthogonal group of (Q, κ) is isomorphic to the semidirect product of W and $Aut(\Delta)$, where $Aut(\Delta)$ denotes the symmetry group of the diagram Δ.*

On the side of singularities, this result is parallelled by the following.

LEMMA 3.4. *Let X be a simple singularity with Coxeter-Dynkin diagram $\Delta \neq A_{even}$. Any maximal reductive subgroup of $Aut(X)_{red} \subset Aut(X)$ is isomorphic to $\mathbb{C}^* \times Aut(\Delta)$, where $\mathbb{C}^*$ is the natural $\mathbb{C}^*$-action. In case $\Delta = A_{even}$, the group $Aut(X)_{red}$ is a non-trivial central extension of $Aut(\Delta) \cong \mathbb{Z}/2\mathbb{Z}$ by $\mathbb{C}^*$.*

One can re-establish the uniformity of this result, in case A_{even}, by then regarding the quotient group $\overline{Aut_{red}(X)} = Aut(X)_{red}/\langle -1 \rangle$ ($-1 \in \mathbb{C}^*$) which is again isomorphic to a product $\mathbb{C}^* \times Aut(\Delta)$.

In the cases $\Delta \neq A_{even}$, the subgroup of $Aut_{red}(X)$ isomorphic to $Aut(\Delta)$ is, of course, not uniquely defined. Henceforth we shall stick to the choice provided by the explicit table (Figure 3 in section 2). This subgroup may be characterized by the fact that it induces $Aut(\Delta)$ on the minimal resolution and that it acts freely on $X \setminus 0$ (cf. [34], 6.2). This last property may also be replaced by the fact that its action on the base $\mathfrak{U} \cong \mathfrak{h}/W$ of the semiuniversal deformation (cf. next section) induces the natural one of $N(W)/W \cong Aut(\Delta)$, cf. e.g. [34, 8.6]. In the case $\Delta = A_{even}$, the 'lifted' subgroup $\mathbb{Z}/4\mathbb{Z} \subset Aut_{red}(X)$ shares similar properties.

Thus, in case of the simple singularities X of type $\Delta \neq A_{odd}$ (resp. A_{odd}), we have an isomorphism of $N(W)/W$ with the component group $Aut_{red}(X)/Aut_{red}(X)^\circ$ (resp. $\overline{Aut_{red}(X)}/\overline{Aut_{red}(X)}^\circ$). Note also that for any $\gamma \in \mathbb{C}^* \subset Aut_{red}(X)$, the class $\mu(\gamma)$ lies in W. In fact, any path from 1 to γ inside $\mathbb{C}^*$ induces, when applied to $u \in \dot{\mathfrak{U}}$, an element in $\pi_1(\dot{\mathfrak{U}}, u)$ whose monodromy coincides with $\mu(\gamma)$.

4. Complex reflection groups

Monodromy groups like finite Weyl groups are generated by reflections in a real orthogonal space. Complex reflection groups are analogues in unitary spaces. Let V be a complex vector space of dimension k. An endomorphism s of V is called a *complex reflection* of order $d \in \mathbb{N}$ if it fixes (pointwise) a hyperplane of V and is of order d. In this case, s is semisimple and its eigenvalues are 1 with multiplicity $k - 1$ and ε where ε is a primitive d-th root of unity. A *complex reflection group* is a subgroup G of $GL(V)$ generated by complex reflections. Assume such a G is finite. It then leaves invariant a hermitian scalar product $\langle \, , \, \rangle$ on V, and any reflection $s \in G$ of order d may be written in the form

$$s(v) = v - (1 - \varepsilon)\langle v, e \rangle e$$

for a suitable vector (a 'root vector') e of unit length and a primitive d-th root of unity ε. Therefore, such an s is also called a *unitary reflection* and G a *unitary reflection group*, cf. e.g. [31]. Any real orthogonal reflection in a euclidean vector space yields a unitary one by extension of scalars. Thus all finite Coxeter groups provide examples of unitary reflection groups. Trivial examples are provided by the cyclic subgroups of $\mathbb{C}^*$ acting on a complex line. The classification of all finite, irreducible complex reflection groups was achieved by Shephard and Todd [31], who listed 37 different classes (denoted G_n). Besides the symmetric and cyclic groups (types G_1 and G_3) there is one triply infinite family $G(m, p, n)$ (type G_2) generalizing the Coxeter groups of type B_k, D_k and $I_2(p)$, the remaining 34 classes consisting each of a single member, 6 of them real Coxeter groups (i.e. G_{23} (resp. $G_{28}, G_{30}, G_{35}, G_{36}, G_{37}$) equals H_3 (resp. F_4, H_4, E_6, E_7, E_8)) and 28 of them proper complex groups acting on spaces of dimensions 2, 3, 4, 5, 6 (19, 4, 3, 1, 1 cases resp.). The most relevant property of unitary reflection groups is stated in the following theorem due to Shephard-Todd and Chevalley (cf. [31], [9], [4]).

THEOREM 4.1. *Let G be a finite subgroup of $GL(V)$, $\dim_{\mathbb{C}}(V) = k$. Then the following properties are equivalent:*

i) G is a complex reflection group.

ii) The quotient variety $V /\!\!/ G$ is smooth.

iii) The algebra of G-invariant polynomials $\mathbb{C}[V]^G$ is a polynomial ring on k algebraically independent generators.

In the situation of the theorem one may choose homogeneous generators $P_i, i = 1, \ldots, k$, for $\mathbb{C}[V]^G$. Their degrees $d_i = \deg(P_i)$ are well defined and one has ([31], [4]):

$$|G| = \prod_{i=1}^{k} d_i \quad .$$

Usually these numbers are called the *degrees* of G; subtracting one one obtains the *exponents*, $m_i = d_i - 1$, of G.

Besides the set of exponents, there is also another set of invariant integers, the *co-exponents* n_i, $i = 1, \ldots, k$, introduced in [25] and tabulated e.g. in [26], p.287. They are the degrees of a basis of the free $\mathbb{C}[V]^G$-module $Mor_G(V, V)$ of polynomial G-covariants $V \longrightarrow V$ or, equivalently, of G-equivariant polynomial vector fields on V.

5. Regular elements in finite reflection groups

Let W be a complex reflection group acting on the complex vector space V.

DEFINITION 5.1. *An element $v \in V$ is called regular if v does not lie on any of the reflection hyperplanes of W in V or, in other words, if $W_v = \{1\}$.*

DEFINITION 5.2. *An element $g \in N(W)$ is called Springer regular if g has a regular eigenvector in V.*

Well known examples of regular elements are given by Coxeter elements and their powers in Weyl groups. Springer gave a complete classification of regular elements in all finite Coxeter groups W and their normalizers $N(W)$. For simplicity, let us first restrict to the case of W itself (for details and generalizations, cf. [37, 19, 20]).

THEOREM 5.3. *Let g be a regular element of order d in a complex reflection group W, and let $E = E(g, \zeta)$ be a regular eigenspace of g, i.e. assume there is a $\zeta \in \mathbb{C}^*$ such that $g(v) = \zeta v$, for all $v \in E$, and such there is a regular vector v in E. Then*

— ζ is a d-th primitive root of unity.

— $dim(E) = e(d)$, where $e(d)$ denotes the number of degrees of W which are divisible by d.

— The normalizer $N_W(E)$ of E in W equals the centralizer $Z_W(g)$ of g in W.

— The centralizer $Z_W(g)$ acts faithfully on E as an irreducible complex reflection group whose degrees are exactly those degrees of W that are divisible by d.

— The eigenvalues of g on V are the numbers ζ^{-m_i} where the m_i run through the exponents of W.

   — Let g' be another regular element and assume $dim E(g', \zeta) = e(d)$
   for the same ζ as for g. Then g and g' are conjugate in W.

   As a consequence, one can show that all regular elements of a given
order d in a Weyl group W are conjugate in W, cf. [37, 4.7, 5]. This is
no longer true for non-crystallographic Coxeter groups [37, 5, Tables 5,
6].

DEFINITION 5.4.   *The numbers d such that W contains a regular
element of order d, are called regular numbers.*

   Let us now look at regular elements in the normaliser $N(W) = \{n \in
GL(V) \mid nW = Wn\}$ of W. An element of finite order in $N(W)$ may be
written as a product $g\sigma$ where $g \in W$ and $\sigma \in N(W) \backslash W$ is an element
of finite order which belongs to a fixed set of representatives for the
elements of the quotient group $N(W)/W$. Note that, for a Weyl group
W, the quotient $N_{O(Q)}W/W$ of the integral normaliser $N_{O(Q)}$ (denoted
simply $N(W)$ in the sections above) may be identified with $Aut(\Delta)$. In
the cases A, D_{odd}, E_6, this normaliser is obtained by adjoining $-id_V$ to
W.
   The element σ acts naturally on V/W as a finite order transformation. It can be linearized, in fact diagonalised, i.e. the fundamental
invariants P_i can be chosen so that σ transforms P_i into $\epsilon_i.P_i$ for suitable
roots of unity ϵ_i, cf. [37, Lemma 6.1]. We fix such a choice from now
on.

THEOREM 5.5.   *Let $g\sigma$ be a regular element of finite order in $N(W)$
where W is a complex reflection group W, and let $E = E(g\sigma, \zeta)$ be a
regular eigenspace of $g\sigma$, i.e. assume there is a primitive d-th root of
unity $\zeta \in \mathbb{C}^*$ such that $g\sigma(v) = \zeta v$, for all $v \in E$, and such there is a
regular vector v inside E. Then*

   — If $\sigma^d = 1$, then $g\sigma$ has order d

   *— $dim(E) = e(d, \sigma)$, where $e(d, \sigma)$ denotes the number of degrees
   d_i of W satisfying $\epsilon_i \zeta^{d_i} = 1$.*

   *— The normalizer $N_W(E)$ of E in W equals the centralizer $Z_W(g\sigma)$
   of $g\sigma$ in W.*

   *— The centralizer $Z_W(g\sigma)$ acts faithfully on E as an irreducible
   complex reflection group whose degrees are exactly those degrees d_i
   of W satisfying $\epsilon_i.\zeta^{d_i} = 1$.*

   *— The eigenvalues of $g\sigma$ on V are the numbers $\epsilon^{-1}\zeta^{-m_i}$ where the
   m_i are running through the exponents of W.*

— Let $g'\sigma$ be another regular element and assume $dim E(g'\sigma, \zeta) = e(d, \sigma)$. Then $g\sigma$ and $g'\sigma$ are conjugate by an element of W.

Note that, in contrast to the case of regular elements in Weyl groups, non-conjugate regular elements of the form $g\sigma$ may happen to have the same order, cf. e.g. [37], 6.12, Table 8. The case $d = 3$ given there corresponds to an element of order 6. Thus there are two non-conjugate regular elements of order 6 in $N(W) \setminus W$.

6. Milnor regularity and Springer regularity

In this section, we want to relate the two concepts of regularity introduced above. Our singularity X will always be a simple singularity of type Δ.

Let us start with an automorphism $\gamma \in Aut_{red}(X) \subset Aut(X)$ of finite order. Besides the action of $\overline{Aut_{red}}$ on $V/W \cong \mathfrak{U} \cong T^1(X)$, there is also a natural action on $V = T^1(\tilde{X}) = H^1(\tilde{X}, T_X)$, normalizing W and covering that on V/W. In case A_{even}, both these actions are effective only for the quotient group $\overline{Aut_{red}(X)} \cong \mathbb{C}^* \times Aut(\Delta)$. To simplify procedures, we shall, occasionally, also write $\overline{Aut_{red}(X)}$ for $Aut(X)$ in the other cases. Note that then $\mathbb{C}^* \subset \overline{Aut_{red}(X)}$ acts on V by scalar transformations ([15, 27]). Due to the type of its induced action on V/W, the fixed subgroup $Aut(\Delta) \subset \overline{Aut_{red}(X)}$ will act on V effectively as a group representing the quotient $N(W)/W$ inside $N(W)$. We shall thus identify its image in $GL(V)$ with $Aut(\Delta)$ as well.

Assume now that γ is Milnor-regular. Then we shall associate with it a Springer-regular element $m(\gamma)$ in $N(W)$ which will be well defined up to W-conjugation.

Let $u = [v] \in (V/W)^\gamma \setminus \mathfrak{D}$ be a 'regular' fixed point of γ. The fixed point equality

$$\gamma([v]) = [v]$$

will hold if and only if the condition

for any $v \in [v]$, there is a $w(\gamma, v) \in W$ such that $\gamma(v) = w(\gamma, v)(v)$

is satisfied. To obtain a Springer regular element in $N(W)$, we split $\gamma \in \overline{Aut_{red}(X)} \cong \mathbb{C}^* \times Aut(\Delta)$ into commuting 'continuous' and 'discrete' parts, i.e. $\gamma = c.d$ with $c \in \mathbb{C}^*$ and $d \in Aut(\Delta)$ (here we assume the splitting of $Aut(X)$ discussed at the end of section 3, and we identify $\overline{Aut_{red}(X)}$ with a subgroup of $GL(V)$ as above). Define $m(\gamma) := d^{-1}.w(\gamma, v)$. Then this is a Springer regular element in $N(W)$ satisfying the 'regular eigenvector' equation

$$m(\gamma, v)(v) = c.v \quad .$$

Up to W-conjugacy, it depends only on the original automorphism γ, and we shall denote (a representative of) that W-class by $m(\gamma)$. Conversely, assume that σg is a Springer regular element in $N(W)$ where σ is chosen in the subgroup $Aut(\Delta) \subset GL(V)$ as above. Then, we have a regular eigenvector equation

$$\sigma g(v) = \zeta.v \ \ ,$$

for some primitive root of unity ζ. Or

$$g(v) = \sigma^{-1}\zeta.v \ \ ,$$

which means that the class $[v]$ of v in the quotient space V/W is fixed under the automorphism $\gamma(g) := \sigma^{-1}\zeta$ in $\overline{Aut_{red}(X)}$. If σ is trivial, this element has the same order as g. In any case, $\gamma(g)$ will be Milnor regular and, up to W-conjugacy, we have

$$m(\gamma(g)) = g \quad \text{and} \quad \gamma(m(\gamma)) = \gamma \ \ .$$

This proves

PROPOSITION 6.1.

  i) Let γ be Milnor regular. Then $m(\gamma)$ is Springer regular.

  ii) Let g be a Springer regular element in $N(W)$. Then there is a Milnor regular automorphism $\gamma \in \overline{Aut_{red}(X)}$ such that $m(\gamma) = g$ (up to W-conjugacy). If g is in W, the orders of g and γ coincide.

A next task will be to identify the W-class $m(\gamma)$ in $N(W)$ with the Milnor class $\mu(\gamma)$ of γ, i.e. with the transformation on the Milnor fibre induced by γ (up to W-conjugacy). This will be achieved, only partially, with the techniques used in the proof of the following Monodromy Theorem. In case γ preserves the defining function f of X, one might also invoke a comparison of the actions of γ on $\mathfrak{U}$ and on $H_2(M, \mathbb{Z})$ (cf. [44]). A complete proof might possibly follow from a thorough investigation of the naturality properties (equivariance with respect to the full group $Aut_{red}(X)$) of suitable period maps for simple singularities, cf. [21, 22].

Let γ be Milnor regular with class $g = m(\gamma) \in N(W)$ as defined above. Furthermore, let $\mathfrak{U}^\gamma$ denote the fixed space of γ in the base $\mathfrak{U}$ of the semiuniversal deformation of X with restricted discriminant locus $\mathfrak{D}^\gamma = \mathfrak{D} \cap \mathfrak{U}$. Choose a base point $u \in \mathring{\mathfrak{U}}^\gamma = \mathfrak{U}^\gamma \setminus \mathfrak{D}^\gamma$. On the other hand, consider a regular eigenspace $E(g) \subset V$ of g with the natural action of $N_W(E(g)) = Z_W(g)$ and quotient $E(g) \to E(g)/Z_W(g)$. Accordingly, we have a 'reflection' discriminant $\mathfrak{D}^g \subset E(g)/Z_W(g)$.

THEOREM 6.2. *By restriction, the monodromy representation*

$$\pi_1(\dot{\mathfrak{U}}, u) \to W$$

induces an epimorphism

$$\pi_1(\dot{\mathfrak{U}}^\gamma, u) \to Z_W(g) ,$$

and the pairs of algebraic varieties $(\mathfrak{U}^\gamma, \mathfrak{D}^\gamma)$ *and* $(E(g)/Z_W(g), \mathfrak{D}^g)$ *are isomorphic.*

For the proof, one needs some generalities on local systems, cf. Lemma, below (the relevant construction was written up long ago in the Bonn thesis [42, §6]). From this, one first gets an inclusion of the 'reflection' discriminant $\mathfrak{D}^g$ into the 'singularity' discriminant $\mathfrak{D}^\gamma$. The coincidence can then be proved by elementary monodromy arguments or, alternatively, by a relevant identification result from [10], Theorem 2.5, or [18], Cor. 5.8, cf. also [19, 20]. Note that this settles, in particular, Conjecture 4.7 in [13], made there with respect to E_8 and G_{31}. The argument also applies to subdeformations not related to automorphisms, like those of Shcherbak [30, 38].

LEMMA 6.3. *Let* Γ *be a group acting freely on a topological space* U, *and let* $\rho : \Gamma \longrightarrow GL(M)$ *be a representation of* Γ, *giving rise to the 'local system'* $\mathcal{L} = U \times^\Gamma M$ *on* U/Γ. *Let* $U' \subset U$ *be a subspace with normalizer* $\Gamma' = N_\Gamma(U') = \{\gamma \in \Gamma \,|\, \gamma(U') = U'\}$. *Assume that* U'/Γ' *embeds as a subspace of* U/Γ. *Then the restricted system* $\mathcal{L}_{|(U'/\Gamma')}$ *is isomorphic to* $U' \times^{\Gamma'} M$, *where* Γ' *acts by the restricted representation on* M.

We now come to the identification of $\mu(\gamma)$ and $m(\gamma)$.

PROPOSITION 6.4. *Let* $\gamma \in \mathbb{C}^* \subset \overline{Aut_{red}(X)}$ *be a Milnor regular automorphism of* X. *Then the Milnor class* $\mu(\gamma)$ *coincides with the class* $m(\gamma)$. *In particular, they have the same order.*

We may assume that $m(\gamma)$ is primitive, i.e. not a power of another regular element in W or, alternatively, γ generates the stabiliser of u in $\mathbb{C}^*$. Let $v \in V$ be a preimage of $u \in \mathfrak{U}^\gamma$ which is a regular eigenvector for $m(\gamma)$. Then the stabiliser of $\mathbb{C}.v$ in W identifies with the subgroup generated by $m(\gamma)$ (use the regularity of v and primitivity). Now we may apply the above Lemma once more to identify the cyclic subgroups of W generated by $m(\gamma)$ and $\mu(\gamma)$. This implies our claim.

7. Examples

In the following, we shall give a list of all regular automorphism of the exceptional simple singularities $X = \{f = 0\}$ (up to conjugation in the automorphism group $Aut(X)$). We should point out that this list is more comprehensive than but consistent with the one in [12] (which concentrates on a minimal list of monodromy groups occurring) and results quite directly from our conceptual approach.

7.1. E_8

Let us start with the singularity $\{x^5 + y^3 + z^2 = 0\}$ of type E_8. There are no symmetries of the Coxeter-Dynkin diagram, and $Aut_{red}(X) \cong \mathbb{C}^*$ acts with weights $(6, 10, 15)$. The degrees are $(2, 8, 12, 14, 18, 20, 24, 30)$, which are the complements in 30 of the degrees of the 8 unfolding monomials

$$x^3 y, x^2 y, x^3, xy, x^2, y, x, 1.$$

One may now use Springer's classification, or a simple check on non-singularity of deformations, to find all regular numbers. In the following table, we list all the regular elements g in the Weyl group $W(E_8)$, their orders, and their attached complex reflection groups (using the numbering of [31]). The identification of these groups can be mainly made by using their degrees. In a few cases, one should also invoke the permanence properties of the co-exponents [10, 19, 20]. Only the case of G_{10} requires a further, more direct geometric argument to distinguish it from the member $G(12, 1, 2)$ of the Shepherd-Todd class G_2. For a very explicit and constructive discussion, see [12].

The corresponding γ-semiuniversal deformations of X are easily written down by unfolding the defining function f of X in the directions of the monomials whose degrees complement the d-divisible degrees to 30. E.g. let γ be a Milnor regular automorphism of order 3. Then the total space of the γ-semiuniversal deformation of X is given by

$$\{x^5 + y^3 + z^2 + ux^3 + vx^2 + wx + t = 0\} \subset \mathbb{C}^7$$

with projection onto the base space $\mathbb{C}^4$ being given by the four functions u, v, w, t.

g	$d = order(g)$	d – divisible degrees	$Z_W(g)$
$c^{30} = id$	1	$2, 8, 12, 14, 18, 20, 24, 30$	E_8
$c^{15} = -id$	2	$2, 8, 12, 14, 18, 20, 24, 30$	E_8
c^{10}	3	$12, 18, 24, 30$	G_{32}
$D_4(a_1)^2$	4	$8, 12, 20, 24$	G_{31}
c^6	5	$20, 30$	G_{16}
$c^5 = -c^{10}$	6	$12, 18, 24, 30$	G_{32}
$D_8(a_3)$	8	$8, 24$	G_9
$c^3 = -c^6$	10	$20, 30$	G_{16}
$E_8(a_3)$	12	$12, 24$	G_{10}
c^2	15	30	$\mathbf{Z}/30\mathbf{Z}$
$E_8(a_2)$	20	20	$\mathbf{Z}/20\mathbf{Z}$
$E_8(a_1)$	24	24	$\mathbf{Z}/24\mathbf{Z}$
c	30	30	$\mathbf{Z}/30\mathbf{Z}$

Fig. 4. Regular automorphisms for E_8

The element c denotes a Coxeter element. Otherwise, we have used Carter's notation, [8], for conjugacy classes in the Weyl group.

Note that the centralizer $Z_W(c)$ of the Coxeter element c is the cyclic group of order 30 generated by c. Therefore, no W-conjugate of the regular element of order 24 can commute with c. In contrast, corresponding geometric automorphisms in $\mathbf{C}^* \subset Aut(X)$ do.

7.2. E_7

Here is the table for the simple singularity $\{x^3 y + y^3 + z^2 = 0\}$ of type E_7. Again, there are no symmetries of the Coxeter-Dynkin diagram and $Aut_{red}(X) \cong \mathbf{C}^*$. The weights are $(4, 6, 9)$ and the degrees $(2, 6, 8, 10, 12, 14, 18)$.

g	$d = order(g)$	$d-$ divisible degrees	$Z_W(g)$
$c^{18} = id$	1	$2, 6, 8, 10, 12, 14, 18$	E_7
$c^9 = -id$	2	$2, 6, 8, 10, 12, 14, 18$	E_7
c^6	3	$6, 12, 18$	G_{26}
c^3	6	$6, 12, 18$	G_{26}
$E_7(a_1)^2 = A_6$	7	14	$\mathbb{Z}/14\mathbb{Z}$
c^2	9	18	$\mathbb{Z}/18\mathbb{Z}$
$E_7(a_1)$	14	14	$\mathbb{Z}/14\mathbb{Z}$
c	18	18	$\mathbb{Z}/18\mathbb{Z}$

Fig. 5. Regular automorphisms for E_7

7.3. E_6

In the case of the simple singularity $\{x^4+y^3+z^2 = 0\}$ of type E_6 we have two tables according to the additional symmetry of the Coxeter-Dynkin diagram. The weights are $(3, 4, 6)$ and the degrees $(2, 5, 6, 8, 9, 12)$. Let us first tabulate the regular automorphisms in $\mathbb{C}^*$.

g	$d = order(g)$	$d-$ divisible degrees	$Z_W(g)$
$c^{12} = id$	1	$2, 5, 6, 8, 9, 12$	E_6
c^6	2	$2, 6, 8, 12$	F_4
c^4	3	$6, 9, 12$	G_{25}
c^3	4	$8, 12$	G_8
c^2	6	$6, 12$	G_5
D_5	8	8	$\mathbb{Z}/8\mathbb{Z}$
$E_6(a_1)$	9	9	$\mathbb{Z}/9\mathbb{Z}$
c	12	12	$\mathbb{Z}/12\mathbb{Z}$

Fig. 6. Regular 'inner' automorphisms for E_6

The group $N(W)$ is obtained by adjoining the transformation $\sigma = -id_V$ to W. Multiplication by that element permutes the two W-cosets of $N(W)$ as well as the respective regular elements. Under that permutation, centralisers of elements are obviously preserved. The fundamental invariants of even (resp. odd) degree are invariant (resp. are transformed into their negatives) under σ. The following table is trivially derived from the table (Fig. 6) of the 'inner' automorphisms.

σg	$d = order(\zeta)$	$degrees\ of\ centraliser$	$Z_W(\sigma g)$
$-c^{12} = -id$	2	$2,5,6,8,9,12$	E_6
$-c^6$	2	$2,6,8,12$	F_4
$-c^4$	3	$6,9,12$	G_{25}
$-c^3$	4	$8,12$	G_8
$-c^2$	6	$6,12$	G_5
$-D_5$	8	8	$\mathbb{Z}/8\mathbb{Z}$
$-E_6(a_1)$	9	9	$\mathbb{Z}/9\mathbb{Z}$
$-c$	12	12	$\mathbb{Z}/12\mathbb{Z}$

Fig. 7. Regular 'outer' automorphisms for E_6

7.4. D_4

Traditionally, the case of D_4 with an automorphism σ of order 3 is considered as belonging to the exceptional types. Without loss of generality, let us fix one such automorphism. Here is the list of the corresponding regular elements in the coset $W\sigma$ denoted by their corresponding elements in $Aut_{red}(X)$. Here, ω (resp. ϵ) will denote a primitive third (resp. twelfth) root of unity. For more explicit information on their form inside $N(W)$, see [37, p.184].

$\sigma\gamma$	$d = order(\zeta)$	$degrees\ of\ centraliser$	$Z_W(\sigma\gamma)$
$\sigma.-1$	2	$2,6$	$I_2(6) = W(G_2)$
$\sigma.\omega$	3	$4,6$	G_4
$\sigma.-\omega$	6	$4,6$	G_4
$\sigma.\epsilon$	12	4	$\mathbb{Z}/4\mathbb{Z}$

Fig. 8. Regular exceptional 'outer' automorphisms for D_4

7.5. A AND D

Let us finally make a few remarks on the cases of type A and D.

In case A_ℓ, all regular elements of W are powers of the Coxeter element c, itself of order $\ell+1$, or a sub-Coxeter element c' of order ℓ, cf. [37], 5.1. Corresponding deformations and monodromy groups (centralisers) have been written down by Goryunov, [11, 13, 12]. Since

$N(W)$ is obtained by adjoining $-id$ to W (for $\ell > 1$), the outer regular elements are easily reduced to the inner ones and no new centralisers (monodromy groups) occur, cf. also [37, 6.9].

In case D_ℓ, the situation is similar. All regular elements in W are powers of two elements, the Coxeter element of order $2\ell - 2$ and a sub-Coxeter element of order ℓ, [37], 5.3. For ℓ odd, the normaliser $N(W)$ is obtained by adjoining $-id$, thus regular elements are easily reduced to those of W. For ℓ even, one has to adjoin another element, however there is no essential effect on the centraliser or monodromy groups. This is most easily seen via the singularities, [12], where the deformations are listed (with respect to outer automorphisms). For the Weyl group theoretic aspects, cf. [37, 6.11].

References

1. V. I. Arnold, Normal forms for functions near degenerate critical points, the Weyl groups A_k, D_k, E_k, and Lagrangian singularities, Funct. Anal. Appl. **27** (1972) 254–272.

2. V. I. Arnold, Critical points of functions on a manifold with boundary, the simple Lie groups B_k, C_k, and F_4, and singularities of evolutes, Russian Math. Surveys **33**:5 (1978) 99–116.

3. V. I. Arnold, S. M. Gusein-Zade and A. N. Varchenko, *Singularities of differentiable maps Vol. I*, Monographs in Math. **82**, Birkhäuser, Basel, Boston, 1985.

4. N. Bourbaki, *Groupes et algèbres de Lie, Chaps. IV, V, VI*, Hermann, Paris, 1968.

5. E. Brieskorn, Singular elements of semisimple algebraic groups, *Actes Congr. Int. Math., Nice 1970*, tome **2** pp 279–284.

6. M. Broué and G. Malle, Zyklotomische Heckealgebren, Astérisque **212** (1993) 119–189.

7. M. Broué and J. Michel, Sur certains élements réguliers des groupes de Weyl et les variétés de Deligne-Lusztig associées, *Finite reductive groups (Luminy 1994)*, Progress in Math. **141**, (Birkhäuser, Basel, 1997) pp 73–139.

8. R.W. Carter, Conjugacy classes in the Weyl groups, Compositio Math. **25** (1972) 1–59.

9. C. Chevalley, Invariants of finite groups generated by reflections, Amer. J. Math. **77** (1955) 778–782.

10. J. Denef and F. Loeser, Regular elements and monodromy of discriminants of finite reflection groups, Indagationes Math. N.S. **6** (1995) 129–143.

11. V. V. Goryunov, Unitary reflection groups associated with singularities of functions with cyclic symmetry, Russ. Math. Surveys **54** (1999) 873-893.

12. V. V. Goryunov, Unitary reflection groups and automorphism groups of simple hypersurface singularities, this volume.

13. V. V. Goryunov and C. E. Baines, Cyclically equivariant function singularities and unitary reflection groups $G(2m, 2, n), G_9, G_{31}$, St. Petersburg Math. J. **11**:5 (2000) 761-774.

14. H. Hauser and G. Müller, Algebraic singularities have maximal reductive automorphism groups, Nagoya Math. J. **113** (1989) 181–186.

15. F. Huikeshoven, On the versal resolutions of deformations of rational double points, Invent. Math. **20** (1973) 15-33.

16. G. Kempken, *Eine Darstellung des Köchers $\tilde{A}_k$*, Bonner Math. Schriften **137** (1982).

17. B. Kostant and S. Rallis, Orbits and representations associated with symmetric spaces, Amer. J. Math. **93** (1971) 753–809.

18. G. Lehrer, Poincaré series for unitary reflection groups, Invent. Math. **120** (1995) 411–425.

19. G. Lehrer and T. A. Springer, Intersection multiplicities and reflection subquotients of unitary reflection groups I, *Geometric group theory down under* (Proceeedings of a special year in Geometric Group Theory, Canberra, Australia 1996), (eds Cossey et al.) (de Gruyter, Berlin, 1999) pp 181–194.

20. G. Lehrer and T. A. Springer, Reflection subquotients of unitary reflection groups, Canadian J. Math. **51** (1999) 1175-1193.

21. E. J. N. Looijenga, A period map for certain semi-universal deformations, Compositio Math. **30** (1975) 299–316.

22. E. J. N. Looijenga, *Isolated singular points on complete intersections*, London Math. Soc. Lecture Notes **77**, Cambridge Univ. Press, 1984.

23. O. V. Lyashko, Classification of critical points of functions on a manifold with singular boundary, Funct. Anal. Appl. **17** (1983) 187–193.

24. J. Milnor, *Singular points of complex hypersurfaces*, Ann. Math. Studies **61**, Princeton University Press, 1974.

25. P. Orlik and L. Solomon, Unitary reflection groups and cohomology, Invent. Math. **59** (1980) 77–94.

26. P. Orlik and H. Terao, Arrangements of hyperplanes, Grundlehren der math. Wiss. **300**, Springer Verlag, Berlin, 1992.

27. H. C. Pinkham, Singularités de Klein I, II; Singularités rationelles de surfaces; Résolution simultanée de points double rationnels, *Séminaire sur les singularités des surfaces* (eds M. Demazure, H. Pinkham and B. Teissier), Springer Lecture Notes in Math **777** (1980), 1-20, 147-178, 179-203.

28. D. S. Rim, Equivariant G-structure on versal deformations, Trans. Amer. Math. Soc. **257** (1980) 217–226.

29. K. Saito, Period mapping associated to a primitive form, Publ. Math. RIMS **19** (1983) 1231–1260.

30. O. P. Shcherbak, Wave fronts and reflection groups, Russian Math. Surveys **43**:3 (1988) 125–160.

31. G. C. Shephard, and J. A. Todd, Finite unitary reflection groups, Canad. J. Math. **6** (1954) 274–304.

32. D. Siersma, Singularities of functions on boundaries, corners, etc., Quart. J. Math. **32** (1981) 119–127.

33. P. Slodowy, Einige Bemerkungen zur Entfaltung symmetrischer Funktionen, Math. Zeits. **158** (1978) 157–170.

34. P. Slodowy, *Simple singularities and simple algebraic groups*, Springer Lecture Notes in Math. **815** (1980).

35. P. Slodowy, *Four lectures on simple groups and singularities*, Communications of the Math. Institute **11** (1980), Rijksuniversiteit Utrecht.

36. P. Slodowy, Singularitäten, Kac-Moody Liealgebren, assoziierte Gruppen und Verallgemeinerungen, Habilitationsschrift, Universität Bonn, 1984.

37. T. A. Springer, Regular elements of finite reflection groups, Invent. math. **25** (1975) 159–198.

38. A. N. Varchenko and S. V. Chmutov, Finite irreducible groups generated by reflections are monodromy groups of appropriate singularities, Funct. Anal. Appl. **18** (1984) 171–183.

39. E. B. Vinberg, On the linear groups associated to periodic automorphisms of semisimple algebraic groups, Soviet Math. Dokl. **16**:2 (1975) 406–409.

40. E. B. Vinberg, On the classification of the nilpotent elements of graded Lie algebras, Soviet Math. Doklady **16**:6 (1975) 1517–1520.

41. E.B. Vinberg, The Weyl group of a graded Lie algebra, Math. USSR – Izvestija **10** (1976) 463–495.

42. E. Voigt, *Ausgezeichnete Basen von Milnorgittern einfacher Singularitäten*, Bonner Math. Schriften **160** (1985).

43. J. Wahl, Derivations, automorphisms, and deformations of quasihomogeneous singularities, *Singularities* (ed Peter Orlik), Proc. Symp. Pure Math. **40**:2, (Amer. Math. Soc., 1983) pp 613–624.

44. C. T. C. Wall, A note on symmetry of singularities, Bull. London Math. Soc. **12** (1980) 169–175.

45. C. T. C. Wall, A second note on symmetry of singularities, Bull. London Math. Soc. **12** (1980) 347–354.

46. T. Yano, Free deformations of isolated singularities, Sci. Rep. Saitama Univ. Ser. A **9** (1980) 61–70.

47. T. Yano, Invariants of finite reflection groups and flat coordinate systems of isolated singularities, (in Japanese), RIMS Kokyuroku **444** (1981) 209–235.

48. T. Yano, Deformations of rational double points associated with unitary reflection groups, Sci. Rep. Saitama Univ. Ser. A **10** (1981) 7–9.

Part C: Singularities of holomorphic maps

Discriminants, vector fields and singular hypersurfaces

Andrew A. du Plessis
Matematisk Institut, Aarhus Universitet
(matadp@imf.au.dk)

C. T. C. Wall
Dept. of Math. Sci., University of Liverpool
(ctcw@liv.ac.uk)

Introduction

The part of singularity theory concerned with geometrical properties of discriminants of mappings is particularly rich, with numerous interrelations between concepts of importance in different applications. The object of this article is to describe some of the basic theory, and to proceed to applications which have arisen in current work of the authors.

We begin with a rather down-to-earth introduction to the discriminant, and to its relations with liftable vector fields, in a fairly concrete case, and describe the now classical algorithm for these. We then summarise the main geometrical properties of discriminants of stable mappings.

As we have explained in the survey article [10] and in [7], vector fields on the discriminant may be used to obtain several insights into the singularities that occur: numerical estimates for Tjurina numbers; in favourable cases, precise determination of the types of these singularities; and certain results on versality and even topological versality of partial unfoldings.

In the main part of this paper we sketch the application of these ideas to hypersurfaces in projective spaces: most of the work is directed at a study of the possible Tjurina numbers, in terms of the lowest degree r of a polynomial vector field leaving the variety invariant.

Many of the arguments are purely algebraic, so we introduce some algebraic terminology, and make precise the relation between the algebraic and geometric contexts. We establish some basic algebraic properties, which underlie our applications. Next we obtain estimates for the lengths of certain modules. We apply these to obtain information about the sum of the Tjurina numbers of a hypersurface. The results establish interesting properties for the cases of curves (first obtained in [11]) and surfaces; for high dimensions the best estimates we have at present

351

D. Siersma et al. (eds.), New Developments in Singularity Theory, 351–377.
© 2001 Kluwer Academic Publishers. Printed in the Netherlands.

are too weak to be really interesting, and we present a preliminary conjecture. The results admit an alternative interpretation as theorems of Cayley-Bacharach type in projective geometry (see [17], [12]).

In the final section we return to the discriminant, but consider the set of unstable points of a non-versal unfolding. This is characterised following [9] by a property closely related to the considerations of the preceding sections.

1. The discriminant

Let $f_0 : \mathbb{C}^n \to \mathbb{C}$ have an isolated singular point at the origin. Then the Jacobian ideal $J f_0 := \langle \partial f_0/\partial x_1, \ldots, \partial f_0/\partial x_n \rangle$ has finite codimension in the ring $\mathcal{O}_x$ of convergent power series in the x variables. This codimension is an important invariant, called the *Milnor number* and denoted $\mu = \mu(f_0)$. The *Tjurina number* $\tau(f_0)$, which is also important for us, is defined to be $\dim_{\mathbb{C}} \mathcal{O}_x/(J f_0 + \langle f_0 \rangle)$.

Choose a basis $\phi_0 = 1, \phi_1, \ldots, \phi_{\mu-1}$ of the quotient vector space $\mathcal{O}_x/J f_0$ consisting of monomials, arranged in some order of non-decreasing degree. Define the unfolding $F(\mathbf{x}, \mathbf{u}) = (y, \mathbf{v}) = (f(\mathbf{x}, \mathbf{u}), \mathbf{u})$, where $\mathbf{x} = (x_1, \ldots, x_n)$, $\mathbf{u} = (u_1, \ldots, u_{\mu-1})$, $\mathbf{v} = (v_1, \ldots, v_{\mu-1})$, and $f(\mathbf{x}, \mathbf{u}) = f_0(\mathbf{x}) + \sum_{i=1}^{\mu-1} u_i \phi_i(\mathbf{x})$.

Write $\mathcal{O}_u, \mathcal{O}_{u,x}$ for the rings of convergent power series in the u variables and u, x variables respectively, and $J_x f := \langle \partial f/\partial x_1, \ldots, \partial f/\partial x_n \rangle$ for the indicated ideal in $\mathcal{O}_{u,x}$. It follows from Nakayama's lemma that the ϕ_i form a free basis of the quotient $\mathcal{O}_u$-module $\mathcal{O}_{u,x}/J_x f$, so there exist uniquely determined elements $a_{i,j} \in \mathcal{O}_u$ such that, for some functions $b_{i,k}(\mathbf{x}, \mathbf{u}) \in \mathcal{O}_{u,x}$, we have equations (for $0 \leq i < \mu$)

$$f(\mathbf{x}, \mathbf{u})\phi_i \equiv \sum_{j=0}^{\mu-1} a_{i,j}(\mathbf{u})\phi_j + \sum_{k=1}^{n} b_{i,k}(\mathbf{x}, \mathbf{u})\frac{\partial f}{\partial x_k}. \tag{1}$$

Then the transpose of $A = (a_{i,j}(\mathbf{v}) - y\,\delta_{i,j})$ is called the discriminant matrix.

The above procedure allows in principle explicit computation of the discriminant matrix: a simple example is given below. However the full matrix soon gets too complicated to be of interest. Many computations for partial unfoldings have been obtained — see, for example, [8, §3.9], and numerous computer calculations in [15].

The vector field ξ_i defined on the source of F by

$$\xi_i = \sum_{k=1}^{n} b_{i,k}(\mathbf{x}, \mathbf{u})\frac{\partial}{\partial x_k} + \sum_{j=1}^{\mu-1}(a_{i,j}(\mathbf{u}) - f(\mathbf{x}, \mathbf{u})\delta_{i,j})\frac{\partial}{\partial u_j}$$

lifts the vector field η_i defined on the target by

$$\eta_i = \sum_{j=0}^{\mu-1} (a_{i,j}(\mathbf{v}) - y\delta_{i,j}) \frac{\partial}{\partial v_j},$$

where $\partial/\partial v_0$ denotes $-\partial/\partial y$; in fact we replace the above y by $-v_0$ from now on.

Here 'lifting' means that for each point $(\mathbf{x}, \mathbf{u})$ in the source, we have $dF(\xi_i)$ at that point equal to the value of η_i at the image point $F(\mathbf{x}, \mathbf{u})$.

Example. Let $f_0(x) = x^3$. Then $Jf_0 = \langle 3x^2 \rangle$. We have $\mu(f_0) = 2$, and take $\phi_0 = 1$, $\phi_1 = x$: thus $f(x, u) = x^3 + u_1 x$. We find the relations
$$x^3 + u_1 x = 0(1) + (\tfrac{2}{3}u_1)(x) + (\tfrac{1}{3}x)(3x^2 + u_1),$$
$$(x^3 + u_1 x)x = (-\tfrac{2}{9}u_1^2)(1) + (0)(x) + (\tfrac{1}{3}x^2 + \tfrac{2}{9}u_1)(3x^2 + u_1),$$
giving discriminant matrix $\begin{pmatrix} -y & \tfrac{2}{3}u_1 \\ -\tfrac{2}{9}u_1^2 & -y \end{pmatrix} = \begin{pmatrix} v_0 & \tfrac{2}{3}v_1 \\ -\tfrac{2}{9}v_1^2 & v_0 \end{pmatrix}$ and vector fields

$$\xi_0 = (\tfrac{1}{3}x)\partial/\partial x + (\tfrac{2}{3}u_1)\partial/\partial u_1,$$
$$\xi_1 = (\tfrac{1}{3}x^2 + \tfrac{2}{9}u_1)\partial/\partial x + (0 - (x^3 + u_1 x))\partial/\partial u_1,$$
$$\eta_0 = v_0\partial/\partial v_0 + \tfrac{2}{3}v_1\partial/\partial v_1,$$
$$\eta_1 = (-\tfrac{2}{9}v_1^2)\partial/\partial v_0 + v_0\partial/\partial v_1.$$

The above construction of F follows a procedure essentially due to Mather [22]: it only yields a germ F, but we may choose a representative map which is stable in the sense defined below. The construction of vector fields is due to Saito [25], with details added by Bruce [2]. It yields a correspondence between the vector fields and functions (in $\mathcal{O}_x/Jf_0$), which allows one to define a multiplication on the space of vector fields. This is the first step in the construction of a Frobenius manifold structure on the unfolding space — see e.g. [20].

We now introduce notations for modules of vector fields in general. Let $G : N \to P$ be a map; write
$$\Sigma(G) := \{\mathbf{x} \in N \mid dG_x \text{ is } not \text{ surjective}\}$$
for the singular set of G and $\Delta(G) := G(\Sigma(G))$ for its discriminant. Write θ_N for the module of vector fields on the source N, θ_P for the module of vector fields on the target P, and $\theta(G)$ for the module of vector fields along G. These definitions may be interpreted in terms of germs of vector fields at a point, of sheaves of such germs, or of their global sections. Here we will mostly think of germs, but later on some of our arguments will be global.

We have a diagram of tangent bundles

$$TN \overset{TG}{\to} TP$$
$$\downarrow \qquad \downarrow$$
$$N \overset{G}{\to} P$$

and interpret θ_N as the set of sections to the projection $TN \to N$, θ_P as the set of sections to $TP \to P$, and $\theta(G)$ as the set of maps from N to TP making the lower triangle commute. Composition (on the right) with TG defines a map $tG : \theta_N \to \theta(G)$; composition (on the left) with G defines a map $\omega G : \theta_P \to \theta(G)$.

In the case when G is the above map-germ F we can identify θ_N with the free $\mathcal{O}_{x,u}$-module having the basis $\partial/\partial x_1, \ldots, \partial/\partial x_n, \partial/\partial u_1, \ldots, \partial/\partial u_{\mu-1}$, θ_P with the free $\mathcal{O}_v$-module with basis $\partial/\partial v_0, \partial/\partial v_1, \ldots, \partial/\partial v_{\mu-1}$, and $\theta(F)$ with the free $\mathcal{O}_{x,u}$-module with basis $\partial/\partial v_0, \partial/\partial v_1, \ldots, \partial/\partial v_{\mu-1}$. In this notation, tF is the $\mathcal{O}_{x,u}$-module map induced by partial differentiation: $\partial/\partial x_i \to \sum_j (\partial v_j/\partial x_i)\partial/\partial v_j$; and ωF is given by substituting u_i for v_i and $-f(\mathbf{x}, \mathbf{u})$ for v_0. The statement that ξ lifts η is expressed by the equation $tF(\xi) = \omega F(\eta)$.

The geometrical properties of the discriminant are simplest in the case when G is a stable map. We refer to our book [15] for a more thorough discussion of stability: for the purposes of this article, as we are concerned with the complex analytic case, we make an *ad hoc* definition. Say that G is stable if it satisfies the local condition that, for the germs at any point $\mathbf{x} \in N$, or more generally at any finite subset $S \subset N$ with a common image, $tG(\theta_N) + \omega G(\theta_P) = \theta(G)$; and the global condition that the restriction of G to $\Sigma(G)$ is a finite, proper map.

A normal form (up to change of local coordinates) for stable map-germs was given by Mather [22]: for the case when TG has rank $\dim P - 1$ this is just the form of F above. The above calculations of vector fields were generalised to the general case (with $\dim N > \dim P$) by Goryunov [18].

We define two points $\mathbf{y}$ and $\mathbf{y}'$ in the target of a stable map $G : N \to P$ to be *equivalent* if the germs of G at the sets $\Sigma(G, \mathbf{y})$ and $\Sigma(G, \mathbf{y}')$ are $\mathcal{K}$- and hence $\mathcal{A}$-equivalent (here $\Sigma(G, \mathbf{y})$ denotes $\Sigma G \cap G^{-1}(\mathbf{y})$). In more detail: there is a bijection $\Sigma(G, \mathbf{y}) \to \Sigma(G, \mathbf{y}')$ such that at corresponding points the local algebras $\mathcal{O}_x/(G^*\mathfrak{m}_y.\mathcal{O}_x)$ are isomorphic. The equivalence classes form a partition of P. Saito calls this the logarithmic stratification of $\Delta(G)$, but it is not a stratification in the usual sense: rather, it is a foliated stratification. We refer to the parts as *leaves*.

We now state a number of major results. We formulate them for stable maps with $\dim N > \dim P$, but they hold much more generally: we give references for the various extensions.

THEOREM 1.1. *(i) The discriminant $\Delta(G)$ is a reduced complex space of dimension $p - 1$ and $G| : \Sigma(G) \to \Delta(G)$ is its normalisation.*
(ii) A vector field on P is liftable under G if and only if it is tangent to the discriminant $\Delta(G)$.
(iii) The leaf L_y containing $\mathbf{y}$ is smooth there, with tangent space given by the values at $\mathbf{y}$ of the liftable vector fields.
(iv) The codimension of L_y equals the sum τ_y of the Tjurina numbers of the singularities of G at the points of $\Sigma(G, \mathbf{y})$.
(v) The space $Derlog(\Delta(G))$ of vector fields tangent to $\Delta(G)$ is a free module over the functions on P. For the above map-germ F, the η_i form a free basis.
(vi) If $G_1 : N_1 \to P$ and $G_2 : N_2 \to P$ have $\Delta(G_1) = \Delta(G_2)$, then there is a holomorphic equivalence $\phi : N_1 \to N_2$ with $G_2 \circ \phi = G_1$.

Proof. Statement (i) is originally due to Teissier [28]: see also [21, 4.7]. It is true much more generally: see [8].

Both (ii) and (iii) are due to Saito [25]: (ii) is generalised in [21, 6.14], with an ultimate generalisation in [3].

The relation (iv) with Tjurina numbers is an elementary consequence of the local normal form, and has been long known. Details are given, for example, in [10].

For assertion (v) the significance of the freeness was first recognised by Saito [25], special cases having been noted previously by Arnold [1]. The proof was given for F as above by Bruce [2] and Terao [27] and in general by Goryunov [18]. A substantial generalisation of the result giving freeness is due to Damon [4]; see also his article [6] in this volume.

Statement (vi) is due to Wirthmüller [29]. It has been generalised several times, more effort being required for results holding in the real (as opposed to complex) case. The final version is given in [8].

Numerous other authors have made important contributions to the theory: a somewhat fuller bibliography is in [10].

A point $\mathbf{v}$ of the target of F corresponds to the map f_v; write τ_v for the sum of the Tjurina numbers at the singular points of $f_v^{-1}(0)$. Applying (iii) and (iv) of Theorem 1.1 to this case, and using the fact that μ_0 is the dimension of the target, we obtain at once

LEMMA 1.2. *The following numbers are equal:*
(i) the difference $\mu_0 - \tau_v$;
(ii) the dimension of the leaf through $\mathbf{v}$;
(iii) the dimension of the span of the vectors η_i at $\mathbf{v}$.

The following result points the way to our applications. First note that we can interpret the statement that the ϕ_i form a free basis over

$\mathcal{O}_u$ of $\mathcal{O}_{u,x}/J_x f$ in terms of the corresponding (coherent) sheaves. It follows that for v in some neighbourhood U of the origin in $\mathbb{C}^\mu$, the ϕ_i form a $\mathbb{C}$-basis of $\mathcal{O}_x/Jf_v$.

LEMMA 1.3. *([9, 1.1]) The linear relation $\sum c_i(a_{i,j} - y\delta_{i,j}) = 0$ between the columns of the discriminant matrix holds at a point $\mathbf{v} \in U$ of the target of F if and only if the function $g = \sum c_i\phi_i$ satisfies the condition $gf_v \in Jf_v$, where f_v is defined by $f_v(\mathbf{x}) = f(\mathbf{x}, \mathbf{v}) + v_0$.*

Proof. The linear relation shows that substituting the values v for u in the corresponding linear combination of equations (1) reduces this to

$$f(\mathbf{x}, \mathbf{v}) \sum c_i \phi_i = \left(\sum c_i \phi_i\right) y + \sum_{k=1}^n c_i b_{i,k}(\mathbf{x}, \mathbf{v}) \frac{\partial f(\mathbf{x}, \mathbf{v})}{\partial x_k},$$

so that g satisfies the stated condition.

Conversely, if $gf_v \in Jf_v$ we must have $\sum c_i(a_{i,j} - y\delta_{i,j})\phi_j \in Jf_v$. Since, by the remark preceding the Lemma, the ϕ_i are independent modulo Jf_v, it follows that $\sum c_i(a_{i,j} - y\delta_{i,j}) = 0$ for each j.

Write Φ for the vector space spanned by the ϕ_i. Consider the ideal in $\mathbb{C}[\mathbf{x}]$ consisting of the g such that $g.f_v \in Jf_v$: we denote this (using the notation introduced below) as the colon ideal $(Jf_v : f_v)$. It follows from Lemma 1.2 and Lemma 1.3 that

COROLLARY 1.4. *We have*

$$\mu_0 - \tau_v = \dim_\mathbb{C} \frac{\Phi}{\Phi \cap (Jf_v : f_v)}.$$

Now assign (positive integer) weights w_i to the variables x_i and suppose f_0 homogeneous of degree D with respect to these weights. Assign weights $D - wt(\phi_i)$ to the unfolding parameters u_i so that $f(\mathbf{x}, \mathbf{u})$ is again homogeneous. These weights may well be negative. We have $\tau_0 = \mu_0 = \prod_{i=1}^n (\frac{D - w_i}{w_i})$. In particular, $wt(\phi_{\mu-1})$ is the degree of the Hessian of f_0, thus equals $nD - 2\sum_1^n w_i$.

Consider that part of the unfolding — let us call it the *positive unfolding* — where the parameters u_i of weight ≤ 0 all vanish. Thus for v in the positive unfolding space, f_v is obtained from f_0 by adding terms of lower weight. The process of solving linear equations which one uses to find relations (1) explicitly on this subspace will produce polynomials in the remaining u variables, as we see by induction on degree.

LEMMA 1.5. *Let v be a point in the positive unfolding space. Then*

$$\tau_0 - \tau_v = \dim_\mathbb{C} \frac{\mathbb{C}[\mathbf{x}]}{(Jf_v : f_v)}.$$

Proof. It will suffice to show that $\mathbb{C}[\mathbf{x}] = \Phi + Jf_v$, for then we have $\mathbb{C}[\mathbf{x}] = \Phi + (Jf_v : f_v)$, and the result follows from Corollary 1.4.

Let $p \in \mathbb{C}[\mathbf{x}]$; we induct on the weight w of p. Write p_0 for the sum of terms of highest weight. Since f_0 is homogeneous, we may write $p_0 = \sum a_i \partial f_0 / \partial x_i + \sum c_j \phi_j$, where a_i is homogeneous of weight $w + w_i - D$ and $c_j = 0$ unless $\mathrm{wt}(\phi_j) = w$. Then the highest weight in $p - \sum a_i \partial f_v / \partial x_i - \sum c_j \phi_j$ is strictly less than w. The assertion thus follows by induction.

2. From geometry to algebra

We will continue to assume weighted homogeneity. More precisely, although much of the algebra to follow makes sense in the weighted case, our best results and also the geometric interpretations assume that all weights are 1. We thus suppose from now on that f_0 is in fact homogeneous.

Suppose, as in the previous section, that f_v arises by deforming f_0 using only unfolding monomials of degree $< D$. Make f_v homogeneous by introducing an additional coordinate x_0, of weight 1, taking $x^+ = (x_0, x)$ as coordinates in $\mathbb{C}^{n+1}$, and setting $F(x_0, x) = x_0^D f_v \left(\frac{x_1}{x_0}, \ldots, \frac{x_n}{x_0} \right)$. Then F is a homogeneous polynomial of degree D defining a hypersurface V in projective space P^n, which meets the hyperplane H given by $x_0 = 0$ transversely. The hypersurface V has isolated singularities and $V \cap H$ is a smooth variety. We also suppose $v \neq 0$: equivalently, f_v is not an equisingular deformation of f_0; i.e. the variety V is not a cone.

The invariant τ_v defined above is equal to the sum $\tau(V)$ of the Tjurina numbers at the singular points of V, since V has no singularities on H. From now on we may drop the subscript v.

We will say that a hypersurface V as above is a *good* hypersurface.

It will be more convenient for our abstract arguments to give an algebraic formulation. First we introduce some standard terminology and establish further notation.

The ideal generated by a collection $g_1, \ldots, g_r$ of elements of a ring R is denoted $\langle g_1, \ldots, g_r \rangle$. Particular ideals in R will be denoted $I, J, \ldots$. If A, B are ideals in R, the *colon ideal* is defined by $(A : B) := \{a \in R : aB \subseteq A\}$. If B is monogenic we simplify the notation and set $(A : f) := (A : \langle f \rangle)$.

A sequence $\{f_1, \ldots, f_r\}$ of elements of a ring R is called *regular* if the ideal $\langle f_1, \ldots, f_r \rangle$ of R is proper and, for each $i = 1, \ldots, r$, f_i is not a zero-divisor in $R/\langle f_1, \ldots, f_{i-1} \rangle$.

We will mainly work in the ring $\mathbb{C}[x_0, x_1, \ldots, x_n]$, and from now on reserve the letter R to denote this ring. The sequence $\{f_1, \ldots, f_r\}$ in R is regular if and only if the locus of their common zeros has dimension $n + 1 - r$ as algebraic subset of $\mathbb{C}^{n+1}$. Thus here regularity does not depend on the ordering of the sequence.

We denote by $\tilde{R}$ the quotient ring

$$R/\langle x_0 - 1 \rangle \cong \mathbb{C}[x_1, \ldots, x_n],$$

and by $\overline{R}$ the quotient ring

$$R/\langle x_0 \rangle \cong \mathbb{C}[x_1, \ldots, x_n].$$

For any ideal I in R we denote by $\tilde{I}, \overline{I}$ its respective images in $\tilde{R}$ and $\overline{R}$; similarly denote the images of $g \in R$ by $\overline{g} \in \overline{R}$, $\tilde{g} \in \tilde{R}$. We will need to know what happens to colon ideals under quotients.

LEMMA 2.1. *For any ideals A, B in R, we have*

$$\widetilde{(A : B)} \subseteq (\tilde{A} : \tilde{B}) \text{ and } \overline{(A : B)} \subseteq (\overline{A} : \overline{B}).$$

If A and B are homogeneous ideals in R, then $\widetilde{(A : B)} = (\tilde{A} : \tilde{B})$.

Proof. If $f \in (A : B)$, then $fB \subseteq A$, so $\tilde{f}\tilde{B} \subseteq \tilde{A}$. Hence $\tilde{f} \in (\tilde{A} : \tilde{B})$. Similarly in $\overline{R}$.

Suppose A and B are homogeneous and $\tilde{g} \in (\tilde{A} : \tilde{B})$. Make g homogeneous: if g has degree d, then set $h(x_0, \ldots, x_n) := x_0^d g(1, \frac{x_1}{x_0}, \ldots, \frac{x_n}{x_0})$. Then h is homogeneous and $\tilde{h} = \tilde{g}$. Thus $\tilde{h}.\tilde{B} \subseteq \tilde{A}$, so $hB \subseteq A + \langle x_0 - 1 \rangle$.

For each homogeneous $b \in B$ we write $gb = a + c(x_0 - 1)$ with $a \in A$. Since A is homogeneous, each graded component a_i of a lies in A. Thus homogenising a produces another element $\alpha = \sum x_0^{D-i} a_i$ of A, and $gb - \alpha$ is still divisible by $x_0 - 1$. If $\deg \alpha - \deg gb = k > 0$, then $x_0^k gb - \alpha$ is a homogeneous polynomial which is divisible by $x_0 - 1$, and hence is identically zero; thus $x_0^k gb \in A$. If $k \leq 0$ then $gb - x_0^{-k}\alpha$ is homogeneous and divisible by $x_0 - 1$, and hence is zero, so $gb \in A$.

Apply this to each of a finite set $\{b_j\}$ of generators of B. If K is the largest of the values of k that appears, then $x_0^K g$ multiplies each b_j into A, so lies in $(A : B)$. Hence $\tilde{g} \in \widetilde{(A : B)}$.

COROLLARY 2.2. *If $f_1, \ldots, f_r$ is a regular sequence of homogeneous elements in R, then unless $\langle \tilde{f}_1, \ldots, \tilde{f}_r \rangle = \tilde{R}$, $\tilde{f}_1, \ldots, \tilde{f}_r$ is a regular sequence in $\tilde{R}$.*

Proof. For each i $(1 \leq i \leq r)$, set $I_i := \langle f_1, \ldots, f_{i-1} \rangle$. Since the sequence is regular we have, for each i, $(I_i : f_i) = I_i$. As I_i is a homogeneous ideal, it follows from the lemma that $(\tilde{I}_i : \tilde{f}_i) = \widetilde{(I_i : f_i)} = \tilde{I}_i$. Since $\tilde{I}_i = \langle \tilde{f}_1, \ldots, \tilde{f}_{i-1} \rangle$, it follows that the sequence $\tilde{f}_1, \ldots, \tilde{f}_r$ is regular.

If I, J are ideals in R with $I \subseteq J$, we will write $\ell(J/I)$ for the length of the quotient indicated. In view of our particular choice of R, this is equal to $\dim_{\mathbb{C}} (J/I)$.

For the next two sections we will work with the following *algebraic hypothesis*.

(A): We have homogeneous elements $f_0, \ldots, f_n$ of R, with $\deg f_i = p$, such that $\overline{f_1}, \ldots, \overline{f_n}$ is a regular sequence in $\overline{R}$. Write $K := \langle f_1, \ldots, f_n \rangle$, and assume $f_0 \notin K$.

LEMMA 2.3. *The function F defines a good hypersurface V if and only if the sequence $f_i := \partial F/\partial x_i$ satisfies (A), with $p = D - 1$.*

Proof. The function F is homogeneous of degree D if and only if the $f_i := \partial F/\partial x_i$ are homogeneous of degree $p = D - 1$.

Next, $\overline{f_1}, \ldots, \overline{f_n}$ is a regular sequence in $\overline{R}$ if and only if $\overline{f_1}, \ldots, \overline{f_n}$ have no common zero in H, i.e. $V \cap H$ is non-singular. Thus the singularities of V are isolated and H is transverse to V.

If f_0 is a linear combination of the other f_i, we have a vector field $\xi = \sum_0^n c_i \partial/\partial x_i$, with c_0 a non-zero constant, annihilating F: there exists such a vector field if and only if V is a cone.

Relaxing the geometric hypothesis to the algebraic hypothesis will give us extra freedom to manipulate our elements.

3. Basic results

In this section we suppose R graded by $\mathrm{wt}(x_i) = w_i$, f_i homogeneous of degree p_i, $f_0 \notin K := \langle f_1, \ldots, f_n \rangle$ and $\overline{f_1}, \ldots, \overline{f_n}$ a regular sequence.

Given a graded ring S and elements $\{\alpha_1, \ldots, \alpha_n\}$ in S, with $deg(\alpha_i) = a_i$, we introduce the exterior algebra Λ over S on n generators θ_i ($1 \le i \le n$), bigraded by $\theta_i \to (1, a_i)$: we will call the first component the *grade* and the second the *weight*. Write Λ_r for the summand where the grade is r: this vanishes unless $0 \le r \le n$. We convert Λ into a differential graded algebra by setting $d\theta_i = \alpha_i \in S = \Lambda_0$: we then call it the Koszul complex $\Lambda(\alpha)$. Thus $\Lambda_r(\alpha)$ is the free S-module on symbols θ_I where I runs over ordered subsets

$$I: \quad (1 \le)i_1 < i_2 < \ldots < i_r(\le n),$$

and

$$\theta_I = \theta_{i_1} \wedge \ldots \wedge \theta_{i_r}; \quad \partial\theta_I = \sum_{s=1}^r (-1)^{s-1}\alpha_{i_s}\theta_{I_s},$$

where I_s is obtained from I by deleting the sth entry, i_s. The differential d lowers grade by 1 and preserves weight.

It is a standard (and easy) fact (see [16, 17.5]) that if $\{\alpha_1, \ldots, \alpha_n\}$ is a regular sequence, the Koszul complex $\Lambda(\alpha)$ has zero homology in positive grades — we will then call it *acyclic*.

By hypothesis, $\{\overline{f_1}, \ldots, \overline{f_n}\}$ is a regular sequence in the ring $\overline{R}$ so its Koszul complex $\Lambda(\overline{f})$ is acyclic, and so is a free resolution over $\overline{R}$ of the module $\overline{R}/K$. Since the generators are homogeneous, we see that the length $\ell(\overline{R}/K) = \tau_0 := \prod_1^n (p_i/w_i)$. It follows that $\{f_1, \ldots, f_n\}$ is a regular sequence in R, so its Koszul complex $\Lambda(f)$ is a free resolution over R of R/K.

LEMMA 3.1. *The module R/K is free of rank τ_0 over the subring* $\mathbb{C}[x_0]$.

Proof. The exact sequence $0 \to R \xrightarrow{x_0} R \to \overline{R} \to 0$ induces an exact sequence

$$0 \to \Lambda(f) \xrightarrow{x_0} \Lambda(f) \longrightarrow \Lambda(\overline{f}) \to 0$$

of chain complexes. This has an exact homology sequence which reduces, since both complexes are acyclic, to

$$0 \to R/K \xrightarrow{x_0} R/K \longrightarrow \overline{R}/\overline{K} \to 0.$$

Thus multiplication by x_0 gives an injective self-map of R/K. But this last is a positively graded $\mathbb{C}[x_0]$-module, where x_0 is homogeneous of grade 1. Since the ring $\mathbb{C}[x_0]$ is euclidean, hence a principal ideal ring, the conclusion follows by a routine argument — indeed, the only graded prime ideal is $\langle x_0 \rangle$.

Factoring out by $\langle x_0 - 1 \rangle$ instead gives an exact sequence

$$0 \to \Lambda(f) \xrightarrow{x_0-1} \Lambda(f) \longrightarrow \Lambda(\widetilde{f}) \to 0$$

of chain complexes. Thus $\Lambda(\widetilde{f})$ is acyclic, and the exact homology sequence reduces to

$$0 \to R/K \xrightarrow{x_0-1} R/K \longrightarrow \widetilde{R}/\widetilde{K} \to 0$$

with $\widetilde{R}/\widetilde{K}$ of dimension τ_0 over $\mathbb{C}$: of course, these modules are not graded.

LEMMA 3.2. *(i) $(K : f_0)/K$ is a free $\mathbb{C}[x_0]$-module,*
(ii) The quotient $R/(K : f_0)$ is a free $\mathbb{C}[x_0]$-module.

Proof. (i) Since $(K : f_0)/K$ is a submodule of the free $\mathbb{C}[x_0]$-module R/K, multiplication on it by x_0 is injective. Thus it is free, as before.

(ii) The ideal $(K : f_0)$ is the kernel of the map $R \to R/K$ induced by multiplication by f_0. Hence $R/(K : f_0)$ is isomorphic to the image of the map, which is a submodule of the free $\mathbb{C}[x_0]$-module R/K. The result again follows.

We obtain greater insight into the role of these modules by considering the Koszul complexes. Write $\Lambda^+(f)$ for the Koszul complex of the sequence $f_0, \ldots, f_n$ (so here f_0 is included). Then (see [16, Cor 17.11]) we have an exact sequence

$$0 \to \Lambda(f) \xrightarrow{i} \Lambda^+(f) \xrightarrow{j} \Lambda(f) \to 0$$

of chain complexes. Here i is the inclusion of the subcomplex of those θ_I with $0 \notin I$ and j is the projection to the quotient by these, composed with 'division' by θ_0: thus j lowers grade by 1 and weight by p_0.

It follows from acyclicity of $\Lambda(f)$ that the only non-zero terms in the homology exact sequence are those that appear in

$$0 \to H_1(\Lambda^+(f)) \to H_0(\Lambda(f)) \to H_0(\Lambda(f)) \to H_0(\Lambda^+(f)) \to 0, \quad (2)$$

where the centre terms may be identified with R/K, and the map between them with multiplication by f_0. In the geometric case, the vanishing of higher homology groups of $\Lambda^+(f)$ is a special case of a well known result: see e.g. [19], [24]. We have seen that the exact sequence (2) is isomorphic to

$$0 \to (K : f_0)/K \longrightarrow R/K \xrightarrow{f_0} R/K \longrightarrow R/(K + \langle f_0 \rangle) \to 0. \quad (3)$$

The final term here is not a free $\mathbb{C}[x_0]$-module; for example, if $f_0, \ldots, f_n$ is a regular sequence in R, it has finite dimension $\prod_0^n \frac{p_i}{w_i}$.

Write T for the rank of the free $\mathbb{C}[x_0]$-module $R/(K : f_0)$. Then the ranks of the terms of (2) as $\mathbb{C}[x_0]$-modules are $\tau_0 - T, \tau_0, \tau_0$ and $\tau_0 - T$ respectively; all but the last are free. Moreover, the module $(K : f_0)/K$ has rank $\tau_0 - T$. Since $R/(K : f_0)$ is free, we have

$$\ell(\widetilde{R}/\widetilde{(K : f_0)}) = \ell(\overline{R/(K : f_0)}) = T$$

and since by Lemma 2.1, $\widetilde{(K : f_0)} = (\widetilde{K} : \widetilde{f_0})$, we also have $\ell(\widetilde{R}/(\widetilde{K} : \widetilde{f_0})) = T$. We will see in Proposition 5.1 that in the geometric situation of a good hypersurface, we have $\tau = \tau_0 - T$, so the estimates of T we obtain in the next section will yield estimates of $\tau(V)$.

We conclude this section by making the connection between (2) and (3) more explicit. We may identify the module Λ_1^+ with the module Vec of sequences $\{\alpha_0, \ldots, \alpha_n\}$ of polynomials, and $d : \Lambda_1^+ \to \Lambda_0^+$ as the map taking the sequence to the sum $\sum_0^n \alpha_i f_i$, so that $Ker\, d : \Lambda_1^+ \to \Lambda_0^+$ is the module of relations between the f_i. The images in Λ_1^+ of the generators $\theta_i \wedge \theta_j \in \Lambda_2^+$ are the trivial relations $f_i \theta_j - f_j \theta_i = 0$, so we may regard $Im\, d : \Lambda_2^+ \to \Lambda_1^+$ as the module of trivial relations.

Now define $\pi_0 : Vec \to \mathbb{C}[x^+]$ by $\pi_0\{\alpha_0, \ldots, \alpha_n\} = \alpha_0$, i.e. as the projection to the first factor. As map $\Lambda_1^+ \to \Lambda_0^+$ it lowers grade by 1 and weight by p_0.

LEMMA 3.3. *We have*
$$\pi_0(Ker\, d : \Lambda_1^+ \to \Lambda_0^+) = (K : f_0), \qquad \pi_0(Im\, d : \Lambda_2^+ \to \Lambda_1^+) = K$$
and π_0 induces an isomorphism of $H_1(\Lambda^+(f))$ onto $(K : f_0)/K$.

Proof. This observation follows by tracing the maps which induce the above identifications.

4. Estimates for the Tjurina number

In this section we will assume hypothesis (A). We begin with two lemmas.

LEMMA 4.1. *Let J be an ideal in a ring S; let $f_1 g_2 + f_2 g_1 \in J$, and suppose neither f_1 nor g_1 is a zero-divisor in S/J. Then $(J + \langle f_1 \rangle :$ $f_2) = (J + \langle g_1 \rangle : g_2)$.*

Proof. Suppose $a \in (J + \langle f_1 \rangle : f_2)$: then af_2 is congruent mod J to a multiple of f_1: say $af_2 \equiv bf_1 \pmod{J}$. We have $f_1 g_2 + f_2 g_1 \equiv 0$: multiplying by a and substituting $af_2 \equiv bf_1$ gives $f_1(bg_1 + ag_2) \equiv 0$. Since f_1 is not a zero divisor mod J, $bg_1 + ag_2 \equiv 0$. Thus $a \in (J + \langle g_1 \rangle : g_2)$.

Since the situation is symmetric, the opposite inclusion follows from the same argument.

We will use this lemma in cases where J has a list of generators which remain a regular sequence when either f_1 or g_1 is appended.

LEMMA 4.2. *([9, 3.1]) Let $f = (f_1, \ldots, f_n)$ be a regular sequence of homogeneous elements of R; let $(h_1, \ldots, h_r)$ be a further regular sequence of homogeneous polynomials. Then there exist $\mathbb{C}$-linear combinations $g_{r+1}, \ldots, g_n$ of $f_1, \ldots, f_n$ such that $(h_1, \ldots, h_r, g_{r+1}, \ldots, g_n)$ is a regular sequence.*

A more general version appears in [12, 2.6].

We define r to be the least degree of a non-zero homogeneous element of $(K : f_0)$ and $\bar{r}$ to be the least degree of a non-zero homogeneous element of $(\overline{K} : \overline{f_0})$. These parameters will play an important role.

LEMMA 4.3. *We have $0 \le \bar{r} \le r \le p$ and $0 < r$.*

Proof. Choose a non-zero homogeneous element α_0 of degree r in $(K : f_0)$. If $\overline{\alpha_0} = 0$, write $\alpha_0 = x_0^k \beta$ with $\beta \neq 0$. By (ii) of Lemma 3.2, $\beta \in (K : f_0)$, contradicting the definition of r. Thus $0 \neq \overline{\alpha_0} \in (\overline{K} : \overline{f_0})$, so $\bar{r} \le r$.

Now $r \leq p$ since $(K : f_0) \supseteq K$, and hence contains the f_i; and $0 < r$ since (A) implies $f_0 \notin K$.

The relation between r and $\bar{r}$ is quite subtle. We can prove that if $n \geq 3$ and $r \leq \frac{p}{2}$, and we make a generic change of coordinates (thus replacing H by a generic hyperplane), then in the new setup $r = \bar{r}$.

We have $\bar{r} = 0$ if and only if $\overline{f_0} \in \overline{K}$. Thus subtracting a linear combination (with constant coefficients) of the other f_i from f_0 will reduce $\overline{f_0}$ to 0.

THEOREM 4.4. *Assume hypothesis* (A). *Then* (i) *if* $\bar{r} \neq 0$, *we have*

$$\bar{r} p^{n-1} \geq T \geq \bar{r}(p - \bar{r}) p^{n-2}.$$

(ii) In all cases

$$r p^{n-1} \geq T \geq r(p - r) p^{n-2}.$$

Proof. (i) By definition of $\bar{r}$ there exists a non-zero $\overline{\alpha_0} \in (\overline{K} : \overline{f_0})$ of degree $\bar{r}$, thus there is a relation $\sum_0^n \overline{\alpha_i f_i} = 0$, where we may suppose each α_i homogeneous of degree $\bar{r}$.

By Lemma 4.2, there exist linear combinations $\overline{f_2'}, \ldots, \overline{f_n'}$ of $\overline{f_2}, \ldots, \overline{f_n}$ (with constant coefficients) such that $\overline{\alpha_0}, \overline{f_2'}, \ldots, \overline{f_n'}$ is a regular sequence (note that this fails if $\bar{r} = 0$ since then $\alpha_0 = 1$). Thus we may adjust $\{f_1, \ldots, f_n\}$ by a linear transformation and suppose $\overline{\alpha_0}, \overline{f_2}, \ldots, \overline{f_n}$ a regular sequence.

Define $L := \langle \alpha_0, f_2, \ldots, f_n \rangle$. Since $\overline{L}$ is generated by a regular sequence of homogeneous elements, we have $\ell(\overline{R}/\overline{L}) = \bar{r} p^{n-1}$. Applying Lemma 4.1 to the relation $\overline{\alpha_0 f_0} + \overline{\alpha_1 f_1} \in \langle \overline{f_2}, \ldots, \overline{f_n} \rangle$ gives $(\overline{K} : \overline{f_0}) = (\overline{L} : \overline{\alpha_1})$. We have an exact sequence

$$0 \to \overline{R}/(\overline{L} : \overline{\alpha_1}) \to \overline{R}/\overline{L} \to \overline{R}/(\overline{L} + \langle \overline{\alpha_1} \rangle) \to 0. \tag{4}$$

Thus we may calculate

$$\ell(\overline{R}/(\overline{K} : \overline{f_0})) = \ell(\overline{R}/(\overline{L} : \overline{\alpha_1})) = \ell(\overline{R}/\overline{L}) - \ell(\overline{R}/(\overline{L} + \langle \overline{\alpha_1} \rangle)). \tag{5}$$

Arguing similarly, we have

$$T = \ell(\tilde{R}/(\widetilde{K} : \widetilde{f_0})) = \ell(\tilde{R}/\tilde{L}) - \ell(\tilde{R}/(\tilde{L} + \langle \widetilde{\alpha_1} \rangle)).$$

Now by Lemma 3.1, R/L is free as a $\mathbb{C}[x_0]$-module. It follows that $\ell(\tilde{R}/\tilde{L}) = \ell(\overline{R}/\overline{L}) = \bar{r} p^{n-1}$. Hence $T \leq \bar{r} p^{n-1}$.

LEMMA 4.5. *In this situation,* $\overline{\alpha_0}$ *and* $\overline{\alpha_1}$ *have no common factor of positive degree, so* $\overline{\alpha_0}, \overline{\alpha_1}$ *is a regular sequence.*

Proof. Suppose h a common factor; write $\overline{\alpha_0} = h\overline{\beta_0}, \overline{\alpha_1} = h\overline{\beta_1}$. Then $\overline{\beta_1}\overline{\alpha_0} = h\overline{\beta_0}\overline{\beta_1} = \overline{\beta_0}\overline{\alpha_1}$, so $\overline{\beta_0} \in (\overline{L} : \overline{\alpha_1})$, which by the lemma equals $(\overline{K} : \overline{f_0})$. But we assumed $\overline{\alpha_0}$ an element of least degree in this ideal.

Continuing the proof of Theorem 4.4, we apply Lemma 4.2 to see that, after making a suitable linear transformation on $\{f_2, \ldots, f_n\}$, we may suppose that $\overline{\alpha_0}, \overline{\alpha_1}, \overline{f_3}, \ldots, \overline{f_n}$ is a regular sequence.

Set $M := \langle \alpha_0, \alpha_1, f_3, \ldots, f_n \rangle$. Although this ideal has no intrinsic significance, we will find that our estimates are obtained by studying properties of M. Now $L + \langle \alpha_1 \rangle = M + \langle f_2 \rangle$. Since $\overline{M}$ is generated by a regular sequence of homogeneous elements, $\ell(\overline{R}/\overline{M}) = \overline{r}^2 p^{n-2}$. We have an exact sequence, similar to (4),

$$0 \to \overline{R}/(\overline{M} : \overline{f_2}) \to \overline{R}/\overline{M} \to \overline{R}/(\overline{M} + \langle \overline{f_2} \rangle) \to 0. \tag{6}$$

Thus we may calculate

$$\ell(\overline{R}/(\overline{L} + \langle \overline{\alpha_1} \rangle)) = \ell(\overline{R}/(\overline{M} + \langle \overline{f_2} \rangle)) = \ell(\overline{R}/\overline{M}) - \ell(\overline{R}/(\overline{M} : \overline{f_2})). \tag{7}$$

Since $T = \ell(\overline{R}/(\overline{K} : \overline{f_0})) \geq \ell(\overline{R}/(\overline{K} : \overline{f_0}))$, assembling our results now gives

$$T \geq \overline{r}p^{n-1} - \ell(\overline{R}/(\overline{L} + \langle \overline{\alpha_1} \rangle)) = \overline{r}p^{n-1} - \overline{r}^2 p^{n-2} + \ell(\overline{R}/(\overline{M} : \overline{f_2})). \tag{8}$$

This proves (i).

The proof of (ii) follows the same pattern. We begin with a (homogeneous) relation $\sum_0^n \alpha_i f_i = 0$ of degree r, adjust to obtain a regular sequence $\alpha_1, f_2, \ldots, f_n$, define $L := \langle \alpha_0, f_2, \ldots, f_n \rangle$, and infer $(K : f_0) = (L : \alpha_1)$. We have exact sequences corresponding to (4) and (5) with the overlines removed.

It follows as in Lemma 4.5 that α_0, α_1 is a regular sequence, so we may adjust again to obtain a regular sequence $\alpha_0, \alpha_1, f_3, \ldots, f_n$, generating an ideal M, so that $L + \langle \alpha_1 \rangle = M + \langle f_2 \rangle$ and we have exact sequences corresponding to (6) and (7). It follows using Lemma 2.1 that all four sequences remain exact when we place tildes over the symbols to replace the overlines.

To complete the proof as before we need to show that $\ell(\tilde{R}/\tilde{L}) = rp^{n-1}$ and $\ell(\tilde{R}/\tilde{M}) = r^2 p^{n-2}$. We no longer know that $\tilde{L}, \tilde{M}$ are generated by regular sequences. Instead we argue as follows. The locus of common zeros of $\alpha_0, \alpha_1, f_3, \ldots, f_n$ in $\mathbb{C}^{n+1}$ has dimension 1, since the sequence is regular: since the sequence consists of homogeneous elements, the locus consists of a finite set of lines through the origin. If we now change coordinates so that $x_0 = 0$ passes through none of these lines, it follows that the sequence $\overline{\alpha_0}, \overline{\alpha_1}, \overline{f_3}, \ldots, \overline{f_n}$ in $\overline{R}$ is regular, so that (as above) by Lemma 3.1 R/L is free over $\mathbb{C}[x_0]$, and

$\ell(\tilde{R}/\tilde{L}) = \ell(\overline{R}/\overline{L}) = rp^{n-1}$. Proceeding similarly with the sequence $\alpha_0, \alpha_1, f_3, \ldots, f_n$ we see that provided $x_0 = 0$ avoids also the finite set of lines which is the zero-locus of M we will have $\ell(\tilde{R}/\tilde{M}) = \ell(\overline{R}/\overline{M}) = r^2 p^{n-2}$. The result now follows.

We can say something in both cases of equality in Theorem 4.4.

LEMMA 4.6. *If $T = rp^{n-1}$, the only common zero of α_0, α_1 and $f_0, \ldots, f_n$ is the origin.*

Proof. The equality implies that $\ell(\overline{R}/(\overline{L} + \langle \overline{\alpha_1} \rangle)) = 0$, so that the ideal $\overline{L} + \langle \overline{\alpha_1} \rangle = \langle \overline{\alpha_0}, \overline{\alpha_1}, \overline{f_2}, \ldots, \overline{f_n} \rangle$ is the whole of $\overline{R}$. The assertion follows.

For the other case of equality we can say rather more.

THEOREM 4.7. *Assume (A); let $0 \neq \alpha_0 \in (K : f_0)$ have degree r. Suppose also $T = r(p - r)p^{n-2}$. Then there exists a non-zero element $\beta_0 \in (K : f_0)$ of degree $p - r$ which is not a multiple of α_0. Moreover, α_0 and β_0 are coprime.*

Proof. It follows from the preceding proof that our hypothesis is equivalent to the vanishing of $\ell(\tilde{R}/(\tilde{M} : \tilde{f_2}))$, and hence to the equality $\tilde{R} = (\tilde{M} : \tilde{f_2})$, or equivalently to the assertion $\tilde{f_2} \in \tilde{M}$, or to $f_2 \in M + \langle x_0 - 1 \rangle$.

Thus f_2 vanishes on every zero of the ideal M outside the hyperplane $x_0 = 1$. Since f_2 is homogeneous and M is a homogeneous ideal, f_2 vanishes on every zero of M. Since M is generated by a regular sequence, it follows from Lasker's Unmixedness Theorem [17, Theorem 8] that $f_2 \in M$.

Thus there is a relation $f_2 = \lambda_0 \alpha_0 + \lambda_1 \alpha_1 + \sum_3^n c_i f_i$. Hence $\lambda_1 \in (L : \alpha_1) = (K : f_0)$. Now λ_1 is not a multiple of α_0 since $\alpha_0, f_2, \ldots, f_n$ is a regular sequence. Since $\deg \lambda_1 = p - r$, we can take $\beta_0 = \lambda_1$.

Finally, suppose the highest common factor h of α_0 and β_0 has degree > 0. Write $\alpha_0 = ha$ and $\beta_0 = hb$. Since $\beta_0 \in (K : f_0)$, we have a homogeneous relation $\sum_0^n \beta_i f_i = 0$. Eliminating f_0 we obtain a relation $\sum_1^n (b\alpha_i - a\beta_i) f_i = 0$ of degree $p - \deg h$ with coefficients of degree $< p$: let c_i be the last of these coefficients which is non-zero. As $\{f_1, \ldots, f_n\}$ is a regular sequence, $c_i \in \langle f_1, \ldots, f_{i-1} \rangle$, so has degree $\geq p$: a contradiction. Hence all coefficients vanish; for each i, $b\alpha_i = a\beta_i$, so α_i is divisible by a, say $\alpha_i = a\gamma_i$. But then we have a non-trivial relation $hf_0 + \sum_1^n \gamma_i f_i = 0$. Since our original relation had the lowest degree possible, $\deg h = \deg \alpha_0$. Thus a is a constant, so α_0 divides β_0, contradicting what we obtained above.

COROLLARY 4.8. *If the theorem applies, we have $p \geq 2r$.*

For since α_0 is of lowest possible degree in $(K; f_0)$, the degree of β_0 is no smaller.

There is a converse to Theorem 4.7, generalising a further result of [11].

THEOREM 4.9. *Assume (A). Suppose $\beta_1, \beta_2 \in (K : f_0)$ are coprime, and are homogeneous of degrees s_1 and s_2 with $s_1 + s_2 = p$. Then*
    (i) The ideal $(K : f_0) = K + \langle \beta_1, \beta_2 \rangle$, so $r = \min(s_1, s_2)$.
    (ii) We have $T = s_1 s_2 p^{n-2}$.

Proof. Let β_3 be a further homogeneous element of $(K : f_0)$. By Lemma 3.3 we can lift β_i $(i = 1, 2, 3)$ to elements $\xi_i \in Ker\, d : \Lambda_1^+ \to \Lambda_0^+$. Set $\xi_i^0 := \xi_i - \beta_i \theta_0$. Then, for (i, j, k) any cyclic permutation of $(1, 2, 3)$, we may write $\xi_i \wedge \xi_j = \xi_i^0 \wedge \xi_j^0 + \eta_k \wedge \theta_0$, where $\eta_k = \beta_j \xi_i^0 - \beta_i \xi_j^0 = \beta_j \xi_i - \beta_i \xi_j$.

Since β_1 and β_2 are coprime, if η_3 vanishes identically, β_1 must divide all the coefficients in ξ_1: say $\xi_1 = \beta_1 \xi$. But then $d(\xi) = 0$. As $\pi_0(\xi) = 1$ we have $1 \in (K : f_0)$, so $f_0 \in K$, contrary to hypothesis.

Now $d(\eta_k) = 0$ and $\pi_0(\eta_k) = 0$, so for some 2-form $\omega_k \in \Lambda_2$ we have $d(\omega_k) = \eta_k$. By inspection, $\sum \beta_k \eta_k = 0$, so the 2-form $\sum_k \beta_k \omega_k$ is a cycle, hence a boundary in the Koszul complex $\Lambda(f)$.

Since η_3 has weight $2p$, so does ω_3. As this is the least weight occurring in Λ_2, ω_3 is a linear combination with constant coefficients $c_{i,j}$, not all zero, of the basic generators $\theta_i \wedge \theta_j$: choose i, j with $C := c_{i,j} \neq 0$. Since $\sum_k \beta_k \omega_k$ is a boundary in $\Lambda(f)$, when we express it in terms of the basic generators, the coefficients all belong to K. Thus if the coefficients of $\theta_i \wedge \theta_j$ in ω_1 and ω_2 are α_1 and α_2 we have $\beta_1 \alpha_1 + \beta_2 \alpha_2 + C\beta_3 \in K$. Hence β_3 belongs to the ideal generated by β_1, β_2 and K.

(ii) It will suffice to show that $(K : f_0)$ is a complete intersection, i.e. admits generators β_1, β_2 and $n - 2$ elements chosen from the f_i $(1 \leq i \leq n)$: we already know that $(K : f_0)$ is generated by these $n + 2$ elements.

We have a relation $\omega_3 = \sum_{i,j} c_{i,j} \theta_i \wedge \theta_j$, where $(c_{i,j})_{0 \leq i, j \leq n}$ is a matrix of constants, not identically zero, which may be supposed skew-symmetric. Thus the rank of this matrix is even, hence at least 2. Now

$$\beta_2 \xi_1 - \beta_1 \xi_2 = \eta_3 = d\omega_3 = 2 \sum_{i,j} c_{i,j} f_i \theta_j,$$

so equating coefficients of θ_j gives a relation $2 \sum_i c_{i,j} f_i \in \langle \beta_1, \beta_2 \rangle$. Since the matrix $c_{i,j}$ has rank at least 2, this shows that we can eliminate at least 2 of the generators f_i.

We believe that the range of values given by Theorem 4.4 for T when $r \leq p/2$ is close to the exact range, but that all the estimates for

$r > p/2$ are susceptible of considerable improvement. We now briefly discuss some improvements, with particular reference to the

CONJECTURE 4.10. *We have $T \le p^{n-1}$ if and only if either $r = 1$ or $p = 4$, $r = 2$ and $T = 4^{n-1}$.*

For example, we have a direct proof that in the case $\bar{r} = 0$ excluded above we have $T > p^{n-1}$. It follows at once from Theorem 4.4 that if $T \le p^{n-1}$ and $p \ne 4$ then r is either 1, $p - 1$ or p. The main difficulty is to exclude the possibility $r = p$.

In view of (8) we may focus on $\ell(\overline{R}/(\overline{M} : \overline{f_2}))$. We need a few preliminaries on Gorenstein ideals.

Let J be a homogeneous ideal of finite colength in R. Then the quotient $A = R/J$ is a $\mathbb{C}$-algebra, and a graded local Artinian ring generated by its first graded part. Writing $H_A(t)$ for the $\mathbb{C}$-dimension of the part of A of degree t defines the *Hilbert function H_A* of A. Clearly, $\ell(A) = \sum_{t \ge 0} H_A(t)$.

Write $\mathfrak{m}$ for the maximal ideal of A generated by elements of positive degree. Its annihilator is called the *socle* of A. The degree $\sigma(A)$ of the socle is called the *socle degree* of A; it is the highest degree possible for non-zero elements of A. If the socle of A is 1-dimensional as a vector space, then A is said to be *Gorenstein* and if $A = R/J$, J is said to be a Gorenstein ideal in R. For example, if J is generated by the regular sequence $\{g_1, \ldots, g_n\}$, with degrees $p_1, \ldots, p_n$, then $\sigma(A) = \sum_1^n (p_i - 1)$; the socle is generated by the Jacobian determinant of $g_1, \ldots, g_n$. The following are well known.

PROPOSITION 4.11. *(i) If A is a graded Gorenstein Artinian ring, then $H_A(t) = H_A(\sigma(A) - t)$ for all $0 \le t \le \sigma(A)$.*

(ii) Let $I \subset J$ be homogeneous Gorenstein ideals of finite colength in R. Then there exists a homogeneous $h \in R \setminus I$ such that $J = (I : h)$; and $\deg h = \sigma(R/I) - \sigma(R/J)$.

(iii) Conversely, let $I \subset R$ be a homogeneous Gorenstein ideal of finite colength, and let $h \in R \setminus I$ be homogeneous. Set $J = (I : h)$. Then R/J is Gorenstein of dimension 0 and $\sigma(R/J) = \sigma(R/I) - \deg h$.

Proof. (i) See [26, 4.1].

(ii), (iii) This is an easy consequence of the Peskine-Szpiro theory of linkage: for this, see for example [16, 21.23, p.541].

We now summarise some properties of the ideal $(\overline{M} : \overline{f_2})$.

PROPOSITION 4.12. *(i) The ideal $(\overline{M} : \overline{f_2})$ is graded Gorenstein.*

(ii) If $(\overline{M} : \overline{f_2})$ is distinct from $\overline{R}$, it has socle degree $\sigma := np - 3p - n + 2\bar{r}$. In this case, $(n-1)(p-1) \ge 2\bar{r} - 1$.

(iii) The ideal $(\overline{M} : \overline{f_2})$ contains no non-zero element of degree less than $2\overline{r} - p$.

Proof. (i) and the first part of (ii) follow from (iii) of Proposition 4.11; the second part of (ii) follows from (i).

(iii) Suppose $(\overline{M} : \overline{f_2})$ contains a homogeneous element $\overline{\beta}_2$ of degree q. Then we have a homogeneous relation $\overline{\beta}_2 \overline{f_2} = \overline{\beta}_0 \overline{\alpha_0} + \overline{\beta}_1 \overline{\alpha_1} + \sum_3^n \overline{\beta}_i \overline{f_i}$, so that $\overline{\beta}_1$ has degree $p + q - \overline{r}$. Thus $\overline{\beta}_1 \in (\overline{L} : \overline{\alpha_1}) = (\overline{K} : \overline{f_0})$, and $\overline{\beta}_1 \neq 0$ since $\overline{\alpha_0}, \overline{f_2}, \ldots, \overline{f_n}$ is a regular sequence. Hence by the definition of $\overline{r}$, $p + q - \overline{r} \geq \overline{r}$, so $q \geq 2\overline{r} - p$.

Write $\overline{P}$ for the ideal $(\overline{M} : \overline{f_2})$: then since $\overline{P}$ is graded Gorenstein, we estimate the length of $A := \overline{R}/\overline{P}$ using the Hilbert function $H_A(t)$. Then by (i) of Proposition 4.12, we have $H_A(t) = \binom{n+t-i}{t}$ for $0 \leq t < 2\overline{r} - p$, and by (iii), together with (i) of Proposition 4.11, we have the symmetry $H_A(np - 3p - n + 2\overline{r} - t) = H_A(t)$. Lower bounds for $H_A(t)$ for intermediate values of t are given if $n = 3$ by the fact that $H_A(t)$ increases up to a maximum and then decreases; for $n \geq 4$ by an estimate due to Macaulay which is complicated to state in detail but implies, for example, that $H_A(m-1) > \frac{m}{m+n} H_A(m)$.

In the case $n = 3$ these remarks yield a rather complicated estimate for T which is however monotone in $\overline{r}$; for higher values of n the estimates are difficult to state. Probably the most interesting is the overall best estimate for T whenever $\overline{r} \geq \frac{p}{2}$. Under this hypothesis we have

n	2	3	4	5
$T >$	$\frac{1}{4}p^2$	$\frac{7}{24}p^3 + \frac{1}{4}p^2 + \frac{4}{3}p - 2$	$p^{3\frac{1}{2}} - p^3$	$p^4 - p^3$

Thus the above conjecture is proved for $n \leq 4$ (but no further), and if $n > 3$ we are a long way from the hoped-for estimate $T > \frac{1}{4}p^n$.

Even without delicate estimates, it follows from Proposition 4.12 that if $\overline{r} > \frac{1}{2}p$, then $\ell(\overline{R}/\overline{P}) \geq 2\binom{n+(2\overline{r}-p-1)}{n}$, and hence $T \geq \overline{r}(p - \overline{r})p^{n-2} + 2\binom{n+(2\overline{r}-p-1)}{n}$. For the difficult case $\overline{r} = p$, if p is large compared to n the final term behaves like $2p^n/n!$, so will eventually exceed p^{n-1}.

5. From algebra to geometry

Recall that a hypersurface V in $P_n(\mathbb{C})$, with equation $F = 0$ of degree D, is said to be good if V has isolated singularities, is transverse to the hyperplane H $(x_0 = 0)$, and is not a cone. We write $f_i := \partial F/\partial x_i$. The sum of the Tjurina numbers of V at the set $\Sigma(V)$ of its singular points is denoted by $\tau(V)$.

PROPOSITION 5.1. *[9] If F defines a good hypersurface V,*

(i) we have $(K : F) = (K : f_0)$,

(ii) the Tjurina number $\tau(V) = (D - 1)^n - T$.

Proof. (i) In view of the Euler relation $DF = \sum_0^n x_i f_i$, and the fact that $f_i \in K$ for $i > 0$, we have $(K : F) = (K : x_0 f_0)$. Thus $(K : f_0) \subseteq (K : F)$. Conversely, if $\phi \in (K : F)$ then $x_0 f_0 \phi \in K$, so $x_0 \phi \in (K : f_0)$. But by Lemma 3.2 (ii), $R/(K : f_0)$ is free over $\mathbb{C}[x_0]$; hence $\phi \in (K : f_0)$.

(ii) By Lemma 1.4, we have $\tau_0 - \tau(V) = \dim_{\mathbb{C}} \mathbb{C}[\mathbf{x}]/(Jf_v : f_v)$. Now in § 2 we renamed the homogenised f_v as F, so that f_v became $\tilde{F}$ and its partial derivatives are the $\tilde{f}_i$, so that $Jf_v = \tilde{K}$. Hence $\tau_0 - \tau(V) = \ell(\tilde{R}/(\tilde{K} : \tilde{F}))$. The result now follows from (i) and Lemma 2.1.

In the geometric case, regarding $\tilde{R}/\tilde{K}$ as $\mathbb{C}[x]$-module defines a point in the Hilbert scheme of $\mathbb{C}^n$, whose support consists of the critical points of F with multiplicities given by the Tjurina numbers at those points. Thus these numbers must sum to τ_0. In general, F will have distinct critical values at them.

We turn to the Koszul complex $\Lambda^+(f)$. We may interpret the abstract symbols θ_i in the definition as the differential operators $\partial/\partial x_i$. The module Λ_1^+ is interpreted as the module Vec of polynomial vector fields (with a degree shift of p, since we define the degree of a homogeneous vector field ξ to be that of $\pi_0(\xi)$), and $Ker\, d : \Lambda_1^+ \to \Lambda_0^+$ as the submodule $Ann(F)$ of those vector fields that annihilate F. The images in Λ_1^+ of the generators $\theta_i \wedge \theta_j \in \Lambda_2^+$ are the basic Hamiltonian vector fields $\frac{\partial F}{\partial x_i}\frac{\partial}{\partial x_j} - \frac{\partial F}{\partial x_j}\frac{\partial}{\partial x_i}$, so the module $Ham(F)$ of Hamiltonian vector fields of F is $Im\, d : \Lambda_2^+ \to \Lambda_1^+$. So the first homology group of $\Lambda^+(f)$ is identified with $Ann(F)/Ham(F)$.

We recall the projection $\pi_0 : Vec \to \mathbb{C}[x^+]$ given by $\pi_0(\sum_0^n \alpha_i \partial/\partial x_i) = \alpha_0$: by Lemma 3.3, we have $\pi_0(Ann(F)) = (K : f_0)$, $\pi_0(Ham(F)) = K$ and π_0 induces an isomorphism of $Ann(F)/Ham(F)$ onto $(K : f_0)/K$.

Of particular interest are the vector fields ξ in $Ann(F)$ of degree 1, for we can exponentiate ξ to obtain a 1-parameter subgroup of $GL_n(\mathbb{C})$ leaving F invariant.

LEMMA 5.2. *Let* ξ, ξ' *be homogeneous vector fields of respective degrees* a, a' *in* $Ann(F)$. *Then if* $\xi \wedge \xi' \neq 0$, *we have* $a + a' \geq D - 1$.

Proof. Since ξ, ξ' are vector fields annihilating F, their product $\xi \wedge \xi'$ is a cycle, hence a boundary, in the Koszul complex $\Lambda^+(f)$. Thus if it is non-zero, it comes from a non-zero element of Λ_3^+. Since the generators θ_i of Λ_1^+ have degrees $D - 1$, the least degree of a generator of Λ_3^+ is $3D - 3$.

The degrees of ξ, ξ' in the Koszul complex are $D + a - 1, D + a' - 1$, so the degree of their product is $2D + a + a' - 2$. Hence $2D + a + a' - 2 \geq 3D - 3$.

This lemma generalises [11, 3.1]. The proof works also for the case of unequal weights; the conclusion becomes $a + a' \geq D + 2 - w_n - w_{n-1} - w_{n-2}$.

Applying Lemma 5.2 to vector fields ξ with linear coefficients, or equivalently, ξ of degree 1, we infer at once

COROLLARY 5.3. *If there exist two independent linear vector fields* $\xi, \xi' \in Ann(F)$, *we have* $D \leq 3$.

Thus if V has a 2-parameter symmetry group in $PGL_{n+1}(\mathbb{C})$ we must have $D \leq 3$ (and the case $D = 2$ is essentially trivial). We will return to this situation below.

5.1. ESTIMATES

We will now apply the main theorems of § 4 to the geometric situation. First observe that it follows from Lemma 3.3 that r is the least degree of a non-zero homogeneous vector field ξ with $\xi(F) = 0$, for K and $Ham(F)$ vanish in degrees below p, so in these degrees π_0 is an isomorphism from $Ann(F)$ onto $(K : f_0)$.

Similarly $\bar{r}$ is the least degree of a non-zero homogeneous vector field ξ such that $\xi(F)$ vanishes on H. For in any homogeneous relation $\sum_0^n \overline{\alpha_i f_i} = 0$ with the α_i of degree $< p$ we have $\overline{\alpha_0} \neq 0$ by (A), and α_0 figures in such a relation if and only if $\overline{\alpha_0} \in (\overline{K} : \overline{f_0})$.

We can now give the geometrical versions of Theorems 4.4, 4.7 and 4.9.

THEOREM 5.4. *Let* V *be a good hypersurface, with* r *as above. Then*
$$(D - 1)^{n-1}(D - r - 1) \leq \tau(V) \leq (D - 1)^{n-2}(D^2 - Dr + r^2 - 2D + r + 1).$$
Moreover $\tau(V)$ *takes the upper value if and only if there exist non-zero vector fields* $\xi, \eta \in Ann(F)$ *of degrees* r, $D - r - 1$, *with* η *not a multiple of* ξ. *In this case,* $Ann(F) = Ham(F) + R\xi_1 + R\xi_2$.

Proof. Since (A) holds by Lemma 2.3, we can apply Theorem 4.4 to deduce that $rp^{n-1} \geq T \geq r(p - r)p^{n-2}$; and $\tau(V) = p^n - T$ by Proposition 5.1. Since $p = D - 1$, these give the values above. Choose ξ of degree r with $\xi F = 0$ and write $\pi_0(\xi) = \alpha_0$, so $\alpha_0 \in (K : f_0)$.

Now by Theorem 4.7, if equality holds there exists a non-zero element $\beta_0 \in (K : f_0)$ of degree $p - r$ which is not a multiple of α_0. By Lemma 3.3, we may choose η with $\pi_0(\eta) = \beta_0$. Conversely, given ξ and η set $\beta_1 := \pi_0(\xi)$, $\beta_2 := \pi_0(\eta)$. These lie in $(K : f_0)$. The argument at the end of the proof of Theorem 4.7 shows that β_1 and β_2 are coprime. Thus by Theorem 4.9(ii) $T = r(p-r)p^{n-2}$.

Now let ζ be any element of $Ann(F)$; set $\pi_0(\zeta) := \beta_3$. Then by Theorem 4.9(i), we may write $\beta_3 = a_1\beta_1 + a_2\beta_2 + \sum_1^n p_i \partial f/\partial x_i$. Then the vector field

$$\xi_3 - (a_1\xi_1 + a_2\xi_2 + \sum_1^n p_i\theta_0 \wedge \theta_i)$$

has zero coefficient of θ_0, so by Lemma 3.3 is Hamiltonian.

For the case when τ takes the lower value, we have

PROPOSITION 5.5. *If V is good, and $\tau(V) = (D-1)^{n-1}(D-r-1)$, then all singularities of V are weighted homogeneous.*

Proof. It follows from Lemma 4.6 that at a singular point P of V, not both of α_0 and α_1 can vanish: suppose α_0 does not; we may also suppose x_0 does not vanish at P. Then in local affine coordinates $x_0 = 1$, using the relations $\sum_0^n \alpha_i f_i = 0$ and $\sum_0^n x_i f_i = DF$, we have $\sum_1^n(\widetilde{\alpha_i} - \widetilde{\alpha_0}x_i)f_i = (D\widetilde{\alpha_0})\widetilde{F}$. Hence the vector field $\xi = \sum_1^n(\widetilde{\alpha_i} - \widetilde{\alpha_0}x_i)\partial/\partial x_i$ satisfies $\xi(\widetilde{F}) = \phi.\widetilde{F}$, with $\phi = D\widetilde{\alpha_0}$; in particular, $\phi(P) \neq 0$. It follows from [10, 5.6] that the singularity at P is weighted homogeneous.

The translation of [10, 5.6] to a projective context was given in [7].

Suppose now that $r = 1$. Then
$$(D^2 - 3D + 2)(D-1)^{n-2} \leq \tau(V) \leq \tau_{max} := (D^2 - 3D + 3)(D-1)^{n-2}.$$
It would follow from Conjecture 4.10 that the left inequality characterises the condition that $r = 1$, i.e. that there is a linear vector field in $Ann(F)$, or a 1-parameter subgroup of $PGL_{n+1}(C)$ under which V is invariant. We have investigated the geometry of such symmetric hypersurfaces in [13] and [14].

In view of Theorem 5.4 this would imply in turn that τ_{max} is the maximum possible value of $\tau(V)$ for good hypersurfaces V, and that $\tau(V) = \tau_{max}$ if and only if there exist vector fields ξ, η, of degrees $1, D-2$, with η not a multiple of ξ, such that $\xi(F) = \eta(F) = 0$. This corollary does not require the full force of the conjecture, and the above estimates establish it in rather more cases: in particular, we have an unconditional proof when $n = 5$.

In the case $n = 2$ of curves, Theorem 5.4 reduces to the main result of [11]. Note that a plane curve is good if and only if it is reduced, and not a union of concurrent lines. Moreover, it follows from the sharper estimates that if $\tau(V) > \frac{3}{4}D^2 - \frac{3}{2}D$, then $r \leq \frac{1}{2}D$. For $\tau(V) > \phi(D)$, where $\phi(D)$ is approximately $D^2 - D^{3/2}$, this shows that $\tau(V)$ determines r. It may be the case that a similar conclusion is true in all dimensions.

5.2. DIMENSIONS OF EQUISINGULAR STRATA

The defining polynomial F of V belongs to the space P_D of all homogeneous polynomials of degree D in $(x_0, \ldots, x_n)$, which is a vector space of dimension $\binom{n+D}{D}$. The group $G = GL_{n+1}(\mathbb{C})$ acts on P_D: write $A(F)$ for the isotropy subgroup of F, and $G.F$ for its orbit. It follows from Corollary 5.3 that if $D > 3$ then dim $A(F) \leq 1$; and from Theorem 4.9 that if $D = 3$ then dim $A(F) \leq 2$.

On the other hand, write $\mathcal{C}$ for the constellation of $\mathcal{K}$-equivalence classes of the singularities of V, and $P_D(\mathcal{C})$ for the set of polynomials in P_D defining hypersurfaces with isolated singularities the set of whose equivalence classes (under $\mathcal{K}$-equivalence) is $\mathcal{C}$. Since the singularities of these hypersurfaces are prescribed, they all have the same total Tjurina number, $\tau_C = \tau(V)$.

Clearly we have $G.F \subseteq P_D(\mathcal{C})$; we now enquire how closely these are related. First dim $G.F = \dim G - \dim A(F)$, so
$$\text{codim } G.F = \dim P_D - \dim G + \dim A(F).$$
Write $Q := \mathbb{C}[x]/\langle F, JF \rangle$ for the coordinate ring of the set $\Sigma(V)$ of singular points of V. This splits as a direct sum $Q = \oplus_{P \in \Sigma(V)} Q_P$.

We can identify P_D with the unfolding space of the unfolding of F by all monomials of degree $\leq D$. For local unfoldings at the singular points we have a semi-universal deformation with parameter space Q. By versality there is an induced map $\Delta : P_D \to Q$: its differential is the reduced Kodaira-Spencer map $\kappa_D : P_D \to Q$, which is surjective if and only if the deformation is versal on the singularities. Since $P_D(\mathcal{C}) = \Delta^{-1}(0)$, we have codim $P_D(\mathcal{C}) \geq \tau_C - \dim \text{Cok } \kappa_D$.

LEMMA 5.6. *([7, 1.4,1.5]) The dimension of* Cok κ_D *is equal to the dimension of the homogeneous part of* Ann$(F)/Ham(F)$ *(or equivalently* $(K : f_0)/K$ *) of degree* $n(D - 2) - 1 - D$.

Combining these remarks, we deduce

PROPOSITION 5.7. *(i) If* $(K : f_0)$ *contains an element* ϵ *of degree 1, then if* $D > 3$, *we have*

$$0 \leq \text{codim } G.f - \text{codim } P_D(\mathcal{C}) \leq \frac{n(n-1)}{2};$$

if $D = 3$ *we must replace the final term by* dim $A(f) + \frac{n^2 - 3n + 2}{2}$.
(ii) If, in addition, $(K : f_0)$ *contains an element of degree* $D - 2$ *not divisible by* ϵ, *then* codim $G.f = \text{codim } P_D(\mathcal{C})$.

The additional hypothesis in (ii) is equivalent, by Theorem 5.4, to the condition $\tau(V) = \tau_{max}$, and Conjecture 4.10 implies that this also implies the hypothesis of (i). It would follow that whenever $\tau_C = \tau_{max}$, $F_D C$ is a finite union of G-orbits.

5.3. HYPERSURFACES WITH $\mathbb{C}^\times$-ACTION

Hypersurfaces V with $\tau(V) \geq (D-2)(D-1)^{n-1}$ admit the action of a 1-parameter algebraic group if $n \leq 4$ and, if Conjecture 4.10 holds, for all n. We now use results from [14] to discuss good varieties V, with $\tau(V) = \tau_{max}$ and equation $F = 0$, which admit an action of the multiplicative group $\mathbb{C}^\times$, or equivalently, such that for some semisimple linear vector field ξ we have $\xi F = 0$.

Choose coordinates in which ξ takes the form $\sum_0^n w_r x_r \partial / \partial x_r$ where the w_r are integers, called the *weights* of the action (N.B. these are not to be confused with the weights occurring in § 1), which may be supposed in non-decreasing order. Write W_λ for the λ-weight space. We call the set of weights *0-symmetric* if $w_{n-r} = -w_r$ for each r. Write, for any m, $s_{m-1}(D) = \frac{(D-1)^m - (-1)^m}{D}$. Then

PROPOSITION 5.8. *[14, 3.5] If 0 is not a weight, then $\mu(V) = s_n(D)$. If 0 occurs $m + 1$ times as a weight, and the set of weights is not 0-symmetric, then*
$$\mu(V) = s_n(D) + (-1)^{m+n+1} s_m(D).$$
If the set of weights is 0-symmetric, we have
$$\mu(V) = s_n(D) + (-1)^{n-m}(\mu(V \cap W_0) - s_m(D)).$$

It is not difficult to deduce

THEOREM 5.9. *If V admits a $\mathbb{C}^\times$-action, then $\mu(V) = \tau(V) \leq \tau_{max}$. Moreover, the following are equivalent:*

(i) $\mu(V) = \tau_{max}$;

(ii) there exist vector fields ξ, η, of degrees $1, D-2$, with η not a multiple of ξ, such that $\xi(f) = \eta(f) = 0$;

(iii) either (a) 0 occurs $(n-2)$ times as weight; or (b) the set of weights is 0-symmetric, 0 occurs $(n-1)$ times (so the set of weights is $[-1, 0, \ldots, 0, 1]$) and $V \cap W_0$ is a cone.

We can also list the systems of weights that occur. We have

PROPOSITION 5.10. *Suppose W is a system of weights containing at least one zero weight; form W^* by adding a further variable x_*, also of zero weight. Then if W^* is the set of weights for an good hypersurface for a given degree D, so is W.*

This can be deduced from [14, Theorem 4.2].

COROLLARY 5.11. *The systems of weights that give the highest value to μ for good hypersurfaces of degree D in P^n are formed from the corresponding systems in P^3 by adding $n - 3$ zero weights.*

The possible systems of weights in 4 variables are listed in [14] (see esp. Table 2): they are the cases of type α in the notation of that paper. For $D \geq 4$ there are $[4\frac{1}{2}D] - 10$ of them. We thus have a complete list of possible types.

Finally we consider the case $D = 3$. We saw in Lemma 5.3 that if there are two linearly independent linear vector fields that annihilate f, or equivalently if and only if V admits a 2-parameter symmetry group, then $D = 3$. We also saw in each of Theorem 5.4 and Theorem 5.9, that (when $D = 3$) V admits a second 1-parameter symmetry group if and only if τ attains its maximal value $\tau_{max} = 3.2^{n-2}$.

If $D = 3$, V admits a $\mathbb{C}^\times$-action, and $\tau(V) = \tau_{max}$, then by Corollary 5.11 the weights arise from those which occur for $n = 3$ by adding zeros ('suspension'); and when $n = 3$ we just have
$$[-1, 0, 1, 0], \quad [-2, 1, 2, 0], \quad [-2, 1, 4, 0].$$
In the [-1,0,0,1] case only we have a semi-simple 2-parameter group. For $n = 3$ we have a surface with 3 A_2 singularities.

In the [-2,0,1,2] case we have a surface with an A_1 and an A_5.

In the [-2,0,1,4] case we have a surface with an E_6 singularity.

For $n = 4$ there is just one case (up to change of coordinates) where V admits a 2-parameter group which contains no semi-simple elements. Such a V has a $T_{2,6,6}$ singular point.

6. Instability loci

The *instability locus Inst g* of a map g is defined to be the set of points y in the target such that the germ of g at $\Sigma(g, y)$ is *not* a stable germ. This notion was independently introduced by Damon [5], who called it the versality discriminant. Continuing the notation of § 1, we consider the partial unfoldings

$$F^k(\mathbf{x}, u_1, \ldots, u_{\mu-1-k}) = (f_u^k(\mathbf{x}), u_1, \ldots, u_{\mu-1-k}),$$

of f_0, where $f_u^k(\mathbf{x}) = f_0(\mathbf{x}) + \sum_1^{\mu-1-k} u_i \phi_i(\mathbf{x})$. Let $c \leq W - D$ be a positive integer, and write $k(c)$ for the number of unfolding variables such that ϕ_i has (weighted) degree $< c$; by duality, there are the same number with degree $> W - c$.

Thus $F^{k(c)}$ is obtained from the miniversal unfolding F^0 of f_0 by omitting all the unfolding monomials of degree $> W - c \geq D$. So the deformations f_v of f_0 have $v_i = 0$ whenever $\deg v_i < 0$.

By Mather's criterion [22] for stable map-germs, for any map $g :$ $N \to P$, $Inst\, g$ is the support of $\theta(g)/tg(\theta_N) + \omega g(\theta_P)$, i.e. of the cokernel of the induced map $\overline{\omega g} : \theta_P \to \theta(g)/tg(\theta_N)$. (N.B. This use of an overline is not related to the convention in preceding sections.) Here θ_P is the $\mathcal{O}_v$-module of vector fields on the target, which is freely generated by the $\partial/\partial v_i$ with $0 \leq i < \mu$.

Write $M(F) := \theta(F)/tF(\theta_N)$; we have an isomorphism of $\mathcal{O}_{x,u}/J_x f$ onto $M(F)$, since $tF(\partial/\partial u_i)$ has one coordinate 1 and all the rest, save the first, 0; so $M(F)$ is generated by the class of $\partial/\partial y$.

Since F is stable, $\overline{\omega F}$ is surjective, so $M(F)$ is a quotient of θ_P. The kernel is the module of liftable vector fields, which is freely generated by the η_i. Thus $M(F)$ is also the cokernel of the map of free $\mathcal{O}_v$-modules defined by the discriminant matrix.

Since $F^{k(c)}$ is the restriction of F where the source variables u_i and target variables v_i are set equal to 0 for $i \geq \mu - k$, we obtain $M(F^{k(c)})$ from $M(F)$ by factoring out by these v_i. The module θ_P generated by all $\partial/\partial v_i$ with $0 \leq i < \mu$ maps onto $M(F^{k(c)})$; $\overline{\omega F^k}$ is the restriction of this surjection to the submodule generated by $\partial/\partial v_i$ with $0 \leq i < \mu - k$. Its cokernel is thus generated by the images of the $\partial/\partial v_i$ with $\mu - k \leq i < \mu$.

Thus $\theta(F^{k(c)})/tF^{k(c)}(\theta_N) + \omega F^{k(c)}(\theta_P)$ is obtained from the cokernel of $A^T : \mathcal{O}_v^\mu \to \mathcal{O}_v^\mu$ by setting $v_i = 0$ for $i \geq \mu - k(c)$ and factoring out the generators corresponding to ϕ_i for $i < \mu - k(c)$: thus (after the substitution) we have the cokernel of the map defined by the last $k(c)$ rows of the discriminant matrix A^T, hence by the last $k(c)$ columns of A. We conclude

PROPOSITION 6.1. *The instability locus $Inst\, F^{k(c)}$ is the support of the quotient of $\mathcal{O}_v^k(c)$ by the submodule spanned by the $k(c)$ columns of A of lowest degree.*

We now use the duality result of Mond and Pellikaan.

THEOREM 6.2. *([23]) Let $f_0 : (\mathbb{C}^n, 0) \to (\mathbb{C}, 0)$ be a weighted homogeneous germ of finite singularity type. Then any homogeneous unfolding of f_0 admits a symmetric homogeneous discriminant matrix.*

PROPOSITION 6.3. *(cf. [15], Theorem 10.5.32) For $F^{k(c)}$ as above, the locus $Inst\, F^{k(c)}$ is the support of the quotient of $\mathcal{O}^{k(c)}$ by the submodule spanned by the $k(c)$ columns of lowest degree of a discriminant matrix for F.*

For the $k(c)$ rows of lowest degree of the discriminant matrix correspond by symmetry to the $k(c)$ columns of lowest degree.

Now by Lemma 1.3, we see that $v \in Inst\ F^{k(c)}$ if and only if there exists a linear combination $g(x)$ of the unfolding monomials of degree $< c$ with $g \in (Jf_v : f_v)$: clearly g may be supposed homogeneous. By Lemma 1.4, this holds if and only if there is a polynomial g with $\deg(g) < c$ and $g \in (Jf_v : f_v)$. In the notation of § 3 we need a non-zero element of $(K : F) = (K : f_0)$ of degree $< c$.

The above remarks, with Lemma 3.3, give an alternative proof of the first main result (Theorem 1.1) of [7].

THEOREM 6.4. *We have $v \in Inst\ F^{k(c)}$ if and only if there is a non-Hamiltonian vector field ξ of weight $< c$ with $\xi f_v = 0$.*

For example, in the equal weight case, if $c \leq D-1$, non-Hamiltonian is equivalent to non-zero. In this case, $v \in Inst\ F^{k(c)}$ if and only if $r < c$, where r is the parameter of Theorem 4.4 (where f_v is denoted F). This gives a geometric interpretation of that parameter, and also implies a relation with the Tjurina number. In the general case, we have

THEOREM 6.5. *[9, 2.1] If $v \in Inst\ F^{k(c)}$ we have*

$$\tau_0 - \tau_v \leq max\left\{\dim_{\mathcal{C}} \frac{\mathcal{O}_n}{Jf_0 + \mathcal{O}_n g}\right\},$$

where the supremum is taken over homogeneous polynomials g not contained in Jf_0 and of degree less than c.

This raises the problem of estimating the maximum. In [9] we give a number of applications of an estimate which applies in easy cases. Work on the general case leads to a well known conjecture in commutative algebra: the Eisenbud-Griffiths-Harris conjecture [17, Conjecture CB12].

References

1. V. I. Arnold, Critical points of functions and the classification of caustics, Uspekhi Mat. Nauk **29** (1974) 243–244.

2. J. W. Bruce, Functions on discriminants, Jour. London Math. Soc. **30** (1984) 551–567.

3. J. W. Bruce, A. A. du Plessis and L. C. Wilson, Discriminants and liftable vector fields, Jour. Alg. Geom. **3** (1994) 725–753.

4. J. Damon, On the legacy of free divisors: discriminants and Morse-type singularities, Amer. J. Math. **120** (1998) 453–492.

5. J. Damon and A. Galligo, Universal topological stratification for the Pham example, Bull. Soc. Math. France **121** (1993) 153–181.

6. J. Damon, Non-linear sections of non-isolated complete intersections, this volume.

7. A. A. du Plessis, Versality properties of projective hypersurfaces, preprint.
    http://www.imf.au.dk/~esn/index.html

8. A. A. du Plessis, T. Gaffney and L. C. Wilson, Map-germs determined by their
    discriminants, *Stratifications, Singularities and Differential Equations I* (eds D.
    Trotman and L. C. Wilson), Travaux en cours **54**, (Hermann, 1997) pp 1–40.

9. A. A. du Plessis and C. T. C. Wall, Versal deformations in spaces of polynomials
    of fixed weight, Compositio Math. **114** (1998) 113–124.

10. A. A. du Plessis and C. T. C. Wall, Discriminants and vector fields, *Singulari-
    ties: the Brieskorn anniversary volume* (eds V. I. Arnold, G.-M. Greuel and J.
    H. M. Steenbrink), (Birkhäuser, 1998) pp 119–140.

11. A. A. du Plessis and C. T. C. Wall, Application of the theory of the discriminant
    to highly singular plane curves, Math. Proc. Camb. Phil. Soc. **126** (1999)
    256–266.

12. A. A. du Plessis and C.T.C. Wall, Singular hypersurfaces, versality and
    Gorenstein algebras, Jour. Alg. Geom. **9** (2000) 309–322.

13. A. A. du Plessis and C. T. C. Wall, Curves in $P^2(C)$ with 1-dimensional
    symmetry, Revista Mat. Complutense **12** (1999) 117–132.

14. A. A. du Plessis and C. T. C. Wall, Hypersurfaces in P^n with one-parameter
    symmetry groups, Proc. Roy. Soc. A **456** (2000) 2515–2541.

15. A. A. du Plessis and C. T. C. Wall, *The geometry of topological stability*, London
    Math. Soc. Monographs new ser. **9**, Oxford Univ. Press, 1995.

16. D. Eisenbud, *Commutative Algebra*, Springer graduate texts **150**, Springer-
    Verlag, 1995.

17. D. Eisenbud, M. Green and J. Harris, Cayley-Bacharach theorems and
    conjectures, Bull. Amer. Math. Soc. **33** (1996) 295–324.

18. V. V. Goryunov, Vector fields and functions on discriminants of complete
    intersections and bifurcation properties of projections, Jour. Soviet Math. **52**
    (1990) 3231–3245.

19. G.-M. Greuel, Der Gauss-Manin Zusammenhang isolierter Singularitäten von
    vollständigen Durchschnitten, Math. Ann. **214** (1975) 235–266.

20. C. Hertling, Frobenius manifolds and variance of the spectral numbers, this
    volume.

21. E. J. N. Looijenga, *Isolated singular points on complete intersections*, London
    Math. Soc. Lecture Notes **77**, Cambridge Univ. Press, 1984.

22. J. N. Mather, Stability of C^∞-mappings IV: classification of stable germs by
    $I\!R$-algebras, Publ. Math. IHES. **37** (1970) 223–248.

23. D. M. Q. Mond and R. Pellikaan, Fitting ideals and multiple points of analytic
    mappings, Springer Lecture Notes in Math. **1414** (1989) 107–161.

24. K. Saito, On a generalisation of de Rham lemma, Ann. Inst. Fourier **26** (1976)
    165–170.

25. K. Saito, Theory of logarithmic differential forms and logarithmic vector fields,
    Jour. Fac. Sci. Tokyo sec. 1A **27** (1980) 265–291.

26. R. P. Stanley, Hilbert functions of graded algebras, Advances in Math. **28**
    (1978) 57–83.

27. H. Terao, The bifurcation set and logarithmic vector fields, Math. Ann. **263**
    (1983) 313–321.

28. B. Teissier, The hunting of invariants in the geometry of discriminants, *Real
    and complex singularities, Oslo 1976* (ed P. Holm), (Sijthoff and Noordhoff,
    1977) pp 565–677.

29. K. Wirthmüller, Singularities determined by their discriminant, Math. Ann.
    252 (1980) 237–245.

The theory of integral closure of ideals and modules: Applications and new developments

Terence Gaffney *
Northeastern University (gaff@neu.edu)

With an appendix by Steven Kleiman and Anders Thorup

Introduction

Many equisingularity conditions such as the Whitney conditions, and their relative versions A_f and W_f, depend on controlling limiting linear structures. The theory of the integral closure of ideals and modules provides a very useful tool for studying these limiting structures. In this paper we illustrate how these tools are used in three case studies, and describe some of the advances in the theory since the survey article of [12], which was written in 1996.

The case studies were chosen because the properties and ideas that they illustrate were the pathways into the problems that we used originally.

In the first section of the paper we recall some of the basic properties of the theory of integral closure of ideals. In our first case study we show how these properties can be used to study equisingularity through weighted homogeneity conditions. We translate a recent condition of Fukui and Paunescu ([3]) which uses weight conditions into integral closure terms. It follows immediately that the set of parameter values in a family of functions for which the condition of Fukui and Paunescu holds is Zariski open, and, with off-the-shelf technology, is non-empty.

In the second section we recall some of the basic properties of the theory of integral closure of modules. In our second case study we show how to use this theory to extend results on functions to mappings. We do this by showing how to extend Malgrange's theorem, which deals with fibers of polynomial functions at infinity, to the fibers of polynomial mappings at infinity ([5]). We also return briefly to case study I to indicate the extension of the results of section 1 to the case of general sets.

In the third section we discuss briefly the theory of the multiplicity of a module and of the Segre numbers of an ideal which extend this notion to ideals of non-finite colength. These ideas are then applied to

---

* Supported in part by NSF grant 9403708-DMS.

D. Siersma et al. (eds.), New Developments in Singularity Theory, 379–404.

our first two case studies as well as to our third, which is the study of mappings from $\mathbb{C}^n$ to $\mathbb{C}^p$, $p \geq 2n$ ([6]).

In the fourth section, we describe some recent work relating the W_f condition to the A_f condition and describe some of the applications of this relation to equisingularity of functions on complex analytic sets, and to the equisingularity of a family of mappings to the plane. We then describe briefly work due to R. Gassler developing a theory of mixed Segre numbers ([13]).

Lastly, in the appendix, S. L. Kleiman and A. Thorup give simple new proofs of two important theorems. These proofs use new ideas of Simis, Ulrich, and Vasconcelos. The theorems concern the "exceptional" fiber F of the generalized conormal space $\mathrm{Projan}\mathcal{O}_X[\mathcal{M}]$ associated to a nested pair $\mathcal{M} \subset \mathcal{N}$ of submodules of a free module on a reduced equidimensional analytic germ X; by definition, F is the preimage of $\mathrm{Supp}(\mathcal{N}/\mathcal{M})$. The first theorem asserts that F has codimension 1 if $\mathcal{N}/\mathcal{M}$ is not integral. It is the key to the generalization in [10], §2 of Teissier's principle of specialization of integral dependence from ideals to modules. The second theorem asserts that F has pure codimension 1 if $\mathcal{N}$ is free. It is due to Massey and the author [12], §5 who used it there to perfect the generalization in [10], §5 of the Lê–Saito theorem.

1.   Ideals

1.1. BASIC PROPERTIES OF THE THEORY OF INTEGRAL CLOSURE FOR IDEALS

Let $X, 0 \subset \mathbb{C}^N, 0$ be a reduced analytic space germ. We denote the local ring of $X, 0$ by $\mathcal{O}_{X,0}$. If $X = \mathbb{C}^n, 0$, we denote the local ring by $\mathcal{O}_n$. Let I be an ideal in $\mathcal{O}_{X,0}$, and f an element in this ring. Then f is integrally dependent on I if one of the following equivalent conditions obtain:

(i) There exists a positive integer k and elements a_j in I^j, so that f satisfies the relation $f^k + a_1 f^{k-1} + \ldots + a_{k-1}f + a_k = 0$ in $\mathcal{O}_{X,0}$.

(ii) There exists a neighborhood U of 0 in $\mathbb{C}^N$, a positive real number C, representatives of the space germ X, the function germ f, and generators $g_1, \ldots, g_m$ of I on U, which we identify with the corresponding germs, so that for all x in X the following equality obtains: $|f(x)| \leq C \max\{|g_1(x)|, \ldots, |g_m(x)|\}$.

(iii) For all analytic path germs $\phi: (\mathbb{C}, 0) \to (X, 0)$ the pull–back $\phi^* f$ is contained in the ideal generated by $\phi^*(I)$ in $\mathcal{O}_1$, the local ring of $\mathbb{C}$ at 0.

If we consider the normalization $\bar{B}$ of the blowup B of X along the ideal I we get another equivalent condition for integral dependence. Denote the pull-back of the exceptional divisor D of B to $\bar{B}$ by $\bar{D}$.

(iv) For any component C of the underlying set of $\bar{D}$, the order of vanishing of the pullback of f to $\bar{B}$ along C is no smaller than the order of the divisor $\bar{D}$ along C.

(Cf. [21] for proofs of these results.)

1.2. EQUISINGULARITY THROUGH WEIGHTS

Now we turn to the study of equisingularity through weights to see how these ideas are used in practice.

Given coordinates $z_1, \ldots, z_n$ on $\mathbb{C}^n$, we can assign weights $w_1, \ldots, w_n$ to the coordinates. We say a monomial $x_1^{k_1} \ldots x_n^{k_n}$ has weight $w_1 k_1 + \ldots + w_n k_n$, and a polynomial is weighted homogeneous of weight d if all of its monomial terms have the same weight d. If $f \in \mathcal{O}_n$, then the initial form of f consists of the weighted homogeneous polynomial made up of the terms in the Taylor expansion of f of lowest weight.

A function germ f is *semi-quasihomogeneous* if the initial form with respect to some system of weights defines a function with an isolated singularity at 0. After the work of Lê and Ramanujam ([19]) it was recognized that families of semi-quasihomogeneous functions whose initial forms have constant weight are topologically trivial ([2]).

It was realized early ([22]) that the weight structure could be used to control vectorfields ensuring the integrability of the fields and producing familes of homomorphisms for trivializing families of functions and sets. This idea of controlling vector fields using metrics defined by weight conditions was used to prove the triviality of families of sets and functions even when the objects were not weighted homogeneous ([27], [28]). In a recent paper Fukui and Paunescu ([3]) showed that if a family of functions or maps satisfied the following inequalities, then vectorfields controlled by metrics based on this choice of weights could be integrated to trivialize the families. The condition of Fukui and Paunescou, translated to the complex analytic setting, is:

Let $w = w_1 \cdot w_2 \cdot \ldots \cdot w_n$, $\underline{w} = (w_1, \ldots, w_n)$, let $||z||_{\underline{w}} = (|z_1|^{2w/w_1} + \ldots + |z_n|^{2w/w_n})^{1/2w}$. Let $f \colon \mathbb{C}^n \times \mathbb{C}^k \to \mathbb{C}^p$ be a k parameter family of maps. If $p = 1$ work on a neighborhood U of the origin in $\mathbb{C}^n \times \mathbb{C}^k$, if $p > 1$ work on $U \cap X$ where X is the set of regular points of $f^{-1}(0)$. Then for all holomorphic $\Psi(z, y) = (\psi_1(z, y), \ldots, \psi_p(z, y))$

$$\left| \sum_{j=1}^{p} \psi_j(z,y) \frac{\partial f_j}{\partial y_s}(z,y) \right| \leq C \sum_{i=1}^{i=n} ||z||_{\underline{w}}^{w_i} \left| \sum_{j=1}^{j=p} \psi_j(z,y) \frac{\partial f_j}{\partial z_i}(z,y) \right|. \quad \text{(FP)}$$

If $p = 1$ this condition simplifies to:

$$\left| \frac{\partial f}{\partial y_s}(z,y) \right| \leq C \sum_{i=1}^{i=n} ||z||_{\underline{w}}^{w_i} \left| \frac{\partial f}{\partial z_i}(z,y) \right|.$$

For the rest of this section we restrict ourselves to this condition. The next section, which deals with the properties of the integral closure of modules, will provide some intuition into the form of the condition when $p > 1$. Since this condition is an analytic inequality, by 1.1 (ii) we should be able to recast it in integral closure terms.

To do this we need some more notation. Assume weights have been ordered so that $w_1 \geq w_2 \geq \ldots \geq w_n$. Denote $\frac{\partial f}{\partial z_i}$ by f_{z_i}, and $(f_{z_i})^w$ by $\tilde{f}_i$. Define ideals I_i by $I_i = (\ldots, z_j^{w \cdot w_i / w_j}, \ldots)$. Define ideals $J_i(f)$ by $J_i(f) = (\tilde{f}_i, \ldots, \tilde{f}_n)$. Then define $J_{\underline{w}}(f)$ by $J_{\underline{w}}(f) := \Sigma_i I_i \cdot J_i$ Since the w_i are monotone decreasing it follows that $I_j \subset I_{j+1}$.

PROPOSITION 1.1. *In the above setting, if $p = 1$ then condition (FP) is equivalent to*

$$f_{y_j}^w \in \overline{J_{\underline{w}}(f)}$$

for $1 \leq j \leq k$, in $\mathcal{O}_n$.

Proof. We use the curve criterion. Condition (FP) implies that

$$\left| f_{y_j} \circ \phi(t) \right| \leq C \sum_i ||z||_{\underline{w}}^{w_i} \circ \phi(t) \, |f_{z_i} \circ \phi(t)|.$$

This implies that

$$\left| f_{y_j} \circ \phi(t) \right| \leq C' \max_{i,l} |z_l|^{(w_i/w_l)} \circ \phi(t) \, |f_{z_i} \circ \phi(t)|.$$

By raising both sides of this inequality to the w power, we get

$$f_{y_j}^w \circ \phi(t) \in \phi^*(z_l^{w w_i / w_l} \tilde{f}_i)$$

which implies the integral closure condition. The reverse implication follows from the analytic inequality implied by the integral closure condition.

COROLLARY 1.2. *In the above setup the set of points in Y at which condition (FP) holds is Z-open.*

Proof. The set of points in Y at which the equivalent integral closure condition fails is Z-closed. To see this, use condition 1.1 (iv); the set of points on the exceptional divisor E of the normalized blow up by $J_{\underline{w}}(f)$ where the pullback of the ideals $J_{\underline{w}}(f)$ and $(J_{\underline{w}}(f), f^w_{y_j})$ are different is a Z-closed subset of E, so the intersection with the inverse image of Y is Z-closed. Then the image of this set in Y is Z-closed also, since the map from E to X is proper. This Z-closed set is exactly the points of Y where (**FP**) fails.

COROLLARY 1.3. *In the above setup the set of points in Y at which condition (**FP**) holds is Z-open and dense.*

Proof. For all weights we have $J_{\underline{w}}(f)$ contains the ideal $(z_1^w \tilde{f}_1, \ldots, z_n^w \tilde{f}_n)$. Now we have for a Z-open subset of Y that

$$f_{y_j} \in \overline{(z_1 f_{z_1}, \ldots, z_n f_{z_n})}$$

This was proved by Teissier ([32]). It follows from the curve criterion that

$$(f_{y_j})^w \in \overline{(z_1^w \tilde{f}_1, \ldots, z_n^w \tilde{f}_n)} \subset \overline{J_{\underline{w}}(f)}$$

2. Modules

2.1. BASIC PROPERTIES OF THE INTEGRAL CLOSURE OF MODULES

Let M be a submodule of $\mathcal{E}$, $\mathcal{E} := \mathcal{O}_X^p$ be a free module of rank p at least 1. One can develop the properties of the integral closure of M by passing from M to an ideal sheaf as follows. Let $\rho : \mathcal{E} \to \mathcal{S}\mathcal{E}$ be the inclusion of $\mathcal{E}$ into its symmetric algebra; then the ideal generated by $\rho(M)$ induces a sheaf of ideals on $X \times \mathbb{P}^{p-1}$, and we can apply (i)-(iv) of section 1 to this context.

One can also develop the analogues of (ii) and (iii) without direct reference to projective space. This gives: $h \in \overline{M}$ iff

(ii') For each choice of generators $\{s_i\}$ of M, there exists a constant $C > 0$ and a neighborhood U of x such that for all $\psi \in \Gamma(Hom(\mathbb{C}^p, \mathbb{C}))$,

$$\|\psi(z) \cdot h(z)\| \leq C \|\psi(z) \cdot s_i(z)\|$$

for all $z \in U$.

(iii') For all analytic path germs $\phi : (\mathbb{C}, 0) \to (X, 0)$ the pull-back $\phi^* h$ is contained in the module generated by $\phi^*(M)$ in $\mathcal{O}_1^p$ at 0.

(Cf. [4] for proofs of these results.)

Although (ii') does not mention projective space directly, the connection with projective space is apparent. For if the section in $\Gamma(Hom(\mathbb{C}^p, \mathbb{C}))$

is non-zero at the origin, then we can think of the section as giving a map from X to $X \times I\!\!P^{p-1}$ by $x \to (x, <\psi(x)>)$. Then the inequality for this ψ follows by pulling back the inequality (ii) from $X \times I\!\!P^{p-1}$ by the map just constructed.

In this paper, if $h \in \mathcal{E}$ we denote the module generated by h and the elements of M by (h, M). The rank of a submodule M of $\mathcal{E}$ will be the rank of a matrix of generators of M. The ideal generated by the $k \times k$ minors of the matrix of generators is denoted $J_k(M)$.

We can now state an important relation between the theory for modules and for ideals:

Jacobian Principle

Suppose the rank of the module (h, M) is k on each component of (X, x). Then $h \in \bar{M}$ if and only if $J_k(h, M) \subset \overline{J_k(M)}$.

(Cf. [4].)

If $f: \mathbb{C}^n, 0 \to \mathbb{C}^p, 0$ is a map germ, and X is its fiber over 0, then the $\mathcal{O}_{X,0}$ module generated by the partial derivatives of f is called the Jacobian module of f, denoted $JM(f)$. We use $JM_z(f)$ and $JM_y(f)$ if we only use the partial derivatives with respect the z or y variables in generating a module.

2.2. EQUISINGULARITY THROUGH WEIGHTS FOR GENERAL ANALYTIC SETS

Recall that the condition of Fukui and Paunescou is: for all holomorphic
$$\Psi(z, y) = (\psi_1(z, y), \dots, \psi_p(z, y))$$

$$\left| \sum_{j=1}^{p} \psi_j(z, y) \frac{\partial f_j}{\partial y_s}(z, y) \right| \leq C \sum_{i=1}^{i=n} ||z||_{\underline{w}}^{w_i} \left| \sum_{j=1}^{j=p} \psi_j(z, y) \frac{\partial f_j}{\partial z_i}(z, y) \right|.$$

Comparing this with condition (ii'), it is clear that they are similar, so an argument similar to that for 1.2 should give an integral closure condition. The big difference is that now $\frac{\partial f}{\partial y_s}(z, y)$ is a vector. To get a condition as we did when $p = 1$, we want to be able to "raise the vector $\frac{\partial f}{\partial y_s}(z, y)$ to a power". To make sense of this we work on $X \times I\!\!P^{p-1}$.

The ideals I_i are defined as before, now $\tilde{f}_i$ is $(\rho(f_{z_i}))^w$. Then the ideals $J_i(f)$ and $J_{\underline{w}}(f)$ are defined as before. Then condition (**FP**) is equivalent to:

$$(\rho(f_{y_j}))^w \in \overline{J_{\underline{w}}(f)}, \quad 1 \leq j \leq k.$$

The analogues to Propositions 1.2 and 1.3 now follow essentially as before.

2.3. FIBERS OF POLYNOMIAL MAPPINGS AT INFINITY

Now we turn to our second case study, *the fibers of a polynomial map at infinity.*

In this study, the starting point is the work of Parusiński ([25], [26]), who studied the singularities at infinity of a polynomial function. The theory of integral closure for modules makes it possible to extend many of the results of these two papers to the case of polynomial mappings without much effort.

Let $f: \mathbb{C}^n \to \mathbb{C}^p$ be a polynomial mapping with $n > p$. A value $t_0 \in \mathbb{C}^p$ of f is called *typical* if f is a C^∞ trivial fibration over a neighborhood of t_0 and *atypical* otherwise.

If f is a polynomial function ($p = 1$), the value t_0 is typical if the following condition holds.

Malgrange's condition $p = 1$
The condition holds at $t_0 \in \mathbb{C}$ if, for $|x|$ large enough and for $f(x)$ close to t_0

$$\exists \, \delta > 0 \quad \text{such that} \quad ||x|| \left(\sum_i ||f_{z_i}(x)||^2 \right)^{1/2} \geq \delta. \qquad \text{(M)}$$

We want to use the properties of integral closure to generalize this condition to polynomial mappings. To study the fibers of f at infinity set:

$$X = \overline{\Gamma(f)} \subset \mathbb{P}^n \times \mathbb{C}^p.$$

We denote the coordinates on $\mathbb{P}^n \times \mathbb{C}^p$ by $(Y_0, Y_1, \ldots, Y_n, t_1 \ldots, t_p)$, $H_\infty =$ hyperplane at infinity of $\mathbb{P}^n$ defined by $Y_0 = 0$.

$$X_\infty = X \cap (H_\infty \times \mathbb{C}^p)$$

$\tilde{f}$ denotes the homogenization of f. Work in a neighborhood of $p_0 = ((0 : \ldots : 0 : 1), 0, \ldots, 0)$, with coordinates $(y_0, \ldots, y_{n-1}, t_1, \ldots, t_p)$, $(y_i = Y_i/Y_n)$. Assume the terms of highest degree, d_i, of the components f_i of f define a complete intersection of codimension p in a neighborhood of p_0. (This set is the *cone at infinity*, denoted C_∞.)

Then X is defined by

$$F_i(y_0, y_1, \ldots, y_{n-1}, t_1 \ldots, t_p) := \tilde{f}_i(y_0, y_1, \ldots, y_{n-1}, 1) - t_i y_0^{d_i} = 0.$$

Working at p_0, we return briefly to the case $p = 1$, because we want to reformulate Malgrange's condition in integral closure terms. We will see that this version extends easily to the complete intersection case.

We start by reformulating the condition in the sup norm. We get:

$$\exists \; \delta' > 0 \;\; \text{such that} \;\; |x_n| \sup_i \{|f_{z_i}(x)|\} \geq \delta'.$$

Here we are assuming x close to p_0, which is why we can replace $\|x\|$ by $|x_n|$.

At this point it is helpful to note some of the relations between the partials of f and F.

$$\frac{1}{y_0^{d-1}} \frac{\partial F}{\partial y_i} = \frac{\partial f}{\partial x_i}, \; 1 \leq i \leq n-1$$

$$\frac{\partial F}{\partial t} = y_0^d$$

$$\frac{1}{y_0^{d-1}} \frac{\partial F}{\partial y_i} = x_1 \frac{\partial f}{\partial x_1} + \ldots + x_n \frac{\partial f}{\partial x_n}$$

Using these relations we can reformulate the condition in terms of F.

$$\exists \; \delta' > 0 \;\; \text{such that} \;\; \sup_i \left\{ \left| \frac{\partial F}{\partial y_i}(y,t) \right| \right\} \geq \delta' \left| y_0^d \right| \qquad 1 \leq i \leq n-1.$$

or

$$\left| y_0 \frac{\partial F}{\partial y_0}(y,t) - \sum_i^{n-1} y_i \frac{\partial F}{\partial y_i}(y,t) \right| \geq \delta' \left| y_0^d \right|.$$

The second inequality comes from the $\partial f / \partial x_n$ term. Since $\partial F / \partial t = y_0^d$ this is equivalent to:

$$\partial F / \partial t \in \overline{\{y_0 \partial F / \partial y_0, \partial F / \partial y_1, \ldots, \partial F / \partial y_{n-1}\}}$$

We can sharpen this a little ([5] prop. 12) to get:

$$(M) \;\; \text{iff} \;\; \partial F / \partial t \in \overline{\{\partial F / \partial y_1, \ldots, \partial F / \partial y_{n-1}\}} \qquad (M_\infty)$$

Let g denote the restriction of y_0 to X. This last condition is equivalent to asking that the limiting tangent planes to the fibers of g are not vertical over C. (The fibers of $y_0|_X$ are defined by $G = (y_0, F)$; if the above inclusion fails, then a limiting kernel of DG is the span of $(\partial / \partial y_1, \ldots, \partial / \partial y_{n-1})$.) If the cone at infinity has isolated singularities, then this is equivalent to the seemingly stronger A_g condition.

In turn this gives us a chance to lift $\partial / \partial t$ over f in a controllable way.

Having rephrased condition $\mathbf{M}$ as condition $\mathbf{M}_\infty$ it is easy to generalize $\mathbf{M}$ to mappings. Both the geometric condition and the integral dependence condition make sense for $p > 1$; therefore we get a condition on polynomial mappings just by replacing the ideal in $\mathbf{M}_\infty$ by the corresponding module. We can apply the Jacobian principle to the integral closure relation for $p > 1$ and convert back to affine space to get a new inequality. The answer is given in terms of various minors of the Jacobian matrix of f. Let $M_I(f)$ with $I = (i_1 < i_2 < \ldots < i_p)$ denote the maximal minor of the Jacobian matrix of f formed from the columns indexed by I. Let $M_J(f, j)$ denote the minor of the Jacobian matrix of size $(p-1) \times (p-1)$ using the columns indexed by J, and all the rows of the Jacobian matrix, except for the j-th row.

For $|x|$ large enough and for $f(x)$ close to t_0, ask

$$\exists \, \delta > 0 \ \text{ such that } \ \|x\| \frac{(\sum_I \|M_I(f)\|^2)^{1/2}}{(\sum_{J,j} \|M_J(f,j)\|^2)^{1/2}} \geq \delta. \qquad \text{(GM)}$$

This condition again implies the smooth triviality of the fibration f locally at t_0, essentially because the levels of y_0 are well-controlled.

3. Multiplicities and Segre numbers

In this section we show how to use invariants and the underlying intersection theory to control equisingularity conditions. In the event that the condition is controlled by a module M of finite colength we can use the Buchsbaum-Rim multiplicity, $e(M)$, if it is controlled by an ideal I of infinite colength, then we can use the Segre numbers $e_i(I)$.

For a discussion of these invariants and the principle of specialization of integral dependence (PSID) which is the main tool in applying them to equisingularity problems Cf. [12].

3.1. MULTIPLICITIES AND EQUISINGULARITY THROUGH WEIGHTS

In case study I, we saw that the equisingularity condition was controlled by $J_{\underline{w}}(f)$, which is an ideal in $\mathcal{O}_{n+k}$ if $p = 1$, and an ideal in SO_X^p if $p > 1$ where X is the zero set of f. Then we can show:

PROPOSITION 3.1. *Suppose f defines a family of complete intersections X_y with isolated singularities, $f^{-1}(0) \supset Y$, and suppose $e(J_{\underline{w}}(f_y))$ is independent of y. Then $Y = \cup \mathrm{Sing}(X_y)$ (resp. $\mathrm{Sing}(f_y)$ if $p = 1$) and the family X_y has a trivialization by a weight controlled vector field (resp. f_y has a trivialization by a weight controlled vector field if $p = 1$).*

Proof. Since $e(J_{\underline{w}}(f_y))$ is upper semicontinous, and its support is $\cup Sing(X_y)$, the independence of parameter implies that $Y = \cup Sing(X_y)$. The second conclusion follows provided $(\rho(f_{y_j}))^w \in \overline{J_{\underline{w}}(f)}$ for $1 \leq j \leq k$ for all points of Y. By Corollary 1.3 it holds generically, hence by the PSID ([31], [10]) it holds at all points of Y.

The question now arises as to the meaning of the invariant, and whether of not its constancy is a necessary as well as sufficent condition for condition (**FP**). Specialize to $p = 1$.

In the case where $p = 1$ and all of the weights are 1, then $J_{\underline{w}}(f_y) = m_n J(f_y)$ and Teissier showed ([31])

$$e(J_{\underline{w}}(f_y)) = \sum_{i=0}^{n} \binom{n}{i} (m(P_i(f_y))).$$

Here $P_i(f_y)$ denotes the relative polar variety of f of codimension i; by convention $P_0(f_y) := C^n$, and $m(P_n(f_y)) := \mu(f)$, the Milnor number of f.

If all of the weights are 1, then (**FP**) is just Whitney equisingularity; and Teissier's decomposition shows that the constancy of $e(J_{\underline{w}}(f_y))$ is necessary because the constancies of the terms in the sum are.

With additional conditions on f a similar decomposition holds for $e(J_{\underline{w}}(f_y))$ when the weights are different from 1. The argument is to show that $J_{\underline{w}}(f_y)$ has a reduction of the form $(m_1 \tilde{f}_1, \ldots, m_n \tilde{f}_n)$ where $m_i \in I_i$, by looking at the exceptional divisor in a certain modification of C^n. Here is a sample of what can be proved.

PROPOSITION 3.2. *Suppose $f: C^n, 0 \to C, 0$ has an isolated singularity at the origin. Let f_0 denote the initial part of f with respect to the choice of weights, assume all weights are unequal. Assume any subset of the partials of f_0 with $n - 1$ elements forms an R-sequence. Then*

$$e(J_{\underline{w}}(f)) = \sum_{I,J} w^{|J|} e(m_{i_1}, \ldots, m_{i_k}, f_{z_{j_1}}, \ldots, f_{z_{j_{n-k}}})$$

Here I and J range over all partitions of $\{1, \ldots, n\}$
Proof. (Sketch) Given

$$\overline{J_{\underline{w}}(f)} = \overline{(m_1(\partial f/\partial z_1)^w, \ldots, m_n(\partial f/\partial z_n)^w)}$$

we have

$$e(J_{\underline{w}}(f)) = e((m_1(\partial f/\partial z_1)^w, \ldots, m_n(\partial f/\partial z_n)^w)) =$$

$$\text{colength } ((m_1(\partial f/\partial z_1)^w, \ldots, m_n(\partial f/\partial z_n)^w)).$$

The result follows by using the additivity of length to calculate the colength term in the last equality, and from the fact that $e(f_1^w, \ldots, f_r^w) = w^k e(f_1, \ldots, f_r)$ in a local ring of dimension k.

With these hypotheses the meaning of the terms in the sum are clear. We have $e(f_{z_1}, \ldots, f_{z_n})$ is the Milnor number, while if $|J| < n$, then

$$e(m_{i_1}, \ldots, m_{i_k}, f_{z_{j_1}}, \ldots, f_{z_{j_{n-k}}}) = e(m_{i_1}, \ldots, m_{i_k}, f_{0,z_{j_1}}, \ldots, f_{0,z_{j_{n-k}}})$$

because of the condition on the initial form of f. These terms can be thought of then as a weighted homogeneous multiplicity of a (non-generic) relative polar variety. In fact, since these are ideals generated by weighted homogeneous forms, their values are determined entirely by the original weights, so we get as a corollary:

COROLLARY 3.3. *Suppose f_y is a family of constant Milnor number, each f_y satisfies the R-sequence condition of Prop. 3.2 for a fixed choice of weights and for these weights $(f_0)_y$ has constant weight, then the family f_y has a trivialization by a weight controlled vector field.*

Proof. Each term in the sum of the previous proposition is constant, so by Proposition 3.1, the result follows.

Notice that under the hypotheses of 3.3, if there exists a weight controlled field, then each term in the above sum must be constant, so $e(J_{\underline{w}}(f))$ must be constant. In this case then, the constancy of $e(J_{\underline{w}}(f))$ is necessary as well as sufficent.

3.2. MULTIPLICITIES AT INFINITY

Now we return to case study II

In order to put ourselves in a situation where our invariants are defined, we restrict ourselves to the case where all of the fibers of f have dimension $n - p$, so that the cone at infinity does not depend on t, is equal to the vanishing of $I_T(f)$ and hence $X_\infty = C_\infty \times C^p$. We also assume that the cone at infinity has isolated singularities.

We also want to work at points $p_t = ((0 : 0 : \ldots : 0 : 1), t)$ at infinity such that in a neighborhood of p_t the fiber over t is smooth at points not in X_∞.

We call a point $p_t \in X_\infty$ *typical* if condition M_∞ holds at p_t. Condition (GM) holds at $t_0 \in C^p$ iff all the points in X_∞ over t_0 are typical. (Cf. Theorems 5,17 of [5].)

Denote the fiber over $t \in \mathbb{C}^p$, by X_t. Denote by $JM(F_t)$ the module obtained by restricting the Jacobian module of F, $JM(F)$, to X_t. Under these conditions the module $JM(F_t)_{Y+} := \{\partial F/\partial y_1, \ldots, \partial F/\partial y_{n-1}\}$ has finite colength ([5] prop 19).

PROPOSITION 3.4. *Suppose p_t as above, p_0 a singular point of the cone at infinity. Suppose the multiplicity of $JM(F_t)_{Y+}$ is independent of t in a neighborhood of p_t, then p_t is typical.*

Proof. We are going to show that the constancy of the multiplicity implies that the the inclusion

$$\partial F/\partial t \in \overline{\{\partial F/\partial y_1, \ldots, \partial F/\partial y_{n-1}\}}$$

holds. The inclusion holds generically along $Y := (0, \ldots, 1) \times \mathbb{C}^p$ because the inclusion is implied by the A_g condition applied to the pair (X_0, Y). Here X_0 denotes the set of points of X where the function g is a submersion. We know that the A_g condition holds generically because g is a function. Now the constancy of the multiplicity, coupled with the fact that the inclusion holds generically, allows us to apply the PSID for modules ([10]). This implies that the inclusion holds at all points close to p_t.

(This condition also appears in the hypersurface case in [35].)

There is an interesting interpretation of the multiplicity of $JM(F_t)_{Y+}$ which we now describe.

PROPOSITION 3.5. *The multiplicity of $JM(F_t)_{Y+}$ at p_t is the sum of the Milnor number of X_t and the Milnor number of C_∞.*

Proof. The number of generators of $JM(F_t)_{Y+}$ is $n-1$. Meanwhile, it is a submodule of $\mathcal{O}_{X_t}^p$ and the dimension of X_t is $n-p$. Now the number of generators of a minimal reduction of $JM(F_t)_{Y+}$ is $dim(X_t) + p - 1 = n - 1$ so $JM(F_t)_{Y+}$ is already a minimal reduction. Hence the multiplicity of $JM(F_t)_{Y+}$ is just its colength ([1]), since the fiber is a complete intersection hence $\mathcal{O}_{X_t}$ is Cohen-Macaulay. By a theorem of Buchsbaum and Rim ([1]), this is the colength of the ideal of maximal minors. Now by a theorem of Lê and Greuel ([16], [18]) the colength of the ideal of maximal minors is the sum of the Milnor number of X_t and the Milnor number of the slice by y_0, which is just the cone at infinity.

COROLLARY 3.6. *Suppose p_t as above, p_0 a singular point of the cone at infinity. Suppose the Milnor number of X_t is independent of t in a neighborhood of p_t, then p_t is typical.*

Proof. Since the Milnor number of the cone at infinity is independent of t, it suffices for the Milnor number of X_t to be independent of t to apply Proposition 4.3.

If $p = 1$, then we can allow the cone at infinity to have non-isolated singularities provided the Segre numbers of $JM(F_t)_{Y+}$ are constant. The argument is the same as 4.3, using the principle of specialization proved in [9].

For an alternate approach to this result in the hypersurface isolated singularity case Cf. [30].

For an approach in the hypersurface case which also uses the theory of integral of ideals, Cf. [35]. For a study of the global topology of the fibers of polynomial mappings Cf. [36].

3.3. DOUBLE POINTS OF MAPS FROM $\mathbb{C}^n \to \mathbb{C}^{2n}$

As we have just seen, we can use the Segre numbers in arguments using the PSID in the case of non-isolated singularities. The Segre numbers can also be used to control geometric quantities which arise as residual intersections.

If f_y is a family of map germs of $\mathbb{C}^n$ to $\mathbb{C}^{2n}$ parametrised by $Y = \mathbb{C}^k$, f_0 injective, then it is necessary to ensure that the map $F(y, z) = (y, f_y(z))$ is injective for the family to be topologically trivial. We do this by controlling the number of double points of f_y at the origin. If f is an injective map-germ of $\mathbb{C}^n$ to $\mathbb{C}^{2n}$, and we take a versal unfolding of f, then the double point set defined in the source of the unfolding has the same dimension as the parameter stratum; so the projection from the double point set to the parameter stratum is a finite map, and the degree of this projection at $(0, 0)$ is the number of double points of f hiding at the origin. It is clear, that as we vary the parameter in the base of the versal unfolding that the number of double points, counted with appropriate multiplicities, of the corresponding map germs remains constant.

How do we calculate this number?

If h is a function in $\mathcal{O}_n$, then we denote $h(z) - h(z')$ in $\mathcal{O}_{2n}$ by h_D. We call $h(z) - h(z')$ the double of h. If $f: \mathbb{C}^n, 0 \to \mathbb{C}^{2n}, 0$, then we can double f by doubling each component function of f. If f injective, the scheme theoretic fiber of the double of f over the origin in $\mathbb{C}^{2n}$ consists of the diagonal of $\mathbb{C}^n \times \mathbb{C}^n$ and an m_{2n} primary ideal at the origin. This primary ideal is related to the double point set of f. Denote the ideal generated by the components of the double of f by $I_D(f)$. We will be interested in $e_{2n}(I_D(f))$, the Segre number of codimension $2n$. Since the double point number is additive for families of map germs $f: \mathbb{C}^n, 0 \to \mathbb{C}^{2n}, 0$, we expect this Segre number to be also.

PROPOSITION 3.7. *Suppose F is an unfolding of $f: \mathbb{C}^n, 0 \to \mathbb{C}^{2n}, 0$, f an injective immersion off the origin. Then, the sum of the Segre numbers of $I_D(f_u)$ of codimension $2n$ is independent of u, for u small.*

Proof. (Sketch) In order to calculate the Segre number of greatest codimension of I, it suffices to take the polar curve of I, and to calculate the multiplicity of the ideal gotten by restricting I to the polar curve. To calculate the polar curve, take $2n - 1$ generic linear combinations of generators of $I_D(f)$, set them equal to zero, and discard any components in the diagonal. In our situation, the restriction of $I_D(f)$ to its polar curve is principal, because $I_D(f)$ has $2n$ generators, and $2n - 1$ of them are zero on the polar curve, by the construction of the polar curve.

To finish the proof it suffices to show that the polar variety of dimension $k + 1$ of a k-parameter family of map germs specializes to the polar curve of f. The problem is that a component of the specialization might lie in the diagonal of $\mathbb{C}^{2n}$. If a component of the specialization lies in the diagonal of $\mathbb{C}^{2n}$, then this must give a curve of singular points of f, which contradicts the fact that f is an injective immersion.

Note that the above proof shows that $e_{2n}(I_D(f))$ is just the colength of $I_D(f)$ in the local ring of $P_{2n-1}(I(f)_D)$ at the origin. This follows because, since $P_{2n-1}(I(f)_D)$ is a reduced curve, hence Cohen-Macaulay, the multiplicity of a principal ideal is just the colength of the ideal.

It seems reasonable to hope that $e_{2n}(I_D(f))$ is the number of double points of f at the origin; unfortunately, this turns out not to be true. Recall that, if $g : \mathbb{C}^n \to \mathbb{C}^{2n-1}$ then the number of Whitney umbrella points of g at the origin is just the colength of the ideal of maximal minors of the Jacobian matrix of g ([24]). Denote the number of Whitney umbrella points of g by $C(g)$. The number of Whitney umbrella points in an unfolding of g is also constant ([24]). Recall also that, if F is a 1-parameter unfolding of f such that for $u \neq 0$, f_u is a stable map-germ on some fixed neighborhood of the origin, then we say that F is a stabilization of f. Then we have the following result:

THEOREM 3.8. *Suppose $f : \mathbb{C}^n, 0 \to \mathbb{C}^{2n}, 0$, f injective and an immersion off the origin, F is a stabilization of f, $\pi : \mathbb{C}^{2n} \to \mathbb{C}^{2n-1}$ a generic linear projection for F, then the sum of $C(\pi \circ f_u)$ and the number of double points of f_u is equal to $e_{2n}(I_D(f))$.*

Proof. The polar curve of $I(f_u)_D$ consists of the double point curve of $\pi \circ f_u$ in $\mathbb{C}^{2n}$. Let g be a generator of the restriction of $I(\bar{f})_D$ to its polar variety of codimension $2n - 1$. Then g_u vanishes at each point where the polar curve of $I(f_u)_D$ intersects the diagonal. These points project to the Whitney umbrella points of $\pi \circ f_u$.

The set of points of the polar curve where g_u vanishes, which are not on the diagonal, are exactly the double points of f_u in $\mathbb{C}^{2n}$. Because f_u is an immersion at the points where $\pi \circ f_u$ has a Whitney umbrella, it follows that the zero set of g is transverse to the polar curve at points where the polar curve intersects the diagonal. (This is easily verified using the normal form for the Whitney umbrella.) This means that each Whitney umbrella point of f_u contributes 1 to the total value of $e_{2n}(I_D(f_u))$. Using local coordinates it is also easy to see that each double point also contributes 1 to the total value of $e_{2n}(I_D(f_u))$.

COROLLARY 3.9. *Suppose f and π are as above, then the number of double points of f at the origin is equal to $e_{2n}(I_D(f)) - C(\pi \circ f)$.*

Proof. All three terms in the above corollary are constant in a stabilization; since equality holds for the general member of a stabilization of f, it holds for f.

In this setting, the other Segre numbers are just the multiplicity at the origin of the set of singular points of the composite of f with various projections to lower dimensional spaces, while the polar multiplicities are just the multiplicities of the corresponding double point loci. In the terminolgy of [9] we can think of the Segre cycles of dimension i, $0 < i < n$ as "moving components", the intersection of the polar varieties with the diagonal which is the cycle of dimension n. From this perspective, $e_{2n}(I_D(f))$ is just the sum of the multiplicity of the "zero dimensional moving component" which comes from the intersection of the polar curve with the diagonal and a "refined Segre number of dimension 0" which is the double point number.

4.   Some Developments in the Theory since '96

4.1. GENERIC SECTIONS, A_f AND W_f

Given a family of germs of complex analytic sets and a family of functions on the sets, when is the family of functions equisingular? This, of course, is a basic question in the field.

The heuristic idea "A weaker equisingularity condition which holds for all generic sections of a family implies a stronger equisingularity condition for the original family" is useful in finding a solution.

For example, Lê and Teissier ([20]) showed that topological triviality for plane sections of complex analytic sets implies the Whitney conditions.

Fix a pair (X, Y) consisting of a reduced equidimensional analytic subspace X of C^{n+k} and a linear subspace Y of C^{n+k} contained in X. Assume $C^{n+k} = C^n \times Y^k$; if $X = F^{-1}(0)$, where $F: C^{n+k}, 0 \to C^p, 0$, assume the fibers of X over Y equidimensional.

Fix a function $f: C^{n+k}, 0 \to C, 0$ which takes Y to 0 and assume that there is an open, smooth, analytic subset X_f of X, dense in each fiber of X over Y, such that $f|_{(X_f, 0)}$ is a submersion onto its image. Denote the map germ with components (F, f) by G.

Recall that if every limiting tangent hyperplane to the fibers of f at 0 contains Y, then we say that the A_f condition holds for the pair (X_f, Y) at 0.

The pair (X_f, Y) satisfies the condition W_f at 0 if there exist a neighborhood U of 0 in X and a constant $C > 0$ such that, for all y in $U \cap Y$ and all x in $U \cap X_f$, we have

$$\text{dist}\,(Y, H) \le C\,\text{dist}(x, Y)$$

where H is a hyperplane which contains the tangent space to the fiber of the restriction of $f|_X$.

Using ideas from the previous section it is not hard to re-write this inequality in integral closure terms. We get ([11]):

(X_f, Y) satisfies W_f at 0 if and only if

$$JM_y(G) \subset \overline{m_Y JM_z(G)}$$

For functions, the A_f and W_f conditions are known to hold on some Zariski open and dense subset of Y.

If X has a stratification containing Y as a stratum and every pair of strata satisfies the W_f condition then the pair (X, f) is locally topologically right trivial over Y.

The relation between the A_f and W_f conditions is contained in the next theorem. The theorem shows that *if you can control A_f, you can control W_f*.

THEOREM 4.1. *([8]) Suppose X, Y as in the above setup, f a function on X; then W_f holds at the origin, if and only if A_f holds at the origin for (X_f, Y) and for a generic set of plane sections of X containing Y.*

Proof. (Sketch) The proof is based on the following ideas. Given the infinitesimal formulation of the W_f condition it is easy to give an equivalent formulation in terms of the relative conormal of f denoted $C(X, f)$: along the preimage in $C(X, f)$ of 0, the ideal of $C(Y) \cap C(X, f)$ is integrally dependent on the ideal of the preimage of Y. ([11] Lêmma 2.1.). In turn, since the W_f condition holds generically, this condition will hold, provided we can show that the fiber over Y

of the exceptional divisor of the blow up of $C(X, f)$ along the inverse image of Y has minimal dimension ([8] Corollary 2.1.1). Now using the information about A_f, show that if the dimension of this fiber is too large, then this failure will cause the failure of the A_f condition for the restriction of f to some generic set of plane sections.

In order to study the A_f condition simultaneously for a family of sections, use the Grassmann modification. This space consists of pairs (x, P) where $x \in X \cap P$ and P is a codimension r linear space containing Y. The formulation of the A_f and W_f condition in infinitesimal terms then makes it possible to move information back and forth between the base X and the Grassmann modification, taking advantage of the genericity condition on the Grassman modification to define the necessary set of generic plane sections and to do the analysis needed for the final result.

In the next result $F_{f_t,(t,0)}$ is the Milnor fiber of f_t at $(t, 0)$.

COROLLARY 4.2. *Let W^d be the germ of an equidimensional analytic subset of $\mathbb{C}^n, 0$. Let $X := D \times W$ be the product of an open disk about the origin with W, and let $Y := D \times 0$. Let $f: X, Y \to \mathbb{C}, 0$ be an analytic function, such that f does not vanish on any component X, and let $f_t(z) := f(t, z)$.*

Suppose that the homology of the Milnor fiber of f_0 at points $\neq 0$ with $\mathbb{C}$ coeficients is the same as the homology of a point, and that the reduced Betti number $\tilde{b}_{d-1-i}(F_{f_a|H_i,(a,0)})$ is independent of a for all small a, for $\{H_i\}$ a generic flag.

Suppose $f \in m_{X,0}^2$, then the pair (X_f, Y) satisfies the W_f condition at 0.

Proof. Massey and Green ([15], [23]), showed that under these hypotheses, the pairs $((X \cap H_i)_f, Y)$ satisfy the A_f condition at 0, so by our main result they satisfy the W_f condition as well. In particular, if the origin is an isolated singular point of W, the family $\{f_t\}$ is topologically right trivial.

This theorem also is useful in the study of mappings from $\mathbb{C}^n, 0 \to \mathbb{C}^2$, $n > 2$ done in the framework of [7]; one applies it to the function in the source which defines the inverse image of the discriminant. If this function and its generic sections satisfy the A_f condition then the function satisfies the W_f condition. This implies that the original family of map germs is Whitney equisingular. Since f is a submersion off a curve, the inverse image of the discriminant contains inside it all of the information carried by the discriminant, so this object which sits in the source also controls target behavior. (Cf. the thesis of our student

R. Vohra ([37]).) In turn, since we have an infinitesimal formulation
of this condition in integral closure terms this gives an infinitesimal
formulation in integral closure terms for the Whitney equisingularity
of the original family of mappings.

Progress in the study of the Whitney equisingularity of finitely determined map germs has also been made by V.H.J. Perez ([29]) who
has made a study of map germs $f : \mathbb{C}^3, 0 \to \mathbb{C}^3, 0$.

4.2. Mixed Segre numbers

This is a description of work from the thesis of our student Gassler
([13]).

As we have seen, the Segre numbers, to some extent, are a good
generalization of the notion of the multiplicity of an ideal of finite
colength. In order to see how to develop the theory of Segre numbers,
it is helpful to try to understand the extent to which other properties
of the multiplicity can be extended to the Segre numbers.

As mentioned earlier, Teissier in [31] gave a formula that expresses
the multiplicity of the product of two m–primary ideals I_1 and I_2 in
terms of some mixed multiplicities. These mixed multiplicities were
the multiplicities of ideals formed by selecting some "generic" elements
from each ideal.

In [33], and in [34] Teissier developed further properties of these
mixed multiplicities, using them in [34] to give a numerical criterion
for any two m–primary ideals, not necessarily one contained in the
other, to have the same integral closure. Teissier's condition is that all
of the mixed multiplicities should agree.

Teissier's proof was based on showing that certain inequalities between the mixed multiplicities held; he reduced this to the case of a
normal surface X, then considered the resolution of the blowup of X
by the product of $I_1 I_2$. The inequalities followed from the negative definiteness of the intersection matrix of the components of the exceptional
divisor of a resolution of singularities of the surface. If these inequalities
were equalities, then the negative definiteness property would force the
pullbacks of the ideals to vanish to the same degree on the components
of the exceptional divisor, forcing the equality of the integral closures.

Given two ideals in $\mathcal{O}_{X,x}$ it is natural to try to define mixed Segre
numbers with the expectation that when they agree, the integral closures of the two ideals should agree.

To define mixed Segre numbers it is necessary as well to define mixed
polar varieties.

For $k = 1, \ldots, n$, and positive integers i, j with $i + j \geq k$, we define
the mixed Segre number of codimension k as follows: select i general

elements $f_1, \ldots, f_i$ from I_1 and j general elements $g_1, \ldots, g_j$ from I_2. Let $\mathbf{h}$ be a k–tuple of general linear combinations of these elements $f_1, \ldots, f_i, g_1, \ldots, g_j$. Then, we proceed inductively. Set $P_0^{i,j}(I_1, I_2) = X$, define $P_k^{i,j}(I_1, I_2)$ to be the closure of $V(h_k|P_{k-1}^{i,j}(I_1, I_2)) - V(I_1 + I_2)$ and $\Lambda_k^{i,j}(I_1, I_2)$ to be the part of the cycle $[V(h_k|P_{k-1}^{i,j}(I_1, I_2))]$ supported in $V(I_1 + I_2)$. Finally, let $e_k^{i,j}(I_1, I_2)$ be its multiplicity.

We also define $e_k^{k,0}(I_1, I_2) := e_k(I_1)$ and $e_k^{0,k}(I_1, I_2) := e_k(I_2)$.

For ideals of finite colength, Gassler's definition of mixed multiplicities coincides with Teissier's.

Gassler then showed that the equality of the mixed Segre numbers implied that the integral closures of the ideals agreed. However, in proving the inequalities analogous to those proved in the smooth case, he was forced to define "intrinsic" Segre numbers.

We discuss how to do this for the first step. Let I_1 be an ideal on X of codimension 1, X has dimension at least 2. Let p be the biggest ideal with the property $\Lambda_1(I_1) = \Lambda_1(p)$. Note that in general, the ideal p is not prinicipal, and, also, its polar variety $P_1(p)$ is in general not empty.

Now, consider the normalized blow-up of X by p,

$$\pi : \tilde{X} \to X$$

with exceptional divisor E. We define $\tilde{I}_1 := \pi^* I_1 \mathcal{O}_{\tilde{X}}$. Then $\tilde{p}$ divides $\tilde{I}_1$; we denote the quotients by $\tilde{I}_1' = (\tilde{I}_1 : \tilde{p})$. Furthermore, the line bundle associated with $\tilde{I}_1$ is the product of the line bundles associated with $\tilde{I}_1'$ and $\tilde{p}$. Therefore, using the construction of polar varieties through blowups (see [9],(2.1)), we see

$$P_1(\tilde{I}_1) = P_1(\tilde{I}_1') + P_1(\tilde{p}).$$

Now, we have

$$e_2(I_1) = \text{mult}_0 \pi_* [V(\tilde{p}|P_1(\tilde{I}_1))] + \text{mult}_0 \pi_* \Lambda_2(\tilde{I}_1') + \text{mult}_0 \pi_* [V(\tilde{I}_1'|P_1(\tilde{p}))]$$

$$= \text{mult}_0([V(p|\pi(P_1(\tilde{I}_1')))] + [\Lambda_2(p)] + \pi_* \Lambda_2(\tilde{I}_1') + \pi_* [V(\tilde{I}_1'|P_1(\tilde{p}))]) \qquad *$$

The multiplicity of $\pi_* \Lambda_2(\tilde{I}_1')$, Gassler calls the second intrinsic Segre number of I_1. It is the part of $e_2(I_1)$ which is independent of $\Lambda_1(I_1)$. By successive blowups higher intrinsic numbers could be defined. Because they correspond to the top dimensional cycles of their associated ideal sheaves, they have nice properties. For example, since all components of $V(\tilde{I}_1')$ map to subsets of X of codimension at least 2, we can compute $\text{mult}_0 \pi_* \Lambda_2(\tilde{I}_1')$ by cutting with a general codimension 2 plane H off 0, if X has dimension at least 3.

Using a formula similar to $*$ to define second mixed intrinsic Segre numbers, Gassler also showed that the mixed intrinsic Segre numbers satisfied a generalization of Teissier's mixed multiplicity formula:

$$\hat{e}_2(I_1 I_2) = \sum_{i=0}^{2} \binom{2}{i} \hat{e}_2^{i,2-i}(I_1, I_2)$$

and the Minkowski–type inequality

$$\hat{e}_2(I_1 I_2)^{1/2} \le \hat{e}_2(I_1)^{1/2} + \hat{e}_2(I_2)^{1/2}.$$

Counterexamples are known to both of these results if "ordinary" mixed Segre numbers are used even if there are no "moving" components. In general the interaction between Segre cycles of different dimensions makes it impossible for the fomulas to hold.

The construction of the refined numbers and the arguments that use them are interesting, because they indicate an approach which avoids some of the technical problems associated with trying to extend the Segre numbers in a straight-forward way to modules. A key property of the multiplicity of modules of finite colength is additivity. If N is a submodule of a module M, which is a submodule of a free module P, and the multiplicities $e(M, P)$, $e(N, P)$ are defined, then the multiplicty $e(N, M)$ is defined, and $e(N, P) = e(N, M) + e(M, P)$. So if $e(M, P) = e(N, P)$, then $e(N, M)$ is zero, so M is dependent on N (Cf. [17] for precise statements of these results). This additivity property only holds for the non-zero Segre number of greatest dimension; it does hold for the refined Segre numbers. It fails for the Segre numbers because of the interaction between Segre cycles of different dimension.

References

1. D. A. Buchsbaum and D. S. Rim, A generalized Koszul complex II. Depth and multiplicity, Trans. Amer. Math. Soc. **111** (1963) 197–224.
2. J. Briançon and J. P. Speder, La trivialite topologue n'implique pas les conditions de Whitney, Comptes Rendus ser. A **280** (1975) 365.
3. T. Fukui and L. Paunescu, Stratification theory from the weighted point of view, preprint, 1998.
4. T. Gaffney, Integral closure of modules and Whitney equisingularity, Invent. Math. **107** (1993) 301–322.
5. T. Gaffney, Fibers of polynomial mappings at infinity and a generalized Malgrange condition, Compositio Math. **119** (1999) 157–167.
6. T. Gaffney, $\mathcal{L}^0$ equivalence of maps, Math. Proc. Camb. Phil. Soc. **128** (2000) 479–496.
7. T. Gaffney, Polar multiplicities and equisingularity of map germs, Topology **32** (1993) 185–223.

8. T. Gaffney, Plane Sections, W_f, and A_f, *Real and Complex Singularities* (eds J.W. Bruce and F. Tari), Research Notes in Math. **412** (Chapman and Hall/CRC 2000) pp 16–32.
9. T. Gaffney and R. Gassler, Segre numbers and hypersurface singularities, Jour. Alg. Geom. **8** (1999) 695–736.
10. T. Gaffney and S. Kleiman, Specialization of integral dependence for modules, Invent. math. **137** (1999) 541–574.
11. T. Gaffney and S. Kleiman, W_f and integral dependence, *Real and Complex Singularities* (eds J.W. Bruce and F. Tari), Research Notes in Math. **412** (Chapman and Hall/CRC 2000) pp 33–45.
12. T. Gaffney and D. Massey, Trends in equisingularity theory, *Singularity Theory* (eds Bill Bruce and David Mond), London Math. Soc. Lecture Notes **263** (Cambridge Univ. Press, 1999) pp 207–248.
13. R. Gassler, Segre numbers and hypersurface singularities, thesis, Northeastern University, 1999.
14. M. D. Green, dissertation, Northeastern University, 1997.
15. M. D. Green and D. B. Massey, Vanishing cycles and Thom's A_f conditions, preprint, Northeastern University, 1996.
16. G. M. Greuel, Der Gauss–Manin Zusammenhang isolierter Singularitäten von vollständigen Durchschnitten, dissertation, Göttingen, 1973; also Math. Ann. **214** (1975) 235–66.
17. S.L. Kleiman and A. Thorup, A geometric theory of the Buchsbaum–Rim multiplicity, J. Algebra **167** (1994) 168–231.
18. D. T. Lê, Calculation of Milnor number of isolated singularity of complete intersection, Funct. Anal. Appl. **8** (1974) 127–31.
19. D. T. Lê and C. P. Ramanujam, The invariance of Milnor's number implies the invariance of the topological type, Amer. J. Math. **98** (1976) 67–78.
20. D. T. Lê and B. Teissier, Cycles évanescents, sections planes et conditions de Whitney II, *Singularities* (ed Peter Orlik), Proc. Symp. Pure Math. **40**:2, (Amer. Math. Soc., 1983) pp 65–103.
21. M. Lejeune-Jalabert and B. Teissier, Clôture intégrale des idéaux et equisingularité, chapitre 1, Publ. Inst. Fourier, 1974.
22. E. J. N. Looijenga, Semi-universal deformation of a simple elliptic singularity: Part I unimodularity, Topology **16** (1977) 257–262.
23. D. Massey, Critical points of functions on singular spaces, Topology and its Applications, to appear.
24. D. Mond, Some remarks on the geometry and classification of germs of maps from surfaces to 3-space, Topology **26** (1987) 361–383.
25. A. Parusiński, On the bifurcation set of a complex polynomial with isolated singularities at infinity, University of Sydney preprint 93-46, 1993.
26. A. Parusiński, A note on singularities at infinity of complex polynomials, *Symplectic Singularities and Geometry of Gauge Fields*, Banach Center Publications **39** (Institute of Mathematics, Polish Academy of Sciences, 1997) pp 131–141.
27. L. Paunescu, A weighted version of the Kuiper-Kuo-Bochnak-Lojasiewicz Theorem, J. Alg. Geom. **2** (1993) 69–79.
28. L. Paunescu, V-sufficiency from the weighted point of view, J. Math. Soc. Japan **46** 1994.
29. V. H. J. Perez, Polar multiplicities and equisingularity of map germs from $C^3 \to C^3$, thesis, Univ. de Sao Paulo, 2000.

30. D. Siersma and M. Tibăr, Singularities at infinity and their vanishing cycles, Duke Math. J. **80** (1995) 771–783.

31. B. Teissier, Cycles évanescents, sections planes et conditions de Whitney, *Singularités à Cargèse*, Astérisque **7–8** (1973) 285–362.

32. B. Teissier, The hunting of invariants in the geometry of the discriminant, *Real and complex singularities, Oslo 1976* (ed P. Holm) (Sijthoff & Noordhoff, 1977) pp 565–678.

33. B. Teissier, Sur une inégalité à la Minkowski pour les multiplicités, Ann. of Math. **106** (1978) 40–44.

34. B. Teissier, On a Minkowski–type inequality for Multiplicities II, *C. P. Ramanujam–a tribute* Studies in Math. **8** (Tata Institute, 1978) pp 347–361.

35. M. Tibăr, Asymptotic equisingularity and topology of complex hypersurfaces, Internat. Math. Research Notices **18** (1998) 979–990.

36. M. Tibăr, Topology at infinity of polynomial mappings and Thom regularity condition, Compositio Math. **111** (1998) 89–109.

37. R. Vohra, Equisingularity of map germs from $C^n \to C^2$, thesis, Northeastern University, 2000.

Appendix
The exceptional fiber of a generalized conormal space

Steven Kleiman
Department of Mathematics, Cambridge, MA
(kleiman@math.mit.edu)

Anders Thorup
Department of Mathematics, University of Copenhagen
(thorup@math.ku.dk)

Let X be the germ of a reduced analytic space of pure dimension d. Let $\mathcal{I}$ be a free $\mathcal{O}_X$-module, and $\mathcal{M} \subset \mathcal{N} \subset \mathcal{I}$ two nested submodules with $\mathcal{M} \neq \mathcal{N}$. Assume $\mathcal{M}$ and $\mathcal{N}$ are generically equal and free of rank e. Set $r := d + e - 1$.

Set $C := \mathrm{Projan}(\mathcal{O}_X[\mathcal{M}])$ where $\mathcal{O}_X[\mathcal{M}] \subset Sym\,\mathcal{I}$ is the subalgebra generated by $\mathcal{M}$ in the symmetric algebra on $\mathcal{I}$. Then C has pure dimension r, see below. Let $c: C \to X$ denote the structure map. Let W be the closed set where $\mathcal{N}$ is not integral over $\mathcal{M}$, and set $E := c^{-1}W$. Set $Y := \mathrm{Supp}(\mathcal{N}/\mathcal{M})$ and $F := c^{-1}Y$.

We will give simple new proofs of the following two theorems. These proofs are geometric versions of some remarkable algebraic proofs given by Simis, Ulrich, and Vasconcelos in [6]. After we complete our proofs, we will discuss some history.

THEOREM A.1. *If $\mathcal{N}$ is not integral over $\mathcal{M}$, then E has dimension $r - 1$.*

THEOREM A.2. *If $\mathcal{N}$ is free, then $E = F$, and F has pure dimension $r - 1$.*

To prove these theorems, form the following two natural commutative diagrams:

$$
\begin{array}{ccc}
B \xrightarrow{\ b\ } P \\
q\downarrow \quad\quad p\downarrow \\
C \xrightarrow{\ c\ } X
\end{array}
\quad\text{and}\quad
\begin{array}{ccc}
D \longrightarrow Z \\
\downarrow \quad\quad \downarrow \\
F \longrightarrow Y
\end{array}
$$

Here $P := \mathrm{Projan}(\mathcal{O}_X[\mathcal{N}])$, and $Z \subset P$ has ideal $\mathcal{M} \cdot \mathcal{O}_X[\mathcal{N}]$. Also B is the blowup of P along Z, and D is the exceptional divisor.

Since $\mathcal{M}$ and $\mathcal{N}$ are submodules of the free module $\mathcal{I}$, the components of C and of P map bijectively onto those of X. Since $\mathcal{M}$ and $\mathcal{N}$ are free of rank e on a dense open subset of X, every component of C and of P has dimension r. Since $\mathcal{M}$ and $\mathcal{N}$ are equal on a dense open set, F is nowhere dense in C, and Z is nowhere dense in P. Hence, $\dim F < r$. Furthermore, the components of B map bijectively onto those of P and of C, and every component has dimension r. So every component of D has dimension $r - 1$. We will prove the theorems by analyzing the image qD in F.

The diagrams above are not product squares, but the first does provide a closed embedding $B \hookrightarrow C \times_X P$. Hence, $D \to C$ will be a closed embedding if $Z \to X$ is one. Now, $Z = \mathrm{Projan}(\mathcal{O}_X[\mathcal{N}]/\mathcal{E})$ where $\mathcal{E} := \mathcal{M} \cdot \mathcal{O}_X[\mathcal{N}]$. If $\mathcal{N}/\mathcal{M}$ is a cyclic $\mathcal{O}_X$-module, then $\mathcal{O}_X[\mathcal{N}]/\mathcal{E}_Z$ is a quotient of a polynomial ring in one variable over $\mathcal{O}_X$; hence, then $Z \to X$ is a closed embedding, and so $D \to C$ is one too.

Let $x \in C$. Then $x \notin qD$ if and only if there is a Zariski neighborhood of x on which the map $f \colon (P - Z) \to C$ is finite, that is, proper with finite fibers; see [3], (2.6), p.183 for example. The latter condition can be put algebraically, in terms of the ring of the germ of $P - Z$ localized along $f^{-1}x$ and of the (local) ring of the germ of C at x; namely, the condition holds if and only if the first ring is integral over the second. Let $\mathbf{p} \subset \mathcal{O}_X[\mathcal{M}]$ be the homogeneous prime of x. Then the second ring is just $\mathcal{O}_X[\mathcal{M}]_{(\mathbf{p})}$, namely, the ring of elements of degree 0 in the homogeneous localization at $\mathbf{p}$ of $\mathcal{O}_X[\mathcal{M}]$, and the first ring is just $\mathcal{O}_X[\mathcal{N}]_{(\mathbf{p})}$. So, if $cx \notin W$, then $x \notin qD$; thus $qD \subset E$.

Write $\mathcal{N}$ as a finite sum, $\mathcal{N} = \sum \mathcal{N}_i$, of submodules $\mathcal{N}_i$ that contain $\mathcal{M}$. Let q_i and D_i be the analogues of q and D. Then $qD = \bigcup q_i D_i$. Indeed, in view of the discussion in the preceding paragraph, this equation holds because $\mathcal{O}_X[\mathcal{N}]_{(\mathbf{p})}$ is, obviously, integral over $\mathcal{O}_X[\mathcal{M}]_{(\mathbf{p})}$ if an only if every $\mathcal{O}_X[\mathcal{N}_i]_{(\mathbf{p})}$ is so.

Take the $\mathcal{N}_i$ so that $\mathcal{N}_i/\mathcal{M}$ is cyclic; just take a finite system of generators of $\mathcal{N}$, and adjoin the ith to $\mathcal{M}$ to get $\mathcal{N}_i$. Then $D_i \xrightarrow{\sim} qD_i$. Hence every component of qD has dimension $r - 1$. Now, qD is empty if and only if Z is. However, Z is empty if and only if $\mathcal{O}_X[\mathcal{N}]$ is integral over $\mathcal{O}_X[\mathcal{M}]$, in other words, if and only if $\mathcal{N}$ is integral over $\mathcal{M}$. Since $qD \subset E$, Theorem A.1 is thus proved.

To prove Theorem A.2, assume $\mathcal{N}$ is free. We will prove $qD = F$. Then F will have pure dimension $r-1$ since qD does, as we just proved. Moreover, $qD \subset E \subset F$; whence, we will have $E = F$. So assume $x \in F$, but $x \notin qD$, and we will arrive at a contradiction. Since $x \notin qD$, in the notation above, $\mathcal{O}_X[\mathcal{N}]_{(\mathbf{p})}$ is integral over $\mathcal{O}_X[\mathcal{M}]_{(\mathbf{p})}$.

Set $y := c(x) \in Y$. The issue being local at y, replace X by its germ at y, replace $\mathcal{M}$ and $\mathcal{N}$ by their restrictions, and so forth. Consider the map of **C**-vector spaces, $\mathcal{M}(y) \to \mathcal{N}(y)$. Take a **C**-basis of the image, lift these elements back to $\mathcal{M}$, and use them to generate a submodule $\mathcal{F} \subset \mathcal{M}$. Next, complete that **C**-basis to a **C**-basis of $\mathcal{N}(y)$, lift the new elements up to $\mathcal{N}$, and use them to generate a submodule

$\mathcal{N}' \subset \mathcal{N}$. Then $\mathcal{N} = \mathcal{F} \oplus \mathcal{N}'$. Moreover, $\mathcal{F}$ and $\mathcal{N}'$ are free since $\mathcal{N}$ is so. Furthermore, $\mathcal{N}' \neq 0$ since $\mathcal{N} \neq \mathcal{M}$ as $y \in Y$. Set $\mathcal{M}' := \mathcal{M} \cap \mathcal{N}'$. Then $\mathcal{M} = \mathcal{F} \oplus \mathcal{M}'$. Moreover, by construction, $\mathcal{M}' \subset \mathbf{m}_y \mathcal{N}'$ where $\mathbf{m}_y$ denotes the maximal ideal of y. Set $\mathbf{p}' := \mathbf{p} \cap \mathcal{O}_X[\mathcal{M}']$.

Set $\mathcal{A} := \mathcal{O}_X[\mathcal{M}']_{\mathbf{p}'}$ and $\mathcal{B} := \mathcal{O}_X[\mathcal{N}']_{\mathbf{p}'}$. Let's prove that $\mathcal{B}$ is integral over $\mathcal{A}$. Since $\mathcal{N}'$ generates $\mathcal{B}$ as an $\mathcal{A}$-algebra, it suffices to show that each $t \in \mathcal{N}'$ satisfies an equation of integral dependence over $\mathcal{A}$. Let $u \in \mathcal{M}$, but $u \notin \mathbf{p}$. Then t/u satisfies an equation of integral dependence over $\mathcal{O}_X[\mathcal{M}]_{(\mathbf{p})}$. Clearing denominators yields an equation,

$$st^n + a_1 t^{n-1} + \cdots + a_n = 0,$$

where $n \geq 1$ and $s, a_i \in \mathcal{O}_X[\mathcal{M}]$, but $s \notin \mathbf{p}$. Now, $\mathcal{O}_X[\mathcal{N}]$ is a polynomial ring over $\mathcal{O}_X[\mathcal{N}']$ because $\mathcal{N} = \mathcal{F} \oplus \mathcal{N}'$ with $\mathcal{F}$ and $\mathcal{N}'$ free. Hence, since $\mathcal{M}' := \mathcal{M} \cap \mathcal{N}'$, we may view s and the a_i as polynomials with coefficients in $\mathcal{O}_X[\mathcal{M}']$. Some coefficient s' of s is not in $\mathbf{p}'$ since $s \notin \mathbf{p}$. Take the corresponding coefficients of the a_i, and divide them by s'. The quotients are the coefficients of the sought equation of integral dependence for t over $\mathcal{A}$.

Therefore, $\mathcal{B}$ is a finitely generated $\mathcal{A}$-module. Pick a finite set of homogeneous generators from $\mathcal{O}_X[\mathcal{N}']$, say of degree at most $m - 1$. Given k, let $\mathcal{Q}_k$ denote the $\mathcal{A}$-submodule of $\mathcal{B}$ generated by the homogeneous elements of $\mathcal{O}_X[\mathcal{N}']$ of degree k. Then $\mathcal{Q}_m \subset \mathcal{M}' \mathcal{Q}_{m-1}$. Now, $\mathcal{M}' \subset \mathbf{m}_y \mathcal{N}'$ and $\mathcal{N}' \mathcal{Q}_{m-1} = \mathcal{Q}_m$. Hence, $\mathcal{Q}_m \subset \mathbf{m}_y \mathcal{Q}_m$. So $\mathcal{Q}_m$ vanishes by Nakayama's lemma over the local ring $\mathcal{A}$. However, $\mathcal{N}' \neq 0$. Hence we have a contradiction. Thus Theorem A.2 is proved.

Theorem A.1 was proved for the first time by the authors in [3], (10.2). That treatment is rather complicated, involving much of the modern theory of the Buchsbaum–Rim multiplicity, as developed in [3]. Then Theorem A.1 was used by Gaffney and the first author in [1], §2 as a basis on which to generalize Teissier's principle of specialization of integral dependence (PSID) from ideals to modules.

The PSID for modules was employed in [1] to discuss, among other applications, the following generalization to ICIS germs of the celebrated Lê–Saito theorem for $\mathbf{C}^n$: a family satisfies Thom's A_f-condition if and only if certain Milnor numbers remain constant. Indeed, [1], §5 proves that A_f implies constancy, and proves a weak converse. The definitive converse was proved by Gaffney and Massey in [2], (5.8); see also [4], (2.2) and [5], (1.7). The key new ingredient is Theorem A.2.

Theorem A.2 was proved by Gaffney and Massey in [2], (5.7) using two remarkable new ideas. First, they expressed F as the union of closed sets, each isomorphic to the exceptional divisor in a suitable blowup. Second, they used complex analytic paths in a delicate way to identify the points of F as suitable limits. These ideas were developed, in the abstract algebraic-geometric setting of [3], by the authors in [5], and then used to recover Theorem A.1 and to prove Theorem A.2 in this general setting. The new proof of Theorem A.1 in [5] is, to quote [5], (1.1), "substantially shorter, simpler, and more direct than the old."

The brand-new proofs of the two theorems are, in turn, substantially shorter, simpler, and more direct than those in [5]. There is now no use of paths. There are again two remarkable new ideas. The first is to reduce to the case where $\mathcal{N}/\mathcal{M}$ is cyclic so that q embeds D in C. This idea is reminiscent of Gaffney and Massey's idea of expressing F as a union of copies of exceptional divisors, but their blowups are more involved. This new idea appeared in the proofs of Theorems 2.6 and 2.2 in the first version of [6], but was omitted from the second. The second new idea is to reduce the general case of Theorem A.2 to the special case where $\mathcal{M} \subset \mathbf{m}_y \mathcal{N}$, by splitting off a free summand from $\mathcal{M}$ and $\mathcal{N}$. This idea appears in the proof of Theorem 4.4 in the second version of [6]. Although in this appendix the proofs are carried out in the complex analytic case, they work with little change in the general setting of [5] and [6].

References

1. T. Gaffney and S. Kleiman, Specialization of integral dependence for modules, Invent. math. **137** (1999) 541–574.
2. T. Gaffney and D. Massey, Trends in equisingularity theory, *Singularity Theory* (eds Bill Bruce & David Mond), London Math. Soc. Lecture Notes **263** (Cambridge University Press, 1999) pp 207–248.
3. S. L. Kleiman and A. Thorup, A geometric theory of the Buchsbaum–Rim multiplicity, J. Algebra **167** (1994) 168–231.
4. S. L. Kleiman, Equisingularity, multiplicity, and dependence, *Commutative algebra and algebraic geometry* (ed F. Van Oystaeyen), Dekker lect. notes in pure and applied math **206**, 1999.
5. S. Kleiman and A. Thorup, Conormal geometry of maximal minors, J. Algebra **230** (2000) 204–21.
6. A. Simis, B. Ulrich and W. V. Vasconcelos, Codimension, multiplicities and integral extensions, preprints, 1999.

Nonlinear Sections of Nonisolated Complete Intersections

James Damon*
University of North Carolina,
(jndamon@math.unc.edu)

Introduction

By a series of discoveries during the past thirty years, a distinguished list of researchers have provided us with a marvelous vista of singularity theory as it applies to isolated singularities, especially isolated complete intersection singularities. This began with Milnor's seminal monograph on isolated hypersurfaces singularities [72], which introduced as a principal tool in the study of isolated singularities the Milnor fibration of the singularity. The basic results were extended by Hamm [43] to isolated complete intersection singularities (ICIS). There has followed a succession of revelations concerning the topology, local geometry, and deformation theory of ICIS using: De Rham cohomology and Gauss-Manin connection, intersection pairing and monodromy, mixed Hodge structures and spectrum, structure of discriminants, equsingularity via multiplicities; and deformation theory. We refer to e.g. [60] and [7, vol 2] where many of these results are presented.

If we were to seek a comparable view of the more complicated non-isolated singularities, then the results for ICIS provide a virtual "wish list"of types of results to be obtained. However, now the vista is considerably clouded, lacking many of the details so apparent for ICIS, although revealing general features via techniques involving stratification theory, resolution of singularities, etc.

A sample of the kinds of questions involving nonisolated singularities which we will consider involve e.g.: topology of complements of hyperplane arrangements, the topology of boundary singularities of complete intersections, critical points of functions $f_1^{\lambda_1} \cdots f_r^{\lambda_r}$ appearing in hypergeometric functions, minimum $\mathcal{A}_e$-codimenson for germs $f : \mathbb{C}^n, 0 \to \mathbb{C}^p, 0$ in a given contact class, de Rham cohomology of highly singular spaces, higher multiplicities à la Teissier for nonisolated singularities together with Buchbaum–Rim multiplicities of modules; as well as properties of discriminants and bifurcation sets for various notions of equivalence. These examples share no obvious common feature

* Partially supported by a grant from the National Science Foundation

D. Siersma et al. (eds.), New Developments in Singularity Theory, 405–445.
© 2001 Kluwer Academic Publishers. Printed in the Netherlands.

except ultimately they concern properties of highly nonisolated singular spaces.

However, there are two key obstacles to extending results for ICIS to such highly singular spaces. First, the Milnor fibration no longer has the connectivity properties possessed in the ICIS case. By the Theorem of Kato–Matsumoto [53], for a hypersurface singularity $(X, 0)$ of dimension n with singular set $\text{sing}(X)$, the connectivity of the Milnor fiber decreases by $\dim_{\mathbb{C}}(\text{sing}(X))$, i.e. it is in general only $(n - 1) - \dim_{\mathbb{C}}(\text{sing}(X))$ connected. Second, unlike isolated singularities, nonisolated singularities generally do not have a versal deformation. Hence, we lack a relation between the topology of Milnor fibers and the structure of discriminants.

Because of the first complication, there is a range of cohomology groups to be determined to understand the topology of the Milnor fiber. By comparison with the state of affairs reported by Randell [88] at the end of the 70's, considerable advances for low dimensional singular sets has been accomplished through the work led by Siersma [95, 1-4] and the Dutch group of Pellikaan [85, 86], Van Straten [110], de Jong [50], etc. along with coworkers Massey [65, 66] [67], Tibar [108], Nemethi [78], etc. This work will be reported on by Siersma at this conference. However, for highly singular spaces this method becomes increasingly difficult to apply.

We describe an alternate approach, which has its origins in algebraic geometry beginning with Hilbert. It applies to large classes of nonisolated singularities which arise as nonlinear sections of fixed "model nonisolated singularities". We use a Thom-Mather type of group of equivalence, $\mathcal{K}_V$, to analyze the singularities of such sections. This allows us to overcome both obstacles.

Beginning with joint work with Mond on hypersurfaces, we obtain for nonlinear sections of nonisolated singularities, a singular analogue of the Milnor fibration. Using a result of Lê [54], we show it retains the same connectivity properties as for Milnor fibers of an ICIS. A crucial ingredient for further analyzing the topology is the notion of freeness introduced by Saito [93] for divisors. If the model singularities are "free divisors and complete intersections", then the freeness first provides the algebraic condition needed to compute the singular Milnor number as the length of a determinantal module. In the hypersurface case, the module is the normal space for another equivalence $\mathcal{K}_H$. It agrees with the $\mathcal{K}_V$-normal space in the weighted homogeneous case, yielding a "$\mu \geq \tau$" result.

Second, in the complete intersection case, there is a generalization of the Lê–Greuel formula which computes the relative singular Milnor number of a divisor on an ICIS as the length of a determinantal module. When the divisor has an isolated singularity, we recover the Lê–Greuel formula as a special case. However, it applies more generally to include "arrangements of hypersurfaces"on complete intersections, singular Milnor fibers for "boundary singularities"(or equivalently complete intersection singularities on a divisor at infinity), projections of discriminants, etc. It further extends to general nonisolated complete intersections, where there is also an alternate approach, which computes the singular Milnor number of a nonisolated complete intersection as an alternating sum of singular Milnor numbers of unions of hypersurfaces.

These results allow us to introduce and compute higher multiplicities à la Teissier for nonisolated complete intersections. For hyperplane arrangements and their nonlinear generalizations, these multiplicities have surprising relations with other invariants of arrangements and with the topology of the complement.

To further explore the topology of nonlinear sections, we describe how methods using de Rham cohomology can be introduced. These include results of Alexandrov and Mond and coworkers which compute cohomology using the complex of logarithmic forms. Besides being used to compute the cohomology of the complement of a free divisor [17], and certain local cohomology [2, 3], the logarithmic complex is used by Mond to construct the correct complex for computing the de Rham cohomology of the singular Milnor fiber and defining a Gauss–Manin connection for the singular Milnor fibration [75].

Lastly, we turn in Part IV to consider why discriminants and bifurcation sets so frequently are free divisors. Except for results concerning discriminants for space curves and for functions on them by Van Sraten [111], Goryunov [38], Mond-Van Straten [77], results on freeness of discriminants can be summarized by the motto "Cohen–Macaulay of codim 1 + genericity of Morse–type singularities implies freeness".

We indicate how this specifically applies for various equivalences, as well as when it does not. When either condition fails this leads to the introduction of the notion of a "Cohen–Macaulay reduction"and a weaker *Free* Divisor structure*. However, the Free* Divisor structure still can be used to determine the topological structure as above.

One final point concerns the role that the structure of modules, especially determinantal modules, and (implicitly) Buchbaum–Rim multiplicities play in all of this work. This changes the emphasis from invariants associated to local rings to invariants of modules of logarithmic vector fields. This parallels the increasingly important role played by modules, their integral closures, reductions, and Buchbaum–Rim multiplicities in the work of Gaffney on Whitney equisingularity [32, 33], [34], as well as its influence on the work of Kirby and Rees [48] [90], Kleiman-Thorup [49], and Henry-Merle [45] (as well as the many other references to be given in Gaffney's lectures). Because of this we raise several natural questions regarding the intrinsic nature of the results we describe and the form they might take as we move "beyond freeness".

I Nonisolated Singularities as Nonlinear Sections

1.   Singularities arising as Nonlinear Sections and $\mathcal{K}_V-$ equivalence

We consider the approach to singularities which represents certain singularities and their deformations as sections of standard model singularities. For example, by the Hilbert–Burch theorem [46], [16], Cohen–Macaulay singularities of codimension 2 can be represented by the $n \times n$ minors of an $n \times (n+1)$ matrix of holomorphic germs. This was extended to their deformations by Schaps [94]. Buchsbaum–Eisenbud characterized codimension 3 Gorenstein singularities [13] as the Pfaffians of principal minors of skew–symmetric matrices. A general criterion was given by Buchweitz [15, Chap 4, 5] who defined a deformation theory for sections and showed that for certain singularities, their deformations can be represented as nonlinear sections of "very rigid singularities". Not all deformations of singularities can be so represented, as illustrated by the example of Pinkam [87, 8.2] of a surface singularity which is a cone on a rational curve of degree 4 in $\mathbb{P}^4$ and can be represented as the 2×2 minors of either a certain 2×4 matrix or a 3×3 symmetric matrix. Each representation gives rise to distinct components in the base of the versal deformation.

$$
\begin{array}{ccc}
\mathbb{C}^n, 0 & \xrightarrow{\ f_0\ } & \mathbb{C}^p, 0 \\
\uparrow & & \uparrow \\
f_0^{-1}(V) = V_0, 0 & \longrightarrow & V, 0
\end{array}
\tag{1}
$$

We shall be concerned instead with the properties of nonisolated singularities and their deformations obtained from V by nonlinear sections f_0 and their unfoldings (this ignores the problem of missing (flat) deformations). Given $V, 0 \subset \mathbb{C}^p, 0$, we consider holomorphic germs $f_0 : \mathbb{C}^n, 0 \to \mathbb{C}^p, 0$ which we view as nonlinear sections of V so that the singularity $V_0 = f_0^{-1}(V)$ is a pullback as given in (1). Here f_0 does not have to be a germ of an embedding.

EXAMPLE 1.1.   Any germ $f_0 : \mathbb{C}^n, 0 \to \mathbb{C}^p, 0$ of finite singularity type, has a stable unfolding $F : \mathbb{C}^{n'}, 0 \to \mathbb{C}^{p'}, 0$. If $g_0 : \mathbb{C}^p, 0 \to \mathbb{C}^{p'}, 0$

denotes the inclusion, then the discriminant $D(f_0) = g_0^{-1}(D(F))$, with $D(F)$ the discriminant of F. If $n < p$, then $D(F)$ is the image of F. How can the properties of $D(f_0)$ be deduced from those of g_0 and $D(F)$?

For properties of nonlinear sections f_0 of $V, 0 \subset \mathbb{C}^p, 0$, we follow the Thom–Mather approach by defining a group of equivalences $\mathcal{K}_V$ [19] acting on sections f_0 and their unfoldings to capture the ambient equivalence of $V_0, 0 = f_0^{-1}(V), 0$. $\mathcal{K}_V$ is a subgroup of the contact group $\mathcal{K}$, introduced by Mather [68] (also see Tougeron [109]), and consists of germs of diffeomorphisms of $\mathbb{C}^{p+n}, 0$ of the form $\Psi(x, y) = (\Psi_1(x), \Psi_2(x, y))$.

$$\mathcal{K}_V = \{\Psi \in \mathcal{K} : \Psi(\mathbb{C}^n \times V) \subseteq \mathbb{C}^n \times V\}. \tag{2}$$

It acts on sections by the restriction of the action of $\mathcal{K}$: graph($\Psi \cdot f_0$) = Ψ(graph(f_0)).

We can extend $\mathcal{K}_V$ to the group of unfolding–equivalences $\mathcal{K}_{V,un}$ acting on unfoldings with any fixed number of unfolding parameters $u \in \mathbb{C}^q$. These groups (together with the associated unfolding groups) are "geometric subgroups of $\mathcal{A}$ or $\mathcal{K}$" and satisfy the basic theorems of singularity theory, especially the versality and finite determinacy theorems [18]. In fact, it is shown in [19] that such results also hold for $\mathcal{K}_V$–equivalence in the real case for smooth germs provided that V is real analytic and coherent (in the sense of Malgrange [63]).

To define the associated extended tangent spaces (which are the deformation theoretic tangent spaces), we must introduce the module of "logarithmic vector fields". We let θ_p denote the module of germs of vector fields on $\mathbb{C}^p, 0$. If $V, 0 \subset \mathbb{C}^p, 0$ is a germ of an analytic set, let $I(V)$ denote the ideal of germs vanishing on V. Then (following Saito [93]) we define

$$\text{Derlog}\,(V) = \{\zeta \in \theta_p : \zeta(I(V)) \subseteq I(V)\}.$$

This is the module of *logarithmic vector fields*, which are vector fields on $\mathbb{C}^p, 0$ tangent to V. If Derlog (V) is generated by $\zeta_0, \ldots, \zeta_r$, the extended $\mathcal{K}_V$ tangent space is computed [19]

$$T\mathcal{K}_{V,e} \cdot f_0 = \mathcal{O}_{\mathbb{C}^n,0}\left\{\frac{\partial f_0}{\partial x_1}, \ldots, \frac{\partial f_0}{\partial x_n}, \zeta_0 \circ f_0, \ldots, \zeta_r \circ f_0\right\} \tag{3}$$

(the R–module generated by $\varphi_1, \ldots, \varphi_k$ is denoted by $R\{\varphi_1, \ldots, \varphi_k\}$, or just $R\{\varphi_i\}$ if k is understood). The analogue of the deformation tangent space T^1 is the extended $\mathcal{K}_V$ normal space

$$N\mathcal{K}_{V,e} \cdot f_0 = \theta(f_0)/T\mathcal{K}_{V,e} \cdot f_0 \simeq \mathcal{O}_{\mathbb{C}^n,0}^{(p)}/T\mathcal{K}_{V,e} \cdot f_0$$

We give an important property of this equivalence (see [20, §1]).

EXAMPLE 1.2. *Invariance under suspension and projection*   If $i : \mathbb{C}^p, 0 \hookrightarrow \mathbb{C}^{p+r}, 0$ denotes inclusion and $\pi : \mathbb{C}^{p+r}, 0 \to \mathbb{C}^p, 0$ projection, then as $\mathcal{O}_{\mathbb{C}^n,0}$–modules,

$$NK_{V,e} \cdot f_0 \simeq NK_{V\times\mathbb{C}^r,e} \cdot i \circ f_0 \quad \text{and} \quad NK_{V\times\mathbb{C}^r,e} \cdot f \simeq NK_{V,e} \cdot \pi \circ f_0$$

Moreover, suspension and projection preserve equivalence classes for K_V and $K_{V\times\mathbb{C}^r}$. Hence, for investigating nonlinear sections we may replace V by $V \times \mathbb{C}^q$ and retain both the topological and deformation theoretic properties of the germ V_0.

Algebraic and Geometric Transversality

Although any analytic germ $V_0, 0 \subset \mathbb{C}^n, 0$ is the zero set ($= f_0^{-1}\{0\}$) for some analytic germ f_0, in general such a germ f_0 will not be transverse to 0 off $0 \in \mathbb{C}^n$. In order to ensure that properties of V are passed on to V_0, we require that f_0 be transverse to V outside 0. The notion of transversality we use depends upon the interpretation of "$T_y V$" at a singular point $y \in V$. If we use $T_y S_i$, where S_i is the stratum of the canonical Whitney stratification of V, then we obtain "geometric transversality". However, for algebraic considerations, the appropriate version of transversality is more subtle, and invokes Derlog (V).

For an $\mathcal{O}_{\mathbb{C}^p,0}$–submodule $M \subset \theta_p$ generated by $\{\zeta_1, \dots, \zeta_r\}$, we let $\langle M \rangle_y$ be the subspace of $\mathbb{C}^p$ generated by $\{\zeta_{1(y)}, \dots, \zeta_{r(y)}\}$. This is well–defined for y in a neighborhood of 0. Then, we define the "logarithmic tangent spaces" $T_{log} V_{(y)} = \langle \text{Derlog}\,(V)\rangle_y$. Then, f_0 is algebraically transverse to V at $x \in \mathbb{C}^n$ if

$$df_0(T_x\mathbb{C}^n) + T_{log}V_{(f_0(x))} = T_{(f_0(x))}\mathbb{C}^p$$

Then, just as for K equivalence, by [19] there is a geometric characterization for f_0 having *finite* K_V–*codimension* (i.e. $\dim_\mathbb{C}(NK_{V,e} \cdot f_0) < \infty$); namely, f_0 has finite K_V codimension iff f_0 is algebraically transverse to V off 0, i.e. at all x in a punctured neighborhood of 0. We can analogously define algebraic (or geometric) transversality of germs of singular varieties, as well as algebraic (or geometric) general position. We always have

$$T_{log}V_{(y)} \subseteq T_y S_i \tag{4}$$

Hence,

algebraic transversality $\Longrightarrow$ geometric transversality.

However, (4) may be strict inclusion, and then the algebraic tangent spaces are tangent to a (possibly singular) foliation of a canonical statum. The strata S_i for which (4) is equality at all points $y \in S_i$ are called *holonomic strata*, and the codimension of the complement of the set of holonomic strata is called the *holonomic codimension* and denoted $hn(V)$. If $n < hn(V)$, then the reverse implication in (4) holds.

Low $\mathcal{K}_V$-codimension germs

f_0 has $\mathcal{K}_{V,e}$-codimension 0 iff it is algebraically transverse to V at 0. By the versality theorem, such an f_0 is already $\mathcal{K}_V$-versal, so any unfolding of f_0 is $\mathcal{K}_V$-trivial. If moreover f_0 is a germ of an embedding, then by the versality theorem, $V \simeq V_0 \times \mathbb{C}^{p-n}$.

For $\mathcal{K}_{V,e}$-codimension 1 we give a definition.

DEFINITION 1.3. *Given $V, 0 \subset \mathbb{C}^p, 0$ and an integer $n > 0$, then a germ $g : \mathbb{C}^n, 0 \to \mathbb{C}^p, 0$ is a Morse-type singularity in dimension n if g has $\mathcal{K}_{V,e}$-codim $= 1$ and is $\mathcal{K}_V$-equivalent to a germ f_0, so that for a common choice of local coordinates, both f_0 and V are weighted homogeneous. We furthermore say V has a Morse-type singularity in dimension n at x if there is a germ $g : \mathbb{C}^n, 0 \to \mathbb{C}^p, x$ which is a Morse-type singularity (using $\mathcal{K}_{(V,x)}$-equivalence).*

Although V is unspecified, such singularities can be precisely classified using $\mathcal{K}_V$-equivalence [25, Lemma 4.12] and [26, Lemma 7.2].

LEMMA 1.4 (Local Normal Form). *Let $f_0 : \mathbb{C}^n, 0 \to \mathbb{C}^p, 0$ be a Morse-type singularity for $V, 0 \subset \mathbb{C}^p, 0$. Then, up to $\mathcal{K}_V$-equivalence, we may assume $V, 0 = \mathbb{C}^r \times V', 0$ for $V', 0 \subset \mathbb{C}^{p'}, 0$ with $T_{log}V'_{(0)} = 0$, and with respect to coordinates for which $V', 0$ is weighted homogeneous, f_0 has the form*

$$f_0(x_1, \ldots, x_n) = (0, \ldots, 0, x_1, \ldots, x_{p'-1}, \sum_{j=p'}^{n} x_j^2) \qquad (5)$$

Remark. The condition of weighted homogeneity in Definition 1.3 is not needed to obtain the normal form (5). However, it is essential so that V will still be weighted homogeneous in (5) for questions involving freeness of discriminants in part IV.

The actual singularity obtained as a nonlinear section depends upon V and can vary considerably, see e.g. the possible Morse–type singularities in dimension 2 for discriminants of stable multigerms [25, §4]. Nonetheless, we see in §3 that the topology (i.e. homotopy type) of a Morse–type singularity exactly matches ordinary Morse singularities.

Relation between $\mathcal{A}$ and $\mathcal{K}_V$-equivalence

To see the relevance of sections for other equivalences, we reconsider
(1) except now allow f_0 to be a multigerm with F its stable unfolding.

THEOREM 1.5 ([20]). *There are the following relations between $\mathcal{A}$-
equivalence for f_0 and $\mathcal{K}_{D(F)}$-equivalence for g_0:*

 1. *f_0 has finite $\mathcal{A}$-codimension iff g_0 has finite $\mathcal{K}_V$-codimension;*

 2. *if either is finite, then there is the isomorphism of $\mathcal{O}_{\mathbb{C}^p,0}$-modules*

$$N\mathcal{A}_e \cdot f_0 \quad \simeq \quad N\mathcal{K}_{V,e} \cdot g_0$$

   Let g be an unfolding of g_0, with pullback of F denoted by f.

 3. *If g is a $\mathcal{K}_V$-trivial unfolding (resp. family) then f is an $\mathcal{A}$-trivial
    unfolding (resp. family);*

 4. *g is $\mathcal{K}_V$-versal iff f is $\mathcal{A}$-versal.*
    Remark. Such a classification by sections has several extensions
such as for $\mathcal{K}$-equivalence of unfoldings for Mond-Montaldi [76], $\mathcal{A}$-
equivalence of unfoldings of hypersurface germs [22, §11], etc.

The preceding theorem relates the $\mathcal{A}$ properties and classification of
germs with those for $\mathcal{K}_V$-equivalence of sections. For a stable simple multigerm F, Morse-type singularities for sections of $V = D(F)$
provide the $\mathcal{A}_e$-codimension 1 multigerms in the contact class of F.
The classification of $\mathcal{A}_e$-codimension 1 germs for simple $\mathcal{K}$ classes follows from Goryunov [37]; and for unfoldings of simple hypersurface
germs and ICIS surface singularities, using local duality and results
from Wahl, Looijenga, and this author (described in [25, §6]). These
yield the $\mathcal{K}_{V,e}$-codimension 1 germs for discriminants of simple stable
germs. Then, a construction using the product union [25] gives the
$\mathcal{K}_{V,e}$-codimension 1 germs for the discriminants of multigerms. In turn,
this gives the $\mathcal{A}_e$-codimension 1 multigerms. Other approaches to their
classification are given in Rieger [91] and by Wik Atique, Cooper, and
Mond [8].

2. Role of Freeness for Divisors and Complete Intersections

To proceed further with properties of nonlinear sections, we need further information about $(V,0)$. Rather than consider its local ring, we
seek instead conditions on Derlog (V) which reveal the properties and

topology of V, and of the singularities arising as nonlinear sections of V. In fact, $\text{Derlog}(V)$ is a Lie algebra; and Hauser and Müller [44] prove that the isomorphism class of $\text{Derlog}(V)$ in an appropriate sense uniquely determines $(V, 0)$.

Among local rings, regular local rings are the simplest and correspond to smooth submanifolds. The simplest structure for $\text{Derlog}(V)$ occurs when $V, 0 = \mathbb{C}^{p-1}, 0 \subset \mathbb{C}^p, 0$ is a smooth hypersurface. Then, $\text{Derlog}(V)$ is a free module generated by $\{\frac{\partial}{\partial y_1}, \ldots, \frac{\partial}{\partial y_{p-1}}, y_p \frac{\partial}{\partial y_p}\}$. However, as discovered by Saito [93], and independently by Arnold [5], this property does not characterize smooth hypersurfaces $(V, 0)$. There are many important highly singular hypersurface singularities for which $\text{Derlog}(V)$ is free of rank p. It is still true that the freeness reveals much about their topology and that of nonlinear sections. This leads to Saito's definition [93].

DEFINITION 2.1. *A hypersurface $V, 0 \subset \mathbb{C}^p, 0$ is called a Free Divisor if* $\text{Derlog}(V)$ *is a free $\mathcal{O}_{\mathbb{C}^p, 0}$-module. (necessarily of rank p)*

Initially three basic classes of free divisors were identified.

THEOREM 2.2 (Three "original" classes of free divisors).

  1. *Discriminants of versal unfoldings of isolated hypersurface and complete intersection singularities are free divisors (Saito [93] and Looijenga [60]);*

  2. *Bifurcation sets of (versal unfoldings of) isolated hypersurface singularities are free divisors (Bruce [11] and Terao [106]);*

  3. *Coxeter arrangements (union of reflecting hyperplanes for a Coxeter group) are free divisors (Terao [105]).*

These examples illustrate how free divisors arise among fundamental objects, but (as we shall later see) this list only scratches the surface.

Saito also recognized for free divisors the important properties which follow for the complex of "logarithmic differential forms". Let $\Omega^k_{\mathbb{C}^p, 0}$ denote the module of germs of holomorphic k-forms on $\mathbb{C}^p, 0$. For a hypersurface germ $V, 0 \subset \mathbb{C}^p, 0$, with reduced defining equation h we follow Saito [93] and define the logarithmic k-forms

$$\Omega^k(\log V) \quad = \quad \{\omega \in \Omega^k_{\mathbb{C}^p \setminus V, 0} : h\omega, h d\omega \in \Omega^k_{\mathbb{C}^p, 0}\}$$

Let $\Omega^\bullet(\log V)$ denote the corresponding complex of logarithmic forms.

THEOREM 2.3 ([93, §1]).

1. *The complex $\Omega^\bullet(logV)$ is an exterior algebra closed under exterior differentiation. Moreover, it is closed under interior product with, and Lie derivative by, vector fields in* Derlog (V).

2. $V, 0$ *is a free divisor iff $\Omega^1(logV)$ is free of rank p; and then $\Omega^k(logV) = \bigwedge^k \Omega^1(logV)$ and is hence a free $\mathcal{O}_{\mathbb{C}^p,0}$-module.*

Furthermore, Saito gives an extremely useful criterion for freeness of a divisor.

THEOREM 2.4 (Saito's Criterion). *Let $\zeta_i \in$ Derlog $(V), i = 1,\ldots,p$ with $\zeta_i = \sum_j a_{ij}\frac{\partial}{\partial y_j}$. If $\det(a_{ij})$ is a reduced defining equation for $V, 0$, then V is a free divisor and the ζ_i generate* Derlog (V).

This is the basis for many results identifying free divisors. The only other intrinsic characterization of freeness is given by Alexandrov [2], [3].

THEOREM 2.5. *Let h be a reduced defining equation for a hypersurface germ $V, 0 \subset \mathbb{C}^p, 0$. Also let $J(h)$ denote the Jacobian ideal of h. Suppose $Sing(V)$ has codimension 1 in $V, 0$, then each of the following is equivalent to V being a free divisor:*

1. $Sing(V)$ *is a determinantal germ;*

2. $Sing(V)$ *is Cohen–Macaulay (for the structure given by $J(h)$).*

Using this criterion, Alexandrov gives an alternate proof that discriminants of versal unfoldings of ICIS are free divisors [3].

The parenthetical condition in 2) is crucial, as the intrinsic geometric structure of $Sing(V)$ does not determine whether V is free.

EXAMPLE 2.6.   Consider the surface singularities in $\mathbb{C}^3$ defined by

$$f_1 = x^{10} + y^{10} + zx^6y^6 \quad \text{and} \quad f_2 = x^{10} + y^{10} + z(x^7y^5 + x^5y^7)$$

Both f_i are equisingular deformations (with parameter z) of the plane curve singularity $f_0 = x^{10} + y^{10}$. Using different methods in [29], it is shown that the first defines a free divisor while the second does not. However, each have the same singular set consisting of the z-axis, which is smooth and hence Cohen–Macaulay. Thus, the extra structure of $J(h)$ is required for Alexandrov's criterion. Moreover, both functions, viewed as deformations, are topologically equivalent to the trivial deformation of f_0. Thus, freeness is NOT a topological property of divisors. This contrasts with a conjecture of Terao that for central hyperplane arrangements (viewed as nonisolated singularities) the freeness is determined by the associated lattice.

The suspension of a free divisor $V \times \mathbb{C}^r, 0$ is easily seen to be free; however, a product of free divisors is not even a hypersurface. The product is naturally replaced by the "product union" of free divisors $V_i, 0 \subset \mathbb{C}^{p_i}, i = 1, 2$

$$V_1 \uplus V_2 \quad = \quad V_1 \times \mathbb{C}^{p_2} \cup \mathbb{C}^{p_1} \times V_2$$

(we can repeat the construction inductively). By [22, prop. 3.1] the product union of free divisors is again a free divisor. For example, by Mather's multitransversality characterization of stability [70], the discriminant of a stable multigerm is the product union of the discriminants of the individual stable germs, and so free. Also, the "product" of hyperplane arrangements [82, Chap. 1] is really a product union.

Derlog (H) and Free Complete Intersections

To extend the preceding to complete intersections, we encounter a basic problem. We would like $\{0\} \subset \mathbb{C}^p$ to be a free complete intersection; however Derlog $(\{0\})$ is generated by $\{y_i \frac{\partial}{\partial y_j}\}$ and is far from being free. We change our perspective to circumvent this problem. We recall from [30, §2] that we may always replace a hypersurface $V \subset \mathbb{C}^p$ by $V' = V \times \mathbb{C}$ and find a "good defining equation" H for V', which means there is an "Euler–like vector field" e such that $e(H) = H$. This does <u>not</u> require that V be weighted homogeneous (although if it is and of nonzero weight then the usual Euler vector field suffices). As this does not alter properties of nonlinear sections, we may suppose V already has this property. Then, we introduce the module of vector fields annihilating H.

$$\text{Derlog}(H) \quad = \quad \{\zeta \in \theta_p : \zeta(H) = 0\}. \tag{6}$$

Then, for example by [30, lemma 3.3] (or equivalently see [2])

$$\text{Derlog}(V) \quad = \quad \text{Derlog}(H) \oplus \mathcal{O}_{\mathbb{C}^p, 0}\{e\}$$

Hence, if V is a free divisor with good defining equation, then Derlog (H) is a free $\mathcal{O}_{\mathbb{C}^p, 0}$-module of rank $p - 1$ and conversely.

For the corresponding notion of freeness for a complete intersection $V, 0 \subset \mathbb{C}^p, 0$ defined by $H : \mathbb{C}^p, 0 \to \mathbb{C}^k, 0$, we define Derlog (H) as in (6).

DEFINITION 2.7. *A complete intersection $V, 0 \subset \mathbb{C}^p, 0$ defined by $H : \mathbb{C}^p, 0 \to \mathbb{C}^k, 0$ is a (H–) free complete intersection if* Derlog (H) *is a free $\mathcal{O}_{\mathbb{C}^p, 0}$-module (necessarily of rank $p - k$).*

EXAMPLE 2.8. The most basic example of a free complete intersection is given by the product of free divisors, see [26, §5] (although there

416   James Damon

are other examples given there). Thus, $\{0\} \subset \mathbb{C}^p$ is a free complete intersection (with $\operatorname{Derlog}(H) = 0$).

While free divisors (and free complete intersections) can be thought of as rigid universal objects, a much larger and richer class of divisors arise as nonlinear sections of free divisors by maps algebraically transverse off 0.

DEFINITION 2.9. *A hypersurface $V', 0 \subset \mathbb{C}^n, 0$ is an almost free divisor (AFD) (based on V) if $V' = f_0^{-1}(V)$ where $V, 0 \subset \mathbb{C}^p, 0$ is a free divisor and $f_0 : \mathbb{C}^n, 0 \to \mathbb{C}^p, 0$ is algebraically transverse to V off 0. Similarly $V', 0$ is an almost free complete intersection (AFCI) if instead $V, 0$ is a free complete intersection.*

Each class of free divisors (or free complete intersections) yields a much larger corresponding class of almost free divisors (or almost free complete intersections).

EXAMPLE 2.10. *Examples of AFD and AFCI*
i) By (2.8) any ICIS is an AFCI.
ii) By Theorems 2.2 and 1.5, the discriminant of any finitely $\mathcal{A}$–determined germ which defines an ICIS is an AFD.
iii) A central arrangement of hyperplanes in general position off 0 is an AFD which is the pullback of a Boolean arrangement, consisting of coordinate hyperplanes.

EXAMPLE 2.11. *Key properties of almost free divisors and complete intersections* (see [22, §3 and §7] and [26, §5]).

1. The pullback of an almost free divisor or complete intersection by a finite map germ algebraically transverse off 0 is again an almost free divisor or complete intersection.

2. If $(V_i, 0)$ are almost free divisors "transverse" off 0, i.e. in algebraic general position off 0 then

   i) the "transverse union" $(\cup V_i, 0)$ is again an almost free divisor and;

   ii) the "transverse intersection" $(\cap V_i, 0)$ is an almost free complete intersection.

EXAMPLE 2.12. *Examples resulting from properties* i) The transverse union of isolated hypersurface singularities is not isolated; however, it remains an almost free divisor. ii) (boundary singularities) If $V, 0 \subset \mathbb{C}^p, 0$ is a free divisor and $X, 0 \subset \mathbb{C}^p, 0$ is an ICIS (algebraically) transverse to V off 0, then $(V \cap X, 0)$ is an AFCI. iii) $(V, 0)$, the union of 4 lines through 0 in $\mathbb{C}^3$, is an ICIS; but it is also an AFCI obtained as the pullback by a general 3–dimensional linear section of the free complete intersection obtained as the product of the Boolean arrangement $\{(x, y) : xy = 0\}$ with itself. In this last example, the different ways of viewing $(V, 0)$ are reflected in the different ways of deforming it and obtaining vanishing topology in part II.

II Topology of Nonlinear Sections

3. Topology of Singular Milnor Fibers

For a nonlinear section of a nonisolated complete intersection we introduce a (topological) stabilization. Via this stabilization, we define a "singular Milnor fibration" of f_0 so that the "singular Milnor fiber" will have the correct connectivity properties. In fact, the "singular Milnor fiber" will be homotopy equivalent to a bouquet of spheres of correct dimension. The number of such spheres defines a "singular Milnor number". We then describe how to algebraically compute the singular Milnor number as the length of a determinantal module, which turns out to be the $\mathcal{K}_{H,e}$–normal space of f_0.

Stabilization of Nonlinear sections of Complete Intersections

Suppose $V, 0 \subset \mathbb{C}^p, 0$ is a nonisolated complete intersection of codimension k. Let $f_0 : \mathbb{C}^n, 0 \to \mathbb{C}^p, 0$ be a nonlinear section of $V, 0$ which is geometrically transverse to V off 0. By a *(topological) stabilization of f_0* we mean a holomorphic family of maps $f_t : U \to \mathbb{C}^p$ (for U a neighborhood of 0), such that: f_t is geometrically transverse to V for $t \neq 0$ (i.e. transverse to the canonical Whitney stratification of V); and for $t = 0$ it is a representative of the germ f_0 which is transverse to V on $U \backslash \{0\}$. In the case $n < hn(V)$ we can even ensure the stabilization is algebraically transverse to V.

Then, we can apply a theorem of Lê [54] which extends Milnor's theorem to a function germ f defined on a (possibly nonisolated) $n+1$–dimensional complete intersection $\mathcal{X}, 0 \subset \mathbb{C}^m, 0$, and which has an isolated singularity in an appropriate sense. Lê proves f_0 has a Milnor fibration with fibers which are singular but still $n - 1$ connected and homotopy equivalent to a bouquet of real n–spheres. Then, Lê's theorem can be appropriately applied to a projection on the stabilization and combined with [59] and standard type stratification arguments. Letting B_ϵ denote a ball about 0 in $\mathbb{C}^n$ of radius $\epsilon > 0$. We obtain

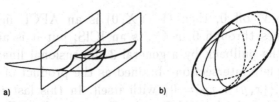

Figure 1. Singular Milnor fibers for: a) hyper/ section of a discriminant and b) an AFCI obtained as the intersection of a braid arrangement and a quadric

for nonlinear sections of hypersurfaces from joint work with Mond [30, Thm 4.6] (and for complete intersections [22, Lemma 7.8], or [23, Thm 3]).

THEOREM 3.1. *Let f_t as above be a stabilization of f_0 as a nonlinear section of the complete intersection $V, 0 \subset \mathbb{C}^p, 0$ of dimension k. For t and $\epsilon > 0$ sufficiently small, $f_t^{-1}(V) \cap B_\epsilon$ is independent of the stabilization f_t and is homotopy equivalent to a bouquet of real $(n-k)$-spheres.*

Then, $f_t^{-1}(V) \cap B_\epsilon$ is called the *singular Milnor fiber* of f_0 (or of V_0). The number of such spheres is called the *singular Milnor number* and denoted by $\mu_V(f_0)$ (or $\mu(V_0)$ if f_0 is understood). The spheres themselves are the *singular vanishing cycles*.

EXAMPLE 3.2.   For the examples in (2.12): i) the singular Milnor fiber is the union of the Milnor fibers of the hypersurfaces, chosen so they are in general position; ii) (boundary singularities) the singular Milnor fiber of $V \cap X, 0$ is the intersection of the boundary singularity V with the usual Milnor fiber of X chosen so it is transverse to V, figure 1 b); iii) for the AFCI of 4 lines in $\mathbb{C}^3$, the singular Milnor fiber is a union of 4 skew lines ℓ_i such that they intersect in cyclic order ℓ_1 and ℓ_2, ℓ_2 and ℓ_3 etc., forming a single singular 1-cycle.

Remark. This extends to images of finite map germs $f_0 : \mathbb{C}^n, 0 \to \mathbb{C}^{n+1}, 0$, see Mond [73] and [74].

Formula for the Singular Milnor number of an Almost Free Divisor

We next give the algebraic formula for the singular Milnor number of an almost free divisor based on the free divisor V with a good defining equation H. This formula is the analogue of Milnor's original formula.

To do so we introduce another equivalence, $\mathcal{K}_H$–equivalence, which is an analog of $\mathcal{K}_V$–equivalence for which Ψ in (2) preserves instead the level sets of the defining equation H. The extended tangent space

$TK_{H,e} \cdot f_0$ is computed by replacing in (3) the generators of Derlog (V) by Derlog (H), i.e. we remove the Euler–like vector field e. (see [30, §3]).

As V is free, we may choose generators $\{\zeta_0, \zeta_1, \ldots, \zeta_{p-1}\}$ for Derlog (V) such that $\zeta_0 = e$ and $\zeta_i, i > 0$. generate Derlog (H).

We let $\nu_V(f_0)$ denote an algebraic codimension $(= \mathcal{K}_{H,e} - \text{codim}\,(f_0))$ which is defined by

$$\nu_V(f_0) = dim_{\mathbb{C}} \; \mathcal{O}_{\mathbb{C}^n,0}\{\frac{\partial}{\partial y_i}\}/\mathcal{O}_{\mathbb{C}^n,0}\{\frac{\partial f_0}{\partial x_1}, \ldots, \frac{\partial f_0}{\partial x_n}, \zeta_1 \circ f_0, \ldots, \zeta_{p-1} \circ f_0\} \quad (7)$$

Just as for Derlog (V), we can define

$$T_{log}(H)_{(y)} \; = \; < \text{Derlog}\,(H) >_{(y)}$$

Equality holds $T_{log}(H)_{(y)} = T_{log}(V)_{(y)}$ if there is an Euler–like vector field defined near y which vanishes at y, e.g. if V is locally weighted homogeneous at y for some choice of local coordinates. Then, we define the H–holonomic codimension $h(V)$ as we did the holonomic codimension $hn(V)$, except using the condition $T_{log}(H)_{(y)} = T_y S_i$, i.e. $h(V)$ is the codimension of the largest stratum which is not H–holonomic.

Then, provided we remain below $h(V)$ we can compute the singular Milnor number using (7).

THEOREM 3.3 ([30, Thm 5]). *Suppose that $V, 0$ is a free divisor with $n < h(V)$, and that f_0 is a germ of an embedding algebraically transverse to V off 0 so that $\nu_V(f_0) < \infty$. Then, the singular Milnor number $\mu_V(f_0) = \nu_V(f_0)$.*

Remark. Originally in [30], the result was stated for embeddings f_0; however, by the graph trick and invariance under suspension, we can apply the theorem to any f_0 algebraically tranverse to V off 0 by reducing to an embedding, see e.g. the discussion in [22, Pt I].

$\mu \geq \tau$ *results* : With the notation of (1.1), suppose f_0 is a finite $\mathcal{A}$-codimension germ with stable unfolding F. Let $\mu = \mu_{D(F)}(g_0)$ and $\tau = \mathcal{K}_{V,e} - \text{codim}\,(f_0)$. Then, as a corollary of Theorem 3.3 (see [30]), provided (n, p) is in the "nice dimensions"in the sense of Mather [71], $\mu \geq \tau$ with equality if f_0 is weighted homogeneous. This is the analogue of the $\mu = \tau$ results obtained by Greuel, Wahl, and Looijenga–Steenbrink see[60, Chap. 8]. However, it can fail if $n \geq h(V)$ (see [30, Thm.6] and [27, §4]). We further consider other $\mu = \tau$ results and their relation to freeness of discriminants in Part IV.

EXAMPLE 3.4. i) In the case of isolated hypersurface singularities, we obtain the usual Milnor fibers and the usual formula for the Milnor numbers.

ii) If $g : \mathbb{C}^n, 0 \to \mathbb{C}^p, 0$ is a Morse type singularity for V, then the weighted homogeneity implies $\mathcal{K}_{H,e}\text{codim}\,(g) = \mathcal{K}_{V,e}\text{codim}\,(g) = 1$. Hence, provided $n < h(V)$, the theorem implies that for Morse–type singularities, there is exactly one singular vanishing cycle in their singular Milnor fibers, as for usual Morse singularities.

iii) For the versal unfolding F of any simple hypersurface singularity, $D(F)$ is weighted homogeneous. A linear section of $D(F)$ defined by the vanishing of unfolding parameter of lowest weight is a Morse–type singularity . A perturbation of the section moves off the origin and creates exactly one vanishing cycle as is usual/tured for the swallowtail singularity figure 1, also see [22, §4]).

Suppose $V, 0 \subset \mathbb{C}^p, 0$ is an almost free divisor based on $V', 0 \subset \mathbb{C}^{p'}, 0$ via $g_0 : \mathbb{C}^p, 0 \to \mathbb{C}^{p'}, 0$, and that $\nu_{V'}(g_0) < \infty$. $\text{Derlog}\,(V)$ is generally not free and Theorem 3.3 does not apply (and, in fact, fails badly) for sections of V. However by the behavior of almost free divisors under pullback by finite map germs, we can still compute the singular Milnor number of a section of V (see [21, Cor. 4.2]).

COROLLARY 3.5. *Suppose that* $f_0 : \mathbb{C}^n, 0 \to \mathbb{C}^p, 0$ *is a finite map germ with* $n < p, h(V')$, *which satisfies* $\nu_{V'}(g_0 \circ f_0) < \infty$. *The singular Milnor number for* f_0 *(which equals the number of* $(n-1)$*–spheres in* $f_t^{-1}(V) \cap B_\epsilon$ *for a stabilization* f_t*) is given by* $\nu_V(g_0 \circ f_0)$.

4. An Extension of the Lê-Greuel Formula

Relative Singular Milnor Number for a Divisor on a Complete Intersection

The Milnor number of an ICIS is computed inductively via the Lê–Greuel formula. Suppose $X, 0 \subset \mathbb{C}^m, 0$ is a positive $n - m$–dimensional ICIS defined by $f_2 = (f_{21}, \ldots, f_{2m})$, and that $f_1 : \mathbb{C}^n, 0 \to \mathbb{C}, 0$ has an isolated singularity restricted to X. Then, $f = (f_1, f_2)$ defines the ICIS $X_0 = X \cap f_0^{-1}(0)$.

THEOREM 4.1 (Lê–Greuel [57]). *In the preceding situation*

$$\mu(f) + \mu(f_2) = \dim{}_{\mathbb{C}}\mathcal{O}_{\mathbb{C}^n,0}/((f_{21}, \ldots, f_{2m}) + J(f)) \qquad (8)$$

$J(f)$ *denotes the ideal generated by the* $(m+1) \times (m+1)$ *minors of* df.

Suppose X_y denotes a Milnor fiber of f_2 over y such that the intersection $X_{(0,y)} = X_y \cap f_1^{-1}(0)$ is transverse giving the Milnor fiber of f. Then, the pair $(X_y, X_{(0,y)})$ has homology only in dimension $n-m$ and the LHS of (8) is the "relative Milnor number" $= \dim_{\mathbb{C}} H^{n-m}(X_y, X_{(0,y)})$. Thus, the RHS of (8) can be alternately viewed as providing an algebraic formula for this relative Milnor number.

We extend this result to the relative case of an almost free divisor V_0 transversely intersecting an almost free complete intersection X off 0. We begin with the case where X is again an ICIS defined by f_2 as in Theorem 4.1. Now we suppose $f_1 : \mathbb{C}^n, 0 \to \mathbb{C}^p, 0$ is algebraically transverse to the free divisor V off 0. Let $\{\zeta_1 \ldots, \zeta_{p-1}\}$ be generators for Derlog (H), with H a good defining equation for V.

If f_{1t} is a stabilization of f_1 and $V_t = f_{1t}^{-1}(V)$ is transverse to $X_y = f_2^{-1}(y) \cap B_\epsilon$ on B_ϵ, then $X_{(t,y)} = V_t \cap X_y$ is the singular Milnor fiber of $V_0 \cap X$. Then, $((X_y, X_{(t,y)})$ is $n-m-1$–connected and the relative singular Milnor number $\mu_{(X,X \cap V_0)}(f) = \dim_{\mathbb{C}} H^{n-m}(X_y, X_{(t,y)})$ is computed via the following module version of the Lê–Greuel formula.

THEOREM 4.2 ([22, Cor. 9.6]). *For the relative case of an almost free divisor V_0 and ICIS X, as above, the relative singular Milnor number is given by*

$$\mu_{(X,X \cap V_0)}(f) = \dim_{\mathbb{C}} \mathcal{O}_{X,0}^{(p+m)} / \mathcal{O}_{X,0}(\frac{\partial f}{\partial x_1}, \ldots, \frac{\partial f}{\partial x_n}, \zeta_1 \circ f_1, \ldots, \zeta_{p-1} \circ f_1)$$

We note several consequences. First, $(X, 0)$ is a complete intersection (and hence Cohen–Macaulay) of positive dimension $n - m$. The RHS in Theorem 4.2 is the quotient of a free $\mathcal{O}_{X,0}$–module of rank $p + m$, on $n + p - 1 = (n - m) + (p + m) - 1$ generators. As the quotient has finite length, by results of Eagon–Northcott the quotient module on the RHS in Theorem 4.2 is Cohen–Macaulay. In particular, its length is also its Buchsbaum–Rim multiplicity. Hence, by results of Buchsbaum–Rim [14], the length equals the length of the quotient algebra of $\mathcal{O}_{X,0}$ by the $(n+p-1) \times (n+p-1)$ minors of the matrix formed from the generators of the quotient. In particular, in the case $V = \{0\}$, there are no ζ_i, and we obtain exactly the Lê– Greuel formula for the ICIS case.

Second, we may apply this result to the case of a fixed free divisor $V, 0 \subset \mathbb{C}^n, 0$ which we view as a "boundary" and consider "boundary singularities", e.g. Arnol'd [6], Lyashko [61], etc. (or equivalently view V as a divisor at infinity and consider locally singularities at infinity, e.g. Siersma-Tibăr [99]). Then, $f_1 = id_{\mathbb{C}^n}$. Let X_y be the Milnor fiber of an ICIS g transverse to V. By projecting off the first m generators, Theorem 4.2 takes the form.

COROLLARY 4.3. *For a ICIS $X, 0$ defined by $g : \mathbb{C}^n, 0 \to \mathbb{C}^m, 0$ for $n > m$, the boundary singularity for the free divisor boundary V has relative singular Milnor number*

$$\dim_{\mathbb{C}} H^{n-m}(X_y, X_y \cap V) = \dim_{\mathbb{C}} \mathcal{O}_{X,0}^{(m)}/\mathcal{O}_{X,0}\{\zeta_1(g), \ldots, \zeta_{n-1}(g)\}$$

In particular, for an isolated hypersurface singularity X, we recover a formula which is similar, although not identical, to one of Bruce–Roberts [12, Prop. 6.4], except they were computing the number of critical points as opposed to the singular Milnor number. As both invariants are computed by counting critical points, these numbers should agree, despite the slight difference in the formulas.

If we consider the more general situation where now we replace the ICIS X by a general AFCI based on a product of free divisors $V' \subset \mathbb{C}^{p_2}$, then we still obtain a formula given by the length of a determinantal module as in (4.2). However, it computes instead the relative Euler characteristic $\chi(X_y, X_{(t,y)})$ where X_y is the usual smooth Milnor fiber of the germ f_2 (which is no longer a bouquet of $n - m$–spheres) and $X_{(t,y)}$ is the intersection of the singular Milnor fiber of f_1 with X_y. We denote this by $\tilde{\chi}(X, X \cap V_0)$. We let V' be defined by $h_2 = (h_{22}, \ldots, h_{2m})$ as a product of free divisors and $V, 0$ by h_1, with each h_{2i} and h_1 good defining equations for the appropriate free divisor. Let $h = (h_1, h_2)$.

THEOREM 4.4 ([22, Thm. 9.4]). *In the preceding situation,*

$$\tilde{\chi}(X, X \cap V_0) = (-1)^{\ell} \dim_{\mathbb{C}} \mathcal{O}_{X,0}^{(p+m)}/\mathcal{O}_{X,0}(\frac{\partial f}{\partial x_1}, \ldots, \frac{\partial f}{\partial x_n}, \zeta_1 \circ f, \ldots, \zeta_{n-m} \circ f)$$

where $\ell = n - m + 1$ and $\zeta_1, \ldots, \zeta_{n-m}$ are generators for Derlog (h).

To relate this relative Euler characteristic with the relative singular Milnor number, we consider an alternate approach when V is given as the transverse intersection off 0 of almost free divisors $V_i, i = 1, \ldots, m$. The transverse union of a subset of the V_i is still an almost free divisor, so the singular Milnor number can be computed by Theorem 3.3. Then, the singular Milnor number of the intersection $\cap_i V_i$ can be expressed as an alternating sum of singular Milnor numbers of various unions of subsets of the V_i, see [22, §8, Thm. 2]. For example, this can be applied to give an alternate derivation of the formula of Guisti [35], Greuel–Hamm [40], and Randell [89] for the Milnor number of a weighted homogeneous ICIS.

In the simplest case the formula takes the form

$$\mu(V_1 \cap V_2) = \mu(V_1 \cup V_2) - \mu(V_1) - \mu(V_2) \tag{9}$$

For example, consider the complete intersection of two free divisors which are suspensions of cusp singularities as shown in figure 2. The

Figure 2. Intersection/union of free divisors, their singular Milnor fibers, with the single vanishing cycles

singular Milnor fiber of both the intersection and union as shown have a single vanishing cycle (and each $\mu(V_i) = 0$). By contrast, Theorem 4.4 gives a relative Euler characteristic $= 9$ (using the smooth Milnor fiber). Hence, it not only counts the vanishing cycle but also adds contributions of 2 for each singularity in the singular Milnor fiber.

REMARK 4.5. [Proof of the relative formulas] As a main step in proving the Theorems 4.3 and 4.7, we apply a generalization of the lemma of Siersma used in [30]. In [95] Siersma shows that the standard Morse theory type argument used by Looijenga in [60, Chap. 5], and in its original form due to Lê [55] [56], can also be extended to nonisolated singularities defined by germs $g : \mathbb{C}^{n+1}, 0 \to \mathbb{C}, 0$. For the relative case, we establish [22, §9], [24, §4] an analogue for nonisolated singularities defined on complete intersections. Note these methods compute the topology of singular spaces without using stratified Morse theory of Goresky-MacPherson [36].

Buchbaum–Rim Multiplicity of a Determinantal module on a Complete Intersection

The formulas for the singular Milnor number or relative singular Milnor number given in Theorems 3.3, 4.2, 4.4 and 4.3 are given by the length of a determinantal module (on a complete intersection). This is the Buchbaum–Rim multiplicity of the module. We next give a formula for this Buchbaum–Rim multiplicity in the (semi-) weighted homogeneous case in terms of the weights.

We suppose that $X, 0 \subset \mathbb{C}^p, 0$ is a weighted homogeneous complete intersection defined by $f = (f_1, \ldots, f_{p-n})$ where $\text{wt}(y_i) = a_i$, $\text{wt}(f_i) = b_i$. Let $F_1, \ldots, F_{n+k-1} \in (\mathcal{O}_{\mathbb{C}^p,0})^k$ be weighted homogeneous of degrees d_i. This means we assign weight c_j to $\epsilon_j = (0, \ldots, 1, 0, \ldots, 0)$, with 1 in the j-th position. Then, $F_i = (F_{i1}, \ldots, F_{ik})$ with each F_{ij} weighted homogeneous of weight $d_i + c_j$ (d_i and c_j are not uniquely determined by F_{ij}). We let $\mathbf{a} = (a_1, \ldots, a_p)$, $\mathbf{b} = (b_1, \ldots, b_{p-n})$, and similarly for $\mathbf{d}$ and $\mathbf{c}$. Then, let

$$M = (\mathcal{O}_{X,0})^k / (\mathcal{O}_{X,0}\{F_1, \ldots, F_{n+k-1}\})$$
$$R = \mathcal{O}_{X,0} / I_k (F_1, \ldots, F_{n+k-1})$$

where $I_k(F_1, \ldots, F_{n+k-1})$ denotes the ideal generated by the $k \times k$ minors of the $k \times (n+k-1)$ matrix (F_{ij}).

There is considerable redundancy in the sets of weights wt (F_{ij}). There is a smaller $n \times k$ degree matrix $\mathbf{D}$ whose ij-th entry is $d_{i+j-1}+c_j$. Then, we define a universal function τ for all matrices [21, §2]; and using properties of determinantal modules due to Macaulay [62] and Northcott [79], we give a formula for the dimensions of M and R in terms of the weights via $\tau(\mathbf{D})$. Moreover, $\tau(\mathbf{D})$ can be expressed using $\sigma_j(\mathbf{d})$, the j-th elementary symmetric function in $d_1, \ldots, d_{p+k-1}$, and $s_j(\mathbf{c})$ which is the sum of all monomials of degree j in $c_1, \ldots, c_k$.

THEOREM 4.6. *Both* $\dim_{\mathbf{C}} M$ *and* $\dim_{\mathbf{C}} R$ *are independent of both* $\{F_i\}$ *and* f *and hence depend only on the weights* $(\mathbf{a}, \mathbf{b}, \mathbf{c}, \mathbf{d})$, *provided the dimensions are finite. Moreover, in this case, these are the Buchsbaum–Rim multiplicity and are given by*

$$\dim_{\mathbf{C}} M = \dim_{\mathbf{C}} R = \frac{b}{a} \cdot \tau(\mathbf{D}) = \frac{b}{a} \cdot \sum_{j=0}^{p} s_{p-j}(\mathbf{c}) \cdot \sigma_j(\mathbf{d}) \qquad (10)$$

where $a = \prod a_i$ *and* $b = \prod b_i$.

Remark. This theorem was proven in [21] for the case of $X = \mathbf{C}^n$. A fairly simple modification of that proof gives this more general result. The theorem extends to the semi–weighted homogeneous case, where for the initial parts of the f_i and F_j, M and R are still finite dimensional.

EXAMPLE 4.7.   The natural properties of the functions σ_j and s_j (see [21, §2]) suggest that formulas as above expressed in terms of them should have analogous forms in the nonweighted homogeneous case where the σ_j and s_j are replaced by appropriate invariants. For example, in the homogeneous case with $X = \mathbf{C}^p$ so $n = p$, there are no f_i, and all $a_i = 0$, $c_i = 0$, so that all F_i are homogeneous with all components of degree d_i. Then, the Buchsbaum–Rim multiplicity is given by $\sigma_n(d_1, \ldots, d_{n+k-1})$. We next see these have interpretations as "higher multiplicities".

5. Higher Multiplicities

Through singular Milnor fibers and numbers for nonlinear sections of complete intersections, we can introduce higher multiplicities á la Teissier for nonisolated complete intersections. Teissier [102] defined for isolated hypersurface singularities a series of higher multiplicities, the μ^*– sequence $\mu^* = (\mu_0, \ldots, \mu_n)$. Given $f_0 : \mathbb{C}^n, 0 \to \mathbb{C}, 0$, if Π is a generic k–dimensional subspace in $\mathbb{C}^n$ then $f_0|\Pi$ has an isolated singularity and Teissier defines $\mu_k(f_0) = \mu(f_0|\Pi)$, where μ denotes the usual Milnor number. Lê–Teissier [58] [104] further considered for general $(V,0) \subset \mathbb{C}^n, 0$, the "k-th vanishing Euler characteristics" $\chi(\pi^{-1}(z) \cap V \cap B_\epsilon)$, where $\pi : V, 0 \to \mathbb{C}^k, 0$ is the restriction of a generic linear projection, z is sufficently general, and $\|z\|$ and $\epsilon > 0$ are sufficiently small. They use the polar multiplicities combined with these vanishing Euler characteristics relative to strata of a Whitney stratification to compute topological invariants of nonisolated singularities.

For nonisolated complete intersections $V, 0 \subset \mathbb{C}^p, 0$, we can define higher multiplicities using the analogue of Teissier's definition for the hypersurface case. A Zariski open subset of k-dimensional subspaces $\Pi \subset \mathbb{C}^p$ are geometrically transverse to V off 0 (which implies algebraic transversality if $k < h(V)$). We view the inclusion $i : \Pi \to \mathbb{C}^p$ as a section of V and define $\mu_k(V) = \mu_V(i)$, the singular Milnor number of the section i (if V is itself a nonlinear section, $\mu_p(V)$ is its singular Milnor number). Then, $\mu_k(V)$ counts the number of singular vanishing cycles for a perturbation i_t of the section i. If k is one less than the codimension of the canonical Whitney stratum of V containing 0, then the singular Milnor fiber of the section is the "complex link" of V as defined by Goresky-MacPherson [36].

In the special case where V is an almost free divisor based on $V', 0 \subset \mathbb{C}^{p'}, 0$ via g_0, Corollary 3.5 computes the higher multiplicities.

PROPOSITION 5.1. *Suppose that $(V,0)$ is an almost free divisor based on $(V', 0)$ via g_0. Let $i : \mathbb{C}^k, 0 \to \mathbb{C}^p, 0$ be a linear section where $k < p, h(V')$. If $\nu_{V'}(g_0 \circ i)$ is finite and minimum among all nearby linear embeddings, then*

$$\mu_k(V) = \mu_{V'}(g_0 \circ i) = \nu_{V'}(g_0 \circ i).$$

This proposition and Theorem 4.3 allow us to compute higher multiplicities for weighted homogeneous free divisors $V, 0 \subset \mathbb{C}^p$. Let wt $(y_i) = a_i$, with $a_1 \leq a_2 \leq \ldots \leq a_p$. Let H be the weighted homogeneous defining equation, and suppose weighted homogeneous generators ζ_i for Derlog (H) have wt $(\zeta_i) = d_i$.

PROPOSITION 5.2. *Suppose that the k–dimensional section of $(V, 0)$ defined by $y_1 = \ldots = y_{p-k} = 0$ is algebraically transverse to V off 0,* *then*

$$\mu_k(V) = \frac{1}{a''} \sum_{j=0}^{k} s_{k-j}(\mathbf{a'}) \cdot \sigma_j(\mathbf{d}) \qquad (11)$$

where $\mathbf{a'} = (a_1, \ldots, a_{p-k})$, $a'' = \prod_{i=p-k+1}^{p} a_i$ *and* $\mathbf{d} = (d_1, \ldots, d_{p-1})$.

Several special cases are of particular interest.

Lower Bound for $\mathcal{A}_e$-codimension

As a consequence, Proposition 5.2 yields the minimum $\mathcal{A}_e$-codimension for germs in the contact class.

COROLLARY 5.3. *Let* $V = D(F)$ *be the discriminant of the versal unfolding* $F : \mathbb{C}^n, 0 \to \mathbb{C}^p, 0$ *of an ICIS* f_0. *Suppose* $k < h(D(F))$ *and the weighted subspace defined by* $y_1 = \ldots = y_{p-k} = 0$ *is algebraically transverse to* V *off* 0. *Then, the minimum* $\mathcal{A}_e$-*codimension of germs* $g : \mathbb{C}^{n'}, 0 \to \mathbb{C}^k, 0$ *in the same contact class as* f_0 *is given by*

$$minimum\ \mathcal{A}_e\text{-}codimension = \mu_k(D(F))$$

EXAMPLE 5.4. Low codimension calculations suggest that for the discriminant $D(A_n)$ of the versal unfolding of an A_n singularity, the sections defined by the $n - k$ lowest weight unfolding parameters are algebraically transverse to $D(A_n)$ off 0. If this generally holds then $\mu_k(D(A_n))$ is given by (11), which can be computed to equal $\binom{n-1}{k}$. Then, by Corollary 5.3 $\binom{n-1}{k}$ is the minimum $\mathcal{A}_e$-codimension of a germ $g : \mathbb{C}^{n'}, 0 \to \mathbb{C}^k, 0$ belonging to the A_n contact class.

Free and Almost Free Hyperplane Arrangements

Freeness plays an extremely important role for hyperplane arrangements, see Orlik-Terao [82]. We consider $A \subset \mathbb{C}^p$ a free hyperplane arrangement. The defining equation for A is homogeneous of the form $Q = \prod \ell_i$ where ℓ_i are linear forms defining the hyperplanes belonging to A. There are homogeneous generators ζ_i for Derlog (A) with ζ_0 the Euler vector field, and $\zeta_i \in$ Derlog $(Q), i > 0$. Let wt $(\zeta_i) = d_i$ and let $e_i = d_i + 1$. Then, $exp(A) = (e_1, \ldots, e_p)$ are called the exponents of A (note $e_1 = 1$). An almost free arrangement $\mathcal{A} \subset$ is obtained as the linear section of a free A. Because we are in the homogeneous case, Proposition 5.2 is always applicable. By properties of $\tau(\mathbf{D})$, we may adjust weights for the generators by decreasing all a_i to zero and replacing d_i by e_i and compute higher multiplicities for almost free arrangements.

COROLLARY 5.5 ([22, Prop. 5.2]). *Let $\mathcal{A} \subset \mathbb{C}^p$ be an almost free arrangement based on the free arrangement A. If $exp'(A) = (e_2, \ldots, e_p)$, then*

$$\mu_k(\mathcal{A}) = \sigma_k(exp'(A))$$

The higher multiplicities have a close connection with other invariants of hyperplane arrangements such as the Mobius function of the arrangement evaluated at A, the Crapo invariant, etc [22, §5] and [28, §2]. We single out a key property:

Betti Numbers and Higher Multiplicities for Arrangements.

If $\mathcal{A} \subset \mathbb{C}^p$ is a central arrangement, then the k-th Betti number of the complement $M(\mathcal{A}) = \mathbb{C}^p \backslash \mathcal{A}$ is given by

$$b_k(M(\mathcal{A})) = \mu_k(\mathcal{A}) + \mu_{k-1}(\mathcal{A}) \tag{12}$$

By this property together with Corollaries 5.5 and 3.5, we can compute algebraically the Poincaré polynomial $P(A', t)$ of the complement $M(A')$ for any almost free arrangement A'. Using this one can find central arrangements in $\mathbb{C}^3$, which are (automatically) locally free in the complement of 0, but which are not almost free. Hence, almost free divisors and complete intersections, like ICIS, can not be characterized by purely local conditions in the complement of a point.

As the preceding includes the free arrangement itself, we deduce as a consequence [22, §5] the factorization theorem of Terao, which generalizes earlier results of Arnold [4] and Brieskorn [9] for Coxeter arrangements.

THEOREM 5.6 (Terao's Factorization Theorem [107]). *If $A \subset \mathbb{C}^p$ is a free arrangement with $exp(A) = (e_1, \ldots, e_p)$, then*

$$P(A, t) = \prod_{i=1}^{p} (1 + e_i t)$$

Considering critical points of functions such as $f_1^{\lambda_1} \cdots f_r^{\lambda_r}$ with the f_i polynomial leads to considerably more complicated "nonlinear arrangements" which can arise by replacing hyperplanes by hypersurfaces or by intersecting a hyperplane arrangement with a smooth affine variety.

Suppose $X, 0 \subset \mathbb{C}^p$ is a homogeneous r–dimensional ICIS defined by a polynomial germ f of multidegree $\mathbf{d} = (d_1, \ldots, d_{n-r})$. Let $X_y = f^{-1}(y)$ be a global smooth Milnor fiber. If the free arrangement A is transverse to X off 0, then $A \cap X$ is an almost free complete intersection at 0. Moreover, for sufficiently general y, A is transverse to X_y, and if both are "nondegenerate at infinity", then $A \cap X_y$ is diffeomorphic to the singular Minor fiber of $(A \cap X, 0)$ [25] or [28]. Then, $A \cap X_y$ defines a nonlinear arrangement on the smooth global complete intersection X_y. The singular vanishing cycles correspond in the real picture to "relative bounding cycles" for regions on X_y determined by the nonlinear arrangement $A \cap X_y$. The generalized Lê–Greuel formula (Thm. 4.3) combined with Corollary 5.5 yields the number of such cycles [22, Theorem 8.19] and [28].

PROPOSITION 5.7. *Suppose $X \subset \mathbb{C}^p$ is a homogeneous ICIS of multidegree* **d**, *and the free arrangement A is transverse to X off* 0. *Then*

$$\text{number of relative bounding cycles of } (X_y, X_y \cap A) = d \cdot \sum_{j=0}^{r} s_j(\mathbf{d}-\mathbf{1})\mu_{r-j}(A)$$

where $d = \prod_{i=1}^{n-r} d_i$ and $\mathbf{d} - \mathbf{1} = (d_1 - 1, \ldots, d_{n-r} - 1)$.

The formula in Proposition 5.7 involves invariants also defined for nonfree A. For the case of a single homogeneous hypersuface, Orlik–Terao [84] obtained an equivalent version of this formula but expressed in terms of the characteristic polynomial of A and valid for any A. Having obtained the formula, it is then possible by different methods to prove that Proposition 5.7 is valid for any A withour regard to freeness and extend it to hypersurface arrangements [28].

   Remark. Besides the number of bounding cycles, such formulas also then apply to determine the number of critical points for holomorphic functions of the form $f_1^{\lambda_1} \cdots f_r^{\lambda_r}$ which appear in hypergeometric functions Aomoto [1], Varchenko [113], Orlik-Terao [83], and [25].

6. Relation with Buchsbaum-Rim Multiplicities

We observed that formulas such as in Proposition 5.7, although arrived at initially as special cases, may remain valid without freeness. As we relax the condition of freeness for the divisor or complete intersection V, we ask what form the formulas will take for the singular Milnor number or its relative version, higher multiplicities, etc. Then, lengths of modules must be replaced by more general invariants. Gaffney's work shows the importance of Buchsbaum–Rim multiplicities as invariants of modules. We note several important connections already established between singular Milnor numbers, the algebraic Buchsbaum–Rim multiplicities, and the geometric higher multiplicities.

1. First, all of the computations of singular Milnor numbers are in terms of lengths of determinantal modules on complete intersections. These modules are extended normal spaces to various groups

$\mathcal{G}$ of equivalences. By results of Buchsbaum–Rim [14], these lengths are Buchsbaum–Rim multiplicities, which we denote by m_{BR}. Hence,

$$\mu_V(f_0) = m_{BR}(N\mathcal{G}_e \cdot f_0) \tag{13}$$

where for complete intersections we replace $\mu_V(f_0)$ by a relative singular Milnor number.

2. Second, Theorem 4.6 gives a general expression for $m_{BR}(M)$ in the (semi-) weighted homogeneous case, where M is a determinantal module on a (possibly nonisolated) complete intersection $X, 0$.

3. Third, in §5 we associated to certain modules $M = \text{Derlog}\,(V)$, geometric higher multiplicities $\mu_k(V)$. For $V, 0$ a weighted homogeneous free divisor, these are again computed as Buchsbaum–Rim multiplicities of certain related modules.

4. Fourth, in certain cases we are able to express Buchsbaum–Rim multiplicities back in terms of higher multiplicities.

When $M = \text{Derlog}\,(V)$ is not free, the preceding no longer hold for lengths of the $\mathcal{K}_{V,e}$ or $\mathcal{K}_{H,e}$ normal spaces. We ask which invariants of the normal spaces will relate to the singular Milnor numbers and higher multiplicities. In particular, can we compute the (relative) singular Milnor number as the Buchsbaum–Rim multiplicity of the original $\mathcal{K}_{V,e}$ or $\mathcal{K}_{H,e}$ normal spaces for general complete intersections?

Alternately, we can consider free submodules $M' \subset \text{Derlog}\,(V)$ which are "good approximations" to $\text{Derlog}\,(V)$ (see §10 and §11). Can we extend the method of computing higher multiplicities for $M = \text{Derlog}\,(V)$ to more general free submodules of M with the same cosupport? When will these then have properties analogous to those for the geometric situation? Then, we seek relations between the invariants for the approximations, the algebraic invariants of $\text{Derlog}\,(V)$, and topological invariants of V.

III Topology of Singular Milnor Fibers via Logarithmic Forms

7. De Rham Cohomology of Free and Almost Free Divisors

Suppose $V_0, 0 \subset \mathbb{C}^n, 0$ is a divisor with reduced defining equation h. There are several natural questions concerning the topology of $(V_0, 0)$ and its complement which can possibly be addressed using differential forms. These include computing the cohomology of:

1. the complement $\mathbb{C}^n \backslash V_0$;

2. the smooth Milnor fiber of h;

3. the singular Milnor fiber of an almost free divisor V_0 based on the free divisor $V, 0$.

In the case of 2) and 3), this would provide a basis as in [10] and [39] for introducing a Gauss–Manin connection.

For 1), by general results of Grothendieck [42] and Griffiths [41] the cohomology of the complement of a divisor can be computed using the complex of meromorphic forms on the complement. For Coxeter arrangements, Arnold [4] and Brieskorn [9] computed the cohomology of the complement of the arrangement using logarithmic forms. This was extended to general hyperplane arrangements, by Orlik–Solomon [80] building on Brieskorn's ideas. For an arrangement $A = \cup H_i \subset \mathbb{C}^m$, let H_i be defined by the linear form ℓ_i. These results express the cohomology of the complement as the quotient of an exterior algebra $R(A)$ on generators $\omega_i = d\ell_i/\ell_i$ by the ideal generated by the relations $r_{ijk} = \omega_i \wedge \omega_j + \omega_j \wedge \omega_k + \omega_k \wedge \omega_i$, for triples of H_i with codim $(H_i \cap H_j \cap H_k) = 2$.

Castro, Mond and Narvaez [17] show that this result has a natural analogue for free divisors using the complex of logarithmic forms. They consider free divisors which are locally weighted homogeneous (i.e. locally weighted homogeneous at each point for some choice of local coordinates).

THEOREM 7.1 ([17]). *Let $V, 0 \subset \mathbb{C}^p$ be a free divisor which is locally weighted homogeneous, then for all k*

$$H^k(\mathbb{C}^p \backslash V : \mathbb{C}) \simeq H^k(\Omega^{\bullet}(logV))$$

Unfortunately, in general this result does not even extend to the simplest almost free divisors such as isolated hypersurface singularities $V, 0 \subset \mathbb{C}^n, 0$. Holland and Mond [47] define an obstruction, the "logarithmic defect", $\delta(logV)_0$, which equals $\dim Gr_n^W H^{n-1}(F; \mathbb{C})$ for the weight filtration for the mixed Hodge structure on the Milnor fiber F of V [101]. If $\delta(logV)_0 \neq 0$, the analogue of Theorem 7.1 fails. They show using results of Steenbrink [101] that even for the simple singularities A_k, D_k, and E_k, there are values of n and k for which it fails.

For a free divisor $V, 0 \subset \mathbb{C}^p$ with singular set $Z = \text{sing}(V)$, Alexandrov has obtained several results computing local cohomology using $\Omega^{\bullet}(logV)$ as described in [3]. He computes the Poincaré polynomials of $H_Z^{\bullet}(\Omega_V^q)$ and $H_Z^{\bullet}(\Omega^q(logV))$ for weighted homogeneous V. Second, he applies ideas of Kunz to obtain a form of Grothendieck local duality for the local cohomology of free divisors [3, Thm. 2.3], obtaining a perfect pairing between the infinite dimensional spaces T_V^1 and the torsion module Tors Ω_V^1 , both of which are isomorphic to $\mathcal{O}_{\mathbb{C}^P,0}/J(h)$ (with $J(h)$ denoting the Jacobian ideal of h). Third, he relates $\Omega^\bullet(\log V)$ to the complex of regular meromorphic differential forms $\omega_V^{\bullet-1}$, obtaining an exact sequence [3, §4] generalizing the classical sequence involving the Poincaré residue for a smooth divisor.

$$0 \longrightarrow \Omega_{\mathbb{C}^P,0}^q \longrightarrow \Omega^q(\log V) \overset{\text{res}}{\longrightarrow} \omega_V^{q-1} \longrightarrow 0 \qquad (14)$$

As a result of (14) he also obtains the Poincaré polynomial of ω_V^{q-1}. In the special case of an isolated hypersurface germ $V, 0$ of dimension ≥ 2, he relates μ and τ to the dimensions of the cohomology of the complex of sections $H^*(\Omega^\bullet(\log V))$ to obtain [3, Cor. 3]

$$\dim_{\mathbb{C}} H^p(\Omega^\bullet(\log V)) - \dim_{\mathbb{C}} H^{p-1}(\Omega^\bullet(\log V)) = \mu - \tau$$

exhibiting the close relation between $H^*(\Omega^\bullet(\log V))$ and other invariants of V.

de Rham Cohomology of the Singular Milnor Fiber

From our point of view, an especially important result was obtained by Mond [75]. He identifies the correct way to use $\Omega^\bullet(\log V)$ to modify the complex of Kähler forms on an almost free divisor to obtain a complex of forms which computes the cohomology of the singular Milnor fiber.

For a divisor $V, 0 \subset \mathbb{C}^p, 0$ defined by h, we have the complex of Kähler forms

$$\Omega_V^\bullet = \Omega_{\mathbb{C}^P,0}^\bullet/(h\Omega_{\mathbb{C}^P,0}^\bullet + dh \wedge \Omega_{\mathbb{C}^P,0}^{\bullet-1}) \qquad (15)$$

Mond observed that $\Omega_V^\bullet$ has torsion and that the torsion can be identified as $h\Omega^\bullet(\log V)/(h\Omega_{\mathbb{C}^P,0}^\bullet + dh \wedge \Omega_{\mathbb{C}^P,0}^{\bullet-1})$. This led him to define $\check{\Omega}_V^\bullet = \Omega_V^\bullet/h\Omega^\bullet(\log V)$. Then, Mond observes that by Theorem 2.3, the complex $\check{\Omega}_V^\bullet$ has properties analogous to those of $\Omega^\bullet(\log V)$. Specifically, exterior differentiation induces a differential on $\check{\Omega}_V^\bullet$ so it is a complex; and it is preserved under both Lie deriviative by and inner product with vector fields in Derlog (V). If $V, 0$ is a free divisor, although $\check{\Omega}_V^k$ is no longer free, he shows $\check{\Omega}_V^k$ is a maximal Cohen–Macaulay module. Moreover, Mond shows

PROPOSITION 7.2. *For a divisor $V, 0 \subset \mathbb{C}^p, 0$, $\check{\Omega}_V^k = 0$ for $k \geq p$; and if $V, 0$ is locally weighted homogeneous, then $\check{\Omega}_V^\bullet$ is a resolution of $\mathbb{C}_V$. Hence, if U is a Stein open subset and $V \subset U$ is locally weighted homogeneous at each point, then $H^k(\Gamma(\check{\Omega}_V^\bullet)) = H^k(V; \mathbb{C})$ for all k.*

To transfer this structure to almost free divisors, we note that a defining germ $f_0 : \mathbb{C}^n, 0 \to \mathbb{C}^p, 0$ for an ICIS is naturally a smoothing, so for $y \notin D(f)$, $f_0^{-1}(y) \cap B_\epsilon$ is the smooth Milnor fiber. In our case of a nonlinear section f_0 of V defining the AFD $(V_0, 0)$, we must instead consider a stabilization of f_0. Let $f : \mathbb{C}^{n+q}, 0 \to \mathbb{C}^p, 0$ be a deformation which is a submersion at 0 (Mond considers other "admissible" deformations, and then further extends them to obtain such a deformation; however, the singular Milnor fibers will be the same). Then, by e.g. the versality theorem for $\mathcal{K}_V$ equivalence, $\mathcal{V} = f^{-1}(V), 0 \subset \mathbb{C}^{n+q}, 0$ is diffeomorphic to $V \times \mathbb{C}^r, 0$, for $r = n - p + q$. If $n < hn(V)$, then algebraic and geometric transversality are the same. Hence, by the parametrized transversality theorem, for almost all values $u \in \mathbb{C}^q$, $f_u(x) = f(x, u)$ is algebraically transverse to V. The set of values u where f_u fails to be transverse form the $\mathcal{K}_V$–discriminant $D_V(f)$, see §8 (Mond calls this the logarithmic discriminant).

Let $\pi : \mathcal{V}, 0 \to \mathbb{C}^q, 0$, be the restriction to $\mathcal{V}$ of the projection $\mathbb{C}^{n+q}, 0 \to \mathbb{C}^q, 0$; this is the analogue of the ICIS germ f_0. For $\epsilon > 0$ and $u \notin D_V(f)$ both sufficiently small, $\pi^{-1}(u) \cap B_\epsilon = f_u^{-1}(V) \cap B_\epsilon$ is a singular Milnor fiber for f_0. Then, Mond introduces the complexes which serve as analogues of those for ICIS (compare [39] or [60])

$$
\begin{aligned}
\check{\Omega}^\bullet_{V_0} &= \Omega^\bullet_{\mathbb{C}^n, 0} / \langle f_0^*(h\Omega^\bullet(\log V)) \rangle \\
\check{\Omega}^\bullet_{\mathcal{V}/\mathbb{C}^q} &= \Omega^\bullet_{\mathbb{C}^{n+q}, 0} / \left(\langle f^*(h\Omega^\bullet(\log V)) \rangle + \sum du_i \wedge \Omega^{\bullet-1}_{\mathbb{C}^{n+q}, 0} \right)
\end{aligned}
\tag{16}
$$

where "$\langle M \rangle$" now denotes the ideal in the exterior algebra generated by M. Moreover, by applying the analogue of the Poincaré Lemma [75, Lemma 3.10] when f is transverse to V, we can replace $\langle f^*(h\Omega^\bullet(\log V)) \rangle$ by $\langle (h \circ f\Omega^\bullet(\log V)) \rangle$.

There are three crucial properties which Mond establishes:

1.
$$
R^k \pi_*(\mathbb{C}_V) \otimes_\mathbb{C} \mathcal{O}_{\mathbb{C}^q} \simeq \mathcal{H}^k(\pi_*(\check{\Omega}^\bullet_{\mathcal{V}/\mathbb{C}^q})) \qquad \text{outside } D_V(f)
$$

2. there is an isomorphism of stalks at 0,
$$
\mathcal{H}^k(\pi_*(\check{\Omega}^\bullet_{\mathcal{V}/\mathbb{C}^q}))_0 \simeq \pi_*(\mathcal{H}^k(\check{\Omega}^\bullet_{\mathcal{V}/\mathbb{C}^q}))_0 \qquad \text{for all } k > 0;
$$

3. the sheaves $\mathcal{H}^k(\pi_*(\check{\Omega}^\bullet_{\mathcal{V}/\mathbb{C}^q}))$ are coherent for all k.

The coherence property 3) is crucial for providing a basis of local sections which generate $\mathcal{H}^k(\pi_*(\check{\Omega}^\bullet_{\mathcal{V}/\mathbb{C}^q}))_u$ at $u \in D_V(f)$. By 1) this explicitly gives a basis for the de Rham cohomology of the singular Milnor fiber over u. The earlier results of Brieskorn and Greuel used the extension of the germ to a proper mapping to obtain coherence. Mond instead applies a result of Van Straten [111]. Then, by an argument which generally follows that for ICIS given in [60, Cor. 8.8], Mond proves

THEOREM 7.3. *Let $V_0, 0 \subset \mathbb{C}^n, 0$ be an almost free divisor defined by a nonlinear section f_0 of $V, 0$. Then, the deRham cohomology of a singular Milnor fiber at u for a stabilization f is computed as the fiber of the sheaf $\mathcal{H}^*(\pi_*(\check{\Omega}^\bullet_{V/\mathbb{C}^q}))$ at u. It is nonzero only in dimension $n-1$ and its rank is given by the rank of the stalk at 0, $\dim_{\mathbb{C}} H^{n-1}(\pi_*(\check{\Omega}^\bullet_{V/\mathbb{C}^q})_0)$.*

The earlier properties imply that $\check{\Omega}^{p-1}_{V_0}$ has torsion and, Theorem 7.3 allows the singular Milnor number to be computed in the weighted homogeneous case as $\dim Tor\check{\Omega}^{p-1}_{V_0}$.

The singular Milnor fiber can be triangulated, so the vanishing cycles can be represented by a union of simplices. Hence, we can still integrate forms over these cycles. This then leads to Mond's construction of the Gauss–Manin connection for a stabilization of an almost free divisor with properties analogous to those of ICIS, especially that it has a regular singularity along the $\mathcal{K}_V$-discriminant [75, §6]. This allows Mond to extend a number of formulae valid for ICIS [39] to the singular Milnor fiber. These results allow for the possibility that many of the earlier results can be extended explicitly using De Rham cohomology.

IV Discriminants

8.   Discriminants for Deformations of Sections

Among the original examples of free divisors in Theorem 2.2, are discriminants for versal unfoldings of ICIS, and bifurcation sets for isolated hypersurface singularities. We explore the general mechanism which produces free divisors in many different situations, which includes both of these examples. As mentioned earlier, there are criteria of Saito and Alexandrov for verifying the a hypersurface singularity is a free divisor. In addition, there is a criterion of Goryunov [38] for functions on space curves: $\mu = \tau$ implies the discriminant of the versal unfolding is a free divisor. Mond and Van Straten then showed that $\mu = \tau$ always holds for functions on space curves, implying that the discriminant is always a free divisor. Van Straten [111] had earlier shown, using Saito's criterion, that the discriminants of versal deformations of space curves are free divisors.

Using a different approach and Van Straten's result, it was shown in [24] that the bifurcation set for smoothings of space curves are free divisors. In fact, this last result applies a general method based on representing various maps, unfoldings, etc. as sections of varieties V and giving sufficient conditions that the $\mathcal{K}_V$-discriminant for the versal unfolding of such a section is a free divisor. The approach is summarized.

Freeness Principle for Discriminants

$$\text{Cohen–Macaulay of codim 1} \; + \; \text{Genericity of Morse Type} \atop \text{Singularities} \implies \text{Freeness of Discriminants} \qquad (17)$$

We explain how this principle applies to a wide variety of situations and discuss what happens as each condition fails.

Given $V, 0 \subset \mathbb{C}^p, 0$ and a germ $f_0 : \mathbb{C}^n, 0 \to \mathbb{C}^p, 0$ which has finite $\mathcal{K}_V$-codimension, let $F : \mathbb{C}^{n+q}, 0 \to \mathbb{C}^{p+q}, 0$ be a $\mathcal{K}_V$-versal unfolding of f_0. Using local coordinates u for $\mathbb{C}^q$, we write $F(x, u) = (\bar{F}(x, u), u)$, and denote $f_u(x) = \bar{F}(x, u) : \mathbb{C}^n, 0 \to \mathbb{C}^p, 0$ as a function of x . Also, we denote by π the projection $\mathbb{C}^{n+q}, 0 \to \mathbb{C}^q, 0$.

DEFINITION 8.1. *The $\mathcal{K}_V$-critical set of F, $C_V(F)$, consists of points (x_0, u_0), such that the germ f_{u_0} is not algebraically transverse to V at x_0. It can equivalently be defined as*

$$C_V(F) = supp(N\mathcal{K}_{V,un,e} \cdot F)$$

(where $N\mathcal{K}_{V,un,e} \cdot F$ is the extended normal space for the action of the unfolding group $\mathcal{K}_{V,un}$). The $\mathcal{K}_V$-discriminant of F is then defined to be $D_V(F) = \pi(C_V(F))$.

We apply Teissier's method [103] for associating a non–reduced structure to the discriminant via the 0–th Fitting ideal.

It can be shown [25, §2] that as f_0 has finite $\mathcal{K}_V$-codimension, $\pi|C_V(F)$ is finite to one. Hence, by Grauert's theorem $D_V(F)) = \pi(C_V(F))$ is the image of an analytic subset $C_V(F)$ under a finite map, hence is also an analytic germ of the same dimension as $C_V(F)$. Second, by [25, Prop. 2.4 and Cor. 2.5] (Mond gives an equivalent formulation), we obtain

PROPOSITION 8.2. *If V is a free divisor and F is a $\mathcal{K}_V$-versal unfolding of f_0 (or at least $\bar{F}$ is algebraically transverse to V at 0) and $n < hn(V)$, then, both $C_V(F)$ and $D_V(F)$ are Cohen–Macaulay of dimension $q - 1$.*

This is half of the condition (17). To obtain the other half , we define

DEFINITION 8.3. *We say that a free divisor $V, 0 \subset \mathbb{C}^p, 0$, generically has Morse-type singularities in dimension n if: all points on canonical* *stata of V of codimension $\leq n+1$ have Morse singularities of nonzero exceptional weight type (which we do not define here); and any stratum of codimension $> n+1$ lies in the closure of a stratum of codimension $= n+1$.*

This condition combined with Proposition 8.2 allows us to apply Saito's criterion to obtain a general criterion for freeness of $\mathcal{K}_V$-discriminants.

THEOREM 8.4 ([25, Thm. 2]). *Let $V, 0 \subset \mathbb{C}^p, 0$ be a free divisor which generically has Morse-type singularities in dimension n where $n < hn(V)$. Then, the $\mathcal{K}_V$-discriminant of the versal unfolding for any $f_0 : \mathbb{C}^n, 0 \to \mathbb{C}^p, 0$ is a free divisor. Moreover,*

$$\text{Derlog}\,(D_V(F)) = \text{module of } \mathcal{K}_V\text{-liftable vector fields.}$$

By a $\mathcal{K}_V$-liftable vector field $\eta \in \theta_q$ we mean there are

$$\xi \in \mathcal{O}_{\mathbb{C}^{n+q},0}\{\frac{\partial}{\partial x_1}, \ldots, \frac{\partial}{\partial x_n}\} \quad \text{and} \quad \zeta \in \mathcal{O}_{\mathbb{C}^{n+q},0}\{\zeta_1, \ldots, \zeta_p\}$$

satisfying:

$$(\xi + \eta)(\bar{F}) \;=\; \zeta \circ \bar{F} \tag{18}$$

9. Morse-type Singularities for Sections and Mappings on Divisors

To apply Theorem 8.4, we must verify for a given V the genericity of Morse–type singularities. We first illustrate the applicability in the case of bifurcation sets of finitely $\mathcal{A}$–determined map germs $f_0 : \mathbb{C}^n, 0 \to \mathbb{C}^p, 0$. By (1.1), f_0 is obtained as the pullback of the stable unfolding F of f_0 by an embedding g_0. Furthermore, by Theorem 1.5, f_0 is a stable multigerm over a point $y \in D(F)$ iff g_0 is transverse to $D(F)$ at y. Hence, for an unfolding $f(x, v) = (\bar{f}(x, v), v)$ of f_0 induced by an unfolding $g(y, v) = (\bar{g}(y, v), v)$ of g_0, $f_v(x) = \bar{f}(x, v)$ is stable iff $g_v(x) = \bar{g}(x, v)$ is transverse to $D(F)$. Thus, the bifurcation set of f is the $\mathcal{K}_{D(F)}$-discriminant of g. Thus, it is sufficient to determine when the discriminants $D(F)$ generically have Morse–type singularities. Using the results on $\mathcal{A}_e$–codimension 1 germs and multigerms, we can identify the following class [25, §6].

DEFINITION 9.1. *We say that a finitely $\mathcal{A}$–determined (multi)germ $f_0 : \mathbb{C}^n, S \to \mathbb{C}^p, 0$ with $n \geq p$ belongs to the distinguished bifurcation class of (multi)germs if it satisfies one of the following:*

1. *"general case"*: $n \neq p+1$ and $p \leq 4$;

2. *"worse case"*: $n = p+1$ and $p \leq 3$;

3. *"best cases"*:

    (i) corank 1 (multi)germs such that :

        (a) $n = p+1$ and $p \leq 6$ or

        (b) $n > p+1$ and $p \leq 5$

    (ii) Σ_{n-p+1} and $\Sigma_{2,(1)}$ (multi)germs without restriction on $n \geq p$.

For example, the Σ_{n-p+1} germs are germs in the $\mathcal{K}$–equivalence class of the A_k germs for $n \geq p$; and the $\Sigma_{2,(1)}$ germs are those in the $\mathcal{K}$–equivalence class of $I_{2,b}$, consisting of the germs $f_0(x,y) = (xy, x^2 + y^b)$ for $n = p$.

Then, the Terao and Bruce result extends as follows [25, Thm. 3].

THEOREM 9.2. *Let $f_0 : \mathbb{C}^n, S \to \mathbb{C}^p, 0$ be a finitely $\mathcal{A}$–determined (multi) germ which belongs to the distinguished bifurcation class. Then, the bifurcation set of its $\mathcal{A}$–versal unfolding is a free divisor.*

David Mond and Andrew DuPlessis obtained an equivalent version of this by directly using $\mathcal{A}$–equivalence.

Also, $\{0\} \subset \mathbb{C}^p$ generically has Morse–type singularities (i.e. the usual Morse singularities) which yields for $p = 1$ the freeness of (the usual) discriminants of versal unfoldings of isolated hypersurface singularities. There are a number of other examples of free divisors which generically have Morse–type singularities. These are given in [25, §6-9], leading to the freeness of various discriminants and bifurcation sets.

One revealing example is the Manin–Schechtman discriminantal arrangement which naturally extends the braid arrangement [64]. For a central general position hyperplane arrangement $A = \cup_{i=1}^n H_i \subset \mathbb{C}^k$, it is obtained as the pullback of the Boolean arrangement $A_n \subset \mathbb{C}^n$ by a linear embedding φ. The associated discriminantal arrangement $B(n, k)$ consists of the set of normal translation vectors for the hyperplanes H_i which when applied, do not give a general position arrangement. However, unlike the braid arrangement, $B(k + 3, k)$ is not free when $k \geq 2$ Orlik–Terao [82, Prop. 5.6.6].

COROLLARY 9.3. *Let $A \subset \mathbb{C}^k$ be a central general position arrangement defined by $\varphi : \mathbb{C}^k \to \mathbb{C}^n$. Then, the $\mathcal{K}_{A_n}$–discriminant of the versal unfolding of φ is a free divisor.*

Then, $B(n, k)$ is the intersection of $D_{A_n}(\varphi)$ with the linear subspace of translation-deformations and the remaining unfolding parameters (at least for low (n, k)) give equisingular $\mathcal{K}_{A_n}$–deformations. Thus, $D_{A_n}(\varphi)$ is topologically equivalent to a suspension of $B(n, k)$, even though $D_{A_n}(\varphi)$ is free while $B(n, k)$ is not.

EXAMPLE 9.4. In the opposite direction, an example of a free divisor without a Morse–type singularity is given by the free hyperplane arrangement $A \subset \mathbb{C}^3$ defined by $Q = xyz(x-y)$. A Morse–type singularity of dimension 2, if it existed, would be obtained as the inclusion of a generic hyperplane section. However, this gives an arrangement of 4 lines in $\mathbb{C}^2$ with singular Milnor fiber having 2 singular cycles (see [25, §4, 9]). This corresponds to the section having $\mathcal{K}_{A,e}$–codimension 2.

These methods can be adjusted to apply to the dual case of the equivalence of mappings $f_0 : \mathbb{C}^n, 0 \to \mathbb{C}^p, 0$ fixing a divisor $V, 0 \subset \mathbb{C}^n, 0$ in the source. We denote the contact group preserving V by $_V\mathcal{K}$. For certain special free divisors V and simple functions f_0, Arnold [6], Lyaschko [61], Zakalyukin [114], and Goryunov [38] prove that the discriminant is a free divisor. What is surprising is that this fails in general even with V free, and it does not depend on whether we consider functions or ICIS map germs. Moreover, the freeness of the $_V\mathcal{K}$–discriminant depends on whether generically V has Morse type singularities. This is because the "Morse type singularities" for $_V\mathcal{K}$–equivalence can be canonically identified with the Morse type singularities for $\mathcal{K}_V$–equivalence [27, §6]. An equivalent version of example (9.3) shows that the discriminant will not be free for the module of $_V\mathcal{K}$–liftable vector fields. We do obtain [27, §6] (with consequences for $_V\mathcal{A}$ and $_V\mathcal{R}$)

THEOREM 9.5. *If $V, 0 \subset \mathbb{C}^n, 0$ is a free divisor which generically has Morse–type singularities in dimension $n - p$, then the $_V\mathcal{K}$–discriminant of the $_V\mathcal{K}$–versal unfolding of an ICIS germ f_0 (of finite $_V\mathcal{K}$–codimension) is a free divisor with*

$$\text{Derlog}\,(D_{_V\mathcal{K}}(F)) = \text{module of }_V\mathcal{K}\text{–liftable vector fields.}$$

10. Beyond Freeness - Free* Divisor Structures

When either of the two conditions in (17) fails for some group of equivalences, we can no longer conclude that the appropriate discriminants are free. We explain how we must introduce new notions reflecting the weaker structure that does remain. We restrict our remarks to the specific question of whether the module of liftable vector fields defines a free divisor structure, reflecting the intrinsic features of the equivalence.

There are always cases such as isolated curve singularities in $\mathbb{C}^2$ which are always free by [93], unrelated to how they are obtained.

First we relax the condition of genericity of Morse type singularities. For example (9.4), the $\mathcal{K}_A$-discriminant has a nonreduced structure resulting from the generic singularities having $\mathcal{K}_{A,e}$-codimension 2 [25, §9]. If the discriminant is Cohen–Macaulay of codimension 1 then at least the discriminant has a "free*" divisor"structure defined by the liftable vector fields.

DEFINITION 10.1.   *By a hypersurface germ $V, 0 \subset \mathbb{C}^p$ having a free* divisor structure on $\mathbb{C}^{p'}$, where $p' = p + m \geq p$, we shall mean: for $V' = V \times \mathbb{C}^m \subset \mathbb{C}^{p'}$ we are given an $\mathcal{O}_{\mathbb{C}^{p'},0}$- submodule $\mathrm{Derlog}^*(V) \subseteq \mathrm{Derlog}(V')$ which satisfies:*

  1. *$\mathrm{Derlog}^*(V)$ is a free $\mathcal{O}_{\mathbb{C}^{p'},0}$-module of rank p'; and*

  2. *$supp(\theta_{p'}/\mathrm{Derlog}^*(V)) = V'$*

In fact there are many ways, including trivial ones, of putting a free* divisor structure on certain hypersurfaces. Their usefulness depends on certain measures of nontriviality and on the intrinsic properties of $\mathrm{Derlog}^*(V)$. By suspending V we can assume $\mathrm{Derlog}^*(V) \subset \mathrm{Derlog}(V)$. This raises the question of how one can understand properties of nonfree submodules of θ_p or more generally $\mathcal{O}_{\mathbb{C}^n,0}^p$ using free submodules which are in an appropriate sense a "good approximation".

For example, we know isolated surface singularities $V, 0 \subset \mathbb{C}^3, 0$ are never free divisors, but are always almost free divisors. Although nonisolated surface singularities need not be free, any weighted homogeneous surface singularity has a natural "pfaffian " free* divisor structure [26, §1].

Also, by the proof of Theorem 8.4 with $n < hn(V)$, but without having genericity of Morse–type singularities, the $\mathcal{K}_V$ discriminant is a free* divisor defined by $\mathrm{Derlog}(V)$ [26, Thm. 1]. For the various consequences mentioned in Example (9), if the conditions are relaxed so genericity of Morse type singularities fails, then at least the discriminants are free* divisors defined by the modules of liftable vector fields. For example, all finitely $\mathcal{A}$–determined germs in the nice dimensions (in the sense of Mather [69]) have bifurcation sets which are free* divisors [26, §1] for the module of $\mathcal{A}$–liftable vector fields.

Importantly, we can still compute the vanishing topology for free* divisors using a modification of Theorem 3.3, where we must correct for "virtual singularities"for the $\mathrm{Derlog}^*(V)$ structure [26, §4].

11. Cohen-Macaulay Reductions for Groups of Equivalences

If we relax instead the first condition in (17), then the group $\mathcal{G}$ of equivalences does not have the correct algebraic structure to ensure that the discriminant is Cohen–Macaulay of codimension 1.

EXAMPLE 11.1.   Several examples where this occurs are for sections of free complete intersections of codimension ≥ 2 (with the exception of $\{0\} \subset \mathbb{C}^p$), the relative case of the intersection of an ADF on an AFCI, and the case of functions on complete intersections where either we allow both the function and complete intersection to deform or we fix the complete intersection.

In these cases which are considered in [26] and [27], we introduce a "Cohen–Macaulay reduction"of the original group, which allows us to keep the same discriminant but with a different structure.

DEFINITION 11.2.   *Given a geometric subgroup $\mathcal{G}$ of $\mathcal{A}$ or $\mathcal{K}$ which has geometrically defined discriminants, a Cohen–Macaulay reduction of $\mathcal{G}$ (abbreviated CM-reduction) consists of a geometric subgroup $\mathcal{G}^* \subset \mathcal{G}$ which still acts on $\mathcal{F}$ (and $\mathcal{F}_{un}$) such that:*

1. *$\mathcal{G}^*$ is Cohen–Macaulay (i.e. for a $\mathcal{G}^*$-versal unfolding F on q parameters, the normal space $N\mathcal{G}^*_{une} \cdot F$ viewed as $\mathcal{O}_{\mathbb{C}^q,0}$-module is Cohen–Macaulay, and whose support, the discriminant $D_{\mathcal{G}^*}(F)$, has codimension 1);*

2. *$f \in \mathcal{F}$ has finite $\mathcal{G}^*$-codimension iff it has finite $\mathcal{G}$-codimension;*

3. *if F is a $\mathcal{G}^*$-versal unfolding on q parameters, then as $\mathcal{O}_{\mathbb{C}^q,0}$-modules,*

$$supp(N\mathcal{G}^*_{un,e} \cdot F) \quad = \quad supp(N\mathcal{G}_{un,e} \cdot F) \qquad (19)$$

As far as we have determined, this is not a reduction in the sense of Rees [90], [48] and Gaffney [34]. In earlier results, the RHS in Theorems 4.2, 4.3 and 4.4 are normal spaces for CM–reductions.

We then establish two key results regarding CM-reduction [27, Thm. 2 and 3].

THEOREM 11.3.   *Suppose a group $\mathcal{G}$ has a CM–reduction, then*

1. *the $\mathcal{G}$-discriminants for $\mathcal{G}$-versal unfoldings are free* divisors for the module of $\mathcal{G}^*$-liftable vector fields; and*

2. *provided $\mathcal{G}$ generically has Morse-type singularities which are $\mathcal{G}^*$-liftable, the $\mathcal{G}$ -discriminants are free divisors.*

In particular, for the examples in (11.1), either they have CM–reductions or are Cohen–Macaulay. Hence, their corresponding $\mathcal{G}$–discriminants are free* divisors. Moreover, we may apply the theorem and conclude that the corresponding $\mathcal{G}$–discriminants of versal unfoldings are free divisors for the following: ICIS, which are sections of $\{0\} \subset \mathbb{C}^p$ (recovering Looijenga's result); the relative cases of the intersection of an ADF on an ICIS; functions on ICIS where we allow both the functions and ICIS to deform (related to a result of Mond–Montaldi [76]); or a complete intersection with boundary singularity a free divisor, with appropriate restrictions on the free divisor generically having Morse–type singularities (9.5).

References

1. K. Aomoto, On the vanishing of cohomology attached to certain many valued meromorphic functions, J. Math. Soc. Japan **27** (1975) 248–255.

2. A. G. Alexandrov, Euler homogeneous singularities and logarithmic differential forms, Ann. Global Anal. Geom. **4** (1986) 225–242.

3. A. G. Alexandrov, Nonisolated hypersurface singularities, Adv. Soviet Math. **1** (1990) 211–245.

4. V. I. Arnold, The cohomology ring of the colored braid group, Math. Notes of Acad. Sci. USSR **5** (1969) 138–140.

5. V. I. Arnold, Wave front evolution and equivariant Morse lemma, Comm. Pure App. Math. **29** (1976) 557–582.

6. V. I. Arnold, Critical points of functions on manifolds with boundaries, simple Lie Groups B_k, C_k, F_4, and singularities of evolutes, Uspehi Mat. Nauk SSR **33**:5 (1978) 91–105.

7. V. I. Arnold, A. N. Varchenko, and S. M. Gusein-Zade, *Singularities of differentiable mappings vol II* Birkhäuser, Basel-Boston, 1988.

8. R. Wik Atique, T. M. Cooper and D. Mond, Vanishing topology of codimension 1 multigerms over $I\!R$ and C, preprint, 1999.

9. E. Brieskorn, Sur les groupes des tresses (d'après V. I. Arnol'd), *Séminaire Bourbaki 1971/72*, Springer Lecture Notes in Math **315** (1973) 21–44.

10. E. Brieskorn, Monodromie der isolertier Singularitäten von Hyperflächen, Manuscripta Math. **2** (1970) 103–161.

11. J. W. Bruce, Vector fields on discriminants and bifurcation varieties, J. London Math. Soc. **17** (1985) 257–262.

12. J. W. Bruce and R. M. Roberts, Critical points of functions on analytic varieties, Topology **27** (1988) 57–91.

13. D. Buchsbaum and D. Eisenbud, Algebra structures for finite free resolutions and some structure theorems for ideals of codimension 3, Amer. J. Math. **99** (1977) 447–485.

14. D. Buchsbaum and D. S. Rim, A generalized Koszul complex II: depth and multiplicity, Trans. Amer. Math. Soc. **111** (1964) 197–224.

15. R. Buchweitz, Contributions à la théorie des singularités, thesis, Univ. Paris VII, 1981.

16. L. Burch, On ideals of finite homological dimension in local rings, Math. Proc. Camb. Phil. Soc. **64** (1968) 941–948.

17. F. J. Castro-Jiménez, D. Mond and L. Narváez- Macarro, Cohomology of the complement of a free divisor, Trans. Amer. Math. Soc. **348** (1996) 3037-3049.

18. J. Damon, *The unfolding and determinacy theorems for subgroups of A and K*, Memoirs Amer. Math. Soc. **50**, no. 306, (1984).

19. J. Damon, Deformations of sections of singularities and Gorenstein surface singularities, Amer. J. Math. **109** (1987) 695-722.

20. J. Damon, *A*-equivalence and the equivalence of sections of images and discriminants, *Singularity Theory and its Applications: Warwick 1989, Part I* (eds D. Mond and J. Montaldi), Springer Lecture Notes in Math. **1462** (1991) 93-121.

21. J. Damon, A Bezout theorem for determinantal modules, Compositio Math. **98** (1995) 117-139.

22. J. Damon, *Higher multiplicities and almost free divisors and complete intersections*, Memoirs Amer. Math. Soc. **123**, no 589, 1996.

23. J. Damon, Singular Milnor fibers and higher multiplicities for nonisolated complete intersections, *Proc. Conf. Sing. and Complex Geom.*, (ed. Q. Lu et al.) AMS/IP Studies in Adv. Math. **5** (1997) 28-53.

24. J. Damon, Critical points of affine multiforms on the complements of arrangements, *Singularity Theory* (eds Bill Bruce and David Mond), London Math. Soc. Lecture Notes **263** (Cambridge Univ. Press, 1999) pp 25-53.

25. J. Damon, On the legacy of free divisors : discriminants and Morse type singularities, Amer. J. Math. **120** (1998) 453-492.

26. J. Damon, The legacy of free divisors II : Free* divisors and complete intersections, preprint.

27. J. Damon, The legacy of free divisors III : Functions and divisors on complete intersections, preprint.

28. J. Damon, On the number of bounding cycles for nonlinear arrangements, *Arrangements-Tokyo*, 1998, (eds M. Falk and H. Terao), Adv. Stud. in Pure Math. **27**, Math. Soc. of Japan (2000) 51-72.

29. J. Damon, On the freeness of equisingular deformations of plane curve singularities, Topology and Appl., to appear.

30. J. Damon and D. Mond, *A*-codimension and the vanishing topology of discriminants Invent. Math. **106** (1991) 217-242.

31. J. A. Eagon and D. Northcott, Ideals defined by matrices and a certain complex associated with them, Proc. Roy. Soc. London **299** (1967) 147-172.

32. T. Gaffney, Integral closure of modules and Whitney equisingularity, Invent. Math. **107** (1992) 301-322.

33. T. Gaffney, Multiplicities and equisingularity of ICIS germs, Invent. Math. **123** (1996) 209-220.

34. T. Gaffney and S. Kleiman, Specialization of integral dependence of modules, Invent. Math. **137** (1999) 541-574.

35. M. Giusti, Intersections complètes quasi-homogènes: calcul d'invariants, thesis, Univ. Paris VII, 1981 (also preprint, Centre Math. de l'Ecole Polytechnique, 1979).

36. M. Goresky and R. MacPherson, *Stratified Morse Theory*, Ergebnisse der Math., Springer-Verlag, Heidelberg–New York, 1988.

37. V. Goryunov, Singularities of projections of full intersections, Jour. Soviet. Math. **27** (1984) 2785-2811.

38. V. Goryunov, Functions on space curves, Jour. London Math. Soc. **61** (2000) 807-822.

39. G.-M. Greuel, Der Gauss-Manin Zusammenhang isolierter Singularitäten von vollständigen Durchschnitten, Math. Ann. **214** (1975) 235-266.

442   James Damon

40. G.-M. Greuel and H. Hamm, Invarianten quasihomogener vollständiger Durchschnitte, Invent. Math. **49** (1978) 67–86.
41. P. Griffiths, On the periods of certain rational integrals I, II, Ann. of Math. **90** (1969), 460–541.
42. A. Grothendieck, On the de Rham cohomology of algebraic varieties, Publ. Math. I.H.E.S. **29** (1966) 95–103.
43. H. Hamm, Lokale topologische Eigenschaften komplexer Räume, Math. Annalen **191** (1971) 235–252.
44. H. Hauser and G. Müller, On the Lie algebra $\Theta(X)$ of vector fields on a singularity, J. Math. Sci. Univ. Tokyo **1** (1994) 239–250.
45. J. H. P. Henry and M. Merle, Conormal spaces and Jacobian modules: A short dictionary, *Singularities* (ed. J.P. Brasselet), London Math. Soc. Lect. Notes **201**, (Cambridge Univ. Press, 1994) pp 147–174.
46. D. Hilbert, Über die Theorie von algebraischen Formen, Math. Ann. **36** (1890) 473–534.
47. M. Holland and D. Mond, Logarithmic differential forms and the cohomology of the complement of a divisor, Math. Scand. **83** (1998) 235–254.
48. D. Kirby and D. Rees, Multiplicities in graded rings II: integral equivalence and the Buchsbaum–Rim multiplicity, Math. Proc. Camb. Phil. Soc. **119** (1996) 425–445.
49. S. Kleiman and A. Thorup, A geometric theory of the Buchsbaum-Rim multiplicity, Jour. Alg. **167** (1994) 168–231.
50. T. de Jong, The virtual number of D_∞ points, Topology **29** (1990) 175–184.
51. T. de Jong and D. van Straten, Disentanglements, *Singularity Theory and its Applications: Warwick 1989, Part I* (eds D. Mond and J. Montaldi), Springer Lecture Notes in Math **1462** (1991) 199–211.
52. T. de Jong and D. van Straten,Deformations of the normalization of hypersurfaces, Math. Ann. **288** (1991) 527–547.
53. M. Kato and Y. Matsumoto, On the connectivity of the Milnor fiber of a holomorphic function at a critical point, *Manifolds Tokyo 1973*, (University of Tokyo Press 1975), pp 131–136.
54. D. T. Lê, Le concept de singularité isolée de fonction analytique, Adv. studies in Pure Math. **8** (1986) 215–227.
55. D. T. Lê, Calcul du nombre de cycles évanouissants d'une hypersurface complexe, Ann. Inst. Fourier **23** (1973) 261–270.
56. D. T. Lê, Calculation of Milnor number of an isolated singularity of complete intersection, Funct. Anal. Appl. **8** (1974) 127–131.
57. D. T. Lê and G.-M. Greuel, Spitzen, Doppelpunkte und vertikale Tangenten in der Diskriminante verseller Deformationen von vollständigen Durchschnitten, Math. Ann. **222** (1976) 71–88.
58. D. T. Lê and B. Teissier, Cycles évanescents, sections planes, et conditions de Whitney II, *Singularities* (ed Peter Orlik), Proc. Symp. pure math. **40**:2, (Amer. Math. Soc., 1983) pp 65–103.
59. D. T. Lê and B. Teissier, Limites d'espaces tangents en géométrie analytique, Comm. Math. Helv. **63** (1988) 540–578.
60. E. J. N. Looijenga, *Isolated singular points on complete intersections*, London Math. Soc. Lecture Notes **77**, Cambridge Univ. Press, 1984.
61. O. Lyashko, Classification of critical points of functions on a manifold with singular boundary, Funct. Anal. Appl. **17** (1984) 187–193.
62. F. S. Macaulay, *The algebraic theory of modular systems*, Cambridge Tracts **19**, Cambridge Univ. Press, 1916.

63. B. Malgrange, *Ideals of Differentiable Functions*, Oxford Univ. Press, 1966.

64. Y. I. Manin and V. V. Schechtman, Arrangements of hyperplanes, higher braid groups, and higher Bruhat orders, *Algebraic Number Theory in honor of K. Iwasawa*, Adv. Stud. in Pure Math. **17** (North-Holland Publ., 1989) pp 289–308.

65. D. Massey, The Lê varieties I, Invent. Math. **99** (1990) 357–376.

66. D. Massey, The Lê varieties II, Invent. Math. **104** (1991), 113–148.

67. D. Massey and D. Siersma, Deformation of polar methods, Ann. Inst. Fourier **42** (1992) 737–778.

68. J. Mather, Stability of C^∞-mappings III. Finitely determined map germs, Publ. Math. IHES. **36** (1968) 127–156.

69. J. Mather, Stability of C^∞-mappings IV. Classification of stable germs by $\mathbb{R}$-algebras, Publ. Math. IHES. **37** (1969) 223–248.

70. J. Mather, Stability of C^∞-mappings V. Transversality, Adv. in Math. **37** (1970) 301–336.

71. J. Mather, Stability of C^∞-mappings VI. The nice dimensions, *Proceedings of Liverpool Singularities Symposium I* (ed C. T. C. Wall, Springer Lecture Notes in Math. **192** (1970) 207–253.

72. J. Milnor, *Singular points on complex hypersurfaces*, Ann. Math. Studies **61**, Princeton Univ. Press, 1968.

73. D. Mond, Some remarks on the geometry and classification of germs of maps from surfaces to 3-space, Topology **26** (1987) 361–383.

74. D. Mond, Vanishing cycles for analytic maps, *Singularity Theory and its Applications: Warwick 1989, Part I* (eds D. Mond and J. Montaldi), Springer Lecture Notes in Math. **1462** (1991) 221–234.

75. D. Mond, Differential forms on free and almost free divisors, Proc. London Math. Soc. **81** (2000) 587-617.

76. D. Mond and J. Montaldi, Deformations of maps on complete intersections, Damon's $\mathcal{K}_V$-equivalence and bifurcations, *Singularities* (ed J.P. Brasselet), London Math. Soc. Lecture Notes **201**, (Cambridge Univ. Press, 1994) pp 263–284.

77. D. Mond and D. van Straten, $\mu = \tau$ for functions on space curves, J. London Math. Soc, to appear.

78. A. Nemethi, The Milnor fiber and the zeta function of singularities of type $f = P(h, g)$, Compositio Math. **79** (1991) 63–97.

79. D. G. Northcott, Semi–regular rings and semi–regular ideals, Quart. J. Math. Oxford **11** (1960) 81–104.

80. P. Orlik and L. Solomon, Combinatorics and the topology of complements of hyperplanes, Invent. Math. **56** (1980) 167–189.

81. P. Orlik and L. Solomon, Coxeter Arrangements, *Singularities* (ed Peter Orlik), Proc. Symp. pure math. **40**:2, (Amer. Math. Soc., 1983) pp 269–292.

82. P. Orlik and H. Terao, *Arrangements of Hyperplanes*, Grundlehren der Math. Wiss. **300**, Springer Verlag, 1992.

83. P. Orlik and H. Terao, The number of critical points of a product of powers of linear functions, Invent. Math. **120** (1995) 1–14.

84. P. Orlik and H. Terao, Arrangements and Milnor fibers, Math. Ann. **301** (1995) 211–235.

85. R. Pellikaan, Hypersurface singularities and resolutions of Jacobi modules, thesis, University of Utrecht, 1985.

86. R. Pellikaan, Deformations of hypersurfaces with a one-dimensional singular locus, Jour. Pure Appl. Algebra **67** (1990) 49–71.

87. H. Pinkham, *Deformations of algebraic varieties with G_m action*, Astérisque **20**, Soc. Math. France, (1974).

88. R. Randell, On the topology of nonisolated singularities, *Geometric Topology*, (Academic Press 1979), pp 445–473.

89. R. Randell, The Milnor number of some isolated complete intersection singularities with C^*-action, Proc. Amer. Math. Soc. **72** (1978) 375–380.

90. D. Rees, Reductions of modules, Math. Proc. Camb. Phil. Soc. **101** (1987) 431–449.

91. J. Rieger, Recognizing unstable equidimensional maps and the number of stable projections of algebraic hypersurfaces, Manuscripta Math. **99** (1999) 73–91.

92. M. Roberts and V. Zakalyukin, Symmetric wave fronts, caustics, and Coxeter groups, *Singularity theory: Proc. College on Singularity Theory (Trieste 1991)* (eds D. T. Lê, K. Saito and B. Teissier), (World Scientific Publ., 1994) pp 594–626.

93. K. Saito, Theory of logarithmic differential forms and logarithmic vector fields, J. Fac. Sci. Univ. Tokyo Sect. Math. **27** (1980) 265–291.

94. M. Schaps, Deformations of Cohen–Macaulay schemes of codimension 2 and nonsingular deformations of space curves, Amer. J. Math. **99** (1977) 669–684.

95. D. Siersma, Isolated line singularities, *Singularities* (ed Peter Orlik), Proc. Symp. pure math. **40**:2, (Amer. Math. Soc., 1983) pp 485–496.

96. D. Siersma, Vanishing cycles and special fibres, *Singularity Theory and its Applications: Warwick 1989, Part I* (ed D. Mond and J. Montaldi), Springer Lecture Notes **1462** (1991) 292–301.

97. D. Siersma, Singularities with critical locus a one–dimensional complete intersection and transversal type A_1, Topology and Appl. **27** (1987) 51–73.

98. D. Siersma, A bouquet theorem for the Milnor fiber, Jour. Alg. Geom. **4** (1995) 51–66.

99. D. Siersma and M. Tibăr, Singularities at infinity and their vanishing cycles, Duke Math. Jour. **80** (1995) 771–783.

100. L. Solomon and H. Terao, A formula for the characteristic polynomial of an arrangement, Adv. in Math. **64** (1987) 305–325.

101. J. Steenbrink, Mixed Hodge structure on the vanishing cohomology, *Real and complex singularities, Oslo 1976* (ed Per Holm), (Sijthoof and Noordhoff, Alphen aan den Rijn, 1977) pp 525–563.

102. B. Teissier, Cycles évanescents, sections planes, et conditions de Whitney, *Singularités à Cargèse*, Asterisque **7-8** (1973) 285–362.

103. B. Teissier, The hunting of invariants in the geometry of the discriminant, *Real and complex singularities, Oslo 1976* (ed Per Holm), (Sijthoof and Noordhoff, Alphen aan den Rijn, 1977) pp 567–677.

104. B. Teissier, Cycles Multiplicités polaires, sections planes, et conditions de Whitney, Springer Lecture Notes in Math. **961** (1982) 314–491.

105. H. Terao, Arrangements of hyperplanes and their freeness I,II, J. Fac. Sci. Univ. Tokyo Sect. Math. **27** (1980) 293–320.

106. H. Terao, The bifurcation set and logarithmic vector fields, Math. Ann. **263** (1983) 313–321.

107. H. Terao, Generalized exponents of a free arrangements of hyperplanes and the Shephard–Todd–Brieskorn formula, Invent. Math. **63** (1981) 159–179.

108. M. Tibăr, Bouquet decomposition for the Milnor fiber, Topology **35** (1996) 227–241.

109. J. Tougeron, Idéaux de fonctions différentiables, I, Ann. Inst. Fourier **18** (1968) 177–240.

110. D. van Straten, Weakly normal surface singularities and their improvements, thesis, Rijksuniversiteit Leiden, 1986.

111. D. van Straten, A note on the discriminant of a space curve, Manuscripta Math. **87** (1995) 167–177.

112. D. van Straten, On the Betti numbers of the Milnor fiber of a certain class of hypersurface singularities, Springer Lecture Notes in Math. **1273** (1987) 203–220.

113. A. N. Varchenko, Critical points of the product of powers of linear functions and families of bases of singular vectors, Compositio Math. **97** (1995) 385–401.

114. V. Zakalyukin, Singularities of circles contact with surfaces and flags, Funct. Anal. Appl. **31** (1997) 67–69.

The Vanishing Topology of Non Isolated Singularities

Dirk Siersma
Mathematisch Instituut, Universiteit Utrecht
(siersma@math.uu.nl)

1. Introduction

We consider holomorphic function germs $f : (\mathbb{C}^{n+1}, O) \to (\mathbb{C}, 0)$ and allow arbitrary singularities (isolated or non-isolated). We are interested in the topology of this situation, especially the so called vanishing homology.

We first recall the definition of the Milnor fibration. For $\epsilon > 0$ small enough there exist an ϵ-ball B_ϵ in $\mathbb{C}^{n+1}$ and an η-disc D_η in $\mathbb{C}$ such that the restriction:

$$f : \quad E := f^{-1}(D_\eta) \cap B_\epsilon \longrightarrow D_\eta$$

is a locally trivial fibre bundle over $D_\eta \setminus \{0\}$. The fibres, mostly denoted by F, are called Milnor fibres. The groups $H_*(E, F)$ are called the vanishing homology groups.

The Milnor fibre F, its homotopy type and homology are interesting topological objects. So is the monodromy operator

$$\mathbb{T}_* \quad : H_*(F) \to H_*(F)$$

of the fibration.

Well known facts are:

- The Milnor fibre is $2n$-dimensional and has the homotopy type of an n-dimensional CW-complex (Milnor [33]),

- The Milnor fibre is $(n - s - 1)$-connected, where s is equal to the dimension of the singular locus of f (Kato-Matsumoto [26]),

- If f has an isolated singularity then the Milnor fibre has the homotopy type of a bouquet of n-dimensional spheres. The number μ of these spheres is called the Milnor number of the isolated singularity (Milnor [33]),

- the eigenvalues of the monodromy operator are roots of unity. See Griffiths [19] for references to four different proofs.

D. Siersma et al. (eds.), New Developments in Singularity Theory, 447–472.
© 2001 Kluwer Academic Publishers. Printed in the Netherlands.

Isolated singularities have been studied in great detail during the last 30 years. They have wonderful properties, which relate different aspects of the singularity. In this paper we want to discuss especially non-isolated singularities. Although the properties (e.g. the topological type of the Milnor fibre) are more complicated than for isolated singularities, there is a lot of interesting structure available.

Let $\Sigma = \Sigma(f)$ be the singular locus of f. For every point of Σ we can do the Milnor construction. So for every $x \in \Sigma$ we have a (local) Milnor fibration (e.g. a space $E_{(x)}$, the local Milnor fibre $F_{(x)}$ and a Milnor monodromy $\mathbb{T}_{(x)}$). We want to investigate the relation between these objects for all $x \in \Sigma$. This is (near to) the study of the sheaf of vanishing cycles [10].

To be more precise: one could try to define a stratification of Σ in such a way that two points of Σ are in the same stratum if they can be joined by a (continuous) path such that there exits a (continuous) family of Milnor fibrations of constant fibration type.

According to a result of Massey, the constancy of Lê numbers (for definition see the contribution of Gaffney [17] in this Volume) implies constancy of the fibration type under certain dimension conditions (more precisely $s \leq n - 2$ for the homotopy-type and $s \leq n - 3$ for the diffeomorphism-type). We refer to Massey's monograph [31] for details and for many other related facts.

Let us suppose that we end up with a situation, where we have stratified Σ (according to the above principle):

$$\Sigma = \Sigma^k \cup \cdots \cup \Sigma^1 \cup \Sigma^0 \ ,$$

where $\Sigma^j \setminus \Sigma^{j-1}$ is j-dimensional and smooth. For every connected component of $\Sigma^j \setminus \Sigma^{j-1}$ we have a monodromy representation of its fundamental group on the homology groups of a typical Milnor fibre at a general point on the statum:

$$\pi_1(\Sigma^j \setminus \Sigma^{j-1}, x) \to \mathrm{Aut}\,[H_*(F_{(x)})] \ .$$

We call these monodromies "vertical". The vertical monodromies contain a lot of extra information about the singularity. The Milnor monodromy is called "horizontal" and commutes with the vertical monodromies.

We intend to discuss this situation in several examples; paying most attention to the situation where Σ is 1-dimensional, where the stratification is rather simple. We also intend to treat some examples of higher dimension.

This paper is organized as follows. In section 2 we recall some facts about isolated singularities. In particular we discuss the relation between variation mapping, monodromy and intersection form.

In section 3 we treat singularities with a 1-dimensional critical set. We follow first [50] and [51], treat several examples, where vertical and horizontal monodromy play a role, and focus at the end on bouquet decompositions of the Milnor fibre. These seem to occur as soon as we stay near to the case of isolated singularities.

Section 4 is about singular sets of higher dimension. We discuss and summarize recent work.

2. About isolated singularities

The theory of variation mappings plays an important role in our discussions. We first repeat some of the well known facts about isolated singularities. They can be found in the literature on several places, e.g. Milnor [33], Lamotke [27], Arnol'd-Gusein Zade-Varchenko [3], Stevens [55].

In the isolated singularity case there exists a geometric monodromy $h : F \to F$ such that $h|_{\partial F}$ is the identity.
Let $\mathbb{T}_q = h_* : H_q(F) \to H_q(F)$ be the algebraic monodromy. The map $\mathbb{T}_q - \mathbb{I} : H_q(F) \to H_q(F)$ factors over:

$$\mathrm{VAR}_q : H_q(F, \partial F) \to H_q(F)$$

which is defined by

$$\mathrm{VAR}_q[x] = [h(x) - x]$$

We have a commutative diagram:

$$
\begin{array}{ccc}
H_q(F) & \xrightarrow{\ \mathbb{T}_q - \mathbb{I}\ } & H_q(F) \\
\downarrow{\scriptstyle j_*} & {\scriptstyle \mathrm{VAR}} & \downarrow{\scriptstyle j_*} \\
H_q(F, \partial F) & \xrightarrow{\ \mathbb{T}_q - \mathbb{I}\ } & H_q(F, \partial F)
\end{array}
$$

LEMMA 2.1. $\mathrm{VAR}_q : H_q(F, \partial F) \to H_q(F)$ is an isomorphism if $q \neq 0$.
Proof. Consider the exact sequence of the pair (S^{2n+1}, F) and the following isomorphism:

$$H_{q+1}(S^{2n+1}, F) \cong H_q(F, \partial F) \otimes H_1(I, \partial I) \cong H_q(F, \partial F).$$

This gives the following exact variation sequence

$$\ldots \to H_{q+1}(S^{2n+1}) \to H_q(F, \partial F) \overset{\text{VAR}_q}{\longrightarrow} H_q(F) \to H_q(S^{2n+1}) \to \ldots$$

The lemma now follows from the fact that $\tilde{H}_q(S^{2n+1}) = 0$ for $q \neq 2n+1$

PROPOSITION 2.2. *For isolated singularities*, $\text{Ker } j_* = \text{Ker } (\mathbf{T}_n - \mathbf{I})$.

Proof. Let E be the total space of the Milnor fibration $f : E \to \partial D_\eta$. The diagram above relates the variation mapping and j_* to the Wang sequence of the fibration:

$$0 \to H_{n+1}(E) \to H_n(F) \overset{\mathbf{T}_n - \mathbf{I}}{\longrightarrow} H_n(F) \to H_n(E) \to 0$$

REMARK 2.3. The intersection form S on $H_n(F)$ is related by Poincaré-duality $[\ ,\ ]$ to j_* by

$$S(x, y) = [j_* x, y].$$

So: $\mathbf{T}$ has eigenvalue $1 \Leftrightarrow S$ is degenerate.
Let $K = f^{-1}(O) \cap \partial B$. A related fact is that for $n \neq 2$:

$$K \text{ is a topological sphere } \Leftrightarrow \det(\mathbf{T}_n - \mathbf{I}) = \pm 1.$$

Because $E \overset{h}{\simeq} S^{2n+1} \backslash K$ and by duality

$$H_{n+1}(E) \cong \tilde{H}^{n-1}(K) \cong H_n(K)$$

$$H_n(E) \cong \tilde{H}^n(K) \cong H_{n-1}(K)$$

the Wang sequence tells:

$$K \text{ is a homology sphere } \Leftrightarrow \det(\mathbf{T}_n - \mathbf{I}) = \pm 1.$$

For the step from homology sphere to topological sphere we refer to Milnor [33].

3. One dimensional singular locus

3.1. INTRODUCTION

In this section we consider singularities with a 1-dimensional critical locus (for short: 1-*isolated singularities*) and study the vanishing homology in a full neighbourhood of the origin. In this case the vanishing homology is concentrated on the 1-dimensional set Σ. We can write

$$\Sigma = \Sigma_1 \cup \ldots \cup \Sigma_r$$

where each Σ_i is an irreducible curve.

At the origin we consider the Milnor fibre F of f and on each $\Sigma_i - \{O\}$ a local system of transversal singularities:
Take at any $x \in \Sigma_i - \{O\}$ the germ of a generic transversal section. This gives an isolated singularity whose μ-class is well-defined. We denote a typical Milnor fibre of this transversal singularity by F_i'. On the level of homology we get in this way a local system with fibre $\tilde{H}_{n-1}(F_i')$.

More precisely we consider in the 1-isolated case the following data:

The **Milnor fibre** F. The vanishing homology is concentrated in dimensions $n - 1$ and n:

$$\begin{cases} H_n(F) = \mathbb{Z}^{\mu_n}, & \text{which is free.} \\ H_{n-1}(F), & \text{which can have torsion.} \end{cases}$$

The Milnor monodromy acts on the fibre F :

$$\mathbb{T}_n \ : H_n(F) \to H_n(F)$$

$$\mathbb{T}_{n-1} : H_{n-1}(F) \to H_{n-1}(F)$$

The **transversal Milnor fibres** F_i'. The vanishing homology is concentrated in dimension $n - 1$:

$$\tilde{H}_{n-1}(F_i') = \mathbb{Z}^{\mu_i'} \text{, which is free.}$$

On this group there act two different monodromies:

1. the vertical monodromy (or local system monodromy)

$$A_i : \tilde{H}_{n-1}(F_i') \to \tilde{H}_{n-1}(F_i')$$

which is the characteristic mapping of the local system over the punctured disc $\Sigma_i - \{O\}$.

2. the horizontal monodromy (or Milnor monodromy)

$$T_i : \tilde{H}_{n-1}(F_i') \to \tilde{H}_{n-1}(F_i')$$

which is the Milnor fibration monodromy, when we restrict f to a transversal slice through $x \in \Sigma_i - \{0\}$.

In fact A_i and T_i are defined over $(\Sigma_i - \{O\}) \times S^1$, which is homotopy equivalent to a torus. So they commute:

$$A_i T_i = T_i A_i$$

One of the topics of this section is to show how the above data enter into a good description of the topology of a 1-isolated singularity. For details we refer to [50, 51].

EXAMPLE 3.1. D_∞-singularity: $f = xy^2 + z^2$.
Σ is given by $y = z = 0$ and is a smooth line. The transversal type is A_1.
It is known that F is homotopy equivalent to S^2 (cf. [45]).
  One can show that:

$$H_2(F) = \mathbb{Z} \quad T_2 = -\mathbb{I} \quad ; \quad H_1(F) = 0$$
$$H_1(F') = \mathbb{Z} \quad T = \mathbb{I} \quad \text{and} \quad A = -\mathbb{I}$$

EXAMPLE 3.2. $T_{\infty,\infty,\infty}$-singularity: $f = xyz$
$\Sigma = \Sigma_1 \cup \Sigma_2 \cup \Sigma_3$ and consists of the three coordinate axes in $\mathbb{C}^3$.
The transversal type is again A_1.
It is known that F is homotopy equivalent to the 2-torus $S^1 \times S^1$ (cf. [46])
One can show that:

$$H_2(F) = \mathbb{Z} \quad\quad T_2 = \mathbb{I}$$
$$H_1(F) = \mathbb{Z} \oplus \mathbb{Z} \quad T_1 = \mathbb{I}$$
$$H_1(F') = \mathbb{Z} \quad\quad T_i = \mathbb{I} \text{ and } A_i = \mathbb{I} \; i = 1, 2, 3.$$

EXAMPLE 3.3. Let f be 1-isolated and homogeneous of degree d. In this case one has the relation:

$$A_i = T_i^{-d}$$

We can assume that all the Σ_i's are straight lines through O. We can suppose that Σ_i is the x_0-axis; the formula follows from $f(sx_0, x_1, \cdots, x_n) = s^d f(x_0, s^{-1}x_1, \cdots, s^{-1}x_n)$, cf [54].

3.2. Series of singularities

Let again $f : (\mathbb{C}^{n+1}, O) \to (\mathbb{C}, 0)$ be a germ of an analytic function. Let f have a 1-dimensional critical locus $\Sigma = \Sigma(f)$. One considers for each $N \in \mathbb{N}$ the series of functions:

$$f_N = f + \epsilon x^N$$

where x is an admissible linear form, which means that $f^{-1}(0) \cap \{x = 0\}$ has an isolated singularity. One calls this series of function germs a Yomdin series of the hypersurface singularity f. Under the above condition all members of the Yomdin series have isolated singularities. Moreover their Milnor numbers can be computed using the so-called Lê-Yomdin formula:

$$\mu(f + \epsilon x^N) = \mu_n(f) - \mu_{n-1}(f) + N e_0(\Sigma).$$

Here μ_n, resp μ_{n-1} are the corresponding Betti-numbers of the Milnor fibre F of the non-isolated singularity f and $e_o(\Sigma)$ is the intersection multiplicity of Σ and $x = 0$. The formula holds for all N sufficiently large. Moreover $e_0(\Sigma) = \sum d_i \mu_i'$, where d_i is the intersection multiplicity of Σ_i (with reduced structure) and x.

The following formula relates the characteristic polynomials of the monodromies of f and f_N. Other ingredients are the horizontal and vertical monodromies. The eigenvalues of the monodromy satisfy Steenbrink's spectrum conjecture, cf [54]. This conjecture was later proved by M. Saito [42], using his theory of Mixed Hodge Modules.

THEOREM 3.4. *Let $f : (\mathbb{C}^{n+1}, O) \to (\mathbb{C}, 0)$ have 1-dimensional critical locus $\Sigma = \Sigma_1 \cup \cdots \cup \Sigma_r$ (irreducible components). Let x be an admissible linear form. Let $M(f)(\lambda)$ be the alternating product of the characteristic polynomials of the monodromy $\mathbf{T}$ of f in dimensions n and $n - 1$. Let $M(f + \epsilon x^N)(\lambda)$ be the characteristic polynomial of the monodromy of $f + \epsilon x^N$ in dimensions n. For all N sufficiently large*

$$M(f + \epsilon x^N)(\lambda) = M(f)(\lambda) \prod \det(\lambda^{N d_i} I - A_i T_i^{N d_i}),$$

where A_i and T_i are the vertical and horizontal monodromy along the branch Σ_i.

Proof. The idea behind the proof is to use polar methods and to consider the map germ

$$\Phi = (f, x) : \mathbb{C}^{n+1} \to \mathbb{C} \times \mathbb{C}.$$

The Milnor fibres F of f, resp F^N of f_N occur as inverse images under Φ of the sets $\{f = t\}$, resp $\{f + \epsilon x^N = t\}$. Next one constructs via a (stratified) isotopy an embedding

$$F \subset F^N.$$

From the corresponding homology sequence one gets the following 4-term exact sequence

$$0 \to H_n(F) \to H_n(F^N) \to H_n(F^N, F) \to H_{n-1}(F) \to 0.$$

The difference $F^N \backslash F$ is (by excision and homotopy equivalence) related to the part of F^N located near the d_i intersection points of Σ_i and F^N. One obtains:

$$H_q(F^N, F) = \oplus_{i=1}^r \oplus_{k=1}^{Nd_i} \tilde{H}_{n-1}(F_{i,k}),$$

where each $F_{i,k}$ is a copy of the Milnor fibre of the transversal singularity F_i'. From this one gets

$$b_n(F) - b_{n-1}(F) = b_n(F^N) - N \sum d_i \mu_i'.$$

To obtain the monodromy statement one constructs a geometric monodromy, which acts on the 4-term sequence. One uses Lê's carrousel method. The monodromy on F^N respects the distance function $|x|$ as nearly as possible. The geometric monodromy gets an x-component, which gives rise to the appearance of the vertical monodromy A_i in $\det(\lambda^{Nd_i} I - A_i T_i^{Nd_i})$. For details cf [50].

REMARK 3.5. $M[f]$ is related to Z_f, the zeta function of the monodromy, which is defined by $Z_f(t) = \prod_{q \geq 0}(\det(\mathbb{I} - t\mathbb{T})^{(-1)^{q+1}})$, cf [33]. For homogeneous singularities of degree d the formula $Z(t) = (1 - t^d)^{-\frac{\chi(F)}{d}}$ is well known and valid in all generality (without assumptions on the dimension of the critical set), cf eg [13].

REMARK 3.6. The theorem can be used in two ways: computing mononodromies for isolated singularities in the series, but also for computing monodromies of certain non-isolated singularities with one dimensional singular sets.

In the case of a homogeneous polynomial one can compute almost all ingredients in the formula by taking N as the degree d of f.

One gets in this way the formula $Z_f(t) = (1-t^d)^{-\frac{\chi(F_{reg})}{d}} \pm \sum \mu_i'$ (where the μ_i' are transversal Milnor numbers), and $\chi(F_{reg}) = 1+(-1)^n(d-1)^n$, the Euler characteristic of an isolated singularity of degree d. Also (in some cases) generalizations to the quasi-homogeneous case can be obtained. See also [12].

Also some questions about the relative monodromy being of finite order can be treated with the above formula, cf [53].

Generalizations of the method are in Tibăr's work [58].

3.3. VARIATION MAPPINGS

It is not possible to construct a geometric monodromy $h : F \to F$ which is the identity on the whole of the boundary ∂F of the Milnor fibre F. However it is possible to make it the identity on a big part $\partial_1 F$ of ∂F but not on its complement $\partial_2 F$, which is situated near the singular locus Σ. We can suppose $\partial F = \partial_1 F \cup \partial_2 F$ and $\partial_2 F = \partial F \cap T$, where T is a tubular neighborhood within S^{2n+1} of the link $L := \Sigma \cap S^{2n+1}$.

The homology sequence of the pair $(F, \partial_2 F)$ gives the following fundamental sequence:

$$0 \to H_{n+1}(F, \partial_2 F) \to H_n(\partial_2 F) \to H_n(F) \to$$

$$\to H_n(F, \partial_2 F) \to H_{n-1}(\partial_2 F) \to H_{n-1}(F) \to 0$$

The homology groups of $\partial_2 F$ play an important role in this sequence. They are related to the local system monodromies $A_i : F_i' \to F_i'$ in the following way:

$\partial_2 F$ is a disjoint union $\partial_2 F = \bigcup\limits_{i=1}^{r} \partial_2 F_i$ concentrated near the components of Σ. Each $\partial_2 F_i$ is fibered over the circle (which is the neighborhood boundary of Σ_i) with fibre F_i'.

The Wang sequence of this fibration is:

$$0 \to H_n(\partial_2 F_i) \to H_{n-1}(F_i') \overset{A_i - \mathbb{I}}{\to} H_{n-1}(F_i') \to H_{n-1}(\partial_2 F_i) \to 0$$

So:

$$H_n(\partial_2 F) = \bigoplus\limits_{i=1}^{r} \mathrm{Ker}(A_i - \mathbb{I})$$

$$H_{n-1}(\partial_2 F) \cong \bigoplus\limits_{i=1}^{r} \mathrm{Coker}\,(A_i - \mathbb{I})$$

The first group is always free, the second can have torsion.

COROLLARY 3.7.

1. dim $H_n(\partial_2 F) = 0 \Leftrightarrow$ *no vertical monodromy* A_i *has an eigenvalue 1* $\Rightarrow$ dim $H_{n-1}(F) = 0$, *but* $H_{n-1}(F)$ *can have torsion!*

2. *Let λ be an eigenvalue of $\mathbf{T}_{n-1} : H_{n-1}(F) \to H_{n-1}(F)$, then λ is also an eigenvalue of some $T_i : H_{n-1}(F_i') \to H_{n-1}(F_i')$; in fact one of the eigenvalues occurring in $\operatorname{Coker}(A_i - \mathbb{I})$.*

EXAMPLE 3.8. If the transversal type of the singularity is A_1 for all the branches then $\mathbf{T}_{n-1}$ can only have eigenvalue 1 if n is even, or eigenvalue -1 if n is odd.

Given a geometric monodromy $h : F \to F$ such that $h|_{\partial_1 F}$ is the identity, we can consider the variation mappings

$$\begin{aligned}
\mathrm{VAR}^I &: H_*(F, \partial_1 F) \to H_*(F) \\
\mathrm{VAR}^{II} &: H_*(F, \partial F) \to H_*(F, \partial_2 F) \\
\mathrm{VAR}^{III} &: H_*(\partial F, \partial_1 F) \to H_*(\partial_2 F).
\end{aligned}$$

One can show that they are isomorphisms near the middle dimensions, along the same lines as in the isolated singularity case. One has a *first variation sequence*

$$\ldots \to H_{q+1}(S^{2n+1}\backslash L) \to H_q(F, \partial_1 F) \overset{\mathrm{VAR}_q^I}{\to} H_q(F) \to H_q(S^{2n+1}\backslash L) \to \ldots$$

and a *second variation sequence*:

$$\ldots \to H_{q+1}(S^{2n+1}, L) \to H_q(F, \partial F) \overset{\mathrm{VAR}_q^{II}}{\to} H_q(F, \partial_2 F) \to H_q(S^{2n+1}, L) \to \ldots$$

The vanishing of $H_q(S^{2n+1}\backslash L)$ and $H_{q+1}(S^{2n+1}, L)$ determine the range of isomorphisms.

Moreover there are Lefschetz type dualities involved (for manifolds with boundary and corners) between

$$\begin{aligned}
H_*(F) \qquad &\text{and} \quad H_*(F, \partial F) \\
H_*(F, \partial_1 F) \quad &\text{and} \quad H_*(F, \partial_2 F)
\end{aligned}$$

See [47] or Dold [16], chapter VII. See also the remark at the end of the paper of Samelson [43].

A lot of homological information about the singularity is contained in a big commutative diagram, which we call the variation ladder. Recall that $\dim \Sigma = 1$. We shall assume $n \geq 3$ in order to avoid special features of low dimensions. In [51] there is a version adapted for the case $n = 2$. Most of the corollaries that follow from the variation ladder are also true in that case.

THEOREM 3.9. [The variation ladder] *The following commutative diagram has as columns the exact sequences of the triple $(F, \partial F, \partial_1 F)$ and the pair $(F, \partial_2 F)$. The maps $\leftarrow$ are induced by inclusion. The maps VAR are isomorphisms.*

$$0 \quad = \quad H_{n+1}(F, \partial_1 F) \xrightarrow{\text{VAR}^I_{n+1}} H_{n+1}(F) \quad = \quad 0$$

$$\mathbb{Z}^{\mu_{n-1}} \quad = \quad H_{n+1}(F, \partial F) \xrightarrow{\text{VAR}^{II}_{n+1}} H_{n+1}(F, \partial_2 F) \quad = \quad \mathbb{Z}^{\mu_{n-1}}$$

$$\mathbb{Z}^{\alpha} \quad = \quad H_n(\partial F, \partial_1 F) \xrightarrow{\text{VAR}^{III}_n} H_n(\partial_2 F) \quad = \quad \mathbb{Z}^{\alpha}$$

$$\mathbb{Z}^{\mu_n} \quad = \quad H_n(F, \partial_1 F) \xrightarrow[(j_1)_*]{\text{VAR}^I_n} H_n(F) \quad = \quad \mathbb{Z}^{\mu_n}$$

$$(i_1)_* \qquad j_* \qquad (j_2)_*$$

$$\mathbb{Z}^{\mu_n} \oplus T = \quad H_n(F, \partial F) \xrightarrow[(i_2)_*]{\text{VAR}^{II}_n} H_n(F, \partial_2 F) \quad = \mathbb{Z}^{\mu_n} \oplus T$$

$$\mathbb{Z}^{\alpha} \oplus T_2 = \quad H_{n-1}(\partial F, \partial_1 F) \xrightarrow{\text{VAR}^{III}_{n-1}} H_{n-1}(\partial_2 F) \quad = \mathbb{Z}^{\alpha} \oplus T_2$$

$$\mathbb{Z}^{\mu_{n-1}} \oplus T = \quad H_{n-1}(F, \partial_1 F) \xrightarrow{\text{VAR}^I_{n-1}} H_{n-1}(F) \quad = \mathbb{Z}^{\mu_{n-1}} \oplus T$$

$$0 \quad = \quad H_{n-1}(F, \partial F) \xrightarrow{\text{VAR}^{II}_{n-1}} H_{n-1}(F, \partial_2 F) \quad = \quad 0$$

where

$$\mu_n = \dim H_n(F)$$

$$\mu_{n-1} = \dim H_{n-1}(F)$$

$$\alpha = \dim H_n(\partial_2 F) = \sum_{i=1}^{r} \dim \ker(A_i - \mathbb{I})$$

$$T_2 = \textit{torsion part of } H_{n-1}(\partial_2 F)$$

$$T = \textit{torsion part of } H_{n-1}(F)$$

It follows from the Lefschetz dualities mentioned above that the rows are "opposite dual" to each other. The composition of $\leftarrow$ and VAR is $\mathbb{T}_* - \mathbb{I}$.

COROLLARY 3.10. *If all vertical monodromies have only eigenvalues $\lambda \neq 1$ then*

$1°$ $(j_2)_*$ *is injective (even bijective over $\mathbb{Q}$)*

$2°$ $H_{n+1}(F, \partial F) = 0$ *and* $H_{n-1}(F)$ *is torsion.*

REMARK 3.11. The variation ladder occurs also in the following way. Consider the sheaf Φ'_f of vanishing cycles, cf [10]. The stalk of this sheaf at a point is the reduced cohomology over $\mathbb{C}$ of the Milnor fibre of f at the point. Our approach works with homology and with coefficients in $\mathbb{Z}$, and we are also interested in torsion.

Let j denote the inclusion of O in $X = f^{-1}(0)$, i be the inclusion of $X - O$ in X, and let $\mathbf{K}^{\cdot}$ denote $\Phi'_f(\mathbb{C})$, Then there exists a distinguished triangle:

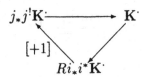

The associated stalk cohomology exact sequence at the origin becomes:

$$0 \to H^{n-1}((\mathbf{K}^{\cdot})_0) \to \overset{r}{\underset{i=1}{\oplus}} \operatorname{Ker}(A_i - \mathbb{I}) \to H^n((j_* j^! \mathbf{K}^{\cdot})_0) \to$$

$$\to H^n((\mathbf{K}^{\cdot})_0) \to \overset{r}{\underset{i=1}{\oplus}} \operatorname{Ker}(A_i - \mathbb{I}) \to H^{n+1}((j_* j^! \mathbf{K}^{\cdot})_0) \to 0$$

where $H^k((\mathbf{K}^{\cdot})_0) = H^k(F, \mathbb{C})$.

Since $H^k((j_* j^! \mathbf{K}^{\cdot})_0)$ can be identified with $H^k(F, \partial_2 F; \mathbb{C})$, this sequence is the cohomology sequence of the pair $(F, \partial_2 F)$ and is the cohomology version (over $\mathbb{C}$) of the right hand side of the variation ladder 3.9, at least for $n \geq 3$.

Other work in this direction was done by M. Saito [42] and D. Barlet [5, 6]. They deal with Hodge theoretical aspects.

3.4. RELATION BETWEEN THE MONODROMY AND THE INTERSECTION FORM

In the isolated singularity case it is known [33] that

$$S \text{ is degenerate} \Leftrightarrow T \text{ has eigenvalue } 1.$$

This follows from $S(x, y) = [j_* x, y]$ by Poincaré Duality and the fact that $\ker j_* = \mathrm{Ker}\,(\mathbf{T}_n - \mathbb{I})$. We discussed this already in section 2. What happens in the case $\dim \Sigma = 1$? Consider the diagram:

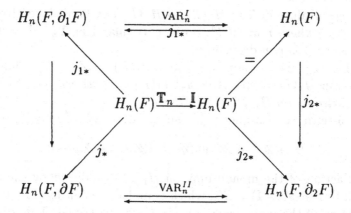

So $\mathbf{T}_n - \mathbb{I} = \mathrm{VAR}^I \circ j_{1*}$; $\mathrm{Ker}\, j_{1*} = \mathrm{Ker}\,(\mathbf{T}_n - \mathbb{I})$ and $(j_2)_* \circ (\mathbf{T}_n - \mathbb{I}) = \mathrm{VAR}^{II} \circ j_*$

PROPOSITION 3.12. *T has eigenvalue* $1 \Rightarrow S$ *is degenerate*
Proof. We have $\mathrm{Ker}\,(\mathbf{T}_n - \mathbb{I}) = \mathrm{Ker}\, j_{1*} \subset \mathrm{Ker}\, j_*$.

In the other direction we have the following two partial results:

PROPOSITION 3.13. *If all vertical monodromies have only eigenvalues* $\lambda \neq 1$, *then:*

$$\mathbf{T}_n \text{ has eigenvalue } 1 \Leftrightarrow S \text{ is degenerate}$$

Proof. Since $(j_2)_*$ is injective we have $\mathrm{Ker}\,(\mathbf{T}_n - \mathbb{I}) = \mathrm{Ker}\, j_*$.

PROPOSITION 3.14.

$$\left.\begin{array}{c} S \text{ is non-degenerate} \\ H_{n-1}(F, \mathbb{Q}) = 0 \end{array}\right\} \Leftrightarrow \left\{\begin{array}{l} \mathrm{Ker}\,(\mathbf{T}_n - \mathbb{I}) = 0 \\ \mathrm{Ker}\,(A_i - \mathbb{I}) = 0 \text{ for all } i. \end{array}\right.$$

EXAMPLE 3.15. We consider the following homogeneous polynomial

$$f = x^2y^2 + y^2z^2 + z^2x^2 - 2xyz(x + y + z)$$

In $\mathbb{P}^2$ this defines a curve, which is known as Zariski's example with 3 cusps. So f has a 1-dimensional singular locus, consisting of 3 complex lines and transversal type A_2.

Since f is homogeneous of degree 4 it is known that $A_i = T_i^{-4}$ where T_i is the monodromy of A_2. Since the eigenvalues of T_i are q and q^2, where $q = e^{2\pi i/6}$ it follows that A_i has eigenvalues q^4 and q^{-4}.

As a corollary: $\begin{cases} \text{Ker } (A_i - \mathbb{I}) = 0 \\ \text{Coker}(A_i - \mathbb{I}) = \mathbb{Z}_3 \text{ torsion} \end{cases}$ and this implies that $H_{n-1}(F)$ is torsion.

The eigenvalues of $\mathbb{T}_n : H_n(F) \to H_n(F)$ can be deduced from 3.4: r, r^2 and r^3 where $r = e^{2\pi i/4}$. So in particular Ker $\mathbb{T}_n - \mathbb{I} = 0$. As a consequence S is non-degenerate.

In this case the fundamental group $\pi_1(F) = \mathbb{Z}_3$, so we have only 3-torsion in $H_1(F) = \mathbb{Z}_3$. Although $b_1(F) = 0$, *we will point out, that the monodromy on $H_1(F)$ is non-trivial!*

The 6-term exact sequence in the variation ladder reduces to:

$$0 \to \mathbb{Z}^3 \to \mathbb{Z}^3 \oplus \mathbb{Z}_3 \to (\mathbb{Z}_3)^3 \to \mathbb{Z}_3 \to 0$$

The action of the monodromy on $H_1(F)$ is induced by the actions on each Coker $(A_i - \mathbb{I}) = \mathbb{Z}_3$, which is multiplication by -1. It follows that action of the monodromy on $H_1(F)$ is non-trivial. This remark is also in [14].

EXAMPLE 3.16.   Proposition (3.12) can be used to construct several examples with totally degenerate intersection forms. Dimca gave $x^a y^{d-a} + yz^{d-1}$ and other examples, where the geometric monodromy can be chosen homotopy equivalent to the identity. This implies that Ker $(\mathbb{T}_n - \mathbb{I}) = \text{Ker } j_* = H_n(F)$.

In case of homogeneous polynomials in x, y, z of degree d it follows from [12] or [50] that a necessary condition for $\mathbb{T}_n = \mathbb{I}$ is $\chi(F) = 0$. Since in the homogeneous case $\chi(F) = d^3 - 3d^2 + 3d - d\sum \mu_i'$, this is equivalent to

$$\Sigma\mu_i' = d^2 - 3d + 3,$$

where μ_i' are the Milnor numbers of the transversal singularities.

Melle [4] pointed out that this condition is not sufficient. First remark that $\chi(F) = 0$ implies that the zeta function of the monodromy is $Z(t) = (1 - t^d)^{-\frac{\chi(F)}{d}} = 1$ (due to homogeneity). It follows that

$T_2 = 1$ is equivalent to $T_1 = 1$; so any eigenvalue of T_1 different from 1 has to cancel against an eigenvalue of T_2 and vice versa. In case $T_2 = 1$ it follows that the 6-term sequence splits into two 3-term sequences, one has

$$0 \to H_n(F, \partial_2 F) \to H_{n-1}(\partial_2 F) \to H_{n-1}(F) \to 0$$

So also the monodromy of the middle term must have only eigenvalues 1.

Melle's example $f = (y^2 + 2xz)(axz + z^2 + ay^2)$ is a polynomial of degree 4, where the corresponding projective curve has only one singularity, which is of type A_7. A straightforward calculation gives $\dim H_{n-1}(\partial_2 F) = \dim \mathrm{Coker}\,(A_* - I) = 3$ and the eigenvalues of T_* on this space are $i, -i, 1$.

Examples of homogeneous singularities with $T = I$ are used in [7] in order to construct complex hypersurfaces which are homology P^n's.

Cancelling of eigenvalues of the monodromy is also the subject of Barlet's studies [5, 6] and Denef's conjectures [11] about the topological zeta function.

3.5. Is ∂F a topological sphere?

In Milnor's book [33], section 8 is entitled: Is K a topological sphere? Remember $K = f^{-1}(0) \cap \partial B$. For isolated singularities Milnor showed:

K is a topological sphere $S^{2n-1} \Leftrightarrow \det T_n - I = \pm 1$

Since ∂F is diffeomorphic to K in this case, we also have

∂F is a topological sphere $S^{2n-1} \Leftrightarrow \det T_n - I = \pm 1$

In this section we study the 1-isolated case. Since K is not smooth, we only study ∂F. The vertical monodromies play an important role in the final result.

PROPOSITION 3.17. *Let $n > 2$. The following are equivalent:*

1. *∂F is a topological sphere S^{2n-1}*

2. $\begin{cases} (a) \ H_{n-1}(F) = 0 \\ (b) \ \textit{The intersection form } S \textit{ on } H_n(F) \textit{ has determinant } \pm 1 \end{cases}$

3. $\begin{cases} (a) \ \det(A_i - I) = \pm 1 \textit{ for all } i = 1, \dots, r \\ (b) \ \det(T_n - I) = \pm 1 \end{cases}$

NB. For $n = 2$ replace (1) by: ∂F is a homology sphere.

REMARK 3.18. ∂F is a topological sphere if and only if K is a homotopy sphere and $\det(A_i - \mathbb{I}) = \pm 1$.

REMARK 3.19.   Let f be 1-isolated with transversal type A_1. We have $A_i = \pm \mathbb{I}$. So in this case ∂F can never be a topological sphere.

REMARK 3.20.   Let f be 1-isolated, and moreover be *homogeneous* of degree d; then ∂F is never a homology sphere.

To show this, consider the exact homology sequence from (3.2) in the case $N = d$:

$$0 \to H_n(F) \to H_n(F^d) \to H_n(F^d, F) \to H_{n-1}(F) \to 0.$$

This sequence compares the homology of the Milnor fibre F with the homology of F^d, the Milnor fibre of the function $f_d = f + \epsilon x^d$. This function is also homogeneous of degree d, but has an isolated singularity.

In case of a homology sphere, we have $H_{n-1}(F) = 0$. Moreover the monodromy actions commute with the sequence. The eigenvalues of $\mathbb{T}_n$ on $H_n(F^d)$ and of $\mathbb{T}_n^{rel}$ on $H_n(F^d, F)$ are explicitly known, so we can compute exactly all the eigenvalues of $\mathbb{T}_n$ on $H_n(F)$. It follows that it is not possible to satisfy the condition: $\det(\mathbb{T}_n - \mathbb{I}) = \pm 1$.

QUESTION 3.21.   Are there examples where ∂F is a topological spheres ?

Up to now only counter-examples are known.

3.6. TRANSVERSAL TYPE A_1.

In this section we study germs of holomorphic functions $f : (\mathbb{C}^{n+1}, O) \to (\mathbb{C}, 0)$, where the critical locus is 1-dimensional and the transversal singularities at points of $\Sigma - \{O\}$ are of type A_1. References are [45], [46], [47].

We know already that the homology of the Milnor fibre F is concentrated in dimensions $n - 1$ and n. The topology seems to depend partly on properties of the critical set. In [47] we showed the following

PROPOSITION 3.22.   *Let $f : (\mathbb{C}^{n+1}, O) \to (\mathbb{C}, 0)$ have a 1-dimensional singular set, which is an ICIS (isolated complete intersection singularity) and let f have transversal type A_1 outside the origin. Then the homotopy type of F is a bouquet of spheres:*

- *Case A:* $F \overset{h}{\simeq} S^n \vee \cdots \vee S^n$ *(general case)*

- *Case B:* $F \overset{h}{\simeq} S^{n-1} \vee S^n \vee \cdots \vee S^n$ *(special case)*

The main idea of the proof of this type of theorem is *deformation with constant topology*. This generalizes the concept of Morsification from the case of isolated singularities. While Morsifications are generic deformations in the isolated case, it is easy to destroy the topology in the non-isolated case, e.g. for a Yomdin series one finds a infinite variety of Milnor fibres. One has to be careful and to study special deformations, which deform both f and Σ in a good way [25].

Let us suppose that we have constructed a deformation (f_s, Σ_s) of (f, Σ), defined on the same neighborhood as (a Milnor representative of) f, with properties:

- the critical set of f_s consists of a curve Σ_s and some isolated points $a_1 \cdots a_\sigma$ inside the Milnor ball,

- during the deformation from f to F_s the fibration is of constant fibration type,

In this case one can use the principle of *additivity of the vanishing homology*, which says:

$$H_*(E, F) = \bigoplus_{i=0}^{\sigma} H_*(E_i, F_i) \ ,$$

where $E =$ Milnor ball of f,
$E_0 =$ tube neighborhood along Σ_s
$E_i =$ small Milnor ball at s_i,
$F =$ Milnor fibre of f,
$F_i =$ restriction of nearby fibre to E_i.
$\mu_i =$ the Milnor number at a_i.
From the properties of isolated singularities it follows that:

$$H_n(F) = H_{n+1}(E, F) = H_{n+1}(E_0, F_0) \oplus \mathbb{Z}^{\sum \mu_i} \ ,$$

$$H_{n-1}(F) = H_n(E, F) = H_n(E_0, F_0) \ .$$

The contribution to $H_{n-1}(F)$ comes entirely from $H_n(E_0, F_0)$ and this is the part related to 1-dimensional part of the deformed critical set.

Returning to the A_1-case with Σ an ICIS one can always produce the above type of deformation, even with the extra properties:

1. all isolated critical points of f_s are of type A_1,

2. Σ_s is smooth, the Milnor fibre of Σ,

3. For the non-isolated critical points of f_s we have only two types:

   –   type A_∞, local formula $w_0^2 + w_1^2 + \cdots w_n^2$ (transversal Morse),

   –   type D_∞, local formula $w_0 w_1^2 + w_2^2 + \cdots w_n^2$ (Whitney umbrella).

Cases A and B in the above proposition are distinguished by $\sharp D_\infty > 0$, in case A, and $\sharp D_\infty = 0$ in case B. In general, $\sharp D_\infty$ is related to the vertical monodromy.

As soon as Σ is not an ICIS it can occur that $b_{n-1}(F) \geq 2$. For example, if $f = xyz$, we have $b_1(F) = 2$.

3.7. BOUQUET THEOREMS

In the last few years different types of 'bouquet theorems' have appeared. Some of them deal with germs $f : (X, x) \to (\mathbb{C}, 0)$ where f defines an isolated singularity. In some cases, F has the homotopy type of a bouquet of $(\dim X - 1)$-spheres, for example when X is an ICIS , or X is a complete intersection. Moreover if both (X, x) and f have isolated singularities, then F has a bouquet decomposition

$$F \overset{h}{\simeq} F_0 \vee S^n \vee \cdots \vee S^n,$$

where F_0 is the complex link of (X, x), cf [52]. Later Tibăr proved a more general bouquet theorem for the case when (X, x) is a stratified space and f defines an isolated singularity (in the sense of the stratified spaces) for details, cf [57]. Related results are in [23].

In the case of non-isolated singularities the bouquet situation is no longer standard, e.g. the torus is the Milnor fibre of $f = xyz$. At the other hand in several special cases (eg many cases discussed in this paper), we still encounter bouquets of spheres (sometimes in different dimensions).

Némethi treated as part of his paper [37] the question: *When is a CW-complex a bouquet of spheres ?* A necessary condition is of course that all homology groups are free (over $\mathbb{Z}$). In the 1-connected case he added the condition, that the Hurewicz map (from homomotopy to homology groups) is surjective in all dimensions. These two conditions together are sufficient.

As a special corollary he showed:

PROPOSITION 3.23. *Let $f : (\mathbb{C}^{n+1}, O) \to (\mathbb{C}, 0)$ $(n \geq 3)$ be a germ of analytic function with a 1-dimensional critical locus. Then its Milnor fibre F has the homotopy type of a bouquet of spheres if and only if $H_*(F, \mathbb{Z})$ is free.*

REMARK 3.24.   The condition $n \geq 3$ is important, the statement does not work in the surface case. For example, $f = xyz$ gives a 2 torus; but $g = xyz + w^2$ (its suspension) gives a bouquet $S^2 \vee S^3 \vee S^3$. In general, the case $n = 2$ is more difficult, due to the influence of the fundamental group. Results in special cases are treated in [45, 46, 47].

REMARK 3.25.   It seems that the following question is relevant: If f has a deformation with constant topology, such that the homology splitting discussed above is valid, does there exist space F_0 and a decomposition:

$$F \overset{h}{\simeq} F_0 \vee S^n \vee \cdots \vee S^n,$$

where the spheres correspond exactly to the vanishing cycles at the isolated critical points ? Is the splitting 'forced' by the topology of CW-complexes or by properties of singularities ?

3.8. OTHER DIRECTIONS

Several developments in the case of 1-dimensional singular sets are not discussed here. We mention them below and give some references.

- The topology of line singularities (smooth 1-dimensional singular set) with transversal type A, D or E studied by De Jong [24]. Several data of the vanishing topology have been recently computed by J. Fernandez de Bobadilla.

- The fundamental studies of Pellikaan about the algebraic aspects of the theory. From his thesis, there originated a series of papers. We mention [39, 40].

- The relation between the deformation theories of weakly normal hypersurface singularities and normal surface singularities given by De Jong and Van Straten [25].

- Van Straten's [56] description of the De Rham complex of 1-isolated singularities and an algebraic description for the highest Betti number. See also [47].

- Jiang's [21, 22] study of functions on an isolated complete intersection with 1-dimensional singular set.

- Relation with Mond's [34, 35] work on map germs from $\mathbb{C}^2$ to $\mathbb{C}^3$ of finite codimension. See also the contribution of Damon [9] in this volume.

- Lê numbers, studied by Massey [31], where polar methods are used to study the structure of the singular set. For the 1-isolated case, cf [32].

- Aleksandrov's study of differential forms and vector fields tangent to a hypersurface germ D. In particular Saito singularities, where the module of vector fields tangent to D is locally free. Also non-isolated singularities occur [1, 2].

- Grandjean's approach to residual discriminants and bifurcation sets [18] for function germs of finite codimension within a given singular set.

4. Higher dimensional singular sets

4.1. 2 DIMENSIONAL SINGULAR SET

We restrict the discussion below to a two dimensional singular set Σ, which is an ICIS and with (generic) transversal type A_1.

The thesis of Zaharia [59] deals with this case; part of it is more general. He considers the situation where f has finite codimension in the space of functions with the set Σ as part of the critical set. This condition is equivalent to having $\Sigma(f) = \Sigma$ and the germ of f at every point of $\Sigma \setminus \{O\}$ equivalent to a so-called $D(k, p)$ singularity [39]. In our two dimensional case we encounter for dimension reasons only:

$$D(2,0) : x_2^2 + \cdots + x_n^2 \qquad ; \ A_\infty \times \mathbb{C},$$
$$D(2,1) : x_1 x_2^2 + x_3^2 + \cdots + x_n^2; \ D_\infty \times \mathbb{C}$$

(coordinates are $x_0, \cdots, x_n$).

Outside the origin one only has to deal with two types of local Milnor fibers, with homotopy type S^{n-2}, respectively S^{n-1}. The closure of the set of $D(2,1)$ types forms an ICIS -curve Δ inside Σ. In exceptional cases, Δ can be void. The stratification according to vanishing homology type consists of the following strata:

$$\Sigma, \Sigma \setminus \Delta, \Delta \setminus \{O\}, \{O\}.$$

Zaharia [59, 60] studied especially the topology of the Milnor fibre. This was later improved by Némethi [37]. The following statement shows that the homotopy type is still a bouquet of spheres, indeed of

n-spheres, in certain cases extended by one sphere of dimension $n - 1$ or $n - 2$. This is very close to the case of isolated singularities.

PROPOSITION 4.1. *Let* $f : (\mathbb{C}^{n+1}, O) \to (\mathbb{C}, 0)$ *have a 2-dimensional singular set, which is an ICIS , and let* f *have transversal type* A_1 *outside a curve* Δ *in* Σ; *then the homotopy type of* F *is a bouquet of spheres:*

- *Case A:* $F \overset{h}{\simeq} S^n \vee \cdots \vee S^n$

- *Case B:* $F \overset{h}{\simeq} S^{n-1} \vee S^n \vee \cdots \vee S^n$

- *Case C:* $F \overset{h}{\simeq} S^{n-2} \vee S^n \vee \cdots \vee S^n$ *(special case).*

If $n = 2$, *statement C should read* $F \overset{h}{\simeq} S^0 \times (S^2 \vee \cdots \vee S^2)$.

This theorem is due to Némethi [37], especially the homotopy part. The results of Zaharia played a crucial role in the proof.

The idea behind the proof is similar to the 1-dimensional case: construct a deformation with constant topology, which has only certain elementary singularity types as 'building blocks'. More precisely Zaharia constructed a deformation $(f_s, \Sigma_s, \Delta_s)$ of (f, Σ, Δ) with constant topology and if $s \neq 0$:

- the singular set of f_s consists of Σ_s and finitely many points, where f_s is Morse,

- Σ_s and Δ_s are Milnor fibres of Σ, resp Δ (recall: they are both ICIS),

- f_s has only points of type $D(2,0)$ and $D(2,1)$ on Σ_s,

- Δ_s is the $D(2,1)$-locus in Σ_s.

Remark next that above each point of Δ_s one can consider the Milnor fibre of the $D(2,1)$-singularty, which has the homotopy type of S^{n-1}. These induces a monodromy map:

$$\Xi : \pi_1(\Delta_s, \star) \to \mathrm{Aut}\, (H_{n-1}(S^{n-1}, \mathbb{Z})) = \mathbb{Z}_2$$

This local system of vertical monodromies plays an important role in the determination of the toplogy of the Milnor fibre. It seems to be 'deeper' than the vertical monodromy on the Δ-stratum, since the system is defined on its Milnor fibre Δ_s. Then cases A, B and C in the above proposition are distinguished by:

- case A: Δ_s is non void and Ξ non trivial,

- case B: Δ_s is non void and Ξ trivial,

- case C: Δ_s is void.

Zaharia showed that the Euler-characteristic of the Milnor fibre is equal to:

$$1 + (-1)^n (2\mu_\Delta + \mu_\Sigma + \sigma - 1),$$

where σ is the number of A_1-points which appear in the deformation, μ_Σ is the Milnor number of the ICIS Σ, and μ_Δ is the Milnor number of the ICIS Δ. (We use the convention $\mu_\emptyset = -1$; moreover in case C, if $n = 2$, one has $\chi(F) = 2 + 2\mu_\Sigma$).

4.2. CODIMENSION 1 CASE

Shubladze [44] studied the case, where the singular set is a hypersurface Σ. Generically there is only one transversal type, which must be an A_k singularity. The function f can be written as $f = h^k g$, where $h = 0$ defines the singular set. Considering the case when f has finite codimension in the set of these functions he showed that if $g(O) = 0$

$$F \simeq S^1 \vee S^n \vee \cdots \vee S^n.$$

NB The case $g(O) \neq 0$ gives k copies of the Milnor fibre of h. The finite codimension condition is equivalent to the conjunction of two conditions (a) g defines an isolated singularity, (b) the pair (g, h) defines an ICIS . Later Némethi [36], unaware of the results of Shubladze, recovered the Shubladze result as a by-product of his theory of composed singularities. The number of n-spheres is related to the Milnor numbers as follows:

$$b_n(F) = (k+1)\mu(h, g) + k\mu(h)) + \mu(g).$$

4.3. COMPOSED SINGULARITIES

The method of composed singularities can give rise to non-isolated singularities of codimension 2 or 1. We discuss here the work carried out by A. Némethi in his paper [36]. The situation is as follows. One considers the sequence of mappings:

$$f : \mathbb{C}^{n+1} \xrightarrow{(g,h)} \mathbb{C}^2 \xrightarrow{P} \mathbb{C},$$

where the pair (g, h) defines an ICIS Y and P is any germ. If P has an isolated singularity at the origin, then the singular set $\Sigma(f)$ is exactly the ICIS Y and has dimension $n - 1$. The local system of vanishing homology groups is defined over (each component) of $Y - \{O\}$. The transversal type is equal to the singularity type of P.

If P is not reduced then the singular set of f is the inverse image of the non-reduced locus of P and has dimension n. The situation becomes more complicated.

Let D be the (reduced) discriminant locus of the ICIS (g, h).

THEOREM 4.2. *Let $P^{-1}(0) \cap D = \{O\}$ then the Milnor fibre F of the composed mapping f has the homotopy type of the disjoint union of π_0 copies of a bouquet of spheres*

$$S^1 \vee \cdots \vee S^1 \vee S^n \vee \cdots \vee S^n,$$

where π_0 is the number of connected components of the Milnor fibre of P.

The number of 1-spheres in a connected component of F is equal to the Milnor number of P, the number of n-spheres is determined by topological data.

Némethi treats also the 'bad case' where $P^{-1}(0) \cap D$ contains 1-dimensional components. His work has some interesting consequences, e.g. the relation to series of singularities.

The theorem also allows P to be regular. In those cases f has an isolated singularity, at least if the condition $P^{-1}(0) \cap D = \{O\}$ is satisfied. In special cases well known situations occur:

- $P = z$ (a generic coordinate), Lê attaching formula,

- $P = z + w^k$: formula for Iomdin series if f has a 1-dimensional singular set,

- $P = z + w$ and g and h have separate variables and isolated singularities: Sebastiani-Thom formula.

A second theorem in the same paper [36] gives a formula for the zeta function of the monodromy of f as a product of other zeta functions related to the topological data of the composed singularity.

4.4. OTHER CASES

Above we studied more or less the summit of the iceberg of non-isolated singularities, those which seem to be close to isolated singularities. There are natural candidates for further investigation of the full system of vanishing homology: central arrangements of hyperplanes, and discriminant spaces of Coxeter arrangements. In both cases, there is also combinatorial and geometric structure around. For the homology of the Milnor fibre of an arrangement we refer to Orlik-Terao [38]. The zeta function is just $Z(t) = (1 - t^d)^{\chi(M^*)}$, where M^* is the complement of the arrangement, modulo the natural $\mathbb{C}^*$-action. The zeta function of the discriminant hypersurface of a Coxeter arrangement is studied in geometric terms by [15].

References

1. A. G. Aleksandrov, The Milnor numbers of non-isolated Saito singularities, (Russian) Funkts. Anal. i Prilozhen **21**:1 (1987) 1–10.
2. A. G. Aleksandrov, Nonisolated Saito singularities, (Russian) Mat. Sb. (N.S.) **137**:4 (179) (1988) 554–567, 576. Translation in Math. USSR-Sb. **65**:2 (1990) 561–574.
3. V. I. Arnol'd, S. M. Gusein Zade and A. N. Varchenko, *Singularities of Differentiable Maps I and II*, Birkhäuser, 1985 and 1988.
4. A. Artal Bartolo, P. Cassou-Noguès, I. Luengo and A. Melle Hernández, Monodromy conjecture for some surface singularities, preprint, September 2000.
5. D. Barlet, Interaction de strates consécutives pour les cycles évanescents, Comptes Rendus Acad. Sci. Paris sér. I, **306** (1988) 473–478.
6. D. Barlet, Interaction de strates consécutives pour les cycles évanescents, Ann. Sci. Ecole Norm. Sup. **24**:4 (1991) 401–505.
7. G. Barthel, A. Dimca, On complex projective hypersurfaces which are homology $\mathbb{P}^n$'s, *Singularities* (ed. J.P. Brasselet), London Math. Soc. Lect. Notes **201**, (Cambridge Univ. Press, 1994), 1–27.
8. E. Brieskorn, Die monodromie der isolierten Singularitäten von Hyperflächen, Manuscripta Math. **2** (1970) 103–161.
9. J. Damon, Nonlinear sections of nonisolated complete intersections, this volume.
10. P. Deligne et. al. *Groupes de monodromie en géométrie algébrique, II, Séminaire de Géometrie Algébrique du Bois-Marie 1967-69* (SGA 7 II), Springer lecture notes in Math. **340** 1973, esp. pp 82–164.
11. J. Denef, F. Loeser, Geometry on arc spaces of algebraic varieties, Proceedings of the Third European Congress of Mathematics, Barcelona 2000 (to appear).
12. A. Dimca, On the Milnor fibration of weighted homogeneous polynomials, Compositio Math. **76** (1990) 19–47.
13. A. Dimca, *Singularities and topology of hypersurfaces*, Universitext, Springer-Verlag, New York, 1992.
14. A. Dimca and A. Némethi, On the monodromy of complex polynomials, preprint, 2000. MathAG 9912072.
15. J. Denef and F. Loeser, Character sums associated to finite Coxeter groups, Trans. Amer. Math. Soc. **350** (1998) 5047–5066.
16. A. Dold, *Lectures on Algebraic Topology*, Grundlehren der Math. Wissenschaften **200**, Springer-Verlag, 1972.
17. T. Gaffney, The theory of integral closure of ideals and modules: applications and new developments, this volume.

18. V. Grandjean, Residual discriminant of a function germ singular along an Euler free divisor, preprint, Rennes, 2000.

19. P. A. Griffiths, Periods of integrals on algebraic manifolds, summary of main results and discussion of open problems, Bull. Amer. Math. Soc. **76** (1970) 228–296.

20. I. N. Iomdin, Local topological properties of complex algebraic sets, Sibirsk. Mat. Z. **15**:4 (1974) 784–805.

21. G. Jiang, Functions with non-isolated singularities on singular spaces, Dissertation, University of Utrecht, 1998.
    http//www.math.uu.nl/people/siersma/

22. G. Jiang, Algebraic descriptions of non-isolated singularities, Hokkaido Math. J. **27**:1 (1998) 233–243.

23. G. Jiang and M. Tibăr, Splitting of Singularities, preprint 2000, No 2, Tokyo Metropolitan University. MathAG 0010035.

24. T. de Jong, Some classes of line singularities, Math. Zeits. **198** (1998) 493–517.

25. T. de Jong and D. van Straten, Deformations of the normalization of hypersurfaces, Math. Ann. **288** (1990) 527–547.

26. M. Kato and Y. Matsumoto, On the connectivity of the Milnor fibre of a holomorphic function at a critical point, *Manifolds Tokyo 1973*, (University of Tokyo Press 1975), pp 131–136.

27. K. Lamotke, Die Homologie isolierter Singularitäten, Mat. Zeits. **143** (1975) 27–44.

28. D. T. Lê, Ensembles analytiques complexes avec lieu singulier de dimension un (d'après I. N. Jomdin), *Séminaire sur les Singularités*, (Publ. Math. Univ. Paris VII) pp 87–95.

29. D. T. Lê, F. Michel and C. Weber, Courbes polaires et topologie des courbes planes, Ann. Sci. cole Norm. Sup. (4) **24** (1991) 141–169.

30. D. Massey, A quick use of the perversity of the vanishing cycles, preprint, Northeastern University, Boston MA, 1989.

31. D. Massey, *Lê cycles and hypersurface singularities*, Springer Lecture Notes in Math. **1615**, 1995.

32. D. Massey and D. Siersma, Deformation of polar methods, Ann. Inst. Fourier **42**:4 (1992) 737–778.

33. J. Milnor, *Singular points of complex hypersurfaces*, Ann. Math. Studies, Princeton Univ. Press, 1968.

34. D. Mond, Some remarks on the geometry and classification of germs of maps from surfaces to 3-space, Topology **26** (1987) 361–383.

35. D. Mond, Vanishing cycles for analytic maps, *Singularity Theory and its Applications: Warwick 1989, Part I* (eds D. Mond and J. Montaldi), Springer Lecture Notes in Math. **1462** (1991) 221–234.

36. A. Némethi, The Milnor fibre and the zeta function of the singularities of the type $f = P(h, g)$, Compositio Math. **79** (1991) 63–97.

37. A. Némethi, Hypersurface singularities with 2-dimensional critical locus, J. London Math. Soc. (2) **59** (1999) 922–938.

38. P. Orlik, H. Terao, Arrangements and Milnor fibers, Math. Ann. **301** (1995) 211–235.

39. R. Pellikaan, Finite determinacy of functions with non-isolated singularities, Proc. London Math. Soc. (3) **57** (1988) 357–382.

40. R. Pellikaan, Series of isolated singularities, *Singularities* (ed R. Randell) Contemp. Math. **90**, (Amer. Math. Soc. Providence, RI, 1989) pp 241–259.

Dirk Siersma

41. R. Randell, On the topology of non-isolated singularities, Proc. 1977 Georgia Topology Conference, 445–473.

42. M. Saito, On Steenbrink's Conjecture, Math. Ann. **289** (??) 703–716.

43. H. Samelson, On Poincaré duality, Jour. d'Analyse Math. **14** (1965) 323–336.

44. M. Shubladze, Hyperplane singularities of transversal type A_k, (Russian) Soobshch. Akad. Nauk Gruzin. SSR **136**:3 (1990) 553–556.

45. D. Siersma, Isolated line singularities, *Singularities* (ed Peter Orlik), Proc. Symp. pure math. **40**:2, (Amer. Math. Soc., 1983) pp 485–496.

46. D. Siersma, Hypersurfaces with singular locus a plane curve and transversal type A_1, *Singularities*, Banach Center Publ. **40** (Warsaw 1988) pp 397–410.

47. D. Siersma, Singularities with critical locus a 1-dimensional complete intersection and transversal type A_1, Topology and its Applications **27** (1987) 51–73.

48. D. Siersma, Quasi-homogeneous singularities with transversal type A_1, Contemporary Math. **90** (1989) 261–294.

49. D. Siersma, Vanishing cycles and special fibres, *Singularity Theory and its Applications: Warwick 1989, Part I* (eds D. Mond and J. Montaldi), Springer Lecture Notes in Math. **1462** (1991) 292–301.

50. D. Siersma, The monodromy of a series of hypersurface singularities, Comment. Math. Helvet. **65** (1990) 181–197.

51. D. Siersma, Variation mappings on singularities with a 1-dimensional critical locus, Topology **30** (1991) 445–469.

52. D. Siersma, A bouquet theorem for the Milnor fibre, J. Alg. Geom. **4** (1995) 51–66.

53. D. Siersma and M. Tibăr, Is the polar relative monodromy of finite order? An example, Chinese Quart. J. Math. **10**:4 (1995) (Proc. 1994 workshop on Topology and Geometry, Zhanjiang) 78–85.

54. J. H. M. Steenbrink, The spectrum of hypersurface singularities, *Actes du Colloque de Thorie de Hodge (Luminy, 1987)*, Astérisque **179-180** (1989) 163–184.

55. J. Stevens, Periodicity of branched cyclic covers of manifolds with open book decomposition, Math. Ann. **273** (1986) 227–239.

56. D. van Straten, On the Betti numbers of the Milnor fibre of a certain class of hypersurface singularities, *Singularities, representation of algebras, and vector bundles, Lambrecht, 1985*, Springer Lecture Notes in Math. **1273** (1987) 203–220.

57. M. Tibăr, Bouquet decomposition of the Milnor fibre, Topology **35** (1996) 227–241.

58. M. Tibăr, Embedding non-isolated singularities into isolated singularities, *Singularities: the Brieskorn anniversary volume* (eds. V. I. Arnold, G.-M. Greuel and J. H. M. Steenbrink), Progress in Math. **162** (Birkhäuser, 1998) pp 103–115.

59. A. Zaharia , A study about singularities with non-isolated critical locus, thesis, Rijksuniversiteit Utrecht, 1993.

60. A. Zaharia, Topological properties of certain singularities with critical locus a 2-dimensional complete intersection, Topology and Appl. **60** (1994) 153–171.